Handbook of Numerical Analysis Applications

Handbook of Numerical Analysis Applications

WITH
PROGRAMS FOR ENGINEERS AND SCIENTISTS

JAROSLAV PACHNER

McGRAW-HILL BOOK COMPANY
New York St. Louis San Francisco Auckland
Bogotá Hamburg Johannesburg London Madrid
Mexico Montreal New Delhi Panama Paris
São Paulo Singapore Sydney Tokyo Toronto

Library of Congress Cataloging in Publication Data

Pachner, Jaroslav.
  Handbook of numerical analysis applications.

  Includes index.
  1. Numerical analysis. 2. Numerical analysis—Data processing. I. Title.
  QA297.P26   983   519.4   82-22843
  ISBN 0-07-048057-5

Copyright © 1984 by McGraw-Hill, Inc. All rights reserved. Printed in the United States of America. Except as permitted under the United States Copyright Act of 1976, no part of this publication may be reproduced or distributed in any form or by any means, or stored in a data base or retrieval system, without the prior written permission of the publisher.

1234567890   KGP/KGP   89876543

ISBN 0-07-048057-5

The editors for this book were Patricia Allen-Browne and Elizabeth P. Richardson, the designer was Mark E. Safran, and the production supervisor was Sally Fliess. It was set in Times Roman by Santype-Byrd.

Printed and bound by The Kingsport Press.

TO MY WIFE JARMILA

Contents

Preface xv

PART ONE   THEORY

1. Linear Algebra I   1.1
   1.1 Systems of Linear Algebraic Equations   1.1
   1.2 Matrices   1.2
   1.3 Determinants and the Inverse Matrix   1.5
   1.4 Direct Methods   1.8
   1.5 Error Analysis   1.13
   1.6 Overdetermined System of Equations   1.20
   1.7 Iterative Methods   1.21

2. Interpolation, Approximation, and Numerical Differentiation   2.1
   2.1 Interpolation Techniques   2.1
   2.2 Lagrange Interpolation   2.3
   2.3 Newton Interpolation   2.5
   2.4 Hermite Interpolation   2.8
   2.5 Cubic-Spline Interpolation   2.10
   2.6 Trigonometric Interpolation   2.14
   2.7 Inverse Interpolation   2.15
   2.8 Least-Squares Techniques   2.15
   2.9 Approximation by Chebyshev Polynomials   2.17
   2.10 Approximation by Orthogonal Polynomials with Arbitrarily Distributed Abscissas   2.19
   2.11 Approximation of a Periodic Function   2.21

	2.12	Economization of a Power Series	2.23
	2.13	Approximation by a Rational Function	2.24
	2.14	Numerical Differentiation	2.27
	2.15	Choosing the Method	2.30
3.	**Evaluation of Definite Integrals**	**3.1**
	3.1	Methods of Numerical Integration	3.1
	3.2	Integrals with Numerically Defined Integrands	3.4
	3.3	Integrals over Finite Intervals	3.5
	3.4	Integrals over Semi-Infinite Intervals	3.6
	3.5	Integrals over Infinite Intervals	3.7
	3.6	Romberg Integration	3.8
	3.7	Singular Integrals	3.10
4.	**Ordinary Differential Equations**	**4.1**
	4.1	Definitions and Analytical Methods of Integration	4.1
	4.2	Differential Equations Defining Special Functions	4.12
	4.3	Euler and Taylor-Series Methods	4.16
	4.4	Runge-Kutta Methods	4.20
	4.5	Predictor-Corrector Methods	4.24
	4.6	Stability and Accuracy of Integration Methods	4.31
	4.7	Choosing the Method	4.32
5.	**Boundary-Value Problems of Ordinary Differential Equations**	**5.1**
	5.1	Analytical Approach to Boundary-Value Problems	5.1
	5.2	Orthogonal Eigenfunctions Defined as Solutions of a Boundary-Value Problem	5.6
	5.3	Numerical Approach to Boundary-Value Problems	5.11
6.	**Nonlinear Equations**	**6.1**
	6.1	Direct Methods for Algebraic Equations	6.1
	6.2	Iterative Methods	6.5
	6.3	Real Roots of Algebraic and Transcendental Equations	6.9
	6.4	Complex Roots of Algebraic Equations	6.11
	6.5	Analysis of Errors in Roots of Algebraic Equations	6.12
	6.6	Real Roots of a System of Nonlinear Equations	6.13
7.	**Linear Algebra II**	**7.1**
	7.1	Algebraic Eigenvalue Problem	7.1
	7.2	Numerical Computation of the Eigenvalues of a Real Matrix	7.3
	7.3	Eigenvectors	7.5
8.	**Special Functions**	**8.1**
	8.1	The Polynomial Function	8.1
	8.2	Orthogonal Polynomials	8.2
	8.3	Hypergeometric Series and Confluent Hypergeometric Functions	8.3
	8.4	Incomplete Elliptic Integrals of the First, Second, and Third Kind	8.5

8.5	The Bessel Function $J_n(x)$ and Modified Bessel Function $I_n(x)$ of Integer Order	8.6
8.6	The Bessel Function $J_v(x)$ of Order $v > -\frac{1}{2}$	8.6
8.7	The Bessel Function $Y_v(x)$ of Order $v > -\frac{1}{2}$	8.7
8.8	The Modified Bessel Function $K_v(x)$	8.7
8.9	The Spherical Bessel Functions $j_n(x)$ and $y_n(x)$	8.7
8.10	The Gamma Function of a Real Argument	8.8
8.11	The Gamma Function $\Gamma(\frac{1}{2} + n)$ for $n = 0, \pm 1, \pm 2, \ldots$	8.9
8.12	The Incomplete Gamma Function	8.9
8.13	The Beta Function	8.9
8.14	The Error Function	8.10
8.15	The Fresnel Integrals $C(x)$ and $S(x)$	8.11
8.16	The Sine Integral and Cosine Integral	8.11
8.17	The Exponential Integral and Logarithmic Integral	8.12
8.18	The Gudermannian and Its Inverse	8.13

9. Selected Problems of Mathematical Statistics — 9.1

9.1	Elements of Combinatorial Analysis	9.1
9.2	Basic Concepts of Mathematical Statistics	9.2
9.3	Curve Fitting	9.5
9.4	Binomial and Negative Binomial Distribution	9.9
9.5	Hypergeometric Distribution	9.10
9.6	Poisson Distribution	9.11
9.7	Normal Distribution and Inverse Normal Distribution	9.11
9.8	Chi-Square Distribution	9.13
9.9	t Distribution	9.14
9.10	F Distribution	9.14

References — R.1

PART TWO   PROGRAMS IN BASIC — P.1

Using the Programs — P.3

1. Linear Algebra I — P.7

P101	Condition Indicator for a System of Linear Algebraic Equations	P.7
P102	Solution of a System of Linear Algebraic Equations by the Doolittle Method with Partial Pivoting and/or Computation of the Determinant	P.11
P103	Solution of a Tridiagonal System of Linear Algebraic Equations	P.17
P104	Solution of a Pentadiagonal System of Linear Algebraic Equations	P.21
P105	Reduction of an Overdetermined System of Linear Algebraic Equations to a Determined System of Normal Equations	P.25
P106	Iterative Methods for a System of Linear Algebraic Equations: Jacobi, Gauss-Seidel, and Successive-Overrelaxation Methods	P.27

2. Interpolation, Approximation, and Numerical Differentiation — P.31

P201	Lagrange Interpolation	P.31

P202	Lagrange Interpolation with Equally Spaced Abscissas	P.34
P203	Newton Interpolation for the Function and Its First and Second Derivatives	P.37
P204	Newton Interpolation with Equally Spaced Abscissas for the Function and Its First and Second Derivatives	P.41
P205	Hermite Interpolation for the Function and Its First and Second Derivatives	P.44
P206	Hermite Interpolation with Equally Spaced Abscissas for the Function and Its First and Second Derivatives	P.48
P207	Hermite Interpolation of a Function Defined at Two Points by the Function and Its First and Second Derivatives for the Function and Its First and Second Derivatives	P.52
P208	Cubic Spline for the Function and Its First and Second Derivatives	P.55
P209	Cubic Spline with Equally Spaced Abscissas for the Function and Its First and Second Derivatives	P.60
P210	Trigonometric Interpolation	P.65
P211	Least-Squares Approximation by Chebyshev Polynomials for the Function and Its First and Second Derivatives	P.68
P212	Least-Squares Approximation by Orthogonal Polynomials with Arbitrarily Spaced Abscissas and a Given Weight Function for the Function and Its First and Second Derivatives	P.73
P213	Least-Squares Approximation by Orthogonal Polynomials with Arbitrarily Spaced Abscissas and a Weight Function Equal to 1 for the Function and Its First and Second Derivatives	P.79
P214	Least-Squares Approximation by Orthogonal Polynomials with Equally Spaced Abscissas and a Given Weight Function for the Function and Its First and Second Derivatives	P.84
P215	Least-Squares Approximation by Orthogonal Polynomials with Equally Spaced Abscissas and a Weight Function Equal to 1 for the Function and Its First and Second Derivatives	P.90
P216	Least-Squares Approximation of a Periodic Function (Fourier Series) for the Function and Its First and Second Derivatives	P.95
P217	Least-Squares Approximation of an Even Periodic Function (Fourier Series) for the Function and Its First and Second Derivatives	P.100
P218	Least-Squares Approximation of an Odd Periodic Function (Fourier Series) for the Function and Its First and Second Derivatives	P.104
P219	Economization of a Power Series	P.108
P220	Padé Approximation of a Truncated Series with 8 Terms for the Function and Its First and Second Derivatives	P.111
P221	Modified Padé Approximation of a Truncated Series with 13 Terms for the Function and Its First and Second Derivatives	P.114
P222	Least-Squares Approximation by a Rational Function with Chebyshev Polynomials for the Function and Its First and Second Derivatives	P.118
P223	Numerical Differentiation: First and Second Derivatives of a Function Defined by 3, 5, or 7 Points with Equally Spaced Abscissas	P.124

3. Evaluation of Definite Integrals — P.128

P301	Integration of a Numerically Defined Integrand by Composite Simpson's Rule	P.128

	P302 Integration of a Numerically Defined Integrand by Modified Composite Simpson's Rule	P.130
	P303 Integration of a Numerically Defined Integrand with Equally Spaced Abscissas by Composite Corrected Trapezoidal Rule	P.132
	P304 Integration of a Numerically Defined Integrand with Arbitrarily Spaced Abscissas by Composite Corrected Trapezoidal Rule	P.134
	P305 Composite Gaussian Integration of an Integral over a Finite Interval	P.136
	P306 Laguerre Integration of an Integral over a Semi-Infinite Interval	P.138
	P307 Composite Gauss-Laguerre Integration of an Integral over a Semi-Infinite Interval	P.140
	P308 Hermite Integration of an Integral over an Infinite Interval	P.143
	P309 Composite Gauss-Laguerre Integration of an Integral over an Infinite Interval	P.144
	P310 Romberg Integration	P.147
	P311 Chebyshev-Gauss Integration of a Singular Integral	P.149
	P312 Integration of an Integral with Logarithmic Singularity	P.150
4.	**Ordinary Differential Equations**	**P.151**
	P401 Fourth-Order Taylor-Series Method for a Set of Differential Equations of First Order	P.151
	P402 Fourth-Order Taylor-Series Method for a Set of Differential Equations of Second Order	P.156
	P403 Fourth-Order Standard Runge-Kutta Method for a Set of Differential Equations of First Order	P.160
	P404 Gill's Version of Fourth-Order Runge-Kutta Method for a Set of Differential Equations of First Order	P.164
	P405 Third-Order Predictor-Corrector Method for a Set of Differential Equations of First Order	P.168
	P406 Fourth-Order Predictor-Corrector Method for a Set of Differential Equations of First Order	P.173
	P407 Fourth-Order Predictor-Corrector Method for a Set of Differential Equations of Second Order	P.178
5.	**Boundary-Value Problems of Ordinary Differential Equations**	**P.183**
	P501 Lagrange Interpolation for Boundary-Value Problems of One Ordinary Differential Equation of Second Order	P.183
6.	**Nonlinear Equations**	**P.186**
	P601 Roots of a Quadratic Equation	P.186
	P602 Roots of a Cubic Equation	P.188
	P603 Roots of a Biquadratic Equation	P.190
	P604 Preliminary Location of Real Roots of a Nonlinear Equation	P.193
	P605 Real Roots of a Nonlinear Equation (with Deflation Subroutine for an Algebraic Equation)	P.196
	P606 Real and Complex Roots of an Algebraic Equation (with Deflation Subroutine)	P.201

xii CONTENTS

	P607 Real Roots of Two Nonlinear Equations by Fixed-Point Iteration	P.209
	P608 Real Roots of Two Nonlinear Equations by the Newton or Modified Newton Method	P.211
7.	**Linear Algebra II**	**P.213**
	P701 Preliminary Location of Eigenvalues by the Gershgorin Method	P.213
	P702 Coefficients of a Secular Equation by the Krylov Method	P.216
	P703 Real and Complex Eigenvalues of a Real Matrix from a Secular Equation	P.221
	P704 Orthonormal Eigenvectors of a Real Matrix with Simple Real Eigenvalues	P.229
8.	**Special Functions**	**P.235**
	P801 The Polynomial Function and Its First and Second Derivatives	P.235
	P802 Orthogonal Polynomials of Legendre, Laguerre, Hermite, and Chebyshev of the First and Second Kind, Their First and Second Derivatives, and Zeros of Chebyshev Polynomials of the First Kind	P.238
	P803 The Hypergeometric Series, Confluent Hypergeometric Function, Their First and Second Derivatives, the Exponential Integral and Logarithmic Integral	P.242
	P804 Incomplete Elliptic Integrals of the First, Second, and Third Kind, Bessel Functions $J_n(x)$ and Modified Bessel Functions $I_n(x)$ of Integer Order, the Incomplete Gamma Function, the Error Function, Fresnel Integrals $C(x)$, $S(x)$, and Sine and Cosine Integrals in Single Precision	P.246
	P805 Incomplete Elliptic Integrals of the First, Second, and Third Kind, Bessel Functions $J_n(x)$ and Modified Bessel Functions $I_n(x)$ of Integer Order, the Incomplete Gamma Function, the Error Function, Fresnel Integrals $C(x)$, $S(x)$, and Sine and Cosine Integrals in Double Precision	P.250
	P806 Bessel Functions $J_\nu(x)$, $Y_\nu(x)$, $K_\nu(x)$ (ν Is Any Real Number $> -\frac{1}{2}$), and Spherical Bessel Functions $j_n(x)$, $y_n(x)$ (n Is Zero or Any Positive Integer) in Single Precision	P.256
	P807 Bessel Functions $J_\nu(x)$, $Y_\nu(x)$, $K_\nu(x)$ (ν Is Any Real Number $> -\frac{1}{2}$), and Spherical Bessel Functions $j_n(x)$, $y_n(x)$ (n Is Zero or Any Positive Integer) in Double Precision	P.260
	P808 Spherical Bessel Functions $j_n(x)$, $y_n(x)$ of Order $n = 0, \ldots, 9$	P.267
	P809 Gamma Function of a Real Argument	P.270
	P810 Gamma Function of Argument $n + \frac{1}{2}$, n an Arbitrary Integer	P.272
	P811 Beta Function	P.273
	P812 Gudermannian, Its Inverse, and Hyperbolic Functions in Single Precision	P.275
9.	**Selected Problems of Mathematical Statistics**	**P.277**
	P901 Permutations, Variations, and Combinations	P.277
	P902 Arithmetic, Geometric, and Harmonic Mean Values, Variances S^2, s^2, Standard Deviations S, s, Standard Errors E, e for a Sample of n Data	P.280
	P903 Arithmetic Mean Value, Variances S^2, s^2, Standard Deviations S, s, Standard Errors E, e for a Sample of Grouped Data	P.282

P904	Linear, Power, Exponential, and Logarithmic Curve Fit	P.284
P905	Binomial-Power Curve Fit	P.287
P906	Parabolic Curve Fit	P.289
P907	Binomial and Cumulative Binomial Distribution	P.291
P908	Negative Binomial and Cumulative Negative Binomial Distribution	P.293
P909	Hypergeometric and Cumulative Hypergeometric Distribution	P.295
P910	Poisson and Cumulative Poisson Distribution	P.297
P911	Normal Distribution, Chi-Square Distribution, t Distribution, F Distribution	P.298
P912	Inverse Normal Distribution	P.303

Appendix — **P.304**

PA1	Derived Elementary Functions in Single Precision	P.304
PA2	Elementary Functions in Double Precision	P.306
PA3	Subroutines for Programming in Double Precision	P.311

Index follows Appendix — I.1

Preface

With the advent of computers, microcomputers, and programmable calculators many mathematical problems important for applications in engineering and sciences but inaccessible by analytical methods can now be treated numerically. Textbooks on numerical analysis usually describe several methods for solving a given problem with more or less detailed deductions, proofs, stability considerations, and statements on the range of applicability which are absolutely necessary in the first study of the discipline. But engineers, scientists, or graduate students working on a research project need compact and precise information about the numerical methods that can be used in the problem and, if possible, an easily adjustable computer program affording full control over its mathematical treatment. The aim of this handbook is to provide just this kind of information, releasing the most valuable and irreplaceable asset, the researcher's time.

The book is divided into two parts. Part One contains nine chapters, each starting with a short survey of the analytical background. The most efficient numerical methods are described in full detail as necessary for applications but without deductions and proofs, which can be found in the literature. In choosing the numerical methods described the emphasis has been on the stability, accuracy, and speed of numerical computation (in that order). The researcher can usually choose between several parallel methods; only one method is discussed when it decisively stands out above the others. Part Two is a set of 85 programs, with 3 additional programs in the Appendix. Examples to each particular program illustrate the applicability of the numerical method applied in the program.

Chapter 1 deals with linear algebra: systems of linear algebraic equations, matrix algebra, evaluation of determinants, and overdetermined systems of equations. Topics include direct methods, gaussian elimination in Doolittle form with optional partial pivoting and with evaluation of the determinant, its modifications for tridiagonal and pentadiagonal matrices, and iterative methods.

Chapter 2 is concerned with the important problem of interpolation and approximation of a function defined either by an analytical expression or by a set of numerical data which can be considered as accurate (data from mathematical tables) or which contain some random errors (data from observation or generated in an earlier numerical computation). The function is interpolated by a polynomial or approximated by a series of orthogonal polynomials or functions (least-squares techniques) or by a rational function. The interpolations and approximations yield not only the function but also its first and second derivatives. Hermite interpolation, the cubic spline, and the differentiation formulas are also included in this chapter.

Chapter 3 deals with the numerical evaluation of integrals over a finite, semi-infinite, or infinite interval, the integrand again being defined by an analytical expression or by a set of equally or arbitrarily spaced numerical data. Problems connected with the evaluation of singular integrals and the corresponding numerical methods are also discussed.

Chapter 4 discusses analytical and numerical methods for the integration of ordinary differential equations. After a short description of the Euler and Taylor-series methods, Runge-Kutta and predictor-corrector methods are extensively examined, including new and powerful predictor-corrector methods with a built-in starter.

The boundary-value problems of ordinary differential equations are discussed from an analytical point of view in Chap. 5; a numerical method is used to reduce the boundary-value problem to an initial-value problem of Chap. 4.

Chapter 6, concerned with the solution of nonlinear algebraic and transcendental equations, starts with the formulas and programs for the roots of a quadratic, cubic, and biquadratic polynomial. Iterative methods compute either the real roots of an algebraic or transcendental equation or the real and complex roots of a polynomial.

Chapter 7 returns to linear algebra again and deals with the computation of eigenvalues and eigenvectors of real matrices. Chapter 8 contains formulas defining special functions and programs evaluating them. Chapter 9 discusses selected problems of mathematical statistics, curve fitting, and both discrete and continuous distributions.

The first program in the Appendix computes 13 derived elementary functions in single precision, and the second computes 18 elementary functions in double precision to be used when they are not included in the standard library of the microcomputer. In that case one or more sections of the third program must be used as subroutines when certain programs are to be run in double precision.

Because the programs are intended to be run on a microcomputer, they are restricted to problems with real matrices or with one real independent variable. The roots of nonlinear algebraic equations with real coefficients may be complex-conjugate, but this is the only exception. Consequently, the eigenvalues of an algebraic problem may be complex-conjugate (because the matrix is assumed to be real), but the computation of eigenvectors is restricted to matrices with simple real eigenvalues.

Whenever a computer is used, an important question is the right choice of programming language. The listing of a source program written in FORTRAN is clearer than when it is programmed in BASIC. On the other hand, a program edited in FORTRAN requires more memory space for the compilation and linkage, which must be rerun when the dimensions of arrays are changed or the program is adjusted to another problem, e.g., to a different integrand or to a new set of differential equations. In BASIC only the source program must be saved in memory; its compilation is automatic and so fast that the user is not aware when it takes nlace. Furthermore, the dimensions of arrays are set in each run by the input data, which need not be formatted (formatting of output data in FORTRAN should be considered an asset, not a liability). The decisive point, however, is the speed of computation. Since BASIC requires only 10 to 12 percent more computer time for running the same program on the same microcomputer, there was no doubt that BASIC should be chosen for the programs in this handbook; its advantages overwhelmingly outweigh its disadvantages.

A microcomputer in the hands of a researcher is the ideal tool for the numerical solution of all the problems dealt with in this handbook; it is always ready and sufficiently fast. The computation is often completed in less than 1 minute and usually takes no more than a few minutes. Of course, the integration of a set of ordinary differential equations carried out in high precision, e.g., Example 3 for Program P406, may take more than 1 hour whereas on a big computer it is completed in seconds, but such a comparison is rather deceptive: the microcomputer is available for an immediate run, while one must usually wait in line not only to run a program on a big computer but also for its output to be printed. The cost of using a big computer may be quite high. The author believes that future generations of microcomputers with increased speed of computation and larger memory will be able to solve all problems occurring in scientific research, including even integration of systems of partial differential equations of the second order.

The present handbook, the result of the author's experience in lecturing on advanced analysis and numerical analysis, is based on his notes and private collection of programs written over several years partly in FORTRAN and partly in RPN for programmable calculators HP-65 and HP-41CV. The inclusion of chapters on partial differential equations of the first order and on integral equations was eventually rejected to restrict the handbook to problems arising repeatedly in everyday research. Nevertheless the author will gratefully receive comments and suggestions of additional problems and programs which should be included in a later edition.

The author cannot conclude this Preface without expressing his sincere thanks to Patricia Allen-Browne of McGraw-Hill Book Company for her cooperation in readying this project for publication and to Elizabeth P. Richardson for copy-editing the manuscript and seeing it through the press.

J. P.

part one
Theory

LINEAR ALGEBRA I

1

There are two fundamental problems of linear algebra, the solution of a system of linear algebraic equations and the eigenvalue problem of a matrix. The first can be solved exactly by a finite number of four basic arithmetic operations, but the solution is difficult when the number of these operations is large enough to give rise to considerable roundoff errors or when the system is ill-conditioned. The present chapter deals with methods of overcoming these difficulties. The eigenvalue problem can be treated by one of several iterative methods, or it can be reduced to the solution of a secular polynomial. Since nonlinear equations are discussed in Chap. 6, the eigenvalue problem is postponed to Chap. 7.

1.1 SYSTEMS OF LINEAR ALGEBRAIC EQUATIONS

A system of m linear algebraic equations in n unknowns

$$a_{11}x_1 + a_{12}x_2 + \cdots + a_{1n}x_n = b_1$$
$$a_{21}x_1 + a_{22}x_2 + \cdots + a_{2n}x_n = b_2$$
$$\dots\dots\dots\dots\dots\dots\dots\dots\dots\dots\dots\dots$$
$$a_{m1}x_1 + a_{m2}x + \cdots + a_{mn}x_n = b_m$$

(1.1.1a)

can be written in matrix form

$$AX = B \tag{1.1.1b}$$

In this chapter Latin capital letters denote matrices. The matrix A is rectangular with m rows and n columns

$$A = \begin{bmatrix} a_{11} & a_{12} & \cdots & a_{1n} \\ a_{21} & a_{22} & \cdots & a_{2n} \\ \dots & \dots & & \dots \\ a_{m1} & a_{m2} & \cdots & a_{mn} \end{bmatrix} \tag{1.1.2}$$

and X and B are single-column matrices

$$X = \begin{bmatrix} x_1 \\ x_2 \\ \vdots \\ x_m \end{bmatrix} \qquad B = \begin{bmatrix} b_1 \\ b_2 \\ \vdots \\ b_m \end{bmatrix} \qquad (1.1.3)$$

If the number of equations equals the number of unknowns, $m = n$, the system is *determined*. If each equation of the determined system represents independent information not contradicting other equations, matrix A has an inverse A^{-1}, defined by

$$A^{-1}A = AA^{-1} = I \qquad (1.1.4)$$

in which I is an identity matrix, i.e., a diagonal matrix with all nonvanishing elements equal to 1

$$I = \begin{bmatrix} 1 & 0 & \cdots & 0 \\ 0 & 1 & \cdots & 0 \\ \multicolumn{4}{c}{\dotfill} \\ 0 & 0 & & 1 \end{bmatrix} \qquad (1.1.5)$$

Equation (1.1.1b) can now be multiplied from the left by A^{-1}; because of (1.1.4), the unknowns x_i are determined exactly by

$$X = A^{-1}B \qquad (1.1.6)$$

Since the computation of the inverse matrix becomes more and more difficult with an increasing number of unknowns, more efficient methods for solving system (1.1.1a) are discussed in the following sections.

If the number of equations in (1.1.1a) is less than the number of unknowns, $m < n$, the system is *underdetermined*. In this case $n - m$ unknowns x_i can be freely chosen and reduced to the right-hand side to make the system determined, provided that the chosen equations are independent and do not contradict each other.

If the number of equations in (1.1.1a) is greater than the number of unknowns, $m > n$, the system is *overdetermined*. This usually occurs when the number of measurements taken is greater than is the number of unknowns (to reduce the influence of observational errors). The overdetermined system cannot be satisfied exactly; one way to find the most satisfactory solution is the least-squares technique (Sec. 1.6).

1.2 MATRICES

A matrix A is an array of $m \times n$ numbers called *elements* a_{ij} and arranged in m rows and n columns, as shown in (1.1.2). If $m = 1$, the matrix is a *single-row matrix*, also called a row *vector*; if $n = 1$, it is a single-column matrix, like those in (1.1.3), called a *column vector*. Every matrix can be made into a square matrix with $m = n$ by putting zeros in the places of missing elements. A square $n \times n$ matrix is said to be *of order n*. A zero or null matrix has all its elements equal to zero. It is denoted by 0.

In a diagonal matrix D all the elements equal zero except those on the main diagonal

$$d_{ii} \neq 0 \qquad d_{ij} = 0 \text{ if } i \neq j \tag{1.2.1}$$

A special case of a diagonal matrix is the identity matrix I, also called a *unit matrix*, with elements equal to Kronecker-delta symbols [cf. (1.1.5)]

$$\delta_{ij} = \begin{cases} 1 & \text{if } i = j \\ 0 & \text{if } i \neq j \end{cases} \tag{1.2.2}$$

In a tridiagonal matrix T only the elements of a narrow band along the main diagonal differ from zero

$$t_{ii} \neq 0 \qquad t_{i, i\pm 1} \neq 0 \qquad \text{all other } t_{ij} = 0 \tag{1.2.3}$$

Similarly the elements of a pentadiagonal matrix F are

$$f_{ii} \neq 0 \qquad f_{i, i\pm 1} \neq 0 \qquad f_{i, i\pm 2} \neq 0 \qquad \text{all other } f_{ij} = 0 \tag{1.2.4}$$

In a lower-triangular matrix L with n rows and n columns the elements on the main diagonal and below it differ from zero while those above the diagonal equal zero

$$\begin{aligned} l_{i, i-k} &\neq 0 & k &= 0, 1, \ldots, i - 1 \\ l_{i, i+k} &= 0 & k &= 1, 2, \ldots, n - 1 \end{aligned} \tag{1.2.5}$$

In an upper-triangular matrix U with n rows and n columns the elements on the main diagonal and above it differ from zero while those below it equal zero

$$\begin{aligned} u_{i, i+k} &\neq 0 & k &= 0, 1, \ldots, n - i \\ u_{i, i-k} &= 0 & k &= 1, 2, \ldots, i - 1 \end{aligned} \tag{1.2.6}$$

If almost all elements of a matrix are nonzeros, the matrix is said to be *full*. On the other hand, a *sparse matrix* is one in which only a small fraction of its elements, usually located in a narrow band along its main diagonal, differs from zero.

A transposed matrix $\tilde{A}$ is formed from the matrix A by interchanging rows and columns

$$\tilde{a}_{ij} = a_{ji} \tag{1.2.7}$$

A complex-conjugate matrix A^* is obtained by taking the complex conjugate of each element of the matrix A

$$a_{ij}^* = (a_{ij})^* \tag{1.2.8}$$

An adjoint matrix $A\dagger$ is generated by transposing a complex-conjugate matrix

1.4 THEORY

A^* or by forming a complex-conjugate matrix of a transposed matrix $\tilde{A}$

$$A\dagger = \tilde{A}^* = \tilde{A^*} \qquad (1.2.9)$$

With the help of these operations and the inverse operation defined by (1.1.4) and described in detail in Sec. 1.3, particular square matrices with distinct properties are defined as follows:

Real: $\qquad A^* = A \qquad (1.2.10)$

Symmetric: $\qquad \tilde{A} = A \qquad (1.2.11)$

Skew-symmetric (antisymmetric): $\quad \tilde{A} = -A \qquad (1.2.12)$

Orthogonal: $\qquad A^{-1} = \tilde{A} \qquad (1.2.13)$

Pure imaginary: $\qquad A^* = -A \qquad (1.2.14)$

Hermitian (self-adjoint): $\qquad A\dagger = A \qquad (1.2.15)$

Skew-hermitian: $\qquad A\dagger = -A \qquad (1.2.16)$

Unitary: $\qquad A^{-1} = A\dagger \qquad (1.2.17)$

If matrix A in (1.2.15) to (1.2.17) is real, the hermitian, skew-hermitian, and unitary matrices reduce to symmetric, skew-symmetric, and orthogonal matrices (1.2.11) to (1.2.13), respectively.

The special properties of these matrices are the result of particular symmetries of their elements which reduce the number of independent elements; e.g., in a symmetric matrix of order n only $n(n+1)/2$ elements are independent. If the magnitudes of matrix elements are taken into consideration, two special matrices are important when a system of linear algebraic equations is being solved. A *strictly* (row) *dominant matrix* is defined by the condition

$$|a_{ii}| \geq \sum_{\substack{j=1 \\ j \neq k}}^{n} |a_{ij}| \qquad i = 1, 2, \ldots, n \qquad (1.2.18)$$

A *symmetric positive definite matrix* is a symmetric matrix (1.2.11) satisfying the additional condition

$$\tilde{X} A X > 0 \qquad (1.2.19)$$

for any vector X. A symmetric matrix of order n is positive definite if and only if

$$\det A_k > 0 \qquad k = 1, 2, \ldots, n \qquad (1.2.20a)$$

where A_k is a $k \times k$ matrix formed by the intersection of the first k rows and columns of A (*Sylvester's criterion*). Consequently, the diagonal elements are all positive

$$a_{ii} > 0 \qquad i = 1, 2, \ldots, n \qquad (1.2.20b)$$

and $\qquad |a_{ij}|^2 \leq a_{ii} a_{jj} \qquad i, j = 1, 2, \ldots, n \qquad (1.2.20c)$

Two matrices A and B are equal to each other if the corresponding elements are equal

$$a_{ij} = b_{ij} \tag{1.2.21}$$

Two matrices A and B can be added or subtracted to form a matrix $C = A \pm B = \pm B + A$ by adding or subtracting the corresponding elements

$$c_{ij} = a_{ij} \pm b_{ij} \tag{1.2.22}$$

A matrix A is multiplied by a scalar to form a matrix $B = gA$ by multiplying each element a_{ij} by the scalar g

$$b_{ij} = ga_{ij} \tag{1.2.23}$$

Two matrices A and B are multiplied to form a matrix $C = AB$ by the formula

$$c_{ij} = \sum_k a_{ik} b_{kj} \tag{1.2.24}$$

Hence, in general,

$$AB \neq BA \qquad AI = IA = A \tag{1.2.25}$$

If $AB = BA$, the matrices are said to *commute*.

The transpose of the product of two matrices AB equals the product of transposed matrices but in reverse order

$$\widetilde{AB} = \widetilde{B}\widetilde{A} \tag{1.2.26}$$

The *trace* of a square matrix A is the sum of its diagonal elements

$$\operatorname{Tr} A = \sum_i a_{ii} \tag{1.2.27}$$

Hence

$$\operatorname{Tr}(AB) = \sum_i \sum_k a_{ik} b_{ki} = \sum_k \sum_i b_{ki} a_{ik} = \operatorname{Tr}(BA) \tag{1.2.28}$$

This holds even if $AB \neq BA$.

1.3 DETERMINANTS AND THE INVERSE MATRIX

A square matrix, i.e., an array of $n \times n$ numbers, can be associated with a single number, the determinant of the order n,

$$\det A = \begin{vmatrix} a_{11} & a_{12} & \cdots & a_{1n} \\ a_{21} & a_{22} & \cdots & a_{2n} \\ \multicolumn{4}{c}{\dotfill} \\ a_{n1} & a_{n2} & \cdots & a_{nn} \end{vmatrix} \tag{1.3.1a}$$

defined by

$$\det A = \sum (-1)^s a_{i1} a_{j2} \cdots a_{pn} \tag{1.3.1b}$$

1.6 THEORY

The sum runs over all permutations $i \neq j \neq \cdots \neq n$, and s denotes the number of exchanges necessary to bring the sequence $(i, j, k, \ldots, p)$ back into the natural order $(1, 2, 3, \ldots, n)$. Accordingly,

$$\det A = \begin{vmatrix} a_{11} & a_{12} \\ a_{21} & a_{22} \end{vmatrix} = a_{11}a_{22} - a_{12}a_{21} \tag{1.3.2}$$

$$\det B = \begin{vmatrix} b_{11} & b_{12} & b_{13} \\ b_{21} & b_{22} & b_{23} \\ b_{31} & b_{32} & b_{33} \end{vmatrix} = b_{11}b_{22}b_{33} - b_{11}b_{32}b_{23}$$
$$+ b_{31}b_{12}b_{23} - b_{31}b_{22}b_{13} + b_{21}b_{32}b_{13} - b_{21}b_{12}b_{33} \tag{1.3.3}$$

The determinant of order n is thus given by the sum of $n!$ terms if evaluated by the formula (1.3.1b). Since its value is usually far smaller than the values of particular terms in the sum which must be computed with highest accuracy to lessen the influence of roundoff errors, it is hopeless to try to evaluate a determinant of order $n > 3$ by formula (1.3.1b).

The best way of computing the determinant of a matrix A is by reducing the matrix to a product of a lower- and an upper-triangular matrix

$$A = LU \tag{1.3.4}$$

and apply the theorem valid for any two matrices of the same order

$$\det (AB) = (\det A)(\det B) \tag{1.3.5}$$

According to (1.3.1b), the determinant of a lower- or upper-triangular matrix equals to the product of its diagonal terms

$$\det L = \prod_{i=1}^{n} l_{ii} \qquad \det U = \prod_{i=1}^{n} u_{ii} \tag{1.3.6}$$

It is shown in Sec. 1.4 that in the LU decomposition all the diagonal terms of L equal 1. Therefore

$$\det A = \prod_{i=1}^{n} u_{ii} \tag{1.3.7}$$

The following theorems may be helpful in evaluating a determinant:

Interchanging two rows (or columns) changes the sign of the determinant.

The value of a determinant does not change if rows and columns are interchanged: $\det A = \det \tilde{A}$.

The value of a determinant does not change if an arbitrary multiple of a row (or column) is added to any parallel row (or column).

The determinant equals zero if all elements in a row (or column) equal zero.

As a consequence of the two preceding theorems, the determinant vanishes if any two rows (or columns) are identical or proportional to each other.

The determinant is multiplied by a scalar if all elements in one row (or column) are multiplied by that scalar.

Definitions

A matrix whose determinant vanishes is called a *singular matrix;* a matrix with a nonvanishing determinant is a *nonsingular matrix*.

If a nonsingular matrix A_r of order r can be formed from a singular matrix A of order n by eliminating $n - r$ rows and columns, the matrix A has rank r and the nonsingular matrix A_r is called the *basis* of A. Obviously $r \leq \min(m, n)$, where m and n are the numbers of rows and columns of a rectangular matrix A.

The minor Min a_{ij} of a square matrix A is the determinant of a matrix obtained by removing all the elements of the ith row and the jth column.

The cofactor Cof a_{ij} of a square matrix A is the minor Min a_{ij} multiplied by $(-1)^{i+j}$

$$\text{Cof } a_{ij} = (-1)^{i+j} \text{ Min } a_{ij} \tag{1.3.8}$$

With the help of these two definitions the determinant can be computed by Laplace expansion

$$\det A = \sum_{k=1}^{n} a_{ik} \text{ Cof } a_{ik} = \sum_{k=1}^{n} a_{ki} \text{ Cof } a_{ki} \tag{1.3.9}$$

where i is freely chosen. According to this formula, the value of the determinant equals the sum of the elements in the ith row (or column) multiplied by the corresponding cofactors.

The elements a_{ik}^{-1} of the inverse matrix A^{-1} defined by (1.1.4) are now given by

$$a_{ij}^{-1} = \frac{\text{Cof } a_{ji}}{\det A} \tag{1.3.10}$$

Thus only a nonsingular matrix has an inverse; it is also called an *invertible matrix*.

Two important properties of an inverse matrix are that

$$(AB)^{-1} = B^{-1}A^{-1} \quad \text{and} \quad (A^{-1})^{-1} = A \tag{1.3.11}$$

The general formula (1.3.10) takes simpler forms for diagonal and lower- or upper-triangular matrices

$$d_{ii}^{-1} = \frac{1}{d_{ii}} \quad \text{other } d_{ij}^{-1} = 0 \tag{1.3.12}$$

$$l_{ii}^{-1} = \frac{1}{l_{ii}} \quad l_{ij}^{-1} = -\frac{\sum_{k=j}^{i-1} l_{ik} l_{kj}^{-1}}{l_{ii}} \quad i = j+1, \ldots, n$$

$$\text{other } l_{ij}^{-1} = 0 \tag{1.3.13}$$

$$u_{ii}^{-1} = \frac{1}{u_{ii}} \qquad u_{ij}^{-1} = -\frac{\sum_{k=i+1}^{j} u_{ik} u_{kj}^{-1}}{u_{ii}} \qquad i = j-1, \ldots, 1$$

$$\text{other } u_{ij}^{-1} = 0 \tag{1.3.14}$$

The inverses of these three matrices preserve their respective diagonal, lower-triangular, or upper-triangular forms.

If the order n of a square matrix is greater than 3, it is extremely difficult to compute the elements of its inverse by formula (1.3.10). The best way of computing it is by the LU decomposition. According to (1.3.4) and (1.3.11),

$$A^{-1} = U^{-1} L^{-1} \tag{1.3.15}$$

This matrix multiplication fills in all elements of the inverse matrix A^{-1} even if U^{-1} and L^{-1} are the upper- and lower-triangular matrices. Therefore systems (1.1.1) are to be solved without using the inverse matrix.

1.4 DIRECT METHODS

Direct methods yield solutions that contain only the roundoff errors. If the numerical data in the equations were known exactly and all the arithmetic computations carried out without roundoff errors, the solution would be exact.

The necessary and sufficient condition for the existence and uniqueness of the solution is the nonsingularity of the matrix A in (1.1.1b). Then the inverse A^{-1} exists, and the unique solution is given by (1.1.6). If the system is homogeneous, $B = 0$, this solution is trivial, $X = 0$. A homogeneous system can have infinitely many nontrivial solutions if the matrix A is singular. If it is of order n and rank r, then those $n - r$ unknowns whose coefficients do not appear in the basis of A can be reduced to the right-hand side of (1.1.1a) and their values freely chosen. In this way the system becomes nonhomogeneous with a nonsingular matrix, the basis A_r, on the left-hand side; it determines the remaining r unknowns uniquely.

The computation of an inverse matrix of order n requires the evaluation of n^2 cofactors, that is, n^2 determinants of order $n - 1$, and of one determinant of order n. According to (1.3.1b), this amounts to the calculation of $[(n - 1)!]^2$ products, each consisting of $n - 1$ terms; the determinant of the system is then computed by Laplace expansion, Eq. (1.3.9). Once the inverse has been determined, the computation of unknowns x_i for each new vector B requires just n^2 multiplications.

The solution of the system of n linear algebraic equations by Cramer's rule requires the evaluation of $n + 1$ determinants of order n, but n determinants must be computed each time the vector B is replaced by a new one. According to this rule,

$$x_i = \frac{\Delta_i}{\Delta} \qquad i = 1, 2, \ldots, n \tag{1.4.1}$$

where $\Delta = \det A$ and Δ_i is the determinant obtained from Δ by replacing its ith column by the single-column matrix B.

Although Cramer's rule substantially reduces the amount of numerical computation, it is still excessive. Neither method should be used in a system with more than three equations.

The most efficient direct method for solving system (1.1.1a) is gaussian elimination and its compact variants. The method eliminates the unknowns systematically to reduce the given system to an upper-triangular one. If $a_{11} \neq 0$, the x_1 is eliminated from the last $n - 1$ equations by subtracting from the ith equation the $m_{i1} = a_{i1}/a_{11}$ multiple of the first equation. These $n - 1$ equations now take the form

$$a_{22}^{(2)}x_2 + \cdots + a_{2n}^{(2)}x_n = b_2^{(2)}$$
$$\cdots\cdots\cdots\cdots\cdots\cdots\cdots\cdots\cdots$$
$$a_{2n}^{(2)}x_2 + \cdots + a_{nn}^{(2)}x_n = b_n^{(2)}$$

In the same way the x_2 is eliminated from the last $n - 2$ equations by subtracting from the ith equation the $m_{i2} = a_{i2}^{(2)}/a_{22}^{(2)}$ multiple of the second equation. Finally the system is reduced to the triangular form ($a_{1i}^{(1)} = a_{1i}$, $b_1^{(1)} = b_1$)

$$a_{11}^{(1)}x_1 + a_{12}^{(1)}x_2 + \cdots + a_{1n}^{(1)}x_n = b_1^{(1)}$$
$$a_{22}^{(2)}x_2 + \cdots + a_{2n}^{(2)}x_n = b_2^{(2)}$$
$$\cdots\cdots\cdots\cdots\cdots\cdots\cdots\cdots$$
$$a_{nn}^{(n)}x_n = b_n^{(n)}$$

Hence

$$x_n = \frac{b_n^{(n)}}{a_{nn}^{(n)}} \qquad x_i = \frac{b_i^{(i)} - \sum_{k=i+1}^{n} a_{ik}^{(k)} x_k}{a_{ii}^{(i)}} \qquad i = n-1, n-2, \ldots, 1$$

The method breaks down if some of the pivotal elements $a_{ii}^{(i)} = 0$. To prevent this and to increase the accuracy of numerical computation, it is necessary to perform *partial pivoting*, i.e., an exchange of rows in such a way that the element $a_{ki}^{(i)}$, $i \leq k \leq n$, of maximum absolute value becomes the divisor in the elimination process. If all the elements $a_{ki}^{(i)}$ equal zero, the matrix is singular.

Partial pivoting is usually quite satisfactory, and there is no need to carry out a complete pivoting, in which the element of largest absolute value in the whole relevant square part of the matrix is taken as a pivot.

If the rows of any nonsingular matrix are properly reordered by partial pivoting, there always exists a lower-triangular matrix L and an upper-triangular matrix U such that $A = LU$. Their nonvanishing elements are related to the elements in the gaussian elimination by

$$\begin{aligned} l_{ii} &= 1 & l_{i,\,i-k} &= m_{i,\,i-k} & k &= 1, 2, \ldots, i-1 \\ & & u_{i,\,i+k} &= a_{i,\,i+k}^{(i)} & k &= 0, 1, \ldots, n-1 \end{aligned} \qquad (1.4.2)$$

The system (1.1.1b) is now solved in two steps. First the vector Y is to be found satisfying the matrix equation

$$LY = B \tag{1.4.3a}$$

Then the unknowns x_i are the components of the vector X, given by

$$UX = Y \tag{1.4.3b}$$

This procedure has the important advantage over gaussian elimination that the LU decomposition is to be carried out only once for the given A. When the vector B is replaced by a new one, only two simple triangular systems (1.4.3) need be solved.

The Doolittle method is a modification of gaussian elimination such that the elements of L and U are computed directly and can be stored in the same memory locations as the elements of the given matrix A (the diagonal elements of L need not be stored because they always equal 1). Partial pivoting should take place before each step of this method to prevent a breakdown of the method and attain the highest possible accuracy of computation. First, the components of an auxiliary vector P are stored; its first component equals $+1$, and the remaining $n-1$ ones equal 0. Then the element of maximum absolute value in the first column of the matrix A is found; if it is not the element a_{11}, all elements of that row are exchanged with those in the first row. If this exchange takes place, the sign of the first component of the auxiliary vector P must be changed and the number of exchanged row put into memory as its second component. After this first step of partial pivoting the components u_{1j} and l_{i1} are computed

$$u_{1j} = a_{1j} \qquad j = 1, 2, \ldots, n \tag{1.4.4a}$$

$$l_{i1} = \frac{a_{i1}}{u_{11}} \qquad i = 2, 3, \ldots, n \tag{1.4.5a}$$

Then, as the second step of partial pivoting, the element of maximum absolute value among the a_{i2}, $i = 2, 3, \ldots, n$, is found; if it is not the element a_{22}, all elements of that row are exchanged with those in the second row. If this exchange takes place, the sign of the first component of P must be changed again and the number of the exchanged row put into memory as the third component of P. After this second step of partial pivoting the components u_{ij}, $i = 2$, and l_{ij}, $j = 2$, are computed

$$u_{ij} = a_{ij} - \sum_{p=1}^{i-1} l_{ip} u_{pj} \qquad j = i, i+1, \ldots, n \tag{1.4.4b}$$

$$l_{ij} = \frac{a_{ij} - \sum_{p=1}^{j-1} l_{ip} u_{pj}}{u_{jj}} \qquad i = j+1, j+2, \ldots, n \tag{1.4.5b}$$

In this way the partial pivoting continues, and all the nonvanishing components are determined. Then the components of the vector B must be exchanged as indicated by the 2nd to nth components of the auxiliary vector P. The components of the unknown vectors Y and X, given by Eqs. (1.4.3), are now easily computed. The explicit formulas are

$$y_1 = b_1 \qquad y_i = b_i - \sum_{k=1}^{n} l_{ik} y_k \qquad i = 2, 3, \ldots, n \qquad (1.4.6a)$$

$$x_n = \frac{y_n}{u_{nn}} \qquad x_i = \frac{y_i - \sum_{k=i+1}^{n} u_{ik} x_k}{u_{ii}} \qquad i = n-1, n-2, \ldots, 1 \qquad (1.4.6b)$$

Once the matrix U has been computed, the determinant of the matrix A is given by (1.3.7), but this value must be multiplied by the first component of the auxiliary vector P, equal to ± 1, to take the exchanges of rows during the partial pivoting into account.

Program P102 uses the Doolittle method for LU decomposition, for solving the determined system of nonhomogeneous equations, and for evaluating the determinant of the system.

The total number of multiplications and divisions in P102 is of the order of $n^3/3$, n being the number of equations. If the matrix is symmetric, it is possible to modify the gaussian elimination to halve the number of multiplications and divisions and the required memory space, but partial pivoting must not be applied since it would destroy the symmetry. If the matrix is diagonally dominant or symmetric and positive definite, gaussian elimination is always stable, even without pivoting. Its compact version is the *Choleski method*. Since this method requires a repeated computation of square roots generating additional numerical errors, the author recommends applying the Doolittle method without partial pivoting and if one is not sure that the symmetric metric is also positive definite, applying the partial pivoting too.

A substantial saving of arithmetic operations and memory space is achieved for tri- and pentadiagonal matrices. The nonvanishing elements of tridiagonal matrices are denoted

$$\begin{aligned} a_{ii} &\to t_{i2} & i &= 1, 2, \ldots, n \\ a_{i, i-1} &\to t_{i1} & i &= 2, 3, \ldots, n \\ a_{i, i+1} &\to t_{i3} & i &= 1, 2, \ldots, n-1 \end{aligned} \qquad (1.4.7)$$

The tridiagonal matrix T is reduced to an upper- and a lower-triangular matrix U and L with the nonvanishing elements

$$\begin{aligned} u_{ii} &\to p_i & i &= 1, 2, \ldots, n \\ u_{i, i+1} &\to t_{i3} & i &= 1, 2, \ldots, n-1 \end{aligned} \qquad (1.4.8)$$

and

$$\begin{aligned} l_{ii} &= 1 & i &= 1, 2, \ldots, n \\ l_{i, i-1} &\to q_i & i &= 2, 3, \ldots, n \end{aligned} \qquad (1.4.9)$$

Thus the U and L matrices preserve the tridiagonal character. The p_i and q_i are successively computed by the formulas

$$p_1 = t_{12} \tag{1.4.10}$$

$$q_i = \frac{t_{i1}}{p_{i-1}} \qquad p_i = t_{i2} - q_i t_{i-1, 3} \qquad i = 2, 3, \ldots, n$$

The solution of the tridiagonal system

$$TX = B \tag{1.4.11}$$

is obtained by forward and backward substitutions

$$y_1 = b_1 \qquad y_i = b_i - q_i y_{i-1} \qquad i = 2, 3, \ldots, n$$
$$x_n = \frac{y_n}{p_n} \qquad x_i = \frac{y_i - t_{i3} x_{i+1}}{p_i} \qquad i = n-1, n-2, \ldots, 1 \tag{1.4.12}$$

The nonvanishing elements of pentadiagonal matrices are denoted

$$\begin{aligned} a_{ii} &\to f_{i3} & i &= 1, 2, \ldots, n \\ a_{i,i-1} &\to f_{i2} & i &= 2, 3, \ldots, n \\ a_{i,i+1} &\to f_{i4} & i &= 1, 2, \ldots, n-1 \\ a_{i,i-2} &\to f_{i1} & i &= 3, 4, \ldots, n \\ a_{i,i+2} &\to f_{i5} & i &= 1, 2, \ldots, n-2 \end{aligned} \tag{1.4.13}$$

The pentadiagonal matrix F is reduced to an upper- and a lower-triangular matrix U and L with the nonvanishing elements

$$\begin{aligned} u_{ii} &\to p_i & i &= 1, 2, \ldots, n \\ u_{i,i+1} &\to s_i & i &= 1, 2, \ldots, n-1 \\ u_{i,i+2} &\to f_{i5} & i &= 1, 2, \ldots, n-2 \end{aligned} \tag{1.4.14}$$

and

$$\begin{aligned} l_{ii} &= 1 & i &= 1, 2, \ldots, n \\ l_{i,i-1} &\to q_i & i &= 2, 3, \ldots, n \\ l_{i,i-2} &\to v_i & i &= 3, 4, \ldots, n \end{aligned} \tag{1.4.15}$$

The U and L matrices preserve the pentadiagonal character. Their elements are successively computed by the formulas

$$p_1 = t_{13} \qquad s_1 = t_{14}$$
$$q_2 = \frac{t_{22}}{p_1} \qquad p_2 = t_{23} - q_2 s_1 \qquad s_2 = t_{24} - q_2 s_1$$
$$v_i = \frac{t_{i1}}{p_{i-2}} \qquad q_i = \frac{t_{i2} - v_i s_{i-2}}{p_{i-1}} \tag{1.4.16}$$
$$\qquad\qquad\qquad\qquad i = 3, 4, \ldots, n$$
$$p_i = t_{i3} - q_i s_{i-1} - v_i t_{i-2, 5}$$
$$s_i = t_{i4} - q_i t_{i-1, 5} \qquad i = 3, 4, \ldots, n-1$$

The solution of the pentadiagonal system

$$FX = B \qquad (1.4.17)$$

is obtained by forward and backward substitutions

$$y_1 = b_1 \qquad y_2 = b_2 - q_2 y_1$$
$$y_i = b_i - q_i y_{i-1} - v_i y_{i-2} \qquad i = 3, 4, \ldots, n$$
$$x_n = \frac{y_n}{p_n} \qquad x_{n-1} = \frac{y_{n-1} - s_{n-1} x_n}{p_{n-1}} \qquad (1.4.18)$$

$$x_i = \frac{y_i - s_i x_{i+1} - t_{i5} x_{i+2}}{p_i} \qquad i = n-2, n-3, \ldots, 1$$

The formulas for tri- and pentadiagonal matrices are special cases of the general formulas (1.4.3) to (1.4.5). The determinant of the system is given by Eq. (1.3.7), which reduces for both tri- and pentadiagonal matrices to

$$\det T = \det F = \prod_{i=1}^{n} p_i \qquad (1.4.19)$$

These formulas are used in Programs P103 and P104 for LU decomposition, for solving tri- and pentadiagonal systems of equations, and for evaluating the determinant of the system. They are faster than P102 because they prevent many multiplications by zero which occur in P0102 when applied to tri- and pentadiagonal systems.

1.5 ERROR ANALYSIS

When the vector $X = X_c$, computed as described in Sec. 1.4, is substituted into

$$R = B - AX_c \qquad (1.5.1)$$

the components of the residual vector R do not vanish because of errors of numerical computation. Even the fact that the components of R are small relative to the components of X does not imply that the solution X_c can always be considered satisfactory.

In order to determine the conditions under which

$$|r_i| \ll |x_i| \qquad i = 1, 2, \ldots, n \qquad (1.5.2)$$

does imply a satisfactory solution, a scalar called a *norm* can be introduced to provide a convenient estimation of the size of a vector and a matrix. The length of a vector X in euclidean n-dimensional space certainly can be considered as a fair measure of its size. It is also called a *euclidean norm* and defined by

$$\|X\|_2 = (\tilde{X}X)^{1/2} = \left(\sum_{i=1}^{n} |x_i|^2\right)^{1/2} \qquad (1.5.3)$$

A generalization of this definition is the weighted p norm of a vector

$$\|X\|_{p,\,w} = \left(\sum_{i=1}^{n} w_i |x_i|^p\right)^{1/p} \tag{1.5.4}$$

where p is some number between 1 and ∞ and the weights $w_1, w_2, \ldots$ are fixed positive numbers. If $p = 2$ and $w_i = 1$ for all i, Eq. (1.5.4) reduces to the euclidean norm (1.5.3). The maximum norm of a vector is the limit of (1.5.4) for $p \to \infty$ and all $w_i = 1$

$$\|X\|_\infty = \lim_{p \to \infty} \left(\sum_{i=1}^{n} |x_i|^p\right)^{1/p} = \max_{1 \leq i \leq n} |x_i| \tag{1.5.5}$$

The vector norms have the following properties, analogous to the usual concept of length:

$$\|X\| \begin{cases} > 0 & \text{if } X \neq 0 \\ = 0 & \text{if } X = 0 \end{cases}$$

$$\|gX\| = |g|\,\|X\| \quad g \text{ any scalar} \tag{1.5.6}$$

$$\|X + Y\| \leq \|X\| + \|Y\|$$

These properties can easily be verified for the maximum norm.

The matrix norms must have similar properties

$$\|A\| \begin{cases} > 0 & \text{if } A \neq 0 \\ = 0 & \text{if } A = 0 \end{cases} \tag{1.5.7a}$$

$$\|gA\| = |g|\,\|A\| \quad g \text{ any scalar} \tag{1.5.7b}$$

$$\|A + B\| \leq \|A\| + \|B\| \tag{1.5.7c}$$

$$\|AB\| \leq \|A\|\,\|B\| \tag{1.5.7d}$$

If a matrix norm and a vector norm are related in such a way that

$$\|AX\| \leq \|A\|\,\|X\|$$

is satisfied for any A and X, the two norms are consistent. A matrix-bound norm subordinated to the vector norm is defined by

$$\|A\| = \max \frac{\|AX\|}{\|X\|} \tag{1.5.8}$$

where the maximum is taken over all nonvanishing vectors X. Since the euclidean norm of a matrix is difficult to calculate, all further considerations are based on the maximum norm of a vector (1.5.5) and the subordinated matrix-bound norm of a matrix

$$\|A\|_\infty = \max \frac{\|AX\|}{\|X\|} = \max_{1 \leq i \leq n} \left(\sum_{j=1}^{n} |a_{ij}|\right) \tag{1.5.9}$$

Thus the maximum norm of a vector (1.5.5) equals the maximum of the absolute values of its components. The maximum norm of a matrix [Eq. (1.5.9)] equals the maximum of the sums of absolute values of all elements in one row. These norms are easy to calculate, and if $|A|$ denotes a matrix with the elements $|a_{ij}|$,

$$\||A|\| = \|A\| \tag{1.5.10}$$

This relation does not hold for the matrix-bound norm subordinated to the euclidean norm.

Some authors prefer the concepts of a *spectral norm* and *spectral radius*. The former is not used in this book, the latter only in Sec. 1.7; their definitions are given in Sec. 7.1.

After this digression, the error analysis can be resumed. If vector B is eliminated from (1.5.1) with the help of (1.1.1b), the residual R is given by

$$R = AE \tag{1.5.11a}$$

where E is the error vector

$$E = X - X_c \tag{1.5.12}$$

Another form of (1.5.11a) is

$$E = A^{-1}R \tag{1.5.11b}$$

The norm of E computed from Eqs. (1.5.11) lies, according to (1.5.7d), within the interval

$$\frac{\|R\|}{\|A\|} \leq \|E\| \leq \|R\| \|A^{-1}\| \tag{1.5.13}$$

Similarly the norm of X computed from (1.1.1b) and (1.1.6) with the help of (1.5.7d) lies within the interval

$$\frac{\|B\|}{\|A\|} \leq \|X\| \leq \|B\| \|A^{-1}\| \tag{1.5.14}$$

Hence the relative error $\|E\|/\|X\|$ expressed in terms of the relative residual $\|R\|/\|B\|$ is bounded within the limits

$$\frac{1}{K(A)} \frac{\|R\|}{\|B\|} \leq \frac{\|E\|}{\|X\|} \leq K(A) \frac{\|R\|}{\|B\|} \tag{1.5.15}$$

where

$$K(A) = \|A\| \|A^{-1}\| \tag{1.5.16}$$

is the condition number of the system. Its numerical value depends to a great extent on the matrix norm used, but it is always greater than 1 (it equals 1 only if $A = I$). Equation (1.5.15) indicates that a small value of the relative residual $\|R\|/\|B\|$ is a reliable indication of a small relative error $\|E\|/\|X\|$ only if the condition number $K(A)$ is close to 1. On the other hand, if $K(A) \gg 1$, a small value of the relative residual $\|R\|/\|B\|$ does not imply that the relative error of the solution $\|E\|/\|X\|$ is small too; the error may be small, but it also may be very large. If $K(A) \gg 1$, the system is said to be *ill-conditioned*.

Evaluation of the condition number is not easy because the inverse matrix A^{-1} must be computed, making this standard technique of restricted practical value. The author has therefore developed a different approach, based on the geometrical interpretation of system (1.1.1a). Each equation of that system, e.g.,

$$a_{i1}x_1 + a_{i2}x_2 + \ldots + a_{in}x_n = b_i \qquad i = 1, 2, \ldots, m$$

defines a hyperplane in an n-dimensional euclidean space. It can be reduced to its normal form

$$c_{i1}x_1 + c_{i2}x_2 + \ldots + c_{in}x_n = d_i \qquad i = 1, 2, \ldots, m \qquad (1.5.17a)$$

with

$$c_{ij} = \frac{a_{ij}}{\Lambda_i} \qquad \begin{array}{l} i = 1, 2, \ldots, m \\ j = 1, 2, \ldots, n \end{array}$$

$$d_i = \frac{b_i}{\Lambda_i} \qquad \Lambda_i = \pm \left(\sum_{k=1}^{n} a_{ik}^2 \right)^{1/2} \qquad i = 1, 2, \ldots, m \qquad (1.5.18)$$

The sign of Λ_i is to be chosen such that $d_i \geq 0$. The left-hand side of (1.5.17a) can now be interpreted as a scalar product of a radius vector $\mathbf{r}$ from the origin of the coordinate system to a point on a hyperplane, its jth component being x_j, and a unit vector $\mathbf{n}_i$ perpendicular to the hyperplane and directed from the origin of the coordinate system, its jth component being c_{ij}. The vector form of Eq. (1.5.17a) is therefore

$$\mathbf{r} \cdot \mathbf{n}_i = d_i \qquad (1.5.17b)$$

where d_i is obviously the distance of the hyperplane from the origin. Because of the particular choice of the positive direction of $\mathbf{n}_i$, d_i is always positive (or zero if the hyperplane goes through the origin). The components c_{ij} of the unit vector $\mathbf{n}_i$ are the directional cosines. The angle ψ_{ij} between two hyperplanes with the unit vectors $\mathbf{n}_i$ and $\mathbf{n}_j$ is given by the well-known formula

$$\cos \psi_{ij} = \mathbf{n}_i \cdot \mathbf{n}_j = \left| \sum_{k=1}^{n} c_{ik} c_{jk} \right| \qquad (1.5.19a)$$

or, with the coefficients a_{ij},

$$\cos \psi_{ij} = \frac{\left| \sum_{k=1}^{n} a_{ik} a_{jk} \right|}{|\Lambda_i \Lambda_j|} \qquad (1.5.19b)$$

In Eqs. (1.5.19) the sign of $\cos \psi_{ij}$ has been chosen positive since the negative sign corresponds to the complementary angle $\pi - \psi_{ij}$.

System (1.1.1a) is ill-conditioned if at least two hyperplanes are almost parallel, i.e., if at least one angle ψ_{ij} is close to 0. Since the cosine of such an angle is close to 1, it is appropriate to compute $1 - \cos \psi_{ij}$ for all the combinations of hyperplanes, call them indicators, and define the condition indicator of the system described by matrix A by

$$C(A) = \min_{\substack{1 \leq k \leq n \\ k \leq l \leq n}} (1 - \cos \psi_{kl}) \qquad (1.5.20)$$

If $C(A) \to 0$, angle ψ_{ij}, measured in radians, is

$$\psi_{ij} \approx [2C(A)]^{1/2} \qquad (1.5.21)$$

Both the condition number $K(A)$, defined by (1.5.16), and the condition indicator $C(A)$, given by (1.5.20), depend only on the elements of matrix A and not on the vector B. The $C(A)$ lies within the interval

$$0 \leq C(A) \leq 1 \qquad (1.5.22)$$

as evident from the definition (1.5.20). A sufficient, but in general not necessary, condition for matrix A to be singular is $C(A) = 0$.

A simple example illustrates the standard and the geometrical approaches. Let the system consist of two equations similar to those constructed by Kahan [9].‡ Its matrices are

$$A = \begin{bmatrix} 1.2968 & 0.8648 \\ 0.2161 & 0.1441 \end{bmatrix} \qquad B = \begin{bmatrix} 0.8640 \\ 0.1440 \end{bmatrix}$$

and its exact solution is

$$X = \begin{bmatrix} 2 \\ -2 \end{bmatrix}$$

All the subsequent computation is carried out in 10-digit floating-point arithmetic. The determinant of A given by (1.3.2) has the value

$$\det A = 0.1868688800 - 0.1868832800 = -1.440 \times 10^{-5}$$

and the inverse matrix, computed by (1.3.10), is

$$A^{-1} = \begin{bmatrix} -1.000694444 \times 10^4 & 6.005555556 \times 10^4 \\ 1.500694444 \times 10^4 & -9.005555556 \times 10^4 \end{bmatrix}$$

Hence the maximum norms are

$$\|A\| = 2.1616 \qquad \|A^{-1}\| = 1.9506 \times 10^5$$

and the condition number given by (1.5.16) is

$$K(A) = 2.2710 \times 10^5$$

The system is thus ill-conditioned, as already indicated by the small value of its determinant relative to the elements of its matrix. The solution computed by

‡ Numbered references appear at the end of Part One.

(1.1.6) is

$$X_c = \begin{bmatrix} 2.000005000 \\ -2.000000000 \end{bmatrix}$$

In spite of the high condition number, the X_c compared with the exact X is very satisfactory.

This is the standard procedure. The computation has been so easy because the system consists of only two equations. In the geometrical approach the condition indicator $C(A)$ and the corresponding angle ψ_{12} computed using (1.5.20) and (1.5.21) take the values

$$C(A) = 7.0 \times 10^{-10} \quad \text{and} \quad \psi_{12} = 3.74 \times 10^{-5}$$

The system has been solved by program P102 (Doolittle method) with the result

$$X_c = \begin{bmatrix} 1.999993994 \\ -1.999990994 \end{bmatrix}$$

The condition indicator is very low, but the solution is nevertheless very good.

If vector B is now replaced by

$$B_1 = \begin{bmatrix} 0.4320 \\ 0.0720 \end{bmatrix} \quad \text{or} \quad B_2 = \begin{bmatrix} 1.2976 \\ 0.2162 \end{bmatrix}$$

the exact solutions are

$$X_1 = \begin{bmatrix} 1 \\ -1 \end{bmatrix} \quad \text{and} \quad X_2 = \begin{bmatrix} -1 \\ 3 \end{bmatrix}$$

Program P102 yields again very satisfactory results

$$X_{1c} = \begin{bmatrix} 0.9999963965 \\ -0.9999945967 \end{bmatrix} \quad \text{and} \quad X_{2c} = \begin{bmatrix} -0.9999939944 \\ 2.999990994 \end{bmatrix}$$

The condition indicator has the same value because matrix A has not been changed.

These results, accurate when rounded off to the first four or five digits, are the consequence of 10-digit floating-point arithmetic. They are surprisingly accurate with regard to the very low value of the condition indicator, but only as long as all the elements in matrix A and the vectors B, B_1, and B_2 can be taken as exact, i.e., as having zeros in all further decimal places. If an element of matrix A, say a_{21}, is perturbed by adding ψ_{12} to it, the perturbed $a'_{21} = 0.2161374000$ still equals its unperturbed value if rounded off to four decimal places, but the solution computed by Program P102

$$X'_c = \begin{bmatrix} 0.6161289233 \\ 0.07516652657 \end{bmatrix}$$

does not even resemble the unperturbed solution X_c. This is the true meaning of an ill-conditioned system: the roundoff errors in the input data completely distort its solution. On the other hand, a well-conditioned system is one in which the roundoff errors in the elements of A and B cause only minor, acceptable changes in the computed vector X_c.

A system

$$A_N X = B_N$$

in which only the a_{11} and b_1 have been changed to $a_{11} = 1.2744$ and $b_1 = 0.8192$ still has the exact solution X, but its condition indicator is 3.28×10^{-5} and $\psi_{12} = 0.810 \times 10^{-2}$. Its solution X_{Nc} computed by Program P102 is identical with the exact X. If element a_{11} is now perturbed by subtracting 5.0×10^{-5} from it, it still equals its unperturbed value if rounded off to four decimal digits, and the solution

$$X'_{Nc} = \begin{bmatrix} 1.995565440 \\ -1.993349698 \end{bmatrix}$$

is quite close to X_{Nc}. But if the element is perturbed by subtracting from it the present $\psi_{12} = 0.810 \times 10^{-2}$, the solution is

$$X''_{Nc} = \begin{bmatrix} 1.470587039 \\ -1.206064254 \end{bmatrix}$$

and if instead of a_{11} the element a_{12} is perturbed by subtracting from it the present ψ_{12}, the solution

$$X'''_{Nc} = \begin{bmatrix} 3.564803048 \\ -4.346661614 \end{bmatrix}$$

differs even more from the exact X_{Nc}

The following conclusions can be drawn:

1. To keep the roundoff errors as small as possible the calculation should be carried out with high precision (double precision if available), especially if the system consists of a large number of equations.

2. If any of the coefficients c_{ik} or c_{jk} or the d_i or d_j of the equations from which the condition indicator has been computed by (1.5.20) are perturbed by adding or subtracting the angle ψ_{ik} determined by (1.5.21), the solution of an ill-conditioned system differs from the unperturbed one to an extent that is not negligible. In other words, the roundoff errors in the input generate negligible changes in the solution only if they are much smaller than ψ_{ik} (at least two orders of magnitude); in this case the system is said to be well-conditioned.

Program P101 computes the Λ_i occurring in Eqs. (1.5.18) to convert a_{ij} and b_i into c_{ij} and d_i and all the indicators of the given system and selects the condition indicator from them. Optionally it also computes the components of the vector B_0 for a chosen vector X_0 (its components are to be of the same order of magnitude as those of the expected solution X_c). If Program P102 with the same matrix A and computed vector B_0 yields a solution X_{0c} that differs negligibly from the chosen vector X_0, the numerical errors in the LU decomposition are acceptable; Program P102 can now be rerun with the given B (matrices L and U being taken from memory, not recomputed) to give a solution that differs from the unknown true solution to the same extent that X_0 and X_{0c} differ from each other.

If the system is ill-conditioned, the solution, even if computed to high precision, is unreliable. The only remedy is a reformulation of those equations which generate low values of the corresponding indicators because they contain bits of information too close to each other.

1.6 OVERDETERMINED SYSTEMS OF EQUATIONS

One way to increase the accuracy of a solution of a system of linear algebraic equations whose data are the result of measurements is to carry out more observations or experiments than the number of unknowns. The system thus becomes overdetermined, consisting of m equations for n unknowns, $m > n$.

The least-squares technique used in solving an overdetermined system effectively reduces the influence of random errors in measurements, i.e. errors which are statistically independent and contain no systematic errors. It minimizes the euclidean norm (1.5.5) of the residual vector (1.5.1).

The technique reduces the overdetermined system of m equations

$$AX - B = 0 \qquad (1.6.1)$$

to a determined system of n normal equations for n unknowns

$$\tilde{A}(AX - B) = 0 \qquad (1.6.2a)$$

After matrix multiplication it takes the form

$$CX = D \qquad (1.6.2b)$$

with the matrices

$$C = \tilde{A}A \qquad D = \tilde{A}B \qquad (1.6.3a)$$

Explicitly, their elements are

$$c_{ij} = \sum_{k=1}^{m} \tilde{a}_{ik} a_{kj} = \sum_{k=1}^{m} a_{ki} a_{kj} \qquad d_i = \sum_{k=1}^{m} \tilde{a}_{ik} b_k = \sum_{k=1}^{m} a_{ki} b_k \qquad (1.6.3b)$$

To obtain the element c_{ij} of the square matrix C all elements of the ith column of the rectangular matrix A are to be multiplied by the elements of its jth column and all these products summed. Similarly, to compute the element d_i of the single-column matrix D all the elements of the ith column of matrix A are to be multiplied by the elements of the single-column matrix B and all these products summed.

Program P105 computes the elements of matrices C and D from the given matrices A and B. Since matrix C is symmetric and positive definite, partial pivoting nccd not be applied when the system of normal equations (1.6.2b) is solved by Program P102.

To prevent loss of significant information in forming the normal equations, which may result in a singular matrix C, the elements c_{ij} should always be computed in double precision. If matrix C still becomes singular, it indicates that the random errors in measurements are so small that $m - n$ equations of the original system (1.6.1) can be dropped and the remaining n equations for n unknowns solved by Program P102 with partial pivoting.

1.7 ITERATIVE METHODS

Large linear systems are advantageously solved not by direct methods but by iteration, especially when matrix A is sparse. These iterative methods start from an initial approximation $X^{(0)}$, often taken to be $X^{(0)} = 0$, and compute the sequence $X^{(1)}, X^{(2)}, \ldots$, which converges for certain types of matrices A to the true solution X.

These approximations are determined in the *Jacobi method* by

$$x_i^{(k+1)} = \frac{b_i}{a_{ii}} - \sum_{j=1}^{i-1} \frac{a_{ij}}{a_{ii}} x_j^{(k)} - \sum_{j=i+1}^{n} \frac{a_{ij}}{a_{ii}} x_j^{(k)} \tag{1.7.1a}$$

in the Gauss-Seidel method by

$$x_i^{(k+1)} = \frac{b_i}{a_{ii}} - \sum_{j=1}^{i-1} \frac{a_{ij}}{a_{ii}} x_j^{(k+1)} - \sum_{j=i+1}^{n} \frac{a_{ij}}{a_{ii}} x_j^{(k)} \tag{1.7.2a}$$

and in the *successive-overrelaxation* (SOR) *method* by

$$x_i^{(k+1)} = (1-\omega)x_i^{(k)} + \frac{b_i}{a_{ii}/\omega} - \sum_{j=1}^{i-1} \frac{a_{ij}}{a_{ii}/\omega} x_j^{(k+1)} - \sum_{j=i+1}^{n} \frac{a_{ij}}{a_{ii}/\omega} x_j^{(k)} \tag{1.7.3a}$$

The SOR method converges only if the relaxation parameter ω lies within the interval

$$0 < \omega < 2 \tag{1.7.4}$$

The Jacobi method uses the improved values $x_i^{(k+1)}$ after one step of iteration has been completed, while the Gauss-Seidel method uses them immediately, as soon as they are computed. In the SOR method the improved values are given by combining the old values $x_i^{(k)}$ with the improved values of the Gauss-Seidel method $x_{i,\,\text{GS}}^{(k+1)}$

$$x_i^{(k+1)} = (1-\omega)x_i^{(k)} + \omega x_{i,\,\text{GS}}^{(k+1)} \tag{1.7.3b}$$

to maximize the rate of convergence by a proper choice of the relaxation parameter ω.

A condition of all three methods is that the diagonal terms a_{ii} not vanish. If some do and matrix A is nonsingular, it is always possible to get all diagonal terms different from zero and as large as possible by permutation of rows and columns.

All three methods can be expressed in the matrix form

$$X^{(k+1)} = MX^{(k)} + C \qquad (1.7.5)$$

which is the general form for a stationary iterative method. Matrix A can be written as a product of the diagonal matrix D with elements equal to a_{jj} and the sum of a lower-triangular, diagonal, and upper-triangular matrix

$$A = D(L + I + U) \qquad (1.7.6)$$

The diagonal matrix I is the identity matrix. Matrices M and C in the Jacobi method are

$$M = -(L + U) \quad \text{and} \quad C = D^{-1}B \qquad (1.7.1b)$$

in the Gauss-Seidel method

$$M = -(I + L)^{-1}U \quad \text{and} \quad C = D^{-1}B \qquad (1.7.2b)$$

and in the SOR method

$$M = (I + \omega L)^{-1}[(1 - \omega)I - \omega U] \quad \text{and} \quad C = \omega D^{-1}B \qquad (1.7.3c)$$

A necessary and sufficient condition for the convergence of a stationary iterative method is that the spectral radius of the matrix M be smaller than 1

$$\rho(M) < 1 \qquad (1.7.7)$$

The spectral radius is defined by Eq. (7.1.8a) in terms of eigenvalues of the matrix M; its upper bound, which can be substituted into the condition (1.7.7), is expressed in terms of elements of the matrix M by Eq. (7.1.8b).

Experience shows that the Jacobi and Gauss-Seidel methods are convergent for strictly (row) diagonally dominant matrices A; the iterations may start from an arbitrary approximation. The Gauss-Seidel method also converges for matrices A that are real, symmetric, and positive definite; usually it converges faster than the Jacobi method. It may converge when the Jacobi method does not, but examples are known of the Gauss-Seidel method's diverging while the Jacobi method is convergent. The SOR method converges for all ω within the interval (1.7.4) when matrix A is real, symmetric, and positive definite. To attain a fast convergence a proper choice of the parameter ω is crucial. This problem is the subject of many papers. For some important classes of matrices the optimal ω is known.

The roundoff errors in iterative methods are less serious than in direct ones because at each iteration the elements of the original matrices A and B are used (there is no propagation of errors), but they can become a difficult problem when the system is ill-conditioned. Iterative methods may not converge, and even when they do the number of iterations may be excessive (large systems may require several hundreds of iterations to obtain a solution correct to four or five digits). Since the number of multiplications in one iteration with a full matrix A is of the order of n^2, and since the number of multiplications and divisions in gaussian elimination is of the order of $n^3/3$, iterative methods are faster if the number of iterations k is

$$k < \frac{n}{3} \tag{1.7.8}$$

Program P106 contains all three iterative methods. The computation stops either when for a chosen δ the condition

$$\delta > \max_{1 \leq i \leq n} \left| 1 - \frac{x_i^{(k+1)}}{x_i^{(k)}} \right| \tag{1.7.9}$$

is met or when the number of iterations surpasses a chosen limit.

Programs P103 and P104 for tri- and pentadiagonal systems using a direct method are usually faster than any iterative method in Program P106.

INTERPOLATION, APPROXIMATION, AND NUMERICAL DIFFERENTIATION

Practically all the numerical methods to be discussed require an approximation of a function defined either analytically or by a discrete set of numerical data. In this chapter the function (and possibly its derivative as well) is given numerically. Either these data can be considered exact (data from mathematical tables) or they are the result of measurements or numerical computation and therefore contain unknown errors. When the data are exact, the approximating function is required to go through all the given points and to have there the prescribed derivatives if they are known. This approximation is called *interpolation*. In the other case the least-squares technique is applied to eliminate (or at least diminish) the random errors in the data defining the function. As a consequence, the approximating function smoothes the given data and does not go through the given points. In both cases the approximating function determines the values of the function and also (optionally) its first and second derivatives at any point within the range of the function. Sometimes only the derivatives are to be computed at a discrete set of points defining the function. Numerical differentiation is one of the most difficult problems of numerical analysis despite the fact that the formulas for derivatives with equally spaced abscissas are so easy to implement.

2.1 INTERPOLATION TECHNIQUES

Let the approximating function $y(x)$ be described on the interval $[a, b]$ by a finite series of properly chosen functions $s_k(x)$

$$y = \sum_{j=1}^{p} A_j s_j(x) \qquad (2.1.1)$$

The function y is given by a discrete set of n data $f(x_i) = f_i$ at arbitrarily spaced

2.2 THEORY

abscissas $x_1 < x_2 < \cdots < x_n$ and possibly also by m derivatives $f'(x_i) = f'_i$, $f''(x_i) = f''_i, \ldots, f^{(m)}(x_i) = f^{(m)}_i$ at each point x_i.

Since at $x = x_i$ the approximating function has to take the value f_i and its derivatives here should equal $f^{(j)}_i$, the unknown coefficients A_j of the series must satisfy a system of the linear algebraic equations

$$\sum_{j=1}^{p} A_j s_j^{(k)}(x_i) = f_i^{(k)} \quad \begin{aligned} i &= 1, 2, \ldots, n \\ k &= 0, 1, \ldots, m \end{aligned} \quad (2.1.2)$$

The superscript k indicates the order of the derivatives. The $s_j^{(0)}(x_i)$ and $f_i^{(0)}$ are the functions $s_j(x_i)$ and f_i, respectively. There are $(1 + m)n$ given data. System (2.1.2) becomes determined if

$$p = (1 + m)n \quad (2.1.3)$$

Once the system has been solved, the approximating function (2.1.1) determines the value $y(x)$ at any point within the interval

$$x_1 \leq x \leq x_n \quad (2.1.4)$$

The jth derivative is to be computed using

$$y^{(k)} = \sum_{j=1}^{p} A_j s_j^{(k)}(x) \quad (2.1.5)$$

If $x < a$ or $x > b$, it is still possible to compute $y(x)$ and $y^{(j)}(x)$ by (2.1.1) and (2.1.5), but this extrapolation rapidly diminishes the accuracy of the computed data.

If the functions $s_j(x)$ are chosen as power functions,

$$s_j(x) = x^{j-1} \quad (2.1.6)$$

the derivatives $s_j^{(k)}(x)$ are easy to calculate but the unknown coefficients A_j are to be determined by solving the system of p linear algebraic equations. This computation may generate roundoff errors in A_j that are not negligible. Therefore the functions $s_j(x)$ should be so chosen that they make solution of system (2.1.2) easy.

In Lagrange interpolation $m = 0$; that is, the discrete set of data consists only of n pairs (x_i, f_i) but contains no derivatives, each function $s_j(x)$ is a parabola of $(j - 1)$th degree called a *Lagrange function*, and the coefficients $A_j = f_j$. The computation of derivatives $s_j^{(k)}(x)$ is rather difficult, and if new points are added to the set of data all the Lagrange functions must be recomputed.

In Newton interpolation again $m = 0$, the function $s_j(x)$ are parabolas of different degrees [$s_j(x)$ is a parabola of $(j - 1)$th degree]. The coefficients A_j must be computed by certain formulas, but they need not be recomputed if one or more points are added to the set of given data.

In both Lagrange and Newton interpolation the roundoff errors are negligible. Far more serious are the errors generated by truncating an infinite series to a finite one (2.1.1). For given n pairs of data (x_i, f_i), the function (2.1.1) is a parabola of $(n - 1)$th degree, exactly the same in Lagrange or Newton interpolation or if

$s_j(x)$ are power functions x^{j-1}. Truncation errors may become unacceptably high if the intervals between x_i and x_{i+1} are large; if the number of data points n is also large, they may give rise to wild oscillations in $y(x)$ and to still greater ones in $y'(x)$ and $y''(x)$, making the interpolation worthless. To avoid this the interpolation interval should be divided into several subintervals and the interpolation carried out piecewise.

Another way to decrease truncation errors is by redistributing the n abscissas x_i across the interpolation interval, e.g., by replacing the equally spaced abscissas by Chebyshev abscissas. A still more efficient way to decrease the truncation errors is to use piecewise Hermite interpolation, in which the function is defined also by its first (and possibly also second) derivative.

If the derivatives are not known and cannot be computed, the function can be interpolated by a cubic spline, in which the function $y(x)$ is always a cubic parabola between two consecutive points x_i and x_{i+1} determined by f_i and by the requirement that the first and second derivatives be continuous when crossing to the neighboring parabolas. Properly applied, the cubic spline gives very satisfactory results.

The trigonometric interpolation is usually applied for periodic functions. In it the functions $s_j(x)$ are not polynomials but, as in Lagrange interpolation, make the system (2.1.2) with $m = 0$ diagonal.

If the function $y(x)$ to be interpolated is approaching the x axis asymptotically and must not cross it [as $y = e^{-x^2}$ or $y = 1/(a + bx^n)$, $n > 0$], it is advisable to interpolate the function $u = \ln y$ and then return to $y = e^u$. If the function is asymptotically approaching the straight line $w = cx$, the interpolated function should be $u = \ln(y - w)$; then $y = w + e^u$. These two examples show that a successful interpolation requires mathematical skill and a particular approach to each given problem.

Valuable indicators of the reliability of an interpolation are the first and second derivatives of the interpolated function computed in most programs. If they vary excessively from one given point to another, it is a positive indication that the interpolation is inadequate and another method should be attempted.

2.2 LAGRANGE INTERPOLATION

In Lagrange interpolation the function is given by n points (x_i, f_i) whose abscissas can be freely chosen. The functions $s_j(x)$ are Lagrange functions $l_j(x)$ given by

$$l_j(x) = \frac{\prod_{i=1, i \neq j}^{n} (x - x_i)}{\prod_{i=1, i \neq j}^{n} (x_j - x_i)} = \begin{cases} 1 & \text{if } x = x_j \\ 0 & \text{if } x = x_i \neq x_j \end{cases} \quad (2.2.1)$$

Thus Lagrange functions make the system (2.1.2) of p linear algebraic equations ($k = 0$) diagonal; its solution is $A_i = f_i$. To reduce the number of multiplications

2.4 THEORY

the series (2.1.1) is written

$$y = \frac{\sum_{j=1}^{n} c_j}{x - x_j} \prod_{i=1}^{n} (x - x_i) \qquad (2.2.2a)$$

where

$$c_j = \frac{f_j}{\prod_{i=1, i \neq j}^{n} (x_j - x_i)} \qquad (2.2.2b)$$

Precautions must be taken to avoid division by zero in (2.2.2a) when $x = x_j$. The easiest way, used in Programs P201 and P202, is to require that

$$\frac{\sum_{j=1}^{n} c_j}{x - x_j} = c_j \qquad \text{if } x = x_j \qquad (2.2.3a)$$

and

$$x - x_j = 1 \qquad \text{if } x = x_j \qquad (2.2.3b)$$

in the product $\prod_{i=1}^{n} (x - x_i)$ in (2.2.2a). The memory space for f_j can now be used for storing the c_j.

The truncation error of Lagrange interpolation for n given points is

$$e_n = \frac{f^{(n)}(\xi)}{n!} \prod_{i=1}^{n} (x - x_i) \qquad (2.2.4a)$$

where ξ is some abscissa within the interpolation interval

$$a < \xi < b \qquad (2.2.4b)$$

Since the nth derivative $f^{(n)}(x)$ and the abscissa ξ are almost always unknown, the formula pinpoints the dangers of interpolation: the numerical value of the derivatives usually decreases very slowly with increasing n, but most likely it grows rapidly and in some cases it may not even exist beyond some n. Therefore a piecewise interpolation should be applied if the interpolation interval is covered by many points. Unless the subintervals between x_i and x_{i+1} are very small, n should not exceed 6 or 8. The interpolation error is decisively influenced by the value of the product in (2.2.4a). For x_i fixed the product becomes smaller if x is close to the center of the interpolation interval $(a + b)/2$. Further, if the interval $b - a$ is short, all the cofactors and the product itself are small. By choosing the x_i to be Chebyshev abscissas, defined on the interval $[a, b]$ by

$$x_i = \tfrac{1}{2}(b + a) + \tfrac{1}{2}(b - a) \cos\left[\frac{\pi}{n}(i - \tfrac{1}{2})\right] \qquad i = 1, 2, \ldots, n \qquad (2.2.5a)$$

the maximum value of the product in (2.2.4a) is minimized

$$\max_{a \leq x \leq b} \prod_{i=1}^{n} (x - x_i) = 2\left(\frac{b - a}{4}\right)^n \qquad (2.2.5b)$$

Since Chebyshev abscissas are shifted to both ends of the interpolation interval, the truncation error for a given function and a given number of points is smaller for x close to the center of interval and equally spaced abscissas but greater for x close to the ends of the interval than for Chebyshev abscissas. In other words, Chebyshev abscissas equalize the truncation error across the interpolation interval.

Program P201 computes the value $y(x)$ of a function defined by an arbitrary number n of points (x_i, f_i) with the help of Lagrange interpolation. A subroutine computes Chebyshev abscissas for the given interval $[a, b]$.

If the abscissas are equally spaced with the interval h, it is appropriate to introduce a new variable X related to the variable x by

$$X = 1 + \frac{x - x_1}{h} \tag{2.2.6a}$$

Then $\quad x - x_i = h(X - i) \quad$ and $\quad x_j - x_i = h(j - i) \tag{2.2.6b}$

Equations (2.2.2) are thus reduced to

$$y = \frac{\sum_{j=1}^{n} c_j}{X - k} \prod_{i=1}^{n} (X - i) \tag{2.2.7a}$$

and

$$c_j = \frac{f_j}{\prod_{i=1, i \neq k}^{n} (j - i)} \tag{2.2.7b}$$

The interval h appears explicitly only in Eq. (2.2.6a) when the new variable X is computed, not in Eqs. (2.2.7). The product in (2.2.7b) is always an integer. The same precaution as in (2.2.2a) must be taken to avoid division by zero in (2.2.7a).

Program P202 computes the value $y(x)$ of a function defined by an arbitrary number n of equally spaced points with the interval h and ordinates f_i by Lagrange interpolation. It requires less memory space than P201, and there are no roundoff errors in computing the products in (2.2.7b).

Both Programs P201 and P202 contain a subroutine that makes piecewise interpolation easy. It shifts all points either to the left or to the right and asks for a new point. The independent variable x can thus always be chosen close to the center of the interpolation subinterval to keep the truncation error as small as possible. The subroutine is especially useful in Program P202 for equally spaced abscissas.

2.3 NEWTON INTERPOLATION

In Newton interpolation the function is again given by n points (x_i, f_i) whose abscissas can be freely chosen. The functions $s_j(x)$ are here defined by

$$s_1(x) = 1 \qquad s_j(x) = \prod_{i=1}^{j-1} (x - x_i) \qquad j = 2, 3, \ldots, n \tag{2.3.1}$$

2.6 THEORY

The coefficients $A_j = A_{1+k}$ are the kth divided difference at the points $x_i, \ldots, x_{i+k}$. The zeroth divided difference is the function itself

$$A_1 = A_{1+0} = f[x_i] = f_i \qquad (2.3.2a)$$

The first divided difference is given by

$$A_2 = A_{1+1} = f[x_i, x_{i+1}] = \frac{f_{i+1} - f_i}{x_{i+1} - x_i} \qquad (2.3.2b)$$

The kth divided difference is a divided difference of the $(k-1)$th divided differences

$$A_{1+k} = f[x_i, \ldots, x_{i+k}] = \frac{f[x_{i+1}, \ldots, x_{i+k}] - f[x_i, \ldots, x_{i+k-1}]}{x_{i+k} - x_i} \qquad (2.3.2c)$$

The divided differences can be generated with the help of a divided-difference table. For $n = 6$ and $x_i = x_1$ it has the form exhibited in Table 2.1. The entries in the table are computed column by column. The first terms in each of the six columns, f_1 to $f[x_1, \ldots, x_6]$, are the coefficients $A_1, \ldots, A_6$ that are to be substituted, together with the functions $s_j(x)$ defined by (2.3.1), into the series (2.1.1)

$$y = f_i + \sum_{k=1}^{n-1} f[x_i, \ldots, x_{i+k}] \prod_{i=1}^{k-1}(x - x_i) = A_1 + \sum_{k=2}^{n} A_k \prod_{i=1}^{k-1}(x - x_i) \qquad (2.3.3)$$

The series can be differentiated. Explicit formulas for the function and its first and second derivatives and $n \leq 6$ are

$$y = A_1 + u_1(A_2 + u_2\{A_3 + u_3[A_4 + u_4(A_5 + u_5 A_6)]\}) \qquad (2.3.4a)$$

$$\frac{dy}{dx} = A_2 + A_3 v_1 + A_4(u_3 v_1 + v_2) + A_5(v_1 v_4 + v_2 v_3)$$

$$+ A_6[u_5 v_1 v_4 + v_2(u_5 v_3 + v_4)] \qquad (2.3.4b)$$

$$\frac{d^2y}{dx^2} = 2\{A_3 + A_4(v_1 + u_3) + A_5(v_1 v_3 + v_2 + v_4)$$

$$+ A_6[v_1 v_4 + v_2 v_3 + u_5(v_1 v_3 + v_2 + v_4)]\} \qquad (2.3.4c)$$

where $\qquad u_i = x - x_i \qquad i = 1, 2, \ldots, 5 \qquad (2.3.5)$

$$v_1 = u_1 + u_2 \qquad v_2 = u_1 u_2 \qquad v_3 = u_3 + u_4 \qquad v_4 = u_3 u_4 \qquad (2.3.6)$$

If $n < 6$, the coefficients A_k with $k > n$ equal zero.

Since n points uniquely determine a parabola of $(n-1)$th degree, the functions (2.2.2a, b) and (2.3.3) are identical. Therefore the error analysis of Lagrange interpolation is valid also for Newton interpolation.

Program P203 computes the value of $y(x)$ and of its first and second derivatives dy/dx and d^2y/dx^2 by Newton interpolation. The function is defined by n points (x_i, f_i), $n \leq 6$. A subroutine computes Chebyshev abscissas for the given interval $[a, b]$.

If the abscissas are equally spaced with the interval h, the kth divided differ-

TABLE 2.1 Divided-difference table

x_1	f_1					
		$f[x_1, x_2]$				
x_2	f_2		$f[x_1, \ldots, x_3]$			
		$f[x_2, x_3]$		$f[x_1, \ldots, x_4]$		
x_3	f_3		$f[x_2, \ldots, x_4]$		$f[x_1, \ldots, x_5]$	
		$f[x_3, x_4]$		$f[x_2, \ldots, x_5]$		$f[x_1, \ldots, x_6]$
x_4	f_4		$f[x_3, \ldots, x_5]$		$f[x_2, \ldots, x_6]$	
		$f[x_4, x_5]$		$f[x_3, \ldots, x_6]$		
x_5	f_5		$f[x_4, \ldots, x_6]$			
		$f[x_5, x_6]$				
x_6	f_6					

ence in the coefficient $A_j = A_{1+k}$ is to be replaced with the kth forward difference $\Delta^k f_i$ divided by $k!$. Hence

$$A_1 = A_{1+0} = f_i$$

$$A_2 = A_{1+1} = \frac{\Delta f_i}{1!} = f_{i+1} - f_i \qquad (2.3.7)$$

$$A_j = A_{1+k} = \frac{\Delta^k f_i}{k!} = \frac{\Delta(\Delta^{k-1} f_i)}{k!} = \frac{\Delta^{k-1} f_{i+1} - \Delta^{k-1} f_i}{k!}$$

The differences can be generated with the help of a difference table similar to that for divided differences. The variable x is now replaced by X, defined by

$$X = 1 + \frac{x - x_1}{h} \qquad (2.3.8)$$

The variables u_i are determined by the formula

$$u_i = X - i \qquad i = 1, 2, \ldots, n - 1 \qquad (2.3.9)$$

Explicit formulas for the function and its first and second derivatives are given for $n \leq 6$ by Eqs. (2.3.4) with A_k defined by (2.3.7) and u_i by (2.3.9), but the right-hand sides of (2.3.4b, c) are to be divided by h and h^2, respectively. Formulas (2.3.6) are still valid. If $n < 6$, the coefficients A_j with $j < n$ equal zero.

Program P204 computes the values of $y(x)$ and of its first and second derivatives by Newton interpolation. The function is defined by n equally spaced points, $n \leq 6$, with the interval h and ordinates f_i. It requires less memory space than P203 and there are fewer roundoff errors in computing A_j.

Programs P203 and P204 both contain a subroutine for a piecewise interpolation, which shifts all points either to the left or to the right and asks for a new point. Hence the independent variable x can be chosen close to the center of the interpolation subinterval to keep the truncation errors smaller. The subroutine is especially useful in Program P204 for equally spaced abscissas.

2.4 HERMITE INTERPOLATION

In Hermite interpolation the function is given by n points (x_i, f_i) and by m derivatives $f_i^{(k)}$ at each point. Formally the interpolation can be reduced to Newton interpolation except that the divided differences are computed here in a slightly different way. In Sec. 2.3 they are defined for $n = 1 + k$ distinct points. If all the points $x_{i+1}, \ldots, x_{i+k}$ are approaching x_i continuously and the function is k-times differentiable, the limit is

$$\lim_{x_{i+1}, \ldots, x_{i+j} \to x_i} f[x_i, \ldots, x_{i+j}] = \frac{f_i^{(j)}}{j!} \qquad j = 1, 2, \ldots, k \qquad (2.4.1)$$

In the extreme case, $n = 1$, $m > 0$; that is, the function is given by one point (x_1, f_1) and m derivatives at this point. All the divided differences are then reduced to derivatives and the series (2.3.3) to a truncated Taylor series.

If the function is given by $n = 3$ points with one derivative at each point, the divided differences are generated as exhibited in Table 2.2. Each point appears here twice because of two pieces of information at each point, f_i and f'_i. In the third column of the table, where the first divided differences are computed, those between x_i and x_i are the first derivatives, while those between x_i and x_{i+1} are the divided differences computed by (2.3.2b). In the rest of the columns all the divided differences are computed by (2.3.2c). The first terms in each of the six columns f_1

TABLE 2.2 Divided-difference table for Hermite interpolation

$x_1 \; f_1$					
	f'_1				
$x_1 \; f_1$		$f[x_1, x_1, x_2]$			
	$\dfrac{f_2 - f_1}{x_2 - x_1}$		$f[x_1, x_1, x_2, x_2]$		
$x_2 \; f_2 Q$		$f[x_1, x_2, x_2]$		$f[x_1, x_1, x_2, x_2, x_3]$	
	f'_2		$f[x_1, x_2, x_2, x_3]$		$f[x_1, x_1, x_2, x_2, x_3, x_3]$
$x_2 \; f_2$		$f[x_2, x_2, x_3]$		$f[x_1, x_2, x_2, x_3, x_3]$	
	$\dfrac{f_3 - f_2}{x_3 - x_2}$		$f[x_2, x_2, x_3, x_3]$		
$x_3 \; f_3$		$f[x_2, x_3, x_3]$			
	f'_3				
$x_3 \; f_3$					

to $f[x_1, x_1, x_2, x_2, x_3, x_3]$ are the coefficients A_k that are to be substituted into the interpolation equations (2.3.4). Each x_i appears in (2.3.5) twice; explicitly

$$u_1 = u_2 = x - x_1 \qquad u_3 = u_4 = x - x_2 \qquad u_5 = x - x_3 \qquad (2.4.2)$$

If six abscissas are equally spaced across the interval $[a, b]$, the subinterval h is given by

$$h = \frac{b - a}{5} \qquad (2.4.3)$$

and the abscissas are

$$x_i = a + h(i - 1) \qquad (2.4.4)$$

These values are now substituted into the preceding equations.

Programs P205 and P206 compute $y(x)$, dy/dx, and d^2y/dx^2 by Hermite interpolation. The function is defined at $n \le 3$ points by the function and its first derivative. The points are arbitrarily spaced in P205 and equally spaced in P206. Both programs contain a subroutine for a piecewise interpolation, which shifts all the points and derivatives either to the left or to the right and asks for another point and the corresponding derivative.

If the function is defined at each point by its ordinate and the first and second derivatives, each point appears in the divided difference table three times, the divided difference of the two first derivatives is the second derivative divided by 2, and each x_i appears in (2.3.3) and in the variables u_i three times. In case the function is defined only at two points by f_i and by first and second derivatives f'_i and f''_i, the interpolating polynomial can be written in the form

$$y = f_1 + f'_1 t + \tfrac{1}{2} f''_1 t^2 + a_3 t^3 + a_4 t^4 + a_5 t^5 \qquad (2.4.5a)$$

where

$$t = x - x_1 \qquad (2.4.6)$$

The unknown coefficients a_3, a_4, and a_5 are determined by the condition that the function y and its derivatives

$$\frac{dy}{dx} = f'_1 + f''_1 t + 3a_3 t^2 + 4a_4 t^3 + 5a_5 t^4 \qquad (2.4.5b)$$

$$\frac{d^2y}{dx^2} = f'' + 6a_3 t + 12a_4 t^2 + 20a_5 t^3 \qquad (2.4.5c)$$

be equal at $t = h = x_2 - x_1$ to the given f_2, f'_2, and f''_2. The system of three linear algebraic equations

$$\begin{aligned} a_3 + a_4 h + a_5 h^2 &= b_1 \\ 3a_3 + 4a_4 h + 5a_5 h^2 &= b_2 \\ 6a_3 + 12a_4 h + 20a_5 h^2 &= b_3 \end{aligned} \qquad (2.4.7)$$

with

$$b_1 = \frac{[(f_2 - f_1)/h - f'_1]/h - \frac{1}{2}f''_1}{h}$$

$$b_2 = \frac{(f'_2 - f'_1)/h - f''_1}{h} \qquad (2.4.8)$$

$$b_3 = \frac{f''_2 - f''_1}{h}$$

has the solution

$$a_3 = 10b_1 - 4b_2 + \tfrac{1}{2}b_3$$

$$a_4 = \frac{-15b_1 + 7b_2 - b_3}{h} \qquad (2.4.9)$$

$$a_5 = \frac{6b_1 - 3b_2 + \tfrac{1}{2}b_3}{h^2}$$

which must be substituted into Eqs. (2.4.5). If the first and second derivatives of a function to be interpolated are known or can be computed, this interpolation should always be preferred because it is simplest and fastest and the interpolation interval is reduced to the distance between two points x and x_{i+1}.

Program P207 computes $y(x)$, dy/dx, and d^2y/dx^2 by Hermite interpolation. The function is defined at two points by the function and its first and second derivatives. A subroutine shifts one point with the corresponding derivatives either to the left or to the right and asks for a further point and both derivatives.

The truncation error in Hermite interpolation is also given by Eqs. (2.2.4), but the n is now the number of points multiplied by $1 + m$, m being the number of derivatives given at each point, and each x_i must appear in the product $1 + m$ times. Since the interpolation interval $[a, b]$ becomes shorter for a given amount of information with increasing number of given derivatives, the numerical value of the product in (2.2.4a) and consequently the truncation error are smaller in Hermite interpolation than in Lagrange or Newton interpolation.

2.5 CUBIC-SPLINE INTERPOLATION

If the first derivatives of the function to be interpolated are neither given nor computable, the cubic-spline interpolation can be applied with advantage. The interpolating function between x_i and x_{i+1} is a cubic parabola

$$y = \frac{f_i}{h_{i+1}^3}(x_{i+1} - x)^2[2(x - x_i) + h_i]$$

$$+ \frac{f_{i+1}}{h_{i+1}^3}(x - x_i)^2[2(x_{i+1} - x) + h_i]$$

$$+ \frac{f'_i}{h_{i+1}^2}(x_{i+1} - x)^2(x - x_i)$$

$$+ \frac{f'_{i+1}}{h_{i+1}^2}(x - x_i)^2(x - x_{i+1}) \qquad (2.5.1a)$$

Its first and second derivatives are

$$\frac{dy}{dx} = 6(x_{i+1} - x)(x - x_i)\frac{f_{i+1} - f_i}{h_{i+1}^3}$$

$$+ \frac{f'_i}{h_{i+1}^2}(x_{i+1} - x)(2x_i + x_{i+1} - 3x)$$

$$+ \frac{f'_{i+1}}{h_{i+1}^2}(x_i - x)(x_i + 2x_{i+1} - 3x) \qquad (2.5.1b)$$

$$\frac{d^2y}{dx^2} = 6(x_i + x_{i+1} - 2x)\frac{f_{i+1} - f_i}{h_{i+1}^3}$$

$$+ 2\frac{f'_i}{h_{i+1}^2}(3x - x_i - 2x_{i+1})$$

$$+ 2\frac{f'_{i+1}}{h_{i+1}^2}(3x - 2x_i - x_{i+1}) \qquad (2.5.1c)$$

where
$$h_{i+1} = x_{i+1} - x_i \qquad (2.5.1d)$$

The parabola goes through the points (x_i, f_i) and (x_{i+1}, f_{i+1}). The unknown first derivatives f'_i and f'_{i+1} are determined by the requirement that the second derivatives of two neighboring parabolas

$$\frac{d^2y}{dx^2} = \frac{6(f_{i+1} - f_i)}{h_{i+1}^2} - \frac{4f'_i}{h_{i+1}} - \frac{2f'_{i+1}}{h_{i+1}}$$

and
$$\frac{d^2y}{dx^2} = -\frac{6(f_i - f_{i-1})}{h_i^2} + \frac{2f'_{i-1}}{h_i} + \frac{4f'_i}{h_i}$$

be continuous. Hence

$$c_i f'_{i-1} + 2f'_i + d_i f'_{i+1} = b_i \qquad i = 2, 3, \ldots, n - 1 \qquad (2.5.2)$$

The right-hand side b_i and the coefficients c_i and d_i are given by

$$b_i = 3c_i \frac{f_i - f_{i-1}}{x_i - x_{i-1}} + 3d_i \frac{f_{i+1} - f_i}{x_{i+1} - x_i}$$

$$c_i = \frac{x_{i+1} - x_i}{x_{i+1} - x_{i-1}} \qquad d_i = \frac{x_i - x_{i-1}}{x_{i+1} - x_{i-1}}$$

(2.5.3)

with
$$c_i + d_i = 1$$

Equations (2.5.2) form a system of $n - 2$ linear algebraic equations for the unknown derivatives f'_i. The derivatives f'_1 and f'_n at both boundaries can be determined in the following different ways:

1. If only the functions f_i are given at n points, the cubic parabola between x_1 and x_2 goes through the points x_1, x_2, x_3 and its derivative at x_2 is f'_2. Its derivative at x_1 satisfies the equation

$$2f'_1 + \frac{2}{c_2} f'_2 = 2(2 + d_2) \frac{f_2 - f_1}{x_2 - x_1} + 2 \frac{d_2^2}{c_2} \frac{f_3 - f_2}{x_3 - x_2} \qquad (2.5.4a)$$

Similarly the cubic parabola between x_{n-1} and x_n goes through the points x_{n-2}, x_{n-1}, x_n, and its derivative at x_{n-1} is f'_{n-1}. Its derivative at x_n satisfies the equation

$$\frac{2}{d_{n-1}} f'_{n-1} + 2f'_n = 2 \frac{c_{n-1}^2}{d_{n-1}} \frac{f_{n-1} - f_{n-2}}{x_{n-1} - x_{n-2}} + 2(2 + c_{n-1}) \frac{f_n - f_{n-1}}{x_n - x_{n-1}} \qquad (2.5.4b)$$

Equation (2.5.4a), the $n - 2$ equations (2.5.2), and Eq. (2.5.4b) form a system of n linear algebraic equations for n unknown f'_i.

2. If the second derivative f''_1 at the boundary x_1 is given, Eq. (2.5.1c), with $i = 1$ and $x = x_1$, yields

$$2f'_1 + f'_2 = \frac{3(f_2 - f_1)}{x_2 - x_1} - \tfrac{1}{2}(x_2 - x_1)f''_1 \qquad (2.5.5a)$$

This equation now replaces (2.5.4a). The n unknown f'_i are to be computed from the system of n linear algebraic equations consisting of (2.5.5a), the $n - 2$ equations (2.5.2), and (2.5.4b).

3. If the second derivative f''_n at the boundary x_n is given, Eqs. (2.5.1c) with $i = n - 1$ and $x = x_{n-1}$, yield

$$f'_{n-1} + 2f'_n = \frac{3(f_n - f_{n-1})}{x_n - x_{n-1}} + \tfrac{1}{2}(x_n - x_{n-1})f''_n \qquad (2.5.5b)$$

This equation now replaces (2.5.4b). The n unknown f'_i are to be computed from the system of n linear algebraic equations consisting of (2.5.4a), the $n - 2$ equations (2.5.2), and Eq. (2.5.5b).

4. If the second derivatives f''_1 and f''_n at both boundaries are given, Eq. (2.5.5a), the $n - 2$ equations (2.5.2), and Eq. (2.5.5b) form a system of n linear algebraic equations for n unknown f'_i.

5. If the first derivative f'_1 at the boundary x_1 is given, Eq. (2.5.4a) is to be dropped and the first equation, $i = 2$, of the system is reduced to

$$2f'_2 + d_2 f'_3 = c_2 \left[\frac{3(f_2 - f_1)}{x_2 - x_1} - f'_1 \right] + 3d_2 \frac{f_3 - f_2}{x_3 - x_2} \qquad (2.5.6a)$$

This equation, the $n - 3$ equations (2.5.2), $i = 3, 4, \ldots, n - 1$, and Eq. (2.5.4b) form a system of $n - 1$ linear algebraic equations for $n - 1$ unknowns f'_i, $i = 2, 3, \ldots, n$.

6. If the first derivatives f'_n at the boundary x_n is given, Eq. (2.5.4b) is to be dropped and the last equation, $i = n - 1$, of the system (2.5.2) is reduced to

$$c_{n-1}f'_{n-2} + 2f'_{n-1} = 3c_{n-1}\frac{f_{n-1} - f_{n-2}}{x_{n-1} - x_{n-2}} + d_{n-1}\left[\frac{3(f_n - f_{n-1})}{x_n - x_{n-1}} - f'_n\right] \quad (2.5.6b)$$

Equation (2.5.4a), the $n - 3$ equations (2.5.2), $i = 2, 3, \ldots, n - 2$, and Eq. (2.5.6b) form a system of $n - 1$ linear algebraic equations for $n - 1$ unknown f'_i, $i = 1, 2, \ldots, -1$.

7. If the first derivatives f'_1 and f'_n at both boundaries x_1 and x_n are given, Eq. (2.5.6a), the $n - 4$ equations (2.5.2), $i = 3, 4, \ldots, n - 2$, and Eq. (2.5.6b) form a system of $n - 2$ linear algebraic equations for $n - 2$ unknown f'_i, $i = 2, 3, \ldots, n - 1$.

8. If the first derivative f'_1 at x_1 and the second derivatives f''_n at x_n are given, the Eq. (2.5.6a), the $n - 3$ equations (2.5.2), $i = 3, 4, \ldots, n - 1$, and Eq. (2.5.5b) form a system of $n - 1$ linear algebraic equations for $n - 1$ unknown f'_i, $i = 2, 3, \ldots, n$.

9. If the second derivative f''_1 at x_1 and the first derivative f'_n at x_n are given, Eq. (2.5.5a), the $n - 3$ equations (2.5.2), $i = 2, 3, \ldots, n - 2$, and Eq. (2.5.6b) form a system of $n - 1$ linear algebraic equations for $n - 1$ unknown f'_i, $i = 1, 2, \ldots, n - 1$.

The system of equations is tridiagonal and, except for Eqs. (2.5.4), strictly diagonally dominant. Therefore its solution is fast. Once the system has been solved, Eqs. (2.5.1a, b, c) determine the function and its first and second derivatives at any point within the interpolation interval.

The cubic spline gives excellent results not only in interpolation but also in numerical differentiation if properly applied. The intervals between two consecutive abscissas can be chosen freely, but they should vary slowly; sudden jumps in their magnitude must always be avoided. The spline exerts a smoothing effect on the computed first derivatives f'_i, but because they are influenced by all points (x_i, f_i), it may sometimes generate unexpected small oscillations in that part of the function which is asymptotically approaching a straight line. In that case, or if the function exhibits a discontinuity in its derivative, the whole interpolation interval must be covered by two or more splines.

When the second derivatives are to be computed, it is usually recommended that the first derivatives be determined by spline and then these data (x_i, f'_i) be used as the input in the second spline. This yields better values than if the second derivatives are computed by (2.5.1c) from the first spline. If the first and second derivatives computed this way are both used in a larger program, however, they may turn out to be incompatible and give rise to numerical instability. If both derivatives are to be determined only at the given abscissas x_i, the values computed by (2.5.1b, c) are usually satisfactory.

Programs P208 and P209 compute $y(x)$, dy/dx, and d^2y/dx^2 by cubic spline for arbitrarily and equally spaced abscissas, respectively. The boundary values f'_1 and f'_n can be determined by any of nine alternative methods. Program P103 for the solution of a triangular system of linear algebraic equations is built in.

2.6 TRIGONOMETRIC INTERPOLATION

If a periodic function defined by n points (x_i, f_i) with equally spaced abscissas on the interval $[0, 2\pi]$ is to be interpolated by (2.1.1), the functions $s_j(x)$ to be substituted into (2.1.2) are $\sin jx$ and $\cos jx$, $j = 0, 1, \ldots$, with $k = 0$ and the abscissas given by

$$x_i = (i-1)\frac{2\pi}{n-1} \qquad i = 1, 2, \ldots, n \qquad (2.6.1)$$

The solution of the system (2.1.2) of n linear algebraic equations for the unknown A_j becomes easy if the functions $s_j(x)$ are chosen so that the system is diagonal. This is the case if for n odd

$$s_j(x) = l_j(x) = \frac{\sin [n(x_j - x)/2]}{n \sin [(x_j - x)/2]} \qquad j = 1, 2, \ldots, n \qquad (2.6.2a)$$

and for n even

$$s_j(x) = l_j(x) = \frac{\sin [n(x_j - x)/2]}{n \sin [(x_j - x)/2]} \cos \frac{x_j - x}{2} \qquad j = 1, 2, \ldots, n \qquad (2.6.2b)$$

This approach is similar to that applied in Lagrange interpolation. All n functions $l_j(x)$ have the same form, but they are shifted with respect to each other by x_j. They are analogous to the Lagrange functions of Sec. 2.2 since $l_j(x_j) = 1$, but if $x = x_i \neq x_j$ then $l_j(x_i) = 0$. The interpolating function now is

$$y = \sum_{j=1}^{n} f_j l_j(x) \qquad (2.6.3)$$

If the function is defined on the interval $[a, b]$, the abscissas t_i corresponding to x_i are

$$t_i = a + \frac{(i-1)(b-a)}{n-1} \qquad i = 1, 2, \ldots, n \qquad (2.6.4a)$$

The variables t and x are related to each other by the linear transformation

$$x = \frac{t-a}{M} \qquad \text{with} \qquad M = \frac{b-a}{2\pi} \qquad (2.6.4b)$$

The x is to be substituted into the interpolating function (2.6.3).

Trigonometric interpolation can also be used for nonperiodic functions that are defined by a set of points with equally spaced abscissas.

There is no simple formula for the truncation error of this interpolation, but it can be said that the terms with x_j close to x contribute most. The error can be decreased only by increasing the number of points defining the function.

Program P210 computes the value of $y(t)$ of a periodic function defined by an arbitrary number n of equally spaced points with the ordinates f_i on the interval $[a, b]$. If n is even and the function is either odd or even, only $(1 + n)/2$ ordinates are to be given on the half interval $[a, b/2]$; the remaining ordinates are determined by a subroutine.

2.7 INVERSE INTERPOLATION

Sometimes a function $\bar{y} = \bar{f}(\bar{x})$ is to be interpolated not at a given value of the independent variable $\bar{x}$ but at a given value of the dependent variable $\bar{y}$; that is, the value $\bar{y}$ is given, and the corresponding $\bar{x}$ is to be determined by an inverse interpolation. In this case the variables are interchanged, $\bar{x}_i \to f_i$ and $\bar{f}_i \to x_i$, and either Lagrange or Newton interpolation for unequally spaced abscissas is used (Program P201 or P203).

The derivatives $d\bar{y}/d\bar{x}$ and $d^2\bar{y}/d\bar{x}^2$ are related to the computed derivatives dy/dx and d^2y/dx^2 by

$$\frac{d\bar{y}}{d\bar{x}} = \frac{1}{dy/dx} \qquad \frac{d^2\bar{y}}{d\bar{x}^2} = -\frac{d^2y/dx^2}{(dy/dx)^3} \qquad (2.7.1)$$

2.8 LEAST-SQUARES TECHNIQUES

There are two groups of problems in which the least-squares approximation can be applied successfully. In problems of the first group the data are the result of observations or measurements with no notion, or merely a vague one of the analytical form of the approximating function. Therefore the function is chosen to be quite simple—a straight line, a quadratic parabola, an exponential, a logarithmic or similar function, whose two or three parameters are to be determined from many data by the least-squares technique. This kind of approximation, called *regression* or *curve fitting*, is discussed in detail in Chap. 9. The numerical data of the other group have a sound theoretical background that sets the analytical form of the approximating function, but they contain random errors generated either by measurements or by previous numerical computation. The function, in the form of a finite series, contains parameters to be determined by the least-squares technique. The number of given numerical data must of course be greater than the number of unknown parameters, but here this ratio is usually not as high as in the problems of curve fitting. If the series converges fast, the results of approximation may be very satisfactory.

The approximating function $y(x)$ given by m points (x_i, f_i) is expressed in the form of a finite series

$$y = \sum_{j=0}^{n} A_j s_j(x) \qquad n \leq m - 1 \qquad (2.8.1)$$

Unlike the interpolation, it does not go through the given set of m points. The deviation at a point with the abscissa x_i is

$$\sum_{j=0}^{n} A_j s_j(x_i) - f_i \qquad i = 1, 2, \ldots, m$$

2.16 THEORY

The parameters A_j are chosen so that the sum δ_n^2 of squares of the deviations at all m points is reduced to a minimum

$$\delta_n^2 \equiv \sum_{i=1}^{m}\left[\sum_{j=0}^{n} A_j s_j(x_i) - f_i\right]^2 = \min \qquad (2.8.2)$$

This condition is satisfied if all the first partial derivatives with respect to each unknown parameter A_k, $k = 0, 1, \ldots, n$, vanish

$$2\sum_{i=1}^{m}\left[\sum_{j=0}^{n} A_j s_j(x_i) - f_i\right] s_k(x_i) = 0$$

In a slightly different form

$$\sum_{j=0}^{n}\left[\sum_{i=1}^{m} s_k(x_i)s_j(x_i)\right] A_j = \sum_{i=1}^{m} s_k(x_i) f_i \qquad k = 0, 1, \ldots, n \qquad (2.8.3)$$

If the $s_j(x_i)$ are interpreted as elements $s_{i,\,j+1}$ of a rectangular matrix S with m rows and $n+1$ columns and the f_i as elements of a single-column matrix F, then

$$\sum_{i=1}^{m} s_k(x_i)s_j(x_i) \quad\text{and}\quad \sum_{i=1}^{m} s_k(x_i)f_i$$

are the elements of the matrix products

$$C = \tilde{S}S \quad\text{and}\quad D = \tilde{S}F$$

Thus the system (2.8.3) of $n+1$ linear algebraic equations for $n+1$ unknown A_j, $j = 0, 1, \ldots, n$, is exactly the determined system of $n+1$ normal equations of Sec. 1.6. If $m = n+1$, the least-squares approximation is reduced to interpolation; the approximating function (2.8.1) goes through each of $n+1$ given points (x_i, f_i).

If the series (2.8.1) contains only two or three terms (as in curve fitting), the system (2.8.3) can be solved for arbitrarily chosen functions $s_j(x)$ by a direct method, but the solution becomes more and more difficult as the number of unknown parameters A_j increases. The solution will be very simple if the system (2.8.3) is reduced to a diagonal one, i.e., when for $k, j = 0, 1, \ldots, m-1$ the scalar product is

$$\sum_{i=1}^{m} s_k(x_i)s_j(x_i) = \begin{cases} N_j \neq 0 & \text{if } k = j \\ 0 & \text{if } k \neq j \end{cases} \qquad (2.8.4)$$

This orthogonality condition can be satisfied by a proper choice of the functions $s_j(x)$. The parameters A_j are then given by

$$A_j = \frac{\sum_{i=1}^{m} s_j(x_i)f_i}{N_j} \qquad 0 \leq j \leq m \qquad (2.8.5)$$

A great majority of orthogonal functions satisfy the orthogonality condition only for the continuous case

$$\int_a^b s_k(x)s_j(x)w(x)\,dx \begin{cases} \neq 0 & \text{if } k = j \\ = 0 & \text{if } k \neq j \end{cases}$$

$w(x) > 0$ being a given weight function, but some of them satisfy it for the discrete case (2.8.4) as well.

If a series of orthogonal functions (2.8.1) whose coefficients are determined by (2.8.5) contains fewer terms than the number m of abscissas x_i defining (together with the ordinates f_i) the function, then the series is a least-squares approximation for any number of terms less than m. If these two numbers are equal, the least-squares approximation goes over to interpolation.

If it is not known from theoretical considerations how many terms should be in the series (2.8.1) to obtain a satisfactory approximation, this number n can be determined on the basis of the null hypothesis of mathematical statistics (applicable if the errors in f_i are normally distributed with zero mean value). The δ_n^2 given by (2.8.2) and the variances σ_n^2

$$\sigma_n^2 = \frac{\delta_n^2}{m - n - 1} \tag{2.8.6}$$

are to be computed for $n = 1, 2, \ldots$. The sequence of σ_n^2 and the series should be terminated for a certain n when σ_n^2 no longer decreases significantly. It is counterproductive to include more terms because the redundant ones describe the noise in f_i that has been eliminated by the smoothing effect of least-squares approximation. On the other hand, if there is no leveling in the sequence of σ_n^2, or if the first and second derivatives computed simultaneously vary excessively from one given point to another, this is an indication that the approximation is inadequate and should be replaced with another having more points defining the function or/and with a shorter interval or with a function more suitable for approximating the given set of data.

In Secs. 2.9 to 2.11 the use of three orthogonal functions is demonstrated: Chebyshev polynomials for which the abscissas are fixed, orthogonal polynomials with arbitrarily spaced abscissas, and the circular functions $\sin x$ and $\cos x$ with equally spaced abscissas for periodic functions (Fourier series).

2.9 APPROXIMATION BY CHEBYSHEV POLYNOMIALS

The only disadvantage of approximation by Chebyshev polynomials lies in the fact that the abscissas x_i cannot be chosen freely. If a function is defined by m points on the interval $[-1, 1]$, the x_i are the m zeros of the Chebyshev polynomial of the first kind and of the mth degree

$$T_m(x_i) = 0 \tag{2.9.1a}$$

2.18 THEORY

Since $T_m(x)$ is defined by

$$T_m(x) = \cos(m \arccos x) \qquad (2.9.2)$$

the zeros are

$$x_i = \cos\left[\frac{\pi}{m}(k - \tfrac{1}{2})\right] \qquad i = 1, 2, \ldots, m \qquad (2.9.1b)$$

Equation (2.9.2) exhibits a valuable property of Chebyshev polynomials, that of oscillating between $+1$ and -1, but their polynomial character becomes obvious upon evaluating them from the recurrence relation

$$T_j = 2xT_{j-1} - T_{j-2} \qquad j \geq 2 \qquad (2.9.2a)$$

with

$$T_0 = 1 \qquad T_1 = x \qquad (2.9.3a)$$

Their first and second derivatives T'_j and T''_j can be computed in a similar way by differentiating (2.9.2a) and (2.9.3a) twice

$$T'_j = 2T_{j-1} + 2xT'_{j-1} - T'_{j-2} \qquad j \geq 2 \qquad (2.9.2b)$$

$$T''_j = 4T'_{j-1} + 2xT''_{j-1} - T''_{j-2} \qquad j \geq 2 \qquad (2.9.2c)$$

$$T'_0 = 0 \qquad T'_1 = 1 \qquad (2.9.3b)$$

$$T''_0 = 0 \qquad T''_1 = 0 \qquad (2.9.3c)$$

Equations (2.9.2a) and (2.9.3a) show that Chebyshev polynomials are even functions for j even and odd functions for j odd.

The orthogonality condition (2.8.4) here takes the form

$$\sum_{i=1}^{m} T_k(x_i) T_j(x_i) = \begin{cases} m & \text{if } k = j = 0 \\ \tfrac{1}{2}m & \text{if } k = j \neq 0 \\ 0 & \text{if } k \neq j \end{cases} \qquad (2.9.4)$$

for $0 \leq k < m$ and $0 \leq j < m$.

A function given by m points (x_i, f_i) with x_i defined by (2.9.1b) can be approximated by the series

$$y = \sum_{j=0}^{n} A_j T_j(x) \qquad n < m - 1 \qquad (2.9.5a)$$

If $n = m - 1$, the least-squares approximation goes over to interpolation. The coefficients A_j are computed by (2.9.5a), which reduces here to

$$A_0 = \frac{1}{m} \sum_{i=1}^{m} f_i \qquad A_j = \frac{2}{m} \sum_{i=1}^{m} T_j(x_i) f_i \qquad 0 < j < m \qquad (2.9.6)$$

The first and second derivatives of the function y are

$$\frac{dy}{dx} = \sum_{j=1}^{n} A_j T'_j(x) \qquad n < m - 1 \qquad (2.9.5b)$$

and
$$\frac{d^2y}{dx^2} = \sum_{j=2}^{n} A_j T_j''(x) \qquad n < m - 1 \qquad (2.9.5c)$$

If a function is defined on the interval $[a, b]$, the abscissas t_i are related to x_i by the linear transformation

$$t = P + Mx_i \qquad \begin{matrix} P = \tfrac{1}{2}(b + a) \\ M = \tfrac{1}{2}(b - a) \end{matrix} \qquad (2.9.7a)$$

The function (t_i, f_i) is now approximated by the series (2.9.5a) with the coefficients A_j computed by (2.9.6) and with x given by the inverse transformation

$$x = \frac{t - P}{M} \qquad (2.9.7b)$$

This x is also to be substituted into (2.9.5b, c); the derivatives with respect to t are

$$\frac{dy}{dt} = \frac{dy/dx}{M} \qquad \frac{d^2y}{dt^2} = \frac{d^2y/dx^2}{M^2} \qquad (2.9.8)$$

Program P211 approximates a function by Chebyshev polynomials. It computes Chebyshev abscissas t_i on an arbitrary interval $[a, b]$ and the function $y(t)$ and its derivatives dy/dt, d^2y/dt^2 at any t within the interval. As an option, for a given number m of points defining the function a subroutine computes the sequence of variances σ_n^2, $n = 1, 2, \ldots$, given by (2.8.6) and (2.8.2) (cf. Sec. 2.8).

2.10 APPROXIMATION BY ORTHOGONAL POLYNOMIALS WITH ARBITRARILY DISTRIBUTED ABSCISSAS

If the abscissas of a function given at m points (x_i, f_i) cannot be the Chebyshev abscissas (2.9.1b) or (2.9.7), it is necessary to use a series of orthogonal polynomials $p_j(x)$ with arbitrarily distributed abscissas. The series developed by Ralston [17] satisfies a generalized least-squares condition requiring that the sum δ_n^2 of products of a weight function $w_i > 0$ given at all m points and of the square of deviation be reduced to a minimum

$$\delta_n^2 \equiv \sum_{i=1}^{m} w_i \left[\sum_{j=0}^{n} A_j p_j(x_i) - f_i \right]^2 = \min \qquad (2.10.1)$$

The polynomials are evaluated by the recurrence relation

$$p_j = (x - a_j)p_{j-1} - b_{j-1}p_{j-2} \qquad j \geq 1 \qquad (2.10.2a)$$

with
$$p_0 = 1 \qquad p_{-1} = 0 \qquad (2.10.3a)$$

and

$$a_j = \frac{1}{N_{j-1}} \sum_{i=1}^{m} w_i x_i [p_{j-1}(x_i)]^2 \qquad b_j = \frac{N_j}{N_{j-1}}$$
(2.10.4)

$$N_j = \sum_{i=1}^{m} w_i [p_j(x_i)]^2$$

where N_j is the scalar product of the orthogonal functions now involving the weight function as well. The first and second derivatives p'_j and p''_j are computed in a similar way by differentiating (2.9.2a) twice

$$p'_j = p_{j-1} + (x - a_j)p'_{j-1} - b_{j-1}p'_{j-2} \qquad j \geq 2 \qquad (2.10.2b)$$

$$p''_j = 2p'_{j-1} + (x - a_j)p''_{j-1} - b_{j-1}p''_{j-2} \qquad j \geq 2 \qquad (2.10.2c)$$

$$p'_0 = 0 \qquad p'_1 = 1 \qquad (2.10.3b)$$

$$p''_0 = 0 \qquad p''_1 = 0 \qquad (2.10.3c)$$

Thus the free choice of abscissas requires far more computation in evaluating the orthogonal polynomials. If the freely chosen abscissas are distributed symmetrically about $x = 0$, all the coefficients a_j vanish. If the abscissas are the Chebyshev abscissas, then $p_j = 2^{1-j}T_j$. The coefficient standing at the highest power of p_j is always equal to 1.

A function given by m points (x_i, f_i) can be approximated by the series

$$y = \sum_{j=0}^{n} A_j p_j(x) \qquad n \leq m - 1 \qquad (2.10.5a)$$

If $n = m - 1$, the least-squares approximation becomes interpolation. The coefficients A_j are computed by a formula similar to (2.8.5)

$$A_j = \frac{\sum_{i=1}^{m} w_i p_j(x_i) f_i}{N_j} \qquad 0 \leq j < m \qquad (2.10.6)$$

The first and second derivatives of the function y are

$$\frac{dy}{dx} = \sum_{j=1}^{n} A_j p'_j(x) \qquad \frac{d^2 y}{dx^2} = \sum_{j=2}^{n} A_j p''_j(x) \qquad n < m \qquad (2.10.5b)$$

There are four programs for approximation of a function by orthogonal polynomials, P212 for a function with arbitrarily distributed abscissas and a given weight function, P213 for a function with arbitrarily distributed abscissas and weight function equal to 1, P214 for a function with equally spaced abscissas and a given weight function, and P215 for a function with equally spaced abscissas and weight function equal to 1. All four programs compute the function $y(x)$ and its derivatives dy/dx and d^2y/dx^2 at any x within the approximation interval. As an option, for a given number m of points defining the function a subroutine in each program computes the sequence of variances σ_n^2 defined by Eqs. (2.8.6) and (2.10.1) for $n = 1, 2, \ldots$ (cf. Sec. 2.8).

2.11 APPROXIMATION OF A PERIODIC FUNCTION

A periodic function defined by $2m$ points (x_i, f_i) with equally spaced abscissas on the interval $[-\pi, \pi]$ can be approximated by a finite Fourier series

$$y = A_0 + \sum_{j=1}^{n} (A_j \cos jx + B_j \sin jx) \qquad n < m \qquad (2.11.1a)$$

The abscissas x_i are fixed by the formula

$$x_i = \frac{i\pi}{m} \qquad i = -(m-1), -(m-2), \ldots, m \qquad (2.11.2)$$

The functions $s_j(x_i) = \cos jx_i$ and $s_j(x_i) = \sin jx_i$ satisfy the orthogonality conditions (2.8.4) with the summation running over all $2m$ values of i given in (2.11.2) and with $N_0 = 2m$ for cosine, $N_j = m$, $1 \leq j \leq m-1$, if both $s_k(x_i)$ and $s_j(x_i)$ are either cosines or sines, and $N_j = 0$ if one of them is a cosine and the other a sine. Both coefficients A_j and B_j can be computed by (2.8.5) with the summation running over all $2m$ points, but since the number is even, the number of arithmetic operations can be halved by determining them from

$$A_0 = \frac{\sum_{i=0}^{m} E_i}{2m}$$

$$A_j = \frac{\sum_{i=0}^{m} E_i \cos jx_i}{m} \qquad (2.11.3)$$

$$B_j = \frac{\sum_{i=1}^{m-1} G_i \sin jx_i}{m} \qquad 1 \leq j < m$$

where [because $\cos x = \cos(-x)$ and $\sin x = -\sin(-x)$]

$$E_0 = f_0 \qquad E_m = f_m$$
$$E_i = f_i + f_{-i} \qquad 1 \leq j < m \qquad (2.11.4)$$
$$G_i = f_i - f_{-i}$$

If the given periodic function is even, it can be defined by $m+1$ points, $f_0, \ldots, f_m$, on the half interval $[0, \pi]$, for $f_{-i} = f_i$, $i = 1, 2, \ldots, m-1$. The $E_i = 2f_i$, all G_i vanish, and the function is expressed only by the cosine terms.

If the given periodic function is odd, it can be defined by $m-1$ points, $f_1, \ldots, f_{m-1}$, on the half interval $[0, \pi]$, for $f_0 = f_m = 0$ and $f_{-i} = -f_i$, $i = 1, 2, \ldots, m-1$. The $G_i = 2f_i$, all the E_i vanish, and the function is expressed only by sine terms.

To speed up the numerical calculation and lower the truncation errors only $\cos x_1$ and $\sin x_i$ are computed using trigonometric subroutines of the computer.

2.22 THEORY

All other $\cos x_i$ and $\sin x_i$ are determined by the recurrence relations

$$\cos x_{i+1} = \cos(x_i + x_1) = \cos x_i \cos x_1 - \sin x_i \sin x_1$$
$$\sin x_{i+1} = \sin(x_i + x_1) = \sin x_i \cos x_1 + \cos x_i \sin x_1$$

These $m + m$ data are stored in memory. Then $\cos(j+1)x_i$ and $\sin(j+1)x_i$ are similarly calculated by

$$\cos(j+1)x_i = \cos jx_i \cos x_i - \sin jx_i \sin x_i$$
$$\sin(j+1)x_i = \sin jx_i \cos x_i + \cos jx_i \sin x_i$$

These formulas are used also for the computation of $\cos jx$ and $\sin jx$ in (2.11.1a).
The first and second derivatives of the function y are

$$\frac{dy}{dx} = \sum_{j=1}^{n}(-A_j j \sin jx + B_j j \cos jx) \qquad n < m \qquad (2.11.1b)$$

$$\frac{d^2y}{dx^2} = \sum_{j=1}^{n}(-A_j j^2 \cos jx - B_j j^2 \sin jx) \qquad n < m \qquad (2.11.1c)$$

If a nonsymmetric function is defined on the interval $[-a, a]$, the $2m$ abscissas t_i are

$$t_i = \frac{ia}{m} \qquad i = -(m-1), -(m-2), \ldots, m \qquad (2.11.5)$$

If an even or odd function is defined on the half interval $[0, a]$, the $m + 1$ abscissas are also given by (2.11.5) but i runs only from 0 to m. The function given by a set of points (t_i, f_i) is now approximated by the series (2.11.1a) with the coefficients A_j, and B_j computed by (2.11.4) and (2.11.3) and with x given by the linear transformation

$$x = \frac{t\pi}{a} \qquad (2.11.6)$$

This x is to be substituted into (2.11.1b, c); the derivatives with respect to t are

$$\frac{dy}{dt} = \frac{dy}{dx}\frac{\pi}{a}$$
$$\frac{d^2y}{dt^2} = \frac{d^2y}{dx^2}\left(\frac{\pi}{a}\right)^2 \qquad (2.11.7)$$

Because the functions $\cos jx$ and $\sin jx$ satisfy the orthogonality conditions, Eqs. (2.8.2) for δ_n^2 and (2.8.6) for the variance σ_n^2 here take a simpler form

$$\sigma_n^2 = \frac{\delta_n^2}{2m - 2n - 1}$$

$$= \frac{E_0^2 + E_m^2 + \frac{1}{2}\sum_{i=1}^{m-1}(E_i^2 + G_i^2) - 2m\left[A_0^2 + \frac{1}{2}\sum_{i=1}^{n}(A_j^2 + B_j^2)\right]}{2m - 2n - 1} \qquad (2.11.8)$$

Program P216 computes the Fourier approximation for a nonsymmetric periodic function defined by $2m$ points (t_i, f_i) on the interval $[-a, a]$. Programs P217 and P218 compute the Fourier approximation for an even and odd function, respectively, defined by $m + 1$ points on the half interval $[0, a]$. All three programs evaluate the function $y(t)$ and its derivatives dy/dt and d^2y/dt^2 at any t. For a given number $2m$ points defining the function a subroutine in each program computes, as an option, the sequence of variances σ_n^2 for $n = 1, 2, \ldots$ (cf. Sec. 2.8).

2.12 ECONOMIZATION OF A POWER SERIES

This technique lowers the truncation error of a given power series

$$y = \sum_{j=0}^{n} A_j x^j \qquad n \leq 12 \qquad (2.12.1)$$

on the interval $[-1, 1]$. It is based on the valuable property of Chebyshev polynomials of never exceeding 1 in absolute value [cf. Eq. (2.9.2)]. The explicit formulas for the first 13 polynomials are

$$T_0 = 1$$
$$T_1 = x$$
$$T_2 = 2x^2 - 1$$
$$T_3 = 4x^3 - 3x$$
$$T_4 = 8x^4 - 8x^2 + 1$$
$$T_5 = 16x^5 - 20x^3 + 5x$$
$$T_6 = 32x^6 - 48x^4 + 18x^2 - 1 \qquad (2.12.2)$$
$$T_7 = 64x^7 - 112x^5 + 56x^3 - 7x$$
$$T_8 = 128x^8 - 256x^6 + 160x^4 - 32x^2 + 1$$
$$T_9 = 256x^9 - 576x^7 + 432x^5 - 120x^3 + 9x$$
$$T_{10} = 512x^{10} - 1280x^8 + 1120x^6 - 400x^4 + 50x^2 - 1$$
$$T_{11} = 1024x^{11} - 2816x^9 + 2816x^7 - 1232x^5 + 220x^3 - 11x$$
$$T_{12} = 2048x^{12} - 6144x^{10} + 6912x^8 - 3584x^6 + 840x^4 - 72x^2 + 1$$

To improve the accuracy of the series (2.12.1) its highest power x^n is replaced by the Chebyshev polynomial T_n and substituted back into (2.12.1). For instance, x^6 is replaced by

$$x^6 = \frac{48x^4 - 18x^2 + 1 + T_6}{32}$$

Since $|T_n| \leq 1$, the T_n divided by the cofactor 2^{n-1} at x^n can be neglected in the economized series; its highest power is thus lowered by 1 and the truncation error decreased. The procedure can be applied successively.

If the interval of the power series with an independent variable t is $[-a, a]$, its coefficients A_j are to be replaced by

$$B_j = A_j a^j \tag{2.12.3}$$

In this way the interval is reduced to $[-1, 1]$, and the series with $x = t/a$ can be economized.

If skillfully applied, the economization technique can give excellent results, improving the accuracy and increasing the rate of convergence, especially if the original series is slowly convergent; however, it must be used with caution. In the neighborhood of $x = 0$ the Taylor series may be more accurate than the economized one.

Program P219 economizes the coefficients of a power series of the twelfth or lower degree on the interval $[-a, a]$.

2.13 APPROXIMATION BY A RATIONAL FUNCTION

In certain cases approximation by a polynomial is not satisfactory because it is not sufficiently accurate or it converges too slowly. An approximation by a rational function may avoid these deficiencies. Three variants of this approximation are discussed in this section; the function is defined in the first two by a truncated power series and in the third by a set of points (x_i, f_i) with Chebyshev abscissas.

The truncated power series defining the function is written in the form

$$f(x) = \sum_{k=0}^{n} c_k x^k \tag{2.13.1}$$

In Padé approximation it is approximated by the rational function

$$y = \frac{\sum_{i=0}^{p} a_i x^i}{\sum_{j=0}^{q} b_j x^j} \qquad b_0 = 1 \tag{2.13.2}$$

The choice $b_0 = 1$ represents no loss of generality because the value of y remains unchanged if both the numerator and denominator are divided by the same number. Since the right-hand sides of Eqs. (2.13.1) and (2.13.2) should be equal, the unknown coefficients a_i, b_i must satisfy the equation

$$\sum_{i=0}^{p} a_i x^i - \left(\sum_{j=0}^{q} b_j x^j\right)\left(\sum_{k=0}^{n} c_k x^k\right) = 0 \tag{2.13.3}$$

and the condition

$$1 + p + q = 1 + n \tag{2.13.4}$$

which states that the total number of unknown coefficients on the left-hand side must equal the total number of coefficients in the given truncated series (2.13.1). Equation (2.13.3) can be satisfied at any x if the cofactors at each power of x cancel each other. This requirement yields two systems of linear algebraic equations

$$\sum_{j=1}^{q} b_j c_{k-j} = -c_k \qquad k = p+1, \ldots, n \tag{2.13.5a}$$

$$a_i = \sum_{j=0}^{i} b_j c_{i-j} \qquad \begin{array}{l} i = 0, 1, \ldots, p \\ b_k = 0 \text{ if } k > q \end{array} \tag{2.13.5b}$$

Experience shows that the choice $p = q$ or $p = q + 1$ gives the smallest maximum error for most functions that are approximated.

The computation of coefficients a_i in (2.13.5a) is easy; difficulty may arise only in solving the system (2.13.5a) if $q > 3$ and the system is ill-conditioned. To increase the number of terms in the truncated series (2.13.1) while maintaining $q \leq 3$ the author modified Padé approximation by expressing the series (2.13.1) as a sum of an even function $E(x)$ and of an odd function $G(x)$

$$f(x) = E(x) + G(x) \tag{2.13.6}$$

with
$$E(x) = \sum_{k=0}^{m'} c_{2k} x^{2k} \quad \text{and} \quad G(x) = \sum_{k=1}^{m''} c_{2k-1} c^{2k-1} \tag{2.13.7}$$

Here $m' = m'' = n/2$ if n is even; $m' = (n-1)/2$ and $m'' = (n+1)/2$ if n is odd. Each of the functions $E(x)$ and $G(x)$ is approximated by a rational function of the form (2.13.2). An equation corresponding to (2.13.3) indicates that the series in the numerator contains only even powers of x if the given power series $f(x)$ is even and only odd powers of x if $f(x)$ is odd. The series in the denominator contains only even powers of x in both cases. Thus the function $f(x)$ is in general approximated by a sum of two rational functions

$$y = \frac{\sum_{i=0}^{p'} a_{2i} x^{2i}}{\sum_{j=0}^{q} b_{2j} x^{2j}} + \frac{\sum_{i=1}^{p''} a_{2i-1} x^{2i-1}}{\sum_{j=0}^{q} d_{2j} x^{2j}} \qquad b_0 = d_0 = 1 \tag{2.13.8}$$

In a way similar to that in Padé approximation the unknown coefficients a_i, b_{2j}, and d_{2j} are determined by four systems of linear algebraic equations

$$\sum_{j=1}^{q} b_{2j} c_{2k-2j} = -c_{2k} \qquad k = p' + 1, \ldots, m' \tag{2.13.9a}$$

$$a_{2i} = \sum_{j=0}^{i} b_{2j} c_{2i-2j} \qquad \begin{array}{l} i = 0, 1, \ldots, p' \\ b_{2k} = 0 \text{ if } k > q \end{array} \tag{2.13.9b}$$

$$\sum_{j=1}^{q} d_{2j} c_{2k-1-2j} = -c_{2k-1} \qquad k = p'' + 1, \ldots, m'' \tag{2.13.9c}$$

$$a_{2i-1} = \sum_{j=1}^{i} d_{2j-2} c_{2i+1-2j} \qquad \begin{matrix} i = 1, 2, \ldots p'' \\ d_{2k} = 0 \text{ if } k > q \end{matrix} \tag{2.13.9d}$$

The solution is unique if

$$1 + p' + p'' + 2q = 1 + n \tag{2.13.10}$$

Since the number of terms in the truncated series (2.13.1) can now be doubled without increasing the number of equations in systems (2.13.9a, c), the accuracy of this approximation is far higher than that of Padé approximation.

In certain cases the power series describing the given function $f(x)$ may converge so slowly [for instance, if $f(x)$ is a rational function] that neither of the preceding approximations gives satisfactory results. If the function is given on the interval $[a, b]$ by a set of m points (t_i, f_i) with abscissas that can be determined by the method, the third variant of approximation by a rational function, also developed by the author, should be applied. First the interval $[-1, 1]$ is reduced to the given interval $[a, b]$ by the usual linear transformation

$$t = P + Mx \qquad \begin{matrix} P = \tfrac{1}{2}(b + a) \\ M = \tfrac{1}{2}(b - a) \end{matrix} \tag{2.13.11}$$

Chebyshev abscissas x_i determined for given m points by (2.9.1b) are related to t_i by (2.13.11). The given function is expressed as a sum of an even function $E(t_i)$ and an odd function $G(t_i)$

$$\begin{matrix} E(t_i) = \tfrac{1}{2}[f(t_i) + f(t_{m+1-i})] \\ G(t_i) = \tfrac{1}{2}[f(t_i) - f(t_{m+1-i})] \end{matrix} \qquad i = 1, 2, \ldots, m \tag{2.13.12}$$

The function is approximated by a sum of two rational functions

$$y = \frac{\sum_{i=0}^{n'} a_{2i} T_{2i}(x)}{\sum_{j=0}^{3} b_{2j} x^{2j}} + \frac{\sum_{i=1}^{n''} a_{2i-1} T_{2i-1}(x)}{\sum_{j=0}^{3} d_{2j} x^{2j}} \qquad \begin{matrix} n', n'' \leq m - 1 \\ b_0 = d_0 = 1 \end{matrix} \tag{2.13.13}$$

Under the assumption that $m = 13$ and $E(0)$ is the value of E at $x = 0$, that is, $t = P$, the unknown coefficients b_{2j} are computed from the condition that

$$\frac{E(0)}{\sum_{j=0}^{3} b_{2j} x^{2j}} = E(x_k) \qquad \text{at } x_k \text{ for } k = 1, 3, \ldots, 13$$

Hence
$$\sum_{j=1}^{3} b_{2j} x_k^{2j} = \frac{E(0)}{E(x_k)} - 1 \qquad k = 9, 11, 13$$

The unknown coefficients a_{2i} are determined by the least-squares technique of

Sec. 2.9 from the condition that the ordinates of the 13 points defining the function $\sum_{i=0}^{n''} a_{2i} T_{2i}(x)$ are

$$E(0) \quad \text{at } x = x_{2i-1} \text{ for } i = 1, 2, \ldots, 7$$

and

$$E(x_i) \sum_{j=0}^{3} b_{2j} x_{2i}^{2j} \quad i = 1, 2, \ldots, 6$$

Similarly the unknown coefficients d_{2j} are computed from the condition that

$$\frac{xG'(0)}{\sum_{j=0}^{3} d_{2j} x^{2j}} = G(x_k) \quad \text{at } x_k \text{ for } k = 1, 3, \ldots, 13$$

Hence

$$\sum_{j=1}^{3} d_{2j} x_k^{2j} = \frac{G'(0) x_k}{G(x_k)} - 1 \quad k = 9, 11, 13$$

Here $G'(0) = (dG/dx)_{x=0}$ and $\lim_{x \to 0} [x/G(x)] = 1/G'(0)$. As a consequence, the preceding equation is satisfied also for $x = x_1 = 0$. The unknown coefficients a_{2i-1} are again determined by the least-squares technique of Sec. 2.9 from the condition that the ordinates of the 13 points defining the function $\sum_{i=1}^{n''} a_{2i-1} T_{2i-1}(x)$ be

$$xG'(0) \quad \text{at } x = x_{2i-1} \text{ for } i = 1, 2, \ldots, 7$$

and

$$G(x_i) \sum_{j=0}^{3} d_{2j} x_{2i}^{2j} \quad i = 1, 2, \ldots, 6$$

If the interval $[a, b]$ is not inordinately long, the accuracy of this approximation is good. It is excellent if the function defined by the set of points (t_i, f_i) is a rational function with an even polynomial in the denominator.

Program P220 computes the Padé approximation of a truncated power series with 8 terms, P221 the modified Padé approximation of a truncated power series with 13 terms, and P222 the rational approximation with Chebyshev polynomials for a function defined by 13 points (t_i, f_i) on the interval $[a, b]$. All three programs evaluate the function and its first and second derivatives for a given value of the independent variable.

2.14 NUMERICAL DIFFERENTIATION

The definition of the first derivative of a function $y = f(x)$

$$\frac{dy}{dx} = \lim_{\Delta x \to 0} \frac{f(x + \Delta x) - f(x)}{\Delta x} \quad (2.14.1)$$

clearly indicates the basic difficulty of numerical differentiation. The definition requires the increment Δx to be chosen as small as possible, but the smaller Δx is the fewer digits in $f(x + \Delta x)$ will differ from the digits in $f(x)$ because both numbers are expressed with a finite precision. As a consequence, the relative errors in $f(x + \Delta x)$ and $f(x)$, which may be quite negligible, are amplified by the subtraction and division by Δx to such an extent that the derivative computed by (2.14.1) from a small but finite increment Δx may contain a very large error.

To improve the accuracy of numerical differentiation Eq. (2.14.1) with Δx finite can be replaced by

$$\frac{dy}{dx} = \lim_{\Delta x \to 0} \frac{f(x + \Delta x) - f(x - \Delta x)}{2\Delta x} \qquad (2.14.2)$$

While (2.14.1) with Δx finite gives a correct value of the derivative only if the function $f(x)$ is linear, (2.14.2) yields a true value for a quadratic function. As long as the given function can be approximated with sufficient accuracy in the interval between $x - \Delta x$ and $x + \Delta x$ by a quadratic parabola, the first derivatives computed by (2.14.2) can be used.

To circumvent the difficulties of numerical differentiation the operation is performed in two steps; first the function given by a set of points is approximated, and then the approximating function is differentiated analytically. This process is called *discretization*, and the error generated by the impossibility in the numerical differentiation of going to the limit $\Delta x \to 0$ is known as *discretization error*. Examples of this approach have been given in the preceding sections.

If a function is defined by a set of m points with equally spaced abscissas and the derivatives are to be determined just at these points, they can be computed directly by very simple formulas; they are the derivatives of a parabola of $(m - 1)$th degree going through those m points, and they are as close to the exact values as the approximating parabola in that range is to the true function. The error is smallest at the central point and increases toward both ends of the interval. Therefore only the formulas for $m = 3, 5, 7$ points are used here. A further increase of their number may lead to disastrous effects, especially at exterior points. The h denotes the interval between two consecutive points

$$h = x_{i+1} - x_i \qquad (2.14.3)$$

Three-point formulas:

$$y'_1 = \frac{-3f_1 + 4f_2 - f_3}{2h}$$

$$y'_2 = \frac{-f_1 + f_3}{2h} \qquad (2.14.4)$$

$$y'_3 = \frac{f_1 - 4f_2 + 3f_3}{2h}$$

$$y''_1 = y''_2 = y''_3 = \frac{f_1 - 2f_2 + f_3}{h^2} \qquad (2.14.5)$$

Five-point formulas:

$$y'_1 = \frac{-25f_1 + 48f_2 - 36f_3 + 16f_4 - 3f_5}{12h}$$

$$y'_2 = \frac{-3f_1 - 10f_2 + 18f_3 - 6f_4 + f_5}{12h}$$

$$y'_3 = \frac{f_1 - 8f_2 + 8f_4 - f_5}{12h} \qquad (2.14.6)$$

$$y'_4 = \frac{-f_1 + 6f_2 - 18f_3 + 10f_4 + 3f_5}{12h}$$

$$y'_5 = \frac{3f_1 - 16f_2 + 36f_3 - 48f_4 + 25f_5}{12h}$$

$$y''_1 = \frac{35f_1 - 104f_2 + 114f_3 - 56f_4 + 11f_5}{12h^2}$$

$$y''_2 = \frac{11f_1 - 20f_2 + 6f_3 + 4f_4 - f_5}{12h^2}$$

$$y''_3 = \frac{-f_1 + 16f_2 - 30f_3 + 16f_4 - f_5}{12h^2} \qquad (2.14.7)$$

$$y''_4 = \frac{-f_1 + 4f_2 + 6f_3 - 20f_4 + 11f_5}{12h^2}$$

$$y''_5 = \frac{11f_1 - 56f_2 + 114f_3 - 104f_4 + 35f_5}{12h^2}$$

Seven-point formulas:

$$y'_1 = \frac{-147f_1 + 360f_2 - 450f_3 + 400f_4 - 225f_5 + 72f_6 - 10f_7}{60h}$$

$$y'_2 = \frac{-10f_1 - 77f_2 + 150f_3 - 100f_4 + 50f_5 - 15f_6 + 2f_7}{60h}$$

$$y'_3 = \frac{2f_1 - 24f_2 - 35f_3 + 80f_4 - 30f_5 + 8f_6 - f_7}{60h}$$

$$y'_4 = \frac{-f_1 + 9f_2 - 45f_3 + 45f_4 - 9f_5 + f_6}{60h} \qquad (2.14.8)$$

$$y'_5 = \frac{f_1 - 8f_2 + 30f_3 - 80f_4 + 35f_5 + 24f_6 - 2f_7}{60h}$$

$$y'_6 = \frac{-2f_1 + 15f_2 - 50f_3 + 100f_4 - 150f_5 + 77f_6 + 10f_7}{60h}$$

$$y'_7 = \frac{10f_1 - 72f_2 + 225f_3 - 400f_4 + 450f_5 - 360f_6 + 147f_7}{60h}$$

$$y_1'' = \frac{1624f_1 - 6264f_2 + 10{,}530f_3 - 10{,}160f_4 + 5940f_5 - 1944f_6 + 274f_7}{360h^2}$$

$$y_2'' = \frac{274f_1 - 294f_2 - 510f_3 + 940f_4 - 570f_5 + 186f_6 - 26f_7}{360h^2}$$

$$y_3'' = \frac{-26f_1 + 456f_2 - 840f_3 + 400f_4 + 30f_5 - 24f_6 + 4f_7}{360h^2}$$

$$y_4'' = \frac{4f_1 - 54f_2 + 540f_3 - 980f_4 + 540f_5 - 54f_6 + 4f_7}{360h^2} \qquad (2.14.9)$$

$$y_5'' = \frac{4f_1 - 24f_2 + 30f_3 + 400f_4 - 840f_5 + 456f_6 - 26f_7}{360h^2}$$

$$y_6'' = \frac{-26f_1 + 186f_2 - 570f_3 + 940f_4 - 510f_5 - 294f_6 + 274f_7}{360h^2}$$

$$y_7'' = \frac{274f_1 - 1944f_2 + 5940f_3 - 10{,}160f_4 + 10{,}530f_5 - 6264f_6 + 1624f_7}{360h^2}$$

Program P223 computes the first and second derivatives at each of 3, 5, or 7 points with equally spaced abscissas which define the given function. As an option, subroutine shifts all points either to the left or to the right and asks for a new point; in this way the derivatives can be repeatedly computed by the formula for the central point.

2.15 CHOOSING THE METHOD

This section contains some hints for making the proper choice of an interpolation or approximation method. If a function that must be interpolated is also given by its first (and possibly second) derivative, or if the derivatives can be computed, Hermite interpolation is superior. If only the data of the function but not its derivatives are known, the cubic spline, applied properly, is a reliable method not only for interpolation but also for numerical differentiation. Lagrange and Newton interpolations are equivalent. The former can be used for any number of points, but the program does not compute the derivatives. In the latter no more than six points define the function; the first and second derivatives are computed too. Trigonometric interpolation must have equally spaced abscissas and is generally used for periodic functions.

If the data defining the function contain errors, a least-squares technique should be used for approximation and numerical differentiation: Chebyshev approximation for points with Chebyshev abscissas, approximation by orthogonal polynomials if the points are arbitrarily distributed, and Fourier approximation for periodic functions with equally spaced abscissas.

The economization technique for a power series and approximation by a rational function can give excellent results in the hands of an experienced numerical analyst.

It must be emphasized once again that numerical differentiation is a very difficult problem of numerical analysis; hence the formulas of Sec. 2.14 should be used with caution, especially at exterior points and for $m = 7$.

EVALUATION OF DEFINITE INTEGRALS

3

A definite integral must be evaluated numerically if the integrand is defined by a set of numerical data or if the integration cannot be carried out exactly by analytical methods. Special techniques are applied for integrals over a semi-infinite or infinite intervals and for singular integrals. The numerical evaluation of an indefinite integral, i.e., of a definite integral whose upper limit is considered as an independent variable, is reduced to the numerical integration of a first-order differential equation and discussed in Chap. 4.

3.1 METHODS OF NUMERICAL INTEGRATION

In the numerical evaluation of definite integrals, i.e., in the process of numerical quadrature, the integrand is approximated by a suitably chosen function that can be integrated analytically. The function is defined by n points (x_i, f_i) and possibly by n first derivatives as well. After the approximating function has been integrated, the general quadrature formula takes the form

$$I_n = \int_a^b f(x)\, dx = \sum_{i=1}^n w_i f_i + \sum_{i=1}^n w'_i f'_i \qquad (3.1.1)$$

In the closed Newton-Cotes formulas the approximating function is a polynomial, the abscissas are equally spaced

$$x_i = x_1 + h(i-1) \qquad i = 1, 2, \ldots, n \qquad (3.1.2)$$

the weights w_i are rational numbers, and all $w'_i = 0$. The best-known of them are the trapezoidal rule ($n = 2$, $b = a + h$)

$$I_2 = \tfrac{1}{2}h(f_1 + f_2) \qquad (3.1.3a)$$

3.1

3.2 THEORY

with an error

$$E_2 = -\frac{f''(\xi)h^2}{12} \quad (3.1.3b)$$

and Simpson's rule ($n = 3$, $b = a + 2h$)

$$I_3 = \frac{h}{3}(f_1 + 4f_2 + f_3) \quad (3.1.4a)$$

with an error

$$E_3 = -\frac{f^{(4)}(\xi)h^5}{90} \quad (3.1.4b)$$

The ξ is some abscissa on the interval $[a, b]$. Higher-order formulas exist but are rarely used.

The open Newton-Cotes formulas differ from the closed ones only by having $w_1 = w_n = 0$. They are used in some methods of numerical integration of ordinary differential equations.

The integrand in the corrected trapezoidal rule is expressed with the help of Hermite interpolation. The approximating function is a cubic parabola defined by two ordinates f_1, f_2 and two derivatives f_1', f_2'. The quadrature formula is

$$I_4 = \tfrac{1}{2}h\left[f_1 + f_2 + \frac{h}{6}(f_1' - f_2')\right] \quad (3.1.5a)$$

with an error

$$E_4 = \frac{f^{(4)}(\xi)h^5}{720} \quad (3.1.5b)$$

In Chebyshev quadrature formulas over the interval $[-1, 1]$ the abscissas x_i are chosen so that all $w_i = 2/n$ and all $w_i' = 0$. The x_i are irrational numbers or zero. The formula with all x_i real exists only for $n = 2, 3, \ldots, 7$ and $n = 9$. If n is even, the formula is exact for polynomials of degree $\leq n + 1$.

To increase the accuracy of integration without increasing the number of arithmetic operations the approximating function in gaussian quadrature over the interval $[-1, 1]$ is a series of Legendre polynomials, the n abscissas are zeros of Legendre polynomial P_n

$$P_n(x_i) = 0 \quad i = 1, 2, \ldots, n \quad (3.1.6a)$$

and the weights are given by

$$w_i = \frac{2}{(1 - x_i^2)[P_n'(x_i)]^2} \quad (3.1.6b)$$

The integration formula (3.1.1) with all $w_i' = 0$ is then exact for polynomials of degree $\leq 2n - 1$. The abscissas x_i and the weights w_i have been computed for $n = 2, 3, \ldots, 10, 12$ in 15 decimal places and for $n = 16, 20, 24, 32, 40, 48, 64, 80, 96$ in 21 decimal places [36].

Since very high accuracy can be achieved by gaussian integration, many special functions can readily be evaluated at a given value of the argument by using the integral representation and evaluating the definite integral by gaussian quadrature.

In gaussian quadrature the external abscissas never equal the endpoints of the integration interval $[-1, 1]$. If there is a reason to emphasize the importance of the region near one or both endpoints, the abscissas -1 or -1 and 1 are included in the integration formulas of Radau and Lobatto.

Radau's formula is

$$I_n = \int_{-1}^{1} f(x)\, dx = \frac{2f(-1)}{n^2} + \sum_{i=1}^{n-1} w_i f_i \qquad (3.1.7a)$$

The x_i abscissa is the ith zero of the equation

$$\frac{P_{n-1}(x_i) + P_n(x_i)}{x_i + 1} = 0 \qquad (3.1.7b)$$

and the corresponding weight is

$$w_i = \{(1 - x_i)[P'_{n-1}(x_i)]^2\}^{-1} \qquad (3.1.7c)$$

The formula is exact for polynomials of degree $\leq 2n - 2$.

Lobatto's formula is

$$I_n = \int_{-1}^{1} f(x)\, dx = \frac{2}{n^2 - n}[f(-1) + f(1)] + \sum_{i=2}^{n-2} w_i f_i \qquad (3.1.8a)$$

The x_i abscissa is the $(i - 1)$th zero of the equation

$$P'_{n-1}(x_i) = 0 \qquad (3.1.8b)$$

and the corresponding weight is

$$w_i = \frac{2}{(n^2 - n)[P_{n-1}(x_i)]^2} \qquad x_i \neq \pm 1 \qquad (3.1.8c)$$

The formula is exact for polynomials of degree $\leq 2n - 3$.

The numerical values of abscissas and weights for Radau's and Lobatto's quadrature can be found in Refs. 17 and 36. Since for the same n the gaussian quadrature may be more accurate than Radau's or Lobatto's formulas, the last two should not be used without a good reason.

Gaussian quadrature with n abscissas requires the integrand to be continuous and its $2n$ derivatives bounded across the interval of integration. Only if this condition is satisfied can any desired accuracy of integration be achieved by choosing the number of abscissas high enough and by using a composite quadrature. On the other hand, the composite trapezoidal rule

$$\int_a^b f(x)\, dx = \frac{b - a}{n - 1}\left[\tfrac{1}{2}(f_1 + f_n) + \sum_{i=2}^{n-1} f_i\right] \qquad (3.1.9)$$

requires only the second derivatives to be bounded; with increasing n only the

3.4 THEORY

roundoff errors limit the accuracy of the formula. In Romberg integration the composite trapezoidal rule is applied in connection with Richardson's extrapolation to the limit. This combination substantially diminishes the number of arithmetical operations, which might be excessive if only the composite trapezoidal rule were used. Romberg quadrature, which gives excellent results and is widely used, is discussed in Sec. 3.6.

In all preceding methods it is tacitly assumed that the integration interval is finite and that the integrand is nonsingular. Numerical integration with semi-infinite and infinite intervals and singular integrals is discussed in subsequent sections.

3.2 INTEGRALS WITH NUMERICALLY DEFINED INTEGRANDS

If the integrand is given numerically by a set of n points (x_i, f_i) with equally spaced abscissas and n is odd, the integral should be evaluated by the composite Simpson's rule

$$\int_a^b f(x)\, dx = \frac{b-a}{3(n-1)} \left(f_1 + f_n + 4 \sum_{i=1}^{n'} f_{2i} + \sum_{i=1}^{n''} f_{1+2i} \right) \qquad (3.2.1a)$$

with

$$n' = \frac{n-1}{2} \qquad n'' = n' - 1 \qquad (3.2.1b)$$

On the other hand, if the integrand is given numerically by a set of n points (x_i, f_i) with arbitrarily distributed abscissas and an arbitrary number of points, the integral should be evaluated by the modification of the composite Simpson's rule developed by the author. The contribution ΔI of the area between two consecutive abscissas $x_{i+1} = p$ and $x_{i+2} = q$ is repeatedly computed by Simpson's rule

$$\Delta I = \frac{q-p}{6} \left[f(p) + f(q) + 4f(\tfrac{1}{2}(p+q)) \right] \qquad (3.2.2)$$

The central ordinate at $x = \tfrac{1}{2}(p+q)$ is determined from four consecutive points by Newton interpolation (cf. Sec. 2.3). At the lower limit of the integral $a = p = x_1, q = x_3$. Similarly, at the upper limit of the integral $p = x_{n-2}, b = q = x_n$. The total number of points defining the integrand must be $n \geq 5$.

If the integrand is given numerically by a set of n points (x_i, f_i) with equally spaced abscissas but n is even, the integration interval can be divided in two and evaluated by the preceding two methods.

If the integrand is given numerically by a set of n points (x_i, f_i) with equally spaced abscissas and by the derivatives at both limits, $f'(a) = f'_1$ and $f'(b) = f'_n$, the integral can be evaluated by the composite corrected trapezoidal rule

$$\int_a^b f(x)\, dx = h \left[\tfrac{1}{2}(f_1 + f_n) + \frac{h}{12}(f'_1 - f'_n) + \sum_{i=2}^{n-2} f_i \right] \qquad (3.2.3a)$$

where the abscissas x_i are given by (3.1.2) and the subinterval h by

$$h = \frac{b-a}{n-1} \tag{3.2.3b}$$

The quadrature formula (3.2.3a) differs from the composite trapezoidal rule (3.1.9) only in two terms involving the derivatives at both integration limits. The interior derivatives cancel each other because the abscissas are equally spaced [cf. Eq. (3.1.5a)].

If the integrand is given numerically by a set of n points (x_i, f_i) with arbitrarily spaced abscissas and by the first derivatives f'_i at each point, the integral can again be evaluated by the composite corrected trapezoidal rule [cf. (3.1.5a)] in which all the derivatives appear explicitly

$$\int_a^b f(x)\,dx = \sum_{i=1}^{n-1} \tfrac{1}{2}(x_{i+1} - x_i)\left[f_i + f_{i+1} + \frac{(x_{i+1} - x_i)(f'_i - f'_{i+1})}{6}\right] \tag{3.2.4}$$

All four quadrature formulas are exact for polynomials of degree ≤ 3. The accuracy of integration depends on the quality of given input data, on the number of points, and on the interval between them; but even if the data contain some random errors, their influence is not as detrimental as in numerical differentiation, which amplifies them, while here they are smoothed in the process of integration.

Program P301 evaluates a definite integral by the composite Simpson's rule, P302 by the modified composite Simpson's rule, P303 by the composite corrected trapezoidal rule for equally spaced abscissas, and P304 by the composite corrected trapezoidal rule for arbitrarily spaced abscissas.

3.3 INTEGRALS OVER FINITE INTERVALS

An integral over a finite interval $[a, b]$ with integrand continuous and defined analytically is here evaluated by composite gaussian quadrature; for a given number of arithmetic operations it is the most accurate integration method. The quadrature formula is

$$\int_a^b f(x)\,dx = \tfrac{1}{2}h \sum_{j=1}^m \sum_{i=1}^n w_i f(t_{ij}) \tag{3.3.1a}$$

The interval $b - a$ is divided into m subintervals

$$h = \frac{b-a}{m} \quad \begin{array}{l} \text{with } m = 2^k \\ k = 0, 1, 2 \ldots, k_m \end{array} \tag{3.3.1b}$$

The abscissas t_{ij} are related to gaussian abscissas x_i by

$$t_{ij} = \tfrac{1}{2}hx_i + (a - \tfrac{1}{2}h) + hj \tag{3.3.1c}$$

3.6 THEORY

The gaussian abscissas x_i and the corresponding weights w_i are defined by Eqs. (3.1.6). Their numerical values can be found in mathematical tables [36].

As long as the integrand $f(x)$ is continuous [if it is not, the integral in (3.3.1a) can be replaced by a finite sum of integrals with continuous integrands] and $2n$ derivatives are bounded, any desired accuracy can be achieved by choosing a sufficiently high number n of abscissas in gaussian quadrature. The error of formula (3.3.1a) is

$$E_n = \frac{(b-a)^{2n+1}}{2n+1} \frac{(n!)^4}{[(2n)!]^3} f^{(2n)}(\xi) \qquad (3.3.2)$$

where ξ is some abscissa within the interval $[a, b]$. If the interval is halved, the remainder decreases $(\frac{1}{2})^{2n}$ times (there are two errors of the same order of magnitude, each in one subinterval), that is, 2.3×10^{-10} times for $n = 16$ and 5.4×10^{-20} times for $n = 32$. The subdivision of the interval in Eqs. (3.3.1) is thus a very efficient way of increasing the accuracy of integration once n has been chosen high enough.

Program P305 evaluates a definite integral over the interval $[a, b]$ by composite gaussian quadrature. The number of abscissas $n = 32$. To increase the accuracy of integration it is possible to divide the interval into m subintervals and to apply gaussian quadrature in each of them. Optionally, the accuracy of quadrature may be chosen. The integral is then evaluated with $m = 1$ and $m = 2$. If the results differ by more than the chosen accuracy of integration, the interval is successively subdivided into 4, 8, 16 subintervals until the required accuracy is achieved or the maximum number of 16 subdivisions is reached.

3.4 INTEGRALS OVER SEMI-INFINITE INTERVALS

An integral over a semi-infinite interval $[0, \infty]$ with integrand continuous and defined analytically is evaluated by Laguerre quadrature

$$\int_0^\infty e^{-x} f(x)\, dx = \int_0^\infty g(x)\, dx = \sum_{i=1}^n (w_i e^{x_i}) g(x_i) \qquad (3.4.1a)$$

The weight function e^{-x} has been introduced to assure the convergence of the integral for a polynomial $f(x)$ of arbitrary degree. The approximating function is a series of Laguerre polynomials, the abscissa x_i is the ith zero of Laguerre polynomial L_n

$$L_n(x_i) = 0 \qquad (3.4.1b)$$

and the corresponding weight is

$$w_i = \frac{(n!)^2 x_i}{[(n+1)L_{n+1}(x_i)]^2} \qquad (3.4.1c)$$

Formula (3.4.1a) is exact for polynomials of degree $\leq 2n - 1$. The abscissas x_i and the weights w_i and $w_i e^{x_i}$ have been computed for $n = 2, 3, \ldots, 10, 12, 15$ and can be found in Ref. 36.

An integral over a semi-infinite interval $[a, \infty]$ can also be evaluated by Laguerre quadrature

$$\int_a^\infty g(t)\, dt = \int_0^\infty g(a + x)\, dx = \sum_{i=1}^n (w_i e^{x_i}) g(a + x_i) \qquad (3.4.2a)$$

Similarly,

$$\int_{-\infty}^a g(t)\, dt = \int_0^\infty g(a - x)\, dx = \sum_{i=1}^n (w_i e^{x_i}) g(a - x) \qquad (3.4.2b)$$

Even with $n = 15$ the Laguerre integration may not be accurate enough. To increase the accuracy the preceding two integrals can be evaluated by the composite Gauss-Laguerre integration

$$\int_a^\infty g(x)\, dx = \int_a^b g(x)\, dx + \int_b^\infty g(x)\, dx \qquad b > 0 \qquad (3.4.3a)$$

and

$$\int_{-\infty}^a g(x)\, dx = \int_{-\infty}^{-b} g(x)\, dx + \int_{-b}^a g(x)\, dx \qquad b > 0 \qquad (3.4.3b)$$

The first integrals on the right-hand sides of Eqs. (3.4.3) are to be evaluated by composite gaussian integration (3.3.1a) and the second integrals by Laguerre quadrature. The greater the chosen absolute value of the limit b the higher the accuracy of Laguerre quadrature. As shown in Sec. 3.3, the accuracy of gaussian quadrature can be achieved as high as required by subdividing the interval $[a, b]$ providing that $2n$ derivatives of the integrand are bounded within the interval of the integration.

Program P306 evaluates an integral over the interval $[a, \infty]$ or $[-\infty, a]$ by Laguerre quadrature with $n = 15$ abscissas. Program P307 evaluates the same integral by the composite Gauss-Laguerre quadrature. To increase the accuracy of gaussian quadrature it is possible to divide the corresponding interval into m subintervals and to apply gaussian quadrature in each of them. As an option the accuracy of gaussian integration can be chosen. Then the integral is evaluated with $m = 1$ and $m = 2$. If the difference in results is greater than the chosen accuracy of quadrature, the interval is successively subdivided into 4, 8, 16 subintervals until the required accuracy is achieved or the maximum number of 16 subdivisions is reached. Of course, the absolute value of the limit b must be chosen sufficiently large.

3.5 INTEGRALS OVER INFINITE INTERVALS

An integral over an infinite interval $[-\infty, \infty]$ with integrand continuous and defined analytically is evaluated by Hermite integration

$$\int_{-\infty}^{+\infty} \exp(-x^2) f(x)\, dx = \int_{-\infty}^{+\infty} g(x)\, dx = \sum_{i=1}^n (w_i \exp x_i^2) g(x_i) \qquad (3.5.1a)$$

3.8 THEORY

The weight function e^{-x^2} has been introduced to assure the convergence of the integral for a polynomial $f(x)$ of arbitrary degree. The approximating function is a series of Hermite polynomials, the abscissa x_i is the ith zero of Hermite polynomial H_n

$$H_n(x_i) = 0 \qquad (3.5.1b)$$

and the corresponding weight is

$$w_i = \frac{2^{n-1} n! \pi^{1/2}}{[nH_{n-1}(x_i)]^2} \qquad (3.5.1c)$$

Formula (3.5.1a) is exact for polynomials of degree $\leq 2n - 1$. The abscissas and weights w_i and $w_i \exp x_i^2$ have been computed for $n = 2, 3, \ldots, 10, 12, 16, 20$ and can be found in Ref. 36.

Even with $n = 20$ Hermite integration may not be accurate enough. To increase the accuracy an integral over an infinite interval can be evaluated by the composite Gauss-Laguerre integration

$$\int_{-\infty}^{+\infty} g(x)\, dx = \int_{-a}^{b} g(x)\, dx + \int_{b}^{+\infty} g(x)\, dx + \int_{-\infty}^{-a} g(x)\, dx \qquad a, b > 0 \quad (3.5.2)$$

The first integral on the right-hand side is to be evaluated by composite gaussian quadrature (3.3.1a) and the other two by Laguerre quadrature (3.4.3a, b). The greater the chosen absolute values of the limits $-a$ and b the higher the accuracy of both Laguerre quadratures. As shown in Sec. 3.3, the accuracy of gaussian quadrature can be as high as required by subdividing the interval $[-a, b]$ providing that $2n$ derivatives of the integrand are bounded.

Program P308 evaluates an integral over an infinite interval by Hermite quadrature with $n = 20$ abscissas. Program P309 evaluates the same integral by composite Gauss-Laguerre quadrature. To increase the accuracy of gaussian quadrature it is possible to divide the corresponding interval into m subintervals and apply gaussian quadrature in each of them. As an option, the accuracy of gaussian quadrature can be chosen. The integral is then evaluated with $m = 1$ and $m = 2$. If the difference is greater than the required accuracy, the interval is successively subdivided into 4, 8, 16 subintervals until the required accuracy is achieved or the maximum number of 16 subdivisions is reached. Of course, the absolute values of the limits $-a$ and b must be chosen sufficiently large.

3.6 ROMBERG INTEGRATION

To keep the number of arithmetic operations as low as possible the composite trapezoidal rule (3.1.9) for a definite integral over the interval $[a, b]$ is reduced to the form

$$T_{0,0} = \tfrac{1}{2} h_0 [f(a) + f(b)] \qquad (3.6.1a)$$

$$T_{0,k} = \tfrac{1}{2} T_{0, k-1} + \Delta T_{0, k} \qquad (3.6.1b)$$

where for $k = 1, 2, \ldots$

$$\Delta T_{0,k} = h_k \sum_{j=1}^{n_k} f(x_{jk}) \tag{3.6.1c}$$

$$h_0 = b - a \qquad h_k = \frac{h_{k-1}}{2} \tag{3.6.1d}$$

$$n_k = 2^{k-1} = 2n_{k-1} \tag{3.6.1e}$$

$$a_0 = a - h_0 \qquad a_k = a_{k-1} + h_k \tag{3.6.1f}$$

$$x_{jk} = a_k + h_{k-1} j \tag{3.6.1g}$$

Formula (3.6.1a) is the simple trapezoidal rule (3.1.3a). To improve its accuracy the number of subintervals is repeatedly doubled, but only the contribution $\Delta T_{0,k}$ generated by adding the new points is computed by (3.6.1c) to get a more accurate value of the integral in (3.6.1b).

The sequence of $T_{0,k}$ converges to the true value of the integral,

$$\int_a^b f(x) \, dx = \lim_{k \to \infty} T_{0,k} \tag{3.6.2}$$

but the number of steps k, and consequently the computer time, may be too high. To increase the rate of convergence, the more accurate values $T_{m,k}$ are determined in Romberg integration by Richardson's extrapolation to the limit. Since the interval in each step is halved, they are computed by the formula

$$T_{m,k} = \frac{4^m T_{m-1, k+1} - T_{m-1, k}}{4^m - 1} \tag{3.6.3}$$

The data $T_{m,k}$ are arranged in the table

$T_{0,0}$
$T_{0,1} \qquad T_{1,0}$
$T_{0,2} \qquad T_{1,1} \qquad T_{2,0}$
$T_{0,3} \qquad T_{1,2} \qquad T_{2,1} \qquad T_{3,0}$
....

Once an improved value $T_{0,m}$ has been computed by the composite trapezoidal rule [Eqs. (3.6.1)], all further data in the same row are calculated by (3.6.3). The data $T_{m,0}$ converge to the true value of the integral,

$$\int_a^b f(x) \, dx = \lim_{m \to \infty} T_{m,0} \tag{3.6.4}$$

much faster. The approximation is completed when $T_{m,k}$ in the new row start to have the same value.

Romberg integration may be considered as an analog of iterative methods of linear algebra: the successive results steadily approach the true solution of the problem. Composite gaussian integration represents a counterpart of direct methods of linear algebra: when the number of abscissas n and the number of subintervals m are properly chosen ($n = 32$, $m = 1$ or 2), one integration yields the correct result.

Program P310 evaluates a definite integral over a finite interval $[a, b]$ by Romberg integration.

3.7 SINGULAR INTEGRALS

A definite integral is *singular* when its integrand or its derivatives become singular within the interval of integration. If the integrand becomes singular, i.e., its value is unbounded, at some point c, the integral is *improper*. If it converges, its value is

$$\int_a^b f(x)\,dx = \lim_{\varepsilon \to 0} \int_a^{c-\varepsilon} f(x)\,dx + \lim_{\eta \to 0} \int_{c+\eta}^b f(x)\,dx \qquad (3.7.1)$$

It may happen that both limits diverge when considered independently, but the integral is still convergent if expressed as one limit

$$\int_a^b f(x)\,dx = \lim_{\varepsilon \to 0} \left[\int_a^{c-\varepsilon} f(x)\,dx + \int_{c+\varepsilon}^b f(x)\,dx \right] \qquad (3.7.2)$$

This value is called the *Cauchy principal value of the integral*.

A singular integrand $f(x)$ can be written as a product of two functions, a regular one $g(x)$ and a singular one $w(x)$

$$\int_a^b f(x)\,dx = \int_a^b g(x)w(x)\,dx \qquad (3.7.3)$$

First, an attempt should be made to remove the singularity analytically. For instance, if $w(x) = (1-x)^{-1/2}$, the substitution $1 - x = t^2$ reduces the singular integral to a regular one

$$\int_a^b g(x)(1-x)^{-1/2}\,dx = 2 \int_a^b g(1-t^2)\,dt$$

Similarly,

$$\int_a^b g(x)(1-x)^{1/2}\,dx = 2 \int_a^b g(1-t^2)t^2\,dt$$

A singular integral can be expressed in another approach as a sum of a regular integral and a singular integral over a very small but finite interval. If the singularity occurs at $x = a$, then

$$\int_a^b f(x)\,dx = \int_{a+\varepsilon}^b f(x)\,dx + \int_a^{a+\varepsilon} g(x)w(x)\,dx \qquad (3.7.4)$$

The first, regular integral on the right-hand side with ε small but finite can be now computed by composite gaussian integration or Romberg integration and the other approximated as

$$g(a) \int_a^{a+\varepsilon} w(x)\, dx$$

and evaluated analytically.

If an integral with unbounded integrand is computed by composite gaussian quadrature and then by Romberg quadrature, the results contain a relatively large error; if the singularity coincides with one of the abscissas used in the method, the computation breaks down. For instance, for $n = 32$

$$\int_0^1 (1-x^2)^{-1/2}\, dx = \begin{cases} 1.551852824 & m=1 \\ 1.557401062 & m=2 \end{cases}$$

when computed by gaussian quadrature in 10-digit floating-point arithmetic. The correct value is $\pi/2 = 1.570796327$ and the errors are 0.018943503 and 0.013395265, respectively. Romberg integration stops at the outset because the integrand is singular at both limits of the integration interval.

If an integral with unbounded derivatives in the integrand is computed by both methods, the results contain errors that are now much smaller. For instance, for $n = 32$

$$\int_0^1 (1-x^2)^{1/2}\, dx = \begin{cases} 0.785402408 & m=1 \\ 0.785399664 & m=2 \end{cases}$$

in gaussian integration. The correct value is $\pi/4 = 0.785398164$ and the errors are 4.244×10^{-6} and 1.501×10^{-6}. Since Romberg integration yields 0.784861688 after $k = 5$ steps, the error 5.36476×10^{-4} is much higher.

These examples show that there is an improvement of accuracy if a higher-precision method is used, but it is much smaller than in the case of regular integrals.

The appropriate solution of the problem of numerical integration of singular integrals is quite simple: let the singular function $w(x)$ be identical with the weight function related to some orthogonal polynomials (as is the case in Laguerre and Hermite integration, Secs. 3.4 and 3.5). Then the quadrature formula is

$$\int_a^b g(x) w(x)\, dx = \sum_{i=1}^n w_i f_i \qquad (3.7.5)$$

In Chebyshev-Gauss quadrature the approximating function is a series of Chebyshev polynomials, the limits of the integral are ± 1, the weight function is

$$w(x) = (1-x^2)^{-1/2} \qquad (3.7.6a)$$

the abscissas are

$$x_i = \cos\left[(i - \tfrac{1}{2})\left(\frac{\pi}{n}\right)\right] \qquad (3.7.6b)$$

and all the weights are

$$w_i = \frac{\pi}{n} \quad (3.7.6c)$$

If this quadrature is applied in the first example (with $n = 5$)

$$\int_0^1 (1 - x^2)^{-1/2} \, dx = \frac{1}{2} \int_{-1}^{+1} (1 - x^2)^{-1/2} \, dx = 1.570796327 = \frac{\pi}{2}$$

the solution is exact. The integral of the other example can also be reduced to a form suitable for Chebyshev-Gauss integration

$$\int_0^1 (1 - x^2)^{1/2} \, dx = \frac{1}{2} \int_{-1}^{+1} (1 - x^2)(1 - x^2)^{-1/2} \, dx = 0.785398164 = \frac{\pi}{4}$$

The solution is also exact.

In gaussian quadrature for an integral with logarithmic singularity the interval of integration is [0, 1] and the weight function is

$$w(x) = \ln x \quad (3.7.7)$$

The abscissas and the corresponding weights to be substituted into (3.7.5) can be found in Ref. 36.

Program P311 computes singular integrals by Chebyshev-Gauss integration, and Program P312 evaluates singular integrals with logarithmic singularity in the integrand by gaussian quadrature.

ORDINARY DIFFERENTIAL EQUATIONS

Before an ordinary differential equation is integrated numerically, its properties and the form of solution should be examined by analytical methods. Therefore the basic definitions and the analytical methods for initial-value problems are discussed first, followed by a short survey of differential equations defining important special functions. To clarify the problems connected with the numerical integration of ordinary differential equations the Euler method is described. The Taylor-series method, Runge-Kutta methods, and predictor-corrector methods for the numerical integration of ordinary differential equations and of systems of ordinary differential equations are then discussed in detail.

4.1 DEFINITIONS AND ANALYTICAL METHODS OF INTEGRATION

An ordinary differential equation of nth order is an equation involving the independent variable x, the dependent variable y, and its derivatives up to nth order

$$F(x, y, y', \ldots, y^{(n)}) = 0 \qquad (4.1.1)$$

The order and the degree of the equation are identical with the order and degree of its highest derivative after the equation has been rationalized. The equation is linear when the dependent variable and all its derivatives are multiplied only by functions of the independent variable. If this condition is not met, the equation is nonlinear.

Equation (4.1.1) is homogeneous if it is satisfied identically not only by the function $y(x)$ but also by $ay(x)$, where a is an arbitrary number. Otherwise the equation is nonhomogeneous.

Equation (4.1.1) is solved (integrated) when a function $y(x)$ has been found which after differentiation and substitution into (4.1.1) satisfies it identically in a specified bounded or unbounded interval. The general solution contains n arbitrary integration constants. If they are chosen such that the function satisfies certain initial or boundary conditions, the solution is called a *particular integral*.

4.1

4.2 THEORY

Many but not all differential equations are satisfied by a function that contains no integration constants at all. This solution is called a *singular integral*.

The differential equation (4.1.1) can also be interpreted as an equation determining the highest, nth, derivative appearing in it in terms of the independent variable, dependent variable, and its derivatives up to $(n-1)$th order

$$y^{(n)}(x) = f(x, y, y', \ldots, y^{(n-1)}) \qquad (4.1.2)$$

From this equation higher derivatives can also be computed as long as its right-hand side is differentiable. For numerical analysis it is important to realize that the derivatives are computed here from the values at the same point; the derivatives computed numerically by the formulas of Sec. 2.14 contain a discretization error because they depend on the values of the differentiated function at neighboring points and are therefore only approximations to the true value. The differentiation of (4.1.2) may lead to rather long expressions. For instance, the second, third, and fourth derivatives computed from the first-order differential equation

$$y' = f(x, y) \qquad (4.1.3a)$$

are

$$y'' = \frac{\partial f}{\partial x} + \frac{\partial f}{\partial y} y' \qquad (4.1.3b)$$

$$y^{(3)} = \frac{\partial^2 f}{\partial x^2} + 2 \frac{\partial^2 f}{\partial x\, \partial y} y' + \frac{\partial^2 f}{\partial y^2} (y')^2 + \frac{\partial f}{\partial y} y'' \qquad (4.1.3c)$$

$$y^{(4)} = \frac{\partial^3 f}{\partial x^3} + 3 \frac{\partial^3 f}{\partial x^2\, \partial y} y' + 3 \frac{\partial^3 f}{\partial x\, \partial y^2} (y')^2$$
$$+ \frac{\partial^3 f}{\partial y^3} (y')^3 + 3 \frac{\partial^2 f}{\partial x\, \partial y} y'' + 3 \frac{\partial^2 f}{\partial y^2} y'y'' + \frac{\partial f}{\partial y} y^{(3)} \qquad (4.1.3d)$$

These formulas may seem formidable, but if a given function (4.1.3a) is differentiated, many of the partial derivatives may vanish.

Every ordinary differential equation of nth order can be replaced by a system of n simultaneous differential equations of first order. For instance, a third-order differential equation

$$F(x, y, y', y'', y^{(3)}) = 0 \qquad (4.1.4a)$$

can be reduced to a system of three first-order differential equations for the unknown functions y, u, w

$$F(x, y, y', u', w') = 0 \qquad (4.1.4b)$$

$$y' - u = 0 \qquad (4.1.4c)$$

$$u' - w = 0 \qquad (4.1.4d)$$

A differential equation in which the dependent variable does not appear explicitly

$$F(x, y', y'', \ldots, y^{(n)}) = 0 \qquad (4.1.5a)$$

can be reduced to a differential equation of $(n-1)$th order for an unknown function $w(x)$

$$F(x, w, w', \ldots, w^{(n-1)}) = 0 \qquad (4.1.5b)$$

After this equation has been integrated, the unknown function $y(x)$ is determined by the first-order differential equation

$$y' = w \qquad (4.1.5c)$$

In a differential equation in which the independent variable x does not appear explicitly

$$F(y, y', \ldots, y^{(n)}) = 0 \qquad (4.1.6a)$$

the y can be considered as a new independent variable and its first derivative y' as a dependent variable $p(y)$

$$p(y) = \frac{dy}{dx} \qquad (4.1.6b)$$

When the differential operator

$$\frac{d}{dx} = \frac{dy}{dx} \frac{d}{dy} \qquad (4.1.6c)$$

is applied to the second and third derivatives, the results are

$$y'' = pp' \qquad (4.1.6d)$$

and

$$y^{(3)} = p \frac{d}{dy}(pp') = p^2 p'' + p(p')^2 \qquad (4.1.6e)$$

In this way Eq. (4.1.6a) is reduced to a differential equation of $(n-1)$th order. For instance, a third-order differential equation is reduced to the second-order differential equation

$$F(y, p, pp', p^2 p'' + p(p')^2) = 0 \qquad (4.1.6f)$$

After this equation has been integrated, the unknown function $y(x)$ is determined by the differential equation

$$\frac{dy}{p(y)} = dx \qquad (4.1.6g)$$

following from (4.1.6b). This equation is solved by quadrature.

If a definite integral must be evaluated at different values of the upper limit but at a constant value of lower limit, or if an indefinite integral must be evalu-

4.4 THEORY

ated at a given set of values of the independent variable, the most efficient way is to reduce the integral to the differential equation

$$y' = f(x) \qquad (4.1.7)$$

and integrate it numerically. This is the simplest differential equation. The initial value at the lower limit x_0 of a definite integral is, of course, $y(x_0) = y_0 = 0$. For an indefinite integral the initial value y_0 can be chosen arbitrarily; in this way the additive integration constant is fixed.

If in a differential equation of first order

$$y' = f(x, y) \qquad (4.1.8a)$$

the function $f(x, y)$ is a product of two functions, each depending on one variable

$$f(x, y) = g(x)h(y) \qquad (4.1.8b)$$

the variables can be separated; the equation

$$\frac{dy}{h(y)} = g(x)\, dx \qquad (4.1.8c)$$

is solved by quadrature.

A first-order differential equation

$$P(x, y)\, dx + Q(x, y)\, dy = 0 \qquad (4.1.9a)$$

is exact if

$$\frac{\partial P}{\partial y} - \frac{\partial Q}{\partial x} = 0 \qquad (4.1.9b)$$

In that case the expression on the left-hand side of (4.1.9a) is an exact differential of a function $u(x, y)$

$$du = \frac{\partial u}{\partial x}\, dx + \frac{\partial u}{\partial y}\, dy$$

Its integral is $u = $ const. Since in an exact differential equation

$$P = \frac{\partial u}{\partial x} \qquad Q = \frac{\partial u}{\partial y}$$

the explicit form of the solution of (4.1.9a) is

$$u \equiv \int P(x, y)\, dx + \int \left[Q(x, y) - \frac{\partial}{\partial y} \int P(x, y)\, dx \right] dy = \text{const} \qquad (4.1.9c)$$

Any ordinary differential equation of first order can be reduced to an exact differential equation by multiplying it by an integrating factor $\mu = e^{\lambda}$. Condition

(4.1.9b) then takes the form

$$P\frac{\partial \lambda}{\partial y} - Q\frac{\partial \lambda}{\partial x} = \frac{\partial Q}{\partial x} - \frac{\partial P}{\partial y} \qquad (4.1.10)$$

If the equation is exact, the right-hand side vanishes and $\lambda = 0$, $\mu = 1$. In general it is not easy to solve the partial differential equation (4.1.10).

A linear differential equation of first order

$$y' + g(x)y = h(x) \qquad (4.1.11a)$$

is not exact for

$$P = yg(x) - h(x) \qquad Q = 1$$

The function $\lambda(x, y)$ must now satisfy the equation

$$[yg(x) - h(x)]\frac{\partial \lambda}{\partial y} - \frac{\partial \lambda}{\partial x} = -g(x) \qquad (4.1.11b)$$

Its solution is

$$\lambda = \int g(x)\, dx \qquad (4.1.11c)$$

where the integration constant has been put equal to 0 without loss of generality. In this way the linear differential equation is reduced to an exact one

$$e^{\lambda}[yg(x) - h(x)]\, dx + e^{\lambda}\, dy = 0$$

The general integral of any linear differential equation of first order can thus be computed by the formula

$$y = \exp\left[-\int g(x)\, dx\right]\left\{C + \int h(x) \exp\left[\int g(x)\, dx\right] dx\right\} \qquad (4.1.11d)$$

where C is an integration constant.

The Bernoulli differential equation

$$y' + g(x)y = h(x)y^n \qquad (4.1.12a)$$

is a nonlinear equation. It is mentioned here not only because it occurs in applications but also to show that the substitution

$$w = y^{1-n} \qquad (4.1.12b)$$

reduces it to a linear equation

$$w' + (1-n)g(x)w = (1-n)h(x) \qquad (4.1.12c)$$

The Clairaut differential equation

$$y - xy' = f(y') \qquad (4.1.13a)$$

is an example of a differential equation with a singular integral. It is a nonlinear differential equation of the first order. After differentiation with respect to x it is

4.6 THEORY

reduced to a nonlinear differential equation of second order

$$y''[f'(y') + x] = 0 \qquad (4.1.13b)$$

Hence
$$y'' = 0 \qquad (4.1.13c)$$

Its integral is a linear function

$$y = ax + b$$

where a and b are two integration constants. Substituting this function into (4.1.13a) yields as one integral of the Clairaut equation

$$y = ax + f(a) \qquad (4.1.13d)$$

This shows that the constant b cannot be chosen freely; it is determined by the differential equation $b = f(a)$. The other equation following from (4.1.13b)

$$f'(y') + x = 0 \qquad (4.1.13e)$$

yields another integral of the Clairaut equation, but it can also be used to eliminate y' from (4.1.13a). This solution is the singular integral since it contains no integration constant. In the special case

$$f(y') = (y')^2 \qquad (4.1.13f)$$

Eq. (4.1.13e) takes the form

$$2y' + x = 0 \qquad (4.1.13g)$$

The singular integral is thus

$$4x^2 + y = 0 \qquad (4.1.13h)$$

It is an envelope of the family of straight lines (4.1.13d). The integral of (4.1.13g) is

$$y = -\tfrac{1}{4}x^2 + c \qquad (4.1.13i)$$

If this function is substituted into the Clairaut equation (4.1.13a), it turns out that the integration constant c cannot be freely chosen either; it must equal zero. Equation (4.1.13i) with $c = 0$ reduces to the singular integral (4.1.13h).

A linear differential equation of nth order

$$a_n(x)y^{(n)} + a_{n-1}(x)y^{(n-1)} + \cdots + a_1(x)y' + a_0(x)y = f(x) \qquad (4.1.14)$$

is a nonhomogeneous equation. It becomes homogeneous if $f(x) = 0$. A general integral of the homogeneous equation consists of n linearly independent solutions y_j, each multiplied by an integration constant C_j

$$y = C_1 y_1 + C_2 y_2 + \cdots + C_n y_n \qquad (4.1.15)$$

The general solution of the nonhomogeneous equation is a sum of the general solution of the homogeneous equation and any solution of the nonhomogeneous equation (called the particular integral). The integration constants C_j of the homogeneous equation are determined by n initial conditions, i.e., by the values y_0,

$y'_0, \ldots, y_0^{(n-1)}$ at $x = 0$. The constants are the unknowns in the system of n linear algebraic equations

$$C_1 y_1(0) + C_2 y_2(0) + \cdots + C_n y_n(0) = y_0$$
$$C_1 y'_1(0) + C_2 y'_2(0) + \cdots + C_n y'_n(0) = y'_0 \qquad (4.1.16)$$
$$\cdots\cdots\cdots\cdots\cdots\cdots\cdots\cdots\cdots\cdots$$
$$C_1 y_1^{(n-1)}(0) + C_2 y_2^{(n-1)}(0) + \cdots + C_n y_n^{(n-1)}(0) = y_0^{(n-1)}$$

The determinant of the system is called the wronskian

$$W = \begin{vmatrix} y_1 & y_2 & \cdots & y_n \\ y'_1 & y'_2 & \cdots & y'_2 \\ \cdots\cdots\cdots\cdots\cdots\cdots \\ y_1^{(n-1)} & y_2^{(n-1)} & \cdots & y_n^{(n-1)} \end{vmatrix} \qquad (4.1.17)$$

It does not vanish in a certain domain if each function is a solution of the differential equation in this domain and is linearly independent. As a consequence, the integration constants are uniquely determined by the initial conditions.

Once the general integral of a homogeneous equation is known, the general solution of a nonhomogeneous equation can be found by variation of constants. In this method the integration constants are considered as unknown functions of x

$$y = C_1(x) y_1(x) + C_2(x) y_2(x) + \cdots + C_n(x) y_n(x) \qquad (4.1.18)$$

They are determined by a system of linear equations

$$C'_1(x) y_1(x) + C'_2(x) y_2(x) + \cdots + C'_n(x) y_n(x) = 0$$
$$C'_1(x) y'_1(x) + C'_2(x) y'_2(x) + \cdots + C'_n(x) y'_n(x) = 0 \qquad (4.1.19)$$
$$\cdots\cdots\cdots\cdots\cdots\cdots\cdots\cdots\cdots\cdots$$
$$C'_1(x) y_1^{(n-1)}(x) + C'_2(x) y_2^{(n-1)}(x) + \cdots + C'_n(x) y_n^{(n-1)}(x) = \frac{f(x)}{a_n(x)}$$

Except for the last, the right-hand sides of these equations equal zero. The determinant of the system is again the wronskian (4.1.17). After the system has been solved for the unknown $C'_j(x)$, the functions $C_j(x)$ are determined by integrating n equations

$$C_j(x) = \int C'_j(x)\, dx + K_j \qquad (4.1.20)$$

where K_j are the new integration constants, computed by solving a system similar to (4.1.16).

Unlike those in system (4.1.16), the unknowns in (4.1.19) and the wronskian are functions of x, not constants. Therefore the system cannot be solved numerically. This fact restricts the applicability of the method of variation of constants

4.8 THEORY

to equations of second and third order. For these problems the numerical treatment is most suitable: the nonhomogeneous equation is reduced to a system of first-order differential equations which are integrated numerically for the given set of initial data.

An important class of linear differential equations comprises the equations with constant coefficients. The solution of a homogeneous equation must then be in form of an exponential function e^{kx}. The substitution of this function and its n derivatives into the differential equation (4.1.14), in which a_j, $j = 0, 1, \ldots, n$, are constants, gives the characteristic equation

$$a_n k^n + a_{n-1} k^{n-1} + \cdots + a_1 k + a_0 = 0 \qquad (4.1.21)$$

Since all the constant coefficients are assumed real, the n roots of the characteristic equations are either real or complex conjugate. If the root k_1 is real and simple, the corresponding term in the solution is

$$C_1 e^{k_1 x} \qquad (4.1.22a)$$

If the root k_1 is real and of multiplicity m, then the corresponding m terms in the solution are

$$(C_1 + C_2 x + \cdots + C_m x^{m-1}) e^{k_1 x} \qquad (4.1.22b)$$

If two roots are complex conjugate and simple

$$k_j = a_j \pm i b_j \qquad i = (-1)^{1/2} \qquad (4.1.22c)$$

the corresponding two terms in the solution are

$$(C_j \cos b_j x + C_{j+1} \sin b_j x) e^{a_j x} \qquad (4.1.22d)$$

If the complex-conjugate roots $a_1 \pm i b_1$ are of multiplicity m, the corresponding $2m$ terms in the solution are

$$[(C_1 + C_2 x + \cdots + C_m x^{m-1}) \cos b_1 x + (C_{m+1} + C_{m+2} x + \cdots$$
$$+ C_{2m} x^{m-1}) \sin b_1 x] e^{a_1 x} \qquad (4.1.22e)$$

The general solution of a homogeneous equation is given by the superposition of these particular solutions. The integration constants C_j, $j = 1, 2, \ldots, n$, are determined from n initial conditions by solving a system of n linear algebraic equations similar to the system (4.1.16).

The method of variation of constants can also be used for solving a nonhomogeneous equation with constant coefficients, but if the function $f(x)$ contains only terms of the form x^m, where m is an integer, or $\cos qx$, $\sin qx$, e^{qx}, or sums or products of two or more such functions, a more powerful method for the equations with constants coefficients is the method of undetermined coefficients. In this method the solution of the nonhomogeneous equation is assumed to be a suitable combination of the four above-mentioned basic functions multiplied by undetermined coefficients. This function with its derivatives is substituted into the nonhomogeneous equation. The coefficients are determined from a set of linear algebraic equations by the requirement that the equations be satisfied identically. The integration constants C_j in the general integral of the homogeneous equation are then computed from n initial conditions by solving a system of n linear algebraic equations similar to the system (4.1.16).

For instance, if $f(x)$ equals $A \cos qx$ or $B \sin qx$ or a sum of these two functions, the solution of the nonhomogeneous equation is assumed to be

$$K_1 \cos qx + K_2 \sin qx$$

This function with its derivatives is substituted into the nonhomogeneous equation and its terms properly arranged. The sums of all the cofactors standing at $\cos qx$ and at $\sin qx$ must vanish to satisfy the equation identically. This requirement yields two linear algebraic equations for the undetermined coefficients K_1, K_2. If the imaginary part of one or more roots of the characteristic equation equals q, the well-known phenomenon of resonance takes place.

A system of m simultaneous nonhomogeneous linear differential equations for m dependent variables $y_j, j = 1, 2, \ldots, m$, can be written in the form

$$L_{11} y_1 + L_{12} y_2 + \cdots + L_{1m} y_m = f_1(x)$$
$$L_{21} y_1 + L_{22} y_2 + \cdots + L_{2m} y_m = f_2(x) \tag{4.1.23}$$
$$\ldots\ldots\ldots\ldots\ldots\ldots\ldots\ldots\ldots\ldots\ldots\ldots\ldots$$
$$L_{m1} y_1 + L_{m2} y_2 + \cdots + L_{mm} y_m = f_m(x)$$

The $L_{jk}, j,k = 1, 2, \ldots, m$, are linear differential operators

$$L_{jk} = \sum_{l=0}^{l_{jk}} a_{jk}(x) \left(\frac{d}{dx}\right)^l \tag{4.1.24}$$

where the integer l_{jk} indicates the highest order of derivatives in the operator. Under the assumption that the linear operators commute, i.e.,

$$L_{jk} L_{pq} = L_{pq} L_{jk}$$

the dependent variables can be eliminated in a way similar to the elimination process for linear algebraic equations. Since L_{jk} are operators, not numbers, Cramer's rule (1.4.1) must be written in the form

$$\Delta y_j = \Delta_j \tag{4.1.25}$$

The determinant of the system Δ must stand on the left-hand side because the operators must operate upon y_j (it is meaningless to divide by a determinant involving operators). The Δ_j are the determinants obtained from Δ by replacing its jth column by the right-hand sides of Eqs. (4.1.23). Of course, the operators in these determinants must stand before the functions operated upon.

A general solution of the system (4.1.23) is again the sum of the integral of the homogeneous system and of the particular integral of the nonhomogeneous system. The homogeneous differential equation for each of m dependent variables y_j

$$\Delta y_j = 0 \quad j = 1, 2, \ldots, m \tag{4.1.26}$$

contains a highest derivative whose order will be denoted q. The general integral of (4.1.26) consists of q linearly independent integrals, each multiplied by an integration constant. There are m such general integrals, each with q integration constants, but these $m \times q$ constants are not all independent. Only q constants are independent; to determine the interdependence of other constants the solution of (4.1.26) for one variable is substituted into whichever of Eqs. (4.1.23) is most convenient for the computation of other unknown variables. The q independent integration constants are given by n initial conditions.

If all the coefficients in (4.1.24) are real constants, the solution of the homogeneous equations (4.1.26) must be in form of an exponential function e^{kx}. The substitution of this function and its q derivatives into (4.1.26) yields a characteristic equation with q roots, either real or complex conjugate. The general solution of (4.1.26) is obtained in the same way as for one differential equation (4.1.21).

Many differential equations important for applications are of second order with variable coefficients, homogeneous or nonhomogeneous. The standard form is

$$y'' + P(x)y' + Q(x)y = f(x) \qquad (4.1.27)$$

A point $x = x_0$ of the homogeneous equation is said to be an *ordinary point* of the differential equation if both $P(x)$ and $Q(x)$ are regular at $x = x_0$; this means that $P(x)$ and $Q(x)$ behave in the neighborhood of x_0 like $(x - x_0)^m$, with m either zero or a positive integer. Near an ordinary point the general integral of the differential equation has two independent solutions, which can be written in form of a Taylor series. If x_0 is not an ordinary point but $(x - x_0)P(x)$ and $(x - x_0)^2 Q(x)$ are both regular at x_0, this point is said to be a *regular singular point* of the differential equation. Near such a point at least one solution is of the form

$$(x - x_0)^s \sum_{j=0}^{\infty} c_j (x - x_0)^j$$

The exponent s may be real or imaginary and need not be an integer. If the point $x = x_0$ is neither an ordinary point nor a regular singular point, it is said to be an *irregular singular point*. A nontrivial solution near such a point may or may not exist, but it is known that in this case the homogeneous equation (4.1.27) cannot have two independent solutions.

A general solution of the differential equation (4.1.27) is given by the superposition of a solution of the homogeneous equation consisting of two independent solutions and of a particular integral of the nonhomogeneous equation. The integral of the homogeneous differential equation of second order cannot be obtained by quadratures as it can for the first-order differential equation [cf. Eqs. (4.1.11a, d]. The differential equation defines, in general, two independent special functions, either polynomials or transcendental functions.

If both independent solutions y_1 and y_2 of the homogeneous differential equation are known, the general integral of the nonhomogeneous equation (4.1.27) can be found by variation of constants. It is worthwhile to write Eqs. (4.1.17) to

(4.1.20) and their solution for a second-order differential equation explicitly. The integration constants, considered now as unknown functions of x, are determined by a system of two linear equations

$$C_1'(x)y_1 + C_2'(x)y_2 = 0$$
$$C_1'(x)y_1' + C_2'(x)y_2' = f(x)$$

After these equations have been solved and C_1', C_2' integrated, the result is

$$C_1(x) = -\int \frac{y_2(x)f(x)}{W(x)} dx + K_1$$
$$C_2(x) = \int \frac{y_1(x)f(x)}{W(x)} dx + K_2 \tag{4.1.28}$$

where K_1, K_2 are two new integration constants and $W(x)$ is the wronskian

$$W(x) = \begin{vmatrix} y_1 & y_2 \\ y_1' & y_2' \end{vmatrix} \tag{4.1.29}$$

The wronskian is a function of x. If its value $W(x_0)$ is known at a point x_0, its value at an arbitrary point x is given by *Abel's formula*

$$W(x) = W(x_0) \exp\left[-\int_{x_0}^{x} P(t)\, dt\right] \tag{4.1.30}$$

The general solution of the nonhomogeneous equation is, according to Eq. (4.1.18),

$$y = K_1 y_1 + K_2 y_2 - y_1 \int \frac{y_2(x)f(x)}{W(x)} dx + y_2 \int \frac{y_1(x)f(x)}{W(x)} dx \tag{4.1.31}$$

The substitution
$$y = w(x)h(x) \tag{4.1.32}$$

replaces the differential equation (4.1.27) with another linear differential equation

$$w'' + \left(P + \frac{2h'}{h}\right)w' + \frac{h'' + Ph' + Qh}{h}w = \frac{f}{h} \tag{4.1.33}$$

If $h(x)$ is chosen such that

$$h = \exp\left[-\frac{1}{2}\int P(x)\, dx\right] \tag{4.1.34}$$

the first derivative w' is eliminated in (4.1.33)

$$w'' + (Q - \tfrac{1}{4}P^2 - \tfrac{1}{2}P')w = f(x) \exp\left[\frac{1}{2}\int P(x)\, dx\right] \tag{4.1.35}$$

This is the canonical form of a second-order differential equation. It is especially useful for approximate solutions of the equation.

If $h(x)$ is chosen as one solution of the homogeneous differential equation, the function w is eliminated in (4.1.33)

$$w'' + \left(P + \frac{2y_1'}{y_1}\right)w' = \frac{f}{y_1}$$

This is a linear differential equation of first order for the unknown w'. If this equation is integrated by (4.1.11d) and then integrated again to determine $w(x)$, which is substituted into (4.1.32), the result is

$$y = C_1 y_1 + C_2 y_1 \int \frac{dx}{Ey_1^2} + y_1 \int \left[\int Ey_1 f(x)\, dx\right] \frac{dx}{Ey_1^2} \quad (4.1.36a)$$

where

$$E = \exp\left[\int P(x)\, dx\right] \quad (4.1.36b)$$

It is thus possible to use quadratures to compute from one given solution y_1 of the homogeneous equation the other independent solution and the particular integral of the nonhomogeneous equation [the second and third terms on the right-hand side of (4.1.36a)].

A homogeneous linear differential equation of second order can be integrated either by the Frobenius method in the form of an infinite series or numerically.

4.2 DIFFERENTIAL EQUATIONS DEFINING SPECIAL FUNCTIONS

Some ordinary differential equations occurring in mathematical physics have been thoroughly investigated and their integrals included in mathematical tables. This section gives a short survey of ordinary differential equations defining special functions of mathematical physics.

The Bessel differential equation

$$x^2 y'' + xy' + (x^2 - v^2)y = 0 \quad (4.2.1a)$$

in which v is a constant called the *order* (also *parameter* or *index*) has two independent solutions

$$y = C_1 J_v(x) + C_2 Y_v(x) \qquad v \text{ arbitrary} \quad (4.2.1b)$$

i.e., the *Bessel function of the first kind* $J_v(x)$ and the *Bessel function of the second kind* $Y_v(x)$ [also called a Neumann function and denoted $N_v(x)$]. The Bessel functions of the third kind, or Hankel functions, $H_v^{(1)}(x)$ and $H_v^{(2)}(x)$ are linear combinations of Bessel functions of the first and second kind

$$H_v^{(1,\,2)}(x) = J_v(x) \pm i Y_v(x) \quad (4.2.1c)$$

If v is not an integer, two independent solutions of the Bessel differential equation can also be written as

$$y = C_1 J_v(x) + C_2 J_{-v}(x) \qquad v \neq 0,\, \pm 1,\, \pm 2,\, \ldots \quad (4.2.1d)$$

Comparing (4.2.1d) with (4.2.1b) makes it obvious that $J_\nu(x)$ and $Y_\nu(x)$ may differ only in a multiplicative constant. In fact

$$Y_{n+1/2}(x) = (-1)^{1-n} J_{-n-1/2}(x) \qquad n = 0, 1, 2, \ldots \qquad (4.2.1e)$$

The modified Bessel functions $I_\nu(x)$ and $K_\nu(x)$ are two independent solutions of the differential equation

$$x^2 y'' + xy' - (x^2 + \nu^2) y = 0 \qquad (4.2.2)$$

These two functions are sometimes referred to as *hyperbolic Bessel functions*.

The spherical Bessel equation

$$x^2 y'' + 2xy' + [x^2 - n(n+1)] y = 0 \qquad n = 0, \pm 1, \pm 2, \ldots \qquad (4.2.3a)$$

has two independent solutions

$$y = C_1 j_n(x) + C_2 y_n(x) \qquad (4.2.3b)$$

called *spherical Bessel functions of the first and second kind*. They are related to Bessel functions by

$$j_n(x) = \left(\frac{\pi}{2x}\right)^{1/2} J_{n+1/2}(x) \qquad (4.2.3c)$$

and

$$y_n(x) = \left(\frac{\pi}{2x}\right)^{1/2} Y_{n+1/2}(x) \qquad (4.2.3d)$$

Spherical Bessel functions of the third kind, $h_n^{(1)}(x)$ and $h_n^{(2)}(x)$, are a linear combination of spherical Bessel functions of the first and second kind

$$h_n^{(1,\,2)}(x) = j_n(x) \pm y_n(x) = \left(\frac{\pi}{2x}\right)^{1/2} H_{n+1/2}^{(1,\,2)}(x) \qquad (4.2.3e)$$

The particular integral of the nonhomogeneous differential equation

$$x^2 y'' + xy' + (x^2 - \nu^2) y = \frac{4\pi^{-1/2} (\tfrac{1}{2} x)^{\nu+1}}{\Gamma(\tfrac{1}{2} + \nu)} \qquad (4.2.4a)$$

is the *Struve function* $\mathbf{H}_\nu(x)$. The general integral of (4.2.4a) is

$$y = C_1 J_\nu(x) + C_2 Y_\nu(x) + \mathbf{H}_\nu(x) \qquad (4.2.4b)$$

The Bessel differential equation and all the related equations have a regular singular point at $x = 0$.

The Legendre differential equation

$$(1 - x^2) y'' - 2xy' + \nu(\nu + 1) y = 0 \qquad (4.2.5a)$$

where ν is an arbitrary constant called a *degree*, has two independent solutions

$$y = C_1 P_\nu(x) + C_2 Q_\nu(x) \qquad (4.2.5b)$$

called *Legendre functions of the first and second kind*. If ν is zero or a positive integer n, the Legendre function of the first kind reduces to a polynomial of degree n.

The associated Legendre differential equation

$$(1 - x^2)y'' - 2xy' + \left[v(v+1) - \frac{\mu^2}{1-x^2}\right]y = 0 \tag{4.2.6a}$$

of degree v and order μ, both arbitrary constants, has two independent integrals

$$y = C_1 P_v^\mu(x) + C_2 Q_v^\mu(x) \tag{4.2.6b}$$

the associated Legendre functions of the first and second kind. If v and μ equal zero or positive integers n and m, $0 \le m \le n$, the associated Legendre functions of the first kind can be computed by differentiating the Legendre polynomial

$$P_n^m(x) = (1 - x^2)^{m/2}\left(\frac{d}{dx}\right)^m P_n(x) \tag{4.2.6c}$$

The Legendre differential equation and the associated Legendre differential equation have two regular singular points at $x = \pm 1$.

The hypergeometric differential equation

$$x(1 - x)y'' + [c - (a + b + 1)x]y' - aby = 0 \tag{4.2.7a}$$

has three regular singular points at $x = 0, 1,$ and ∞. If $c \ne 0$, one solution is the hypergeometric series (or hypergeometric function)

$$y_1 = F(a, b; c; x) = \frac{\Gamma(c)}{\Gamma(a)\Gamma(b)} \sum_{j=0}^{\infty} \frac{\Gamma(a+j)\Gamma(b+j)}{\Gamma(c+j)} \frac{x^j}{j!} \tag{4.2.7b}$$

The series terminates if a or b is a negative integer and $|c| > |a|$ or $|c| > |b|$ if c is a negative integer. If c is not a negative integer, the infinite series converges for $|x| < 1$, at $x = 1$ if $c > a + b$ and at $x = -1$ if $c > a + b - 1$. Many elementary functions and special transcendental functions are special cases of the hypergeometric series. The other independent solution of (4.2.7a) is

$$y_2 = x^{1-c} F(a - c + 1, b - c + 1; 2 - c; x) \qquad c \ne 0 \tag{4.2.7c}$$

The hypergeometric differential equation is a special case of Riemann's differential equation

$$y'' + \left(\frac{1 - \alpha - \alpha'}{x - a} + \frac{1 - \beta - \beta'}{x - b} + \frac{1 - \gamma - \gamma'}{x - c}\right)y'$$
$$+ \left[\frac{(a-b)(a-c)}{x-a} + \frac{(b-c)(b-a)}{x-b}\right.$$
$$\left. + \frac{(c-a)(c-b)}{x-c}\right] \frac{y}{(x-a)(x-b)(x-c)} = 0 \tag{4.2.8a}$$

which has three regular singular points at $x = a, b, c$. The indices $\alpha, \alpha'; \beta, \beta'; \gamma, \gamma'$ are related by

$$\alpha + \alpha' + \beta + \beta' + \gamma + \gamma' = 1 \tag{4.2.8b}$$

The complete set of solutions is denoted by the symbol

$$y = \left\{ \begin{array}{ccc} a & b & c \\ \alpha & \beta & \gamma \\ \alpha' & \beta' & \gamma' \end{array} \; x \right\} \tag{4.2.8c}$$

The solution of Riemann's differential equation can be expressed as a hypergeometric series

$$y = \left(\frac{x-a}{x-b}\right)^{\alpha}\left(\frac{x-c}{x-b}\right)^{\gamma} F\left[\alpha + \beta + \gamma, \alpha + \beta' + \gamma; 1 + \alpha - \alpha'; \frac{(x-a)(c-b)}{(x-b)(c-a)}\right] \tag{4.2.8d}$$

If the constants a, b, c; α, α'; β, β'; γ, γ' are permuted in a suitable manner, Riemann's differential equation remains unchanged. Thus, there exist 24 solutions of the equation provided none of the differences $\alpha - \alpha'$, $\beta - \beta'$, $\gamma - \gamma'$ are integers.

If the dependent variable x in the hypergeometric differential equation (4.2.7a) is replaced by x/b, the regular singular point at $x = 1$ will be shifted to $x = b$. The singularity at $x = b$ will approach the regular singular point at $x = \infty$ as $b \to \infty$ (confluence of singularities). In this way Eq. (4.2.7a) is reduced to Kummer's confluent hypergeometric differential equation

$$xy'' + (c - x)y' - ay = 0 \tag{4.2.9a}$$

with regular singular point at $x = 0$ and irregular singular point at $x = \infty$. If $c \neq 0$, one solution is the confluent hypergeometric function (*Kummer's function*)

$$y_1 = M(a, c; x) = \frac{\Gamma(c)}{\Gamma(a)} \sum_{j=0}^{\infty} \frac{\Gamma(a+j)}{\Gamma(c+j)} \frac{x^j}{j!} \tag{4.2.9b}$$

The series terminates if a is a negative integer, and $|c| > |a|$ if c is a negative integer. The infinite series converges for all values of x, a, and c, provided c is not a negative integer. A second solution of (4.2.9a) is the function

$$y_2 = x^{1-c} M(a + 1 - c, 2 - c; x) \qquad c \neq 2, 3, \ldots \tag{4.2.9c}$$

The standard form of the second solution is a linear combination of (4.2.9b) and (4.2.9c)

$$y_2 = U(a, c; x) = \frac{\Gamma(1-c)}{\Gamma(a-c+1)} M(a, c; x)$$

$$+ \frac{\Gamma(c-1)}{\Gamma(a)} x^{1-c} M(a + 1 - c, 2 - c; x) \tag{4.2.9d}$$

This function is defined even when c is a positive or negative integer. Many special functions are special cases of the confluent hypergeometric function.

The differential equation of orthogonal polynomials has the form

$$g_2(x)y'' + g_1(x)y' + g_0(x)y = 0 \tag{4.2.10a}$$

The functions $g_2(x)$, $g_1(x)$, and $g_0(x)$ are specified for particular functions:

Jacobi polynomials $P_n^{(\alpha, \beta)}(x)$:

$$g_2 = 1 - x^2 \qquad g_1 = \beta - \alpha - (\alpha + \beta + 2)x \qquad g_0 = n(n + \alpha + \beta + 1) \qquad (4.2.10b)$$

Ultraspherical (Gegenbauer) polynomials $C_n^{(\alpha)}(x)$:

$$g_2 = 1 - x^2 \qquad g_1 = -(2\alpha + 1)x \qquad g_0 = n(n + 2\alpha) \qquad (4.2.10c)$$

Chebyshev polynomials of the first kind $T_n(x)$:

$$g_2 = 1 - x^2 \qquad g_1 = -x \qquad g_0 = n^2 \qquad (4.2.10d)$$

Chebyshev polynomials of the second kind $U_n(x)$ (cf. Sec. 8.2):

$$g_2 = 1 - x^2 \qquad g_1 = -3x \qquad g_0 = n^2 - 1 \qquad (4.2.10e)$$

Generalized Laguerre polynomials $L_n^{(m)}(x)$:

$$g_2 = x \qquad g_1 = m + 1 - x \qquad g_0 = n \qquad (4.2.10f)$$

Hermite polynomials $H_n(x)$:

$$g_2 = 1 \qquad g_1 = -2x \qquad g_0 = 2n \qquad (4.2.10g)$$

Hermite polynomials $\text{He}_n(x) = 2^{-n/2} H_n(2^{-1/2}x)$:

$$g_2 = 1 \qquad g_1 = -x \qquad g_0 = n \qquad (4.2.10h)$$

Jacobi polynomials, ultraspherical polynomials, and Chebyshev polynomials of the first and second kind have regular singular points at $x = \pm 1$. Generalized Laguerre polynomials have one regular singular point at $x = 0$. Hermite polynomials have no singular points at any finite value of x.

A powerful and fast method for generating a table of a special function (not to evaluate it just at one or a few given x) is the numerical integration of its differential equation. This technique, however, should not be applied if the function is a polynomial; in this case the recurrence formulas are faster and more accurate because the errors are restricted to roundoff errors.

4.3 EULER AND TAYLOR-SERIES METHODS

Before the decision is made to integrate a given ordinary differential equation numerically, the author recommends consulting Kamke's catalog [33] of more than 1500 ordinary differential equations with their solutions and extensive references to further information. Even when the equation has been solved analytically, one may still decide to integrate it numerically (for instance, when no mathematical tables of the integral of the differential equation are available), but one should have all the accessible information about the equation first in order not to duplicate research already done and to be aware of what to expect from the numerical integration.

A differential equation of the first order

$$y' = f(x, y) \tag{4.3.1}$$

determines the derivative y' from the given x and y. The function $f(x, y)$ is assumed to be continuous, sufficiently differentiable with respect to both x and y, and bounded over the the interval of integration. As long as $\partial f/\partial y$ is continuous, the solution of (4.3.1) for a given initial value of y is unique. In the simplest method of numerical integration of ordinary differential equations the unknown function $y(x)$ is approximated in a very short subinterval h by a straight line. Starting from the initial value y_0 at x_0 the function at $x_0 + h$ is given by

$$y(x_0 + h) = y_0 + hy'(x_0, y_0) \tag{4.3.2}$$

The new values $x_0 + h$ and $y_0 + h$ are now substituted into (4.3.1), and the procedure is repeated. This is the Euler method. Its obvious disadvantage is the approximation of $y(x)$ by a straight line: to keep the error of the linear approximation low the subinterval h must be chosen very short. As a consequence, the integration is much too time-consuming, and because many steps are required, the accumulation of roundoff errors may be considerable. If the function $y(x)$ is monotonically increasing or decreasing in the whole integration domain, the function computed by the Euler method over a longer interval will lag unacceptably behind the true integral of the differential equation.

A differential equation of second order

$$y'' = f(x, y, y') \tag{4.3.3}$$

with $f(x, y, y')$ continuous, sufficiently differentiable with respect to both x and y, and bounded, can always be reduced to a system of two differential equations of first order [cf. Eqs. (4.1.4)] and integrated simultaneously by the Euler method or its integral approximated by a Taylor series truncated after its quadratic term. Starting from the initial data x_0, y_0, y'_0, one has

$$y(x_0 + h) = y_0 + hy'_0 + \tfrac{1}{2}h^2 y''(x_0, y_0, y'_0) \tag{4.3.4}$$

For the computation of the second derivatives in the next step by (4.3.4) the first derivative must also be approximated. A differentiation of Eq. (4.3.4) gives

$$y'(x_0 + h) = y'_0 + hy''(x_0, y_0, y'_0) \tag{4.3.5}$$

The function is now approximated by a quadratic parabola, but its derivative is approximated by a straight line, which lowers the accuracy of the second derivative computed by (4.3.3). If the differential equation (4.3.3) is linear, it is therefore appropriate to reduce it to its canonical form [cf. Eqs. (4.1.32) to (4.1.35)].

A differential equation of first order can be differentiated analytically [cf. Eqs. (4.1.3a, b)] to obtain Eq. (4.3.3). The first derivative computed by (4.3.1) is substituted into (4.3.3) and both these derivatives then used in (4.3.4) to approximate the unknown function $y(x)$; Eq. (4.3.5) is now superfluous. This brings a substantial improvement of accuracy, but unfortunately it is not sufficient in many cases.

To increase the accuracy of integration further the first-order differential equation can be differentiated twice more [cf. Eqs. (4.1.3c, d)], the Taylor series truncated after its biquadratic term

$$y(x_0 + h) = y_0 + hy_0' + \frac{h^2 y_0''}{2} + \frac{h^3 y_0^{(3)}}{6} + \frac{h^4 y_0^{(4)}}{24} \tag{4.3.6a}$$

and the step size h determined by the requirement that the absolute value of the biquadratic term be equal to a small constant P_1

$$\frac{h^4 |y_0^{(4)}|}{24} = P_1$$

Hence the subinterval

$$h = \left(\frac{24 P_1}{|y_0^{(4)}|}\right)^{1/4} \tag{4.3.6b}$$

If P_1 is chosen sufficiently small, the truncation error in (4.3.6a) may become negligible.

To determine that the error in the truncated series (4.3.6a) with the subinterval h determined by (4.3.6b) does not surpass a chosen error bound of integration P_2, the second control of step size can be applied. In the procedure developed by the author the value $y(x_0 + h)$ is substituted into the differential equation (4.3.1) to determine the first derivative at $x = x_0 + h$. This value

$$y'(x_0 + h) = f(x_0 + h, y(x_0 + h))$$

is now used to compute the unknown fifth derivative at $x = x_0$

$$y'(x_0 + h) = y_0' + hy_0'' + \frac{h^2 y_0^{(3)}}{2} + \frac{h^3 y_0^{(4)}}{6} + \frac{h^4 y_0^{(5)}}{24}$$

The right-hand side is the first derivative of a Taylor series truncated after its fifth-order term. Hence the correction term

$$C = \frac{h^5 y^{(5)}}{120}$$

that must be added to the right-hand side of (4.3.6a) is

$$C = \frac{h}{5}\left[y'(x_0 + h) - y_0' - hy_0'' - \frac{h^2 y_0^{(3)}}{2} - \frac{h^3 y_0^{(4)}}{6}\right] \tag{4.3.7a}$$

The corrected $y^C(x_0 + h)$ that is replacing the value $y(x_0 + h)$ computed by (4.3.6a) is

$$y^C(x + h) = y(x + h) + C \tag{4.3.7b}$$

If the correction term C is above the error bound P_2, always chosen smaller than P_1, this constant in formula (4.3.6b) is replaced with a smaller one $\bar{P}_1$, given by

$$\bar{P}_1 = (0.8)^4 P_1 \tag{4.3.8a}$$

and the step recomputed with h determined now by $\bar{P}_1$. In this way the subinterval is reduced to $0.8h$. If the integral of the given differential equation is sharply increasing or decreasing (for instance, when it is approaching a singular point) the constant $\bar{P}_1$, and hence the subinterval h, may be shortened again and again. On the other hand, if in any step

$$C < \frac{P_2}{8} \tag{4.3.8b}$$

the corrected $\bar{P}_1$ is increased to

$$P_1 = \frac{\bar{P}_1}{(0.8)^4} \tag{4.3.8c}$$

but it is not allowed to be greater than the given input value P_1.

The second control of step size acts at the same time as a fifth-order corrector, Eq. (4.3.7b), to the fourth-order predictor (4.3.6a). (Other predictor-corrector methods are discussed in detail in Sec. 4.5.)

A second-order differential equation (4.3.3) is to be differentiated analytically only twice, but during the integration by Eqs. (4.3.6) the first derivative must be computed by

$$y'(x_0 + h) = y'_0 + hy''_0 + \frac{h^2 y_0^{(3)}}{2} + \frac{h^3 y_0^{(4)}}{6} \tag{4.3.9}$$

In the second control of step size the $y(x_0 + h)$ and $y'(x_0 + h)$ are substituted into the differential equation (4.3.3) to determine the second derivatives at $x = x_0 + h$. This value

$$y''(x_0 + h) = f(x_0 + h, y(x_0 + h), y'(x_0 + h))$$

is now used to compute the unknown fifth derivative at $x = x_0$

$$y''(x_0 + h) = y''_0 + hy_0^{(3)} + \frac{h^2 y_0^{(4)}}{2} + \frac{h^3 y_0^{(5)}}{6}$$

The right-hand side is the second derivative of a Taylor series truncated after its fifth-order term. Hence the correction term

$$C = \frac{h^5 y_0^{(5)}}{120}$$

that must be added to the right-hand side of (4.3.6a) is

$$C = \frac{h^2}{20} [y''(x_0 + h) - y''_0 - hy_0^{(3)} - \tfrac{1}{2} h^2 y_0^{(4)}] \tag{4.3.10}$$

The corrected $y^C(x_0 + h)$ that is replacing the value $y(x_0 + h)$ computed by (4.3.6a) is again given by (4.3.7b). The correction of the constant P_1 is also carried out here by relations (4.3.8).

4.20 THEORY

A set of m simultaneous ordinary differential equations of first order

$$y'_j = f_j(x, y_1, \ldots, y_m) \qquad j = 1, 2, \ldots, m \qquad (4.3.11)$$

can be integrated numerically by the Euler method (4.3.2) or, better, by a truncated Taylor series in the same way as Eq. (4.3.6a); the only difference is that the derivatives on the right-hand side of (4.3.6a) depend now on all the functions and their derivatives. From the fourth derivatives of m functions the one which is of maximum absolute value is substituted into the formula (4.3.6b) for the step size. The second control of step size can be applied here too. The correction term must be computed for each equation and the term of maximum absolute value substituted into Eqs. (4.3.8).

Programs P401 and P402 integrate a differential equation (or a system of simultaneous differential equations) of first and second orders, respectively, by a Taylor series truncated after its biquadratic term and with a step size determined by (4.3.6b). As an option the accuracy of integration can be tested and improved by the fifth-order corrector in the second control of step size.

The Taylor-series method is a one-step method: it determines the value of the dependent variable at the end of a subinterval $x = x_0 + h$ from the data at the beginning. The accuracy of integration can be controlled by computing the step size by the formula (4.3.6b) and the solution improved by the fifth-order corrector. This advantage sometimes outweighs the necessity of analytical differentiation of the differential equation which may lead to rather long expressions.

The Runge-Kutta methods are also one-step methods, achieving the same accuracy as Taylor-series methods by evaluating only the first derivative but several times at selected points of each step. Unfortunately, the only control of accuracy of integration is by halving the subinterval, recomputing in two steps, and comparing the results, which requires too much computer time.

The predictor-corrector methods evaluate the right-hand side of the differential equation only once in the predictor and once in the corrector and use the information from previous steps to achieve the required accuracy. The comparison of the result from the predictor with that from the corrector at each step is a realiable indicator of the accuracy of integration. These methods are multistep methods; they need a starter at the beginning of integration.

4.4 RUNGE-KUTTA METHODS

Instead of computing higher derivatives for the truncated Taylor series the higher accuracy in Runge-Kutta methods is obtained by evaluating the first derivative several times at selected points of each subinterval. The basic idea of the method is explained here for a second-order Runge-Kutta method.

The value of the function $y(x)$, defined by the differential equation (4.3.1) and the initial data x_0, y_0, is given at $x = x_0 + h$ in the second-order Runge-Kutta method by

$$y(x_0 + h) = y_0 + w_1 k_1 + w_2 k_2 \qquad (4.4.1a)$$

where
$$k_1 = hf(x_0, y_0) \qquad (4.4.1b)$$
$$k_2 = hf(x_0 + ah, y_0 + bk_1) \qquad (4.4.1c)$$

The constants w_1, w_2 and a, b are to be determined from the requirement that $y(x_0 + h)$ computed by (4.4.1a) agree with the value determined by Taylor series truncated after its quadratic term. The substitution of (4.4.1b, c) into (4.4.1a) and Taylor expansion for the function (4.4.1c) depending on two variables yields

$$y(x_0 + h) = y_0 + w_1 hf + w_2 hf + w_2 h^2 \left(a \frac{\partial f}{\partial x} + bf \frac{\partial f}{\partial y} \right)$$

The truncated Taylor series with the second derivatives given by (4.1.3b) has the form

$$y(x + h) = y + hf + \tfrac{1}{2} h^2 \left(\frac{\partial f}{\partial x} + f \frac{\partial f}{\partial y} \right).$$

Here $f = f(x_0, y_0)$; the partial derivatives are evaluated also at the point (x_0, y_0). Comparing the terms with the same powers of h and the same derivatives results in a set of three equations for four unknowns

$$w_1 + w_2 = 1 \qquad (4.4.1d)$$
$$w_2 a = \tfrac{1}{2} \qquad (4.4.1e)$$
$$w_2 b = \tfrac{1}{2} \qquad (4.4.1f)$$

In the method due to Runge w_1 is chosen to vanish. Then

$$w_1 = 0 \qquad w_2 = 1 \qquad a = b = \tfrac{1}{2} \qquad (4.4.2)$$

This method is now called the *modified Euler method*. In the improved Euler method

$$w_1 = w_2 = \tfrac{1}{2} \qquad a = b = 1 \qquad (4.4.3)$$

When the improved Euler method is applied to a system of two simultaneous differential equatons

$$y'_1 = f_1(x, y_1, y_2) \qquad (4.4.4a)$$
$$y'_2 = f_2(x, y_1, y_2) \qquad (4.4.4b)$$

the integration formulas with the initial data y_{10}, y_{20} at $x = x_0$ are

$$y_1(x_0 + h) = y_{10} + \tfrac{1}{2}(k_{11} + k_{12}) \qquad (4.4.4c)$$

and

$$y_2(x_0 + h) = y_{20} + \tfrac{1}{2}(k_{21} + k_{22}) \qquad (4.4.4d)$$

where

$$k_{11} = hf_1(x_0, y_{10}, y_{20})$$
$$k_{21} = hf_2(x_0, y_{10}, y_{20})$$
$$k_{12} = hf_1(x_0, y_{10} + k_{11}, y_{20} + k_{21})$$
$$k_{22} = hf_2(x_0, y_{10} + k_{11}, y_{20} + k_{21})$$

$$(4.4.4e)$$

4.22 THEORY

In Runge-Kutta methods of third and fourth order the first derivatives are evaluated at three or four points, respectively, but since the fifth-order method requires the evaluation at six points, Runge-Kutta methods or fifth or higher order [2] are seldom used. In all these methods the system of linear algebraic equations for the unknown constants is underdetermined.

The fourth-order Runge-Kutta method is described by

$$y(x_0 + h) = y_0 + \frac{k_1 + 2ak_2 + 2bk_3 + k_4}{6} \qquad (4.4.5a)$$

$$k_1 = hf(x_0, y_0) \qquad (4.4.5b)$$

$$k_2 = hf(x_0 + \tfrac{1}{2}h, y_0 + \tfrac{1}{2}k_1) \qquad (4.4.5c)$$

$$k_3 = hf(x_0 + \tfrac{1}{2}h, y_0 + ck_1 + dk_2) \qquad (4.4.5d)$$

$$k_4 = hf(x_0 + h, y_0 + ek_2 + bk_3) \qquad (4.4.5e)$$

In the standard method

$$a = 1 \qquad b = 1 \qquad c = 0 \qquad d = \tfrac{1}{2} \qquad e = 0 \qquad (4.4.5f)$$

In Gill's method

$$a = 1 - 2^{-1/2} \qquad b = 1 + 2^{-1/2} \qquad c = -\tfrac{1}{2} + 2^{-1/2} \qquad d = a \qquad e = -2^{-1/2}$$
$$(4.4.5g)$$

Gill's method reduces the accumulation of roundoff errors by the choice of the coefficients a to e.

These formulas can be adjusted to a system of m simultaneous differential equations of first order

$$y'_j = f_j(x, y_1, \ldots, y_m) \qquad j = 1, 2, \ldots, m \qquad (4.4.6)$$

The fourth-order Runge-Kutta method with the initial data $y_{10}, \ldots, y_{m0}$ at $x = x_0$ is described by

$$y_j(x_0 + h) = y_{j0} + \frac{k_{j1} + 2ak_{j2} + 2bk_{j3} + k_{j4}}{6} \qquad j = 1, 2, \ldots, m \quad (4.4.7a)$$

where

$$k_{j1} = hf_j(x_0, y_{10}, \ldots, y_{m0})$$

$$k_{j2} = hf_j(x_0 + \tfrac{1}{2}h, y_{10} + \tfrac{1}{2}k_{11}, \ldots, y_{m0} + \tfrac{1}{2}k_{m1})$$

$$k_{j3} = hf_j(x_0 + \tfrac{1}{2}h, y_{10} + ck_{11} + dk_{12}, \ldots, y_{m0} + ck_{m1} + dk_{m2}) \qquad (4.4.7b)$$

$$k_{j4} = hf_j(x_0 + h, y_{10} + ek_{12} + bk_{13}, \ldots, y_{m0} + ek_{m2} + bk_{m3})$$

The coefficients a to e are given by (4.4.5f) for the standard method and by (4.4.5g) for Gill's method.

The differential equations of second or higher order (or a system of higher-order differential equations) can be reduced to a system of first-order differential equations by introducing new dependent variables, as shown for a third-order equation in Eqs. (4.1.4). Runge-Kutta methods also exist for higher-order differential equations [2]. A fourth-order method for a second-order differential equation

$$y'' = f(x, y, y') \tag{4.4.8}$$

with the initial data y_0, y_0' at $x = x_0$ is

$$y(x_0 + h) = y_0 + h\left(y_0' + \frac{k_1 + k_2 + k_3}{6}\right)$$

$$y'(x_0 + h) = y_0' + \frac{k_1 + 2k_2 + 2k_3 + k_4}{6} \tag{4.4.9a}$$

where
$$k_1 = hf(x_0, y_0, y_0')$$

$$k_2 = hf\left(x_0 + \tfrac{1}{2}h,\ y_0 + \tfrac{1}{2}hy_0' + \frac{hk_1}{8},\ y_0' + \tfrac{1}{2}k_1\right)$$

$$k_3 = hf\left(x_0 + \tfrac{1}{2}h,\ y_0 + \tfrac{1}{2}hy_0' + \frac{hk_1}{8},\ y_0' + \tfrac{1}{2}k_2\right) \tag{4.4.9b}$$

$$k_4 = hf(x_0 + h,\ y_0 + hy_0' + \tfrac{1}{2}hk_3,\ y_0' + k_3)$$

Runge-Kutta methods have two serious disadvantages compared with Taylor-series or predictor-corrector methods. The accuracy of integration can be controlled only by halving the subinterval, recomputing in two steps, comparing the results, and when the difference is above the chosen error bound repeating the halving again and again until the difference is below the error bound or until the number of halving surpasses a chosen limit (in this case the result contains an error larger than the error bound). This testing procedure is inefficient and usually requires excessive computer time. The other disadvantage, not mentioned in many textbooks on numerical analysis, is the difficulty of integrating a differential equation with a term involving y/x multiplied by an even function of x. This problem is demonstrated here for the linear differential equation

$$y' = px^{p+q-1} + \frac{qy}{x} \tag{4.4.10a}$$

with positive integers p and q (one of which may equal zero) and initial data $x_0 = y_0 = 0$. Its particular integral is

$$y = x^{p+q} \tag{4.4.10b}$$

If $p + q = 4$, the numerical integration by a fourth-order Runge-Kutta method should be exact, no matter how large the step size. This is the case for $p = 4$, $q = 0$. If $p = 3$, $q = 1$, the result is exact only in the first step (the undetermined term y/x at $x = 0$ has been evaluated analytically: $\lim_{x \to 0} y/x = 0$); then the error steadily increases but in such a way that the relative error after the same number of steps is exactly the same for any step size. After five steps

$$y(5.0) = 6.248376590 \times 10^2 \qquad h = 1.0$$

and
$$y(0.05) = 6.248376590 \times 10^{-6} \qquad h = 0.01$$

If $p = 1$ and $q = 3$, the errors are higher; an error appears even in the first step, but again relative errors are independent of the step size. After five steps

$$y(5.0) = 5.597229717 \times 10^2 \qquad h = 1.0$$

and
$$y(0.05) = 5.597229717 \times 10^{-6} \qquad h = 0.01$$

The absolute errors, of course, decrease when the step size is smaller. If $p = 0$ and $q = 4$, the Runge-Kutta integration never gets started because in the first step, according to (4.4.5b, c), $k_1 = 0$ and as a result $k_2 = k_3 = k_4 = 0$.

When the differential equation (4.4.10a) is integrated by the fourth-order Taylor-series method, Eqs. (4.3.6), the results are satisfactory. The integration starts even for $p = 0$, $q = 4$; in this case the results are exact. It is not difficult to understand this behavior: a general integral of the differential equation (4.4.10a) is

$$y = x^{p+q} + Cx^q \tag{4.4.10c}$$

The integration constant C equals zero in the particular integral (4.4.10b), but the numerical errors in the first step of integration generate initial data for the second step which correspond to $C \neq 0$. The errors in the first step of the fourth-order Runge-Kutta method are relatively large because $k_1 = 0$, while in the fourth-order Taylor-series method they are restricted to roundoff errors [because the function $y(x)$ is a polynomial of the order $p + q = 4$].

An efficient remedy for this instability is eliminating the term y/x from the differential equation by the substitution

$$y = x^q w(x) \tag{4.4.10d}$$

Then the differential equation for the new dependent variable $w(x)$ has the simple form

$$w' = px^{p-1} \tag{4.4.10e}$$

Programs P403 and P404 integrate a differential equation (or a system of differential equations) of first order by the standard fourth-order Runge-Kutta method or Gill's version of it, respectively. As an option, the accuracy of integration can be tested by the usual procedure of halving the subinterval, recomputing, and comparing the results.

4.5 PREDICTOR-CORRECTOR METHODS

A predictor-corrector method has already been applied in the Taylor-series method described in Sec. 4.3, but usually they are multistep methods. As in Runge-Kutta methods, the predictor evaluates the differential equation (not its higher derivatives) only once each step and increases the accuracy by making use of information on the integral of the equation at several earlier steps.

A function defined by its value y_n at x_n and by its first derivative y'_n at this point and at one, two, three preceding equally spaced points $(x_n - h, y'_{n-1})$, $(x_n - 2h, y'_{n-2})$, and $(x_n - 3h, y'_{n-3})$ is interpolated by a parabola of second, third, or fourth degree, which is extrapolated to the next point at $x = x_n + h$. The result of a straightforward computation yields Adams-Bashforth predictors of the second, third, and fourth order

$$y^P_{n+1} = y_n + \frac{h}{2}(3y'_n - y'_{n-1})$$

$$y^P_{n+1} = y_n + \frac{h}{12}(23y'_n - 16y'_{n-1} + 5y'_{n-2}) \qquad (4.5.1)$$

$$y^P_{n+1} = y_n + \frac{h}{24}(55y'_n - 59y'_{n-1} + 37y'_{n-2} - 9y'_{n-3})$$

The derivatives are the right-hand sides of the first-order differential equation

$$y' = f(x, y) \qquad (4.5.2)$$

Milne's predictor of the fourth order is based on numerical evaluation of a definite integral

$$\int_{x_{n-3}}^{x_{n+1}} y' \, dx = \int_{x_{n-3}}^{x_{n+1}} f(x, y) \, dx$$

When the Newton-Cotes formula of open type is used, the predictor takes the form

$$y^P_{n+1} = y_{n-3} + \frac{4h}{3}(2y'_n - y'_{n-1} + 2y'_{n-2}) \qquad (4.5.3)$$

The value y_{n+1} computed by a predictor is substituted into the differential equation (4.5.2), and y'_{n+1} is used to recompute y_{n+1} by an interpolation. The function is defined by its value y_n at x_n and its first derivative y'_n at this point, by the derivative y'_{n+1} and possibly by the derivatives at one or two preceding points $(x_n - h, y'_{n-1})$ and $(x_n - 2h, y'_{n-2})$ and interpolated by a parabola of second, third, or fourth degree at $x = x_n + h$. In this way Adams-Moulton correctors of the second, third, or fourth order are obtained

$$y^C_{n+1} = y_n + \frac{h}{2}(y'_{n+1} + y'_n)$$

$$y^C_{n+1} = y_n + \frac{h}{12}(5y'_{n+1} + 8y'_n - y'_{n-1}) \qquad (4.5.4)$$

$$y^C_{n+1} = y_n + \frac{h}{24}(9y'_{n+1} + 19y'_n - 5y'_{n-1} + y'_{n-2})$$

Milne's corrector of the fourth order is deduced from a definite integral

$$\int_{x_{n-1}}^{x_{n+1}} y' \, dx = \int_{x_{n-1}}^{x_{n+1}} f(x, y) \, dx$$

which is evaluated by a Newton-Cotes formula of closed type (Simpson's rule)

$$y_{n+1}^C = y_{n-1} + \frac{h}{3}(y'_{n+1} + 4y'_n + y'_{n-1}) \qquad (4.5.5)$$

In Hamming's corrector the interpolating polynomial of fourth degree is defined by the data $(x_n + h, y'_{n+1})$, (x_n, y_n, y'_n), $(x_n - h, y'_{n-1})$, and $(x_n - 2h, y_{n-2})$. The ordinate y_{n-1} is not used in the definition of the polynomial to increase its stability. The interpolated value y_{n+1} is now given by

$$y_{n+1}^C = \frac{9y_n - y_{n-2}}{8} + \frac{3h}{8}(y'_{n+1} + 2y'_n - y'_{n-1}) \qquad (4.5.6)$$

There are many more predictor-corrector methods, but in application the most popular are the fourth-order Adams-Bashforth predictor (4.5.1c) with Adams-Moulton corrector (4.5.4c), Milne's predictor (4.5.3) with corrector (4.5.5), and in Hamming's method Milne's predictor (4.5.3) with Hamming's corrector (4.5.6). Milne's predictor (4.5.3) should be used with caution because of possible instabilities.

The data y'_{n+1} and y_{n+1}^C in the corrector are related to each other through the differential equation (4.5.2). Since y'_{n+1} has been computed from an approximate y_{n+1}^P, the corresponding y_{n+1}^C is just a correction, not an exact solution of the corrector equation, and can further be improved by iteration: y_{n+1}^C is substituted into the differential equation to obtain a more accurate y'_{n+1}, which is substituted into the corrector, and so on, until the absolute value of the difference of two successive improvements lies below a chosen error bound of integration. If the subinterval h is not estimated small enough (but not too small!), the convergence of the sequence may be very slow, and sometimes the terms approach the limit in an oscillatory manner. There are other difficulties connected with using the predictor-corrector methods described above: it is obvious that a change of the subinterval is not easy to accomplish, and the method needs a starter, say a Runge-Kutta method, to get a few preceding data for the predictor-corrector formulas.

For these reasons the author has developed three new predictor-corrector methods free of these problems. Their main features are the introduction of a variable step size (determined in each step by the method itself) and a second control of the step size.

The first method is a predictor-corrector of third order for a differential equation of first order. The predictor is of the Adams-Bashforth type but with a variable subinterval related to the abscissas by

$$h_j = x_j - x_{j-1} \qquad (4.5.7)$$

The predictors of first, second, and third order are

$$y_{n+1}^P = y_n + h_{n+1} y'_n \qquad (4.5.8a)$$

$$y_{n+1}^P = y_n + h_{n+1}\left[y'_n + \frac{1}{2}\frac{h_{n+1}}{h_n}(y'_n - y'_{n-1})\right] \qquad (4.5.8b)$$

$$y_{n+1}^P = y_n + h_{n+1}\left[y'_n + \frac{1}{2}\frac{h_{n+1}}{h_n}(y'_n - y'_{n-1})\right]$$

$$+ \frac{h_{n+1}^2}{6h_{n-1}}(3h_n + 2h_{n+1})\left(\frac{y'_n - y'_{n-1}}{h_n} - \frac{y'_n - y'_{n-2}}{h_n + h_{n-1}}\right) \quad (4.5.8c)$$

The corrector is of the Adams-Moulton type but again with a variable subinterval. The correctors of second and third order are

$$y_{n+1}^C = y_n + \frac{h_{n+1}}{2}(y'_{n+1} + y'_n) \quad (4.5.9a)$$

$$y_{n+1}^C = y_n + \frac{h_{n+1}}{2}(y'_{n+1} + y'_n) - \frac{h_{n+1}^2}{6(h_{n+1} + h_n)}\left[y'_{n+1} - y'_n - \frac{h_{n+1}}{h_n}(y'_n - y'_{n-1})\right]$$

$$(4.5.9b)$$

The integration starts with the evaluation of the first derivative y'_n by (4.5.2) and computation of the subinterval h_1 in the first step, which is determined by the condition that the contribution of the linear term in the predictor (4.5.8a) equal a chosen small constant P_0. Hence

$$h_1 = \frac{P_0}{|y'_0|} \quad (4.5.10a)$$

Then the first-order predictor (4.5.8a) and second-order corrector (4.5.9a) are applied. In the second step the subinterval is increased to

$$h_2 = \tfrac{3}{2}h_1 \quad (4.5.10b)$$

and the second-order predictor (4.5.8b) and third-order corrector (4.5.9b) used. These two steps constitute the starter. Henceforth the subinterval is determined by the requirement that the contribution of the term involving cubic h_{n+1}^3 in the third-order predictor equal a chosen small constant P_1

$$h_{n+1} = \left(\frac{3P_1}{|b|}\right)^{1/3} \quad (4.5.10c)$$

where

$$b = \frac{(y'_n - y'_{n-1})/h_n - (y'_n - y'_{n-2})/(h_n + h_{n-1})}{h_{n-1}} \quad (4.5.10d)$$

To avoid deterioration of the accuracy of interpolation by a sudden large increase of the subinterval at the transition from the starter to a normal integration with the third-order predictor and third-order corrector the size of the subinterval is limited by an additional condition

$$h_{n+1} \leq \tfrac{3}{2}h_n \quad (4.5.10e)$$

As a consequence, the step size (which must be very small in the starter because the first predictor is the Euler method) is only gradually increased until it reaches its full size, given by (4.5.10c, d). Thereafter the variation of the step size is gentle.

As an option, a second control of step size can be applied: if the absolute value of the difference between the predictor and corrector, $|y^C_{n+1} - y^P_{n+1}|$, is above the error bound P_2, always chosen smaller than P_1, the constant P_1 in formula (4.5.10c) is to be replaced with a smaller one, $\bar{P}_1$, given by

$$\bar{P}_1 = (0.75)^3 P_1 \tag{4.5.11a}$$

and the step size recomputed with h determined now by $\bar{P}_1$. In this way the subinterval is reduced to $0.75h$. It may happen that the constant $\bar{P}_1$ and the subinterval h are decreased again and again. On the other hand, if in any step

$$|y^C_{n+1} - y^P_{n+1}| < \frac{P_2}{8} \tag{4.5.11b}$$

the corrected $\bar{P}_1$ is increased to

$$P_1 = \frac{\bar{P}_1}{(0.75)^3} \tag{4.5.11c}$$

but it is not allowed to be greater than the chosen input value P_1.

The second method is a combination of a Taylor-series method and a predictor-corrector method of fourth order for a differential equation of the first order (4.5.2) from which a second-order differential equation

$$y'' = g(x, y, y') \tag{4.5.12}$$

has been deduced analytically [cf. Eq. (4.1.3b)]. The third method is a predictor-corrector method of fourth order for a differential equation of second order. In both these methods the predictors of second, third, and fourth order for the function (and in the third method for its first derivative as well) are

$$y^P_{n+1} = y_n + h_{n+1} y'_n + \tfrac{1}{2} h^2_{n+1} y''_n \tag{4.5.13a}$$

$$y'^P_{n+1} = y'_n + h_{n+1} y''_n \tag{4.5.13b}$$

$$y^P_{n+1} = y_n + h_{n+1} y'_n + h^2_{n+1}\left(\tfrac{1}{2} y''_n + \frac{h_{n+1}}{h_n} \frac{y''_n - y''_{n-1}}{6}\right) \tag{4.5.14a}$$

$$y'^P_{n+1} = y'_n + h_{n+1} y''_n + \frac{1}{2} \frac{h^2_{n+1}}{h_n} (y''_n - y''_{n-1}) \tag{4.5.14b}$$

$$y^P_{n+1} = y_n + h_{n+1} y'_n + h^2_{n+1}\left(\tfrac{1}{2} y''_n + \frac{h_{n+1}}{h_n} \frac{y''_n - y''_{n-1}}{6}\right)$$

$$+ \frac{h^3_{n+1}}{12 h_{n-1}} (2h_n + h_{n+1})\left(\frac{y''_n - y''_{n-1}}{h_n} - \frac{y''_n - y''_{n-2}}{h_n + h_{n-1}}\right) \tag{4.5.15a}$$

$$y'^P_{n+1} = y'_n + h_{n+1} y''_n + \frac{1}{2} \frac{h^2_{n+1}}{h_n} (y''_n - y''_{n-1})$$

$$+ \frac{h^2_{n+1}}{6 h_{n-1}} (3h_n + 2h_{n+1})\left(\frac{y''_n - y''_{n-1}}{h_n} - \frac{y''_n - y''_{n-2}}{h_n + h_{n-1}}\right) \tag{4.5.15b}$$

The interpolating quadratic parabola in the predictor of second order is defined by y_n, y'_n, y''_n at x_n. In the predictor of third order the cubic parabola is given by one more piece of information, y''_{n-1} at $x_{n-1} = x_n - h_n$, and in the predictor of fourth order the biquadratic parabola is determined by the additional information y''_{n-2} at $x_{n-2} = x_{n-1} - h_{n-1}$. The correctors of third and fourth order and their first derivatives are

$$y^C_{n+1} = y_n + h_{n+1} y'_n + \frac{h^2_{n+1}}{6} (y''_{n+1} + 2y''_n) \tag{4.5.16a}$$

$$y'^C_{n+1} = y'_n + \frac{h_{n+1}}{3} (y''_{n+1} + 2y''_n) \tag{4.5.16b}$$

$$y^C_{n+1} = y_n + h_{n+1} y'_n + \frac{h^2_{n+1}}{6} (y''_{n+1} + 2y''_n)$$
$$- \frac{h^3_{n+1}}{12(h_{n+1} + h_n)} \left[y''_{n+1} - y''_n - \frac{h_{n+1}}{h_n} (y''_n - y''_{n-1}) \right] \tag{4.5.17a}$$

$$y'^C_{n+1} = y'_n + \tfrac{1}{2} h_{n+1}(y''_{n+1} + y''_n) - \frac{h^2_{n+1}}{6(h_{n+1} + h_n)} \left[y''_{n+1} - y''_n - \frac{h_{n+1}}{h_n} (y''_n - y''_{n-1}) \right]$$
$$\tag{4.5.17b}$$

The interpolating cubic parabola in the corrector of third order is defined by y_n, y'_n, y''_n at x_n and by y''_{n+1} at $x_{n+1} = x_n + h_{n+1}$. In the corrector of fourth order the biquadratic parabola is determined by one more piece of information, y''_{n-1} at $x_{n-1} = x_n - h_n$. The integration starts with the evaluation of the first and second derivatives. In the second method the first derivative is computed from the given differential equation; in the third method it is one of the initial data. The subinterval h_1 in the first step is determined by the condition that the contribution of the quadratic term in the predictor (4.5.13a) equals a chosen small constant P_0. Hence

$$h_1 = \left(\frac{2P_0}{|y''_0|} \right)^{1/2} \tag{4.5.18a}$$

Then the second-order predictor (4.5.13a) [or in the third method (4.5.13a, b)] and third-order corrector (4.5.16a) [or (4.5.16a, b)] are applied. In the second step the interval is increased to h_2, given by (4.5.10b); and third-order predictor (4.5.14a) [or (4.5.14a, b)] and fourth-order corrector (4.5.17a) [or (4.5.17a, b)] are used. These two steps constitute the starter. Henceforth the fourth-order predictor (4.5.15a) [or (4.5.15a, b)] and corrector (4.5.17a) [or (4.5.17a, b)] are applied repeatedly. The only difference between the second method for the first-order differential equation and the third method for the second-order differential equation lies in the computation of the first derivatives: in the former they are computed from the given differential equation; however, the predictors (4.5.13b), (4.5.14b), and (4.5.15b) and the correctors (4.5.16b) and (4.5.17b), which are not used in the former, are used in the latter. The subinterval is determined by the

4.30 THEORY

condition that the contribution of the term involving the biquadratic h_{n+1}^4 in the fourth-order predictor equals a small chosen constant P_1

$$h_{n+1} = \left(\frac{12P_1}{|d|}\right)^{1/4} \qquad (4.5.18b)$$

where

$$d = \frac{(y_n'' - y_{n-1}'')/h_n - (y_n'' - y_{n-2}'')/(h_n + h_{n-1})}{h_{n-1}} \qquad (4.5.18c)$$

For the same reasons as in the first method, the size of the subinterval is limited by the additional condition

$$h_{n+1} \leq \tfrac{3}{2} h_n \qquad (4.5.18d)$$

Thus the subinterval is only gradually increasing from the value in the starter until it reaches its full size, given by (4.5.18c). Thereafter the variation of its size is gentle. As an option, control of step size can be applied. It is very similar to the procedure in the first method, but since the predictor and corrector are of fourth order, the constant P_1 is decreased and increased here by the formulas

$$\bar{P}_1 = (0.75)^4 P_1 \qquad (4.5.19a)$$

and

$$P_1 = \frac{\bar{P}_1}{(0.75)^4} \qquad (4.5.19c)$$

Condition (4.5.11b) is not changed:

$$|y_{n+1}^C - y_{n+1}^P| < \frac{P_2}{8} \qquad (4.5.19b)$$

The differential equation (two differential equations in the second method) is evaluated at the beginning of each predictor and corrector. If the second control of step size decides that the subinterval must be shortened and the step recomputed, the differential equation is evaluated only in the corrector, not in the predictor.

Systems of simultaneous differential equations can be integrated by any of the three methods. The only difference is in the differential equations, each of which now depends on all the dependent variables, and in the computation of the subinterval, which must be determined from the maximum absolute value of y'_{j0}, y''_{j0}, b_j, d_j, $j = 1, 2, \ldots, m$, respectively.

Programs P405 to P407 integrate a first-order differential equation (first method), a first-order differential equation with a second-order differential equation deduced analytically (second method), and a second-order differential equation (third method). They also solve a system of those simultaneous differential equations. An optional second control of step size can be applied in each program.

4.6 STABILITY AND ACCURACY OF INTEGRATION METHODS

The problem of the stability of numerical integration of ordinary differential equations is an object of extensive theoretical investigation, referred to in many papers and textbooks on numerical analysis. The present brief notes are restricted mainly to the methods for which the seven programs had been developed.

One-step methods, i.e., Taylor-series and Runge-Kutta methods, do not exhibit any numerical instability if the step size of integration is sufficiently small. In multistep methods the situation is different. The integration formulas with nonvariable step size are in fact difference equations with constant coefficients. One of the roots of the corresponding characteristic equation tends to the exact solution of the differential equation as the step size $h \to 0$. The other roots generate extraneous solutions. A multistep method is called *strongly stable* if the absolute values of all the extraneous roots are smaller than 1 as $h \to 0$. Then any errors introduced into the numerical integration will decay with the increasing number of integration steps, provided $h \to 0$. If the dependent variable appears in a formula for a multistep method only twice and at two consecutive discrete values of the independent variable (for instance, y_n, y_{n+1}), then one root of the characteristic equation equals 1 and all the others equal 0 as $h \to 0$. Therefore Adams-Bashforth and Adams-Moulton methods are strongly stable for a sufficiently small h. On the other hand, Milne's predictor (4.5.3) and corrector (4.5.5) are weakly unstable: if in the differential equation (4.5.2) $\partial f/\partial y > 0$ on the interval of integration, the solution is stable, while for $\partial f/\partial y < 0$ it is unstable. In the latter case the extraneous solutions are exponentially increasing.

The three methods with a variable step size developed by the author can be considered in the first approximation as methods with constant coefficients because the change of the step size is very small as $h \to 0$. An analysis of their characteristic equations shows that they are strongly stable too.

A general integral of a differential equation is represented by a family of integral curves, a one-parameter family for first-order differential equations and a two-parameter family for equations of second order. The sum of roundoff errors and truncation errors in each step of integration shifts the solution to a neighboring integral curve with slightly different initial data. Roundoff errors are diminished in a higher-precision arithmetic. Truncation errors are smaller in higher-order methods and can be further reduced by decreasing the step size, but a large number of steps increases the accumulation of roundoff errors. Therefore a compromise must be made: there is an optimum value of the step size which for a given precision (double precision if possible) and the given differential equation results in best accuracy. Finding that optimum or at least being close to it depends on the experience of the researcher.

The example in Sec. 4.4 shows that it is difficult to obtain a required accuracy by Runge-Kutta method when the integral curve starts at, or goes through, a point where its first derivative vanishes, while the integration by Taylor-series method is quite successful. Here another example demonstrates the difficulties of

4.32 THEORY

numerical integration of a linear differential equation

$$y' - ay = g(x) \tag{4.6.1}$$

where a is a positive constant and $g(x)$ a polynomial. The general integral is

$$y = h(x) + Ce^{ax} \tag{4.6.2}$$

where $h(x)$ is another polynomial of the same degree as $g(x)$ and C an integration constant. The integral curves are in this case diverging rapidly because of the exponential function. If the initial data are $x = x_0$, $y(x_0) = h(x_0)$, the integration constant $C = 0$. The numerical errors gradually shift the solution to a sequence of integral curves with $C \neq 0$, which rapidly diverges from the solution with $C = 0$. This phenomenon is an instability inherent to the given differential equation. One cannot escape it by choosing another method. The best way to fight it is to use high-precision arithmetic and the proper choice of the step size.

This example is not intended to discourage readers from numerical integration of ordinary differential equations, but to call attention to its pitfalls. Excellent results can be achieved with adequate mathematical skill.

4.7 CHOOSING THE METHOD

In this section the characteristic properties of different methods of numerical integration of ordinary differential equations are summarized to make the choice of method easier.

Of the methods discussed in this chapter the Taylor-series method is the simplest. It is a one-step method approximating the integral curve in each subinterval by a polynomial of higher degree. It easily and automatically adjusts the step size to the differential equation and to the desired accuracy of solution. The only objection to its use is the difficulty of analytical differentiation of the given differential equation. There are equations where any attempt at differentiation is obviously hopeless, but contrary to the general opinion expressed in textbooks on numerical analysis many equations can be differentiated without too much effort. It is important to realize that the derivatives of the first, second, and third order appearing on the right-hand sides of Eqs. (4.1.3b, c, d) should never be replaced by explicit analytical expressions; the computer will substitute for them just one number. Hence whenever it is possible to compute the derivatives up to the fourth order analytically, the Taylor-series method should be taken into serious consideration.

Runge-Kutta methods of the nth order are also one-step methods, but unlike Taylor-series methods they evaluate only first derivatives for n times. In each step they approximate the integral curve by a straight line whose slope is an average of n slopes defined in such a way that the endpoint of the straight line coincides with the endpoint of a truncated Taylor series of the nth degree. They are widely and successfully used for the integration of systems of ordinary differential equations of first order. Aside from the difficulty exhibited in the example in Sec. 4.4, the most serious objection is the need to control the accuracy of integration by repeatedly halving the subinterval and recomputing the step until the desired accuracy is achieved. This procedure may be very time-consuming if the step size has not been estimated adequately for the differential equation and the required accuracy of solution.

The strongly stable predictor-corrector methods of nth order approximate the integral curve by a polynomial of nth degree defined at several consecutive points by derivatives of first or second order. Their great advantage lies in the fact that no (or only one) derivative is to be computed analytically and that they easily and automatically adjust the step size to the differential equation and the desired accuracy of integration. Since the step size is finite, extraneous solutions do exist but they can be kept under control in strongly stable methods by a proper choice of the step size, i.e., by a proper choice of the constant P_1 in Programs P405 to P407. There are fewer extraneous solutions in the second method (Program P406) than in the first (Program P405); therefore the former should be preferred in integrating first-order differential equations whenever it is possible to compute their second derivative analytically. Compared with the Taylor-series method, Programs P405 to P407 for strongly stable predictor-connector methods are at a slight disadvantage because of the existence of extraneous solutions. Nevertheless the author experienced excellent results in their application [15].

BOUNDARY-VALUE PROBLEMS OF ORDINARY DIFFERENTIAL EQUATIONS

For the numerical solution of the boundary-value problems of ordinary differential equations it is important to know the analytical background of the problem, and this topic is therefore discussed first. Some eigenfunctions frequently occurring in applications are then defined as solutions of a boundary-value problem. The chapter closes with the description of a simple but powerful method for the numerical solution of boundary-value problems of one differential equation of second order.

5.1 ANALYTICAL APPROACH TO BOUNDARY-VALUE PROBLEMS

In applications many solutions of ordinary differential equations have to satisfy certain conditions on the boundaries of the integration domain. Since most of these equations are of second order and linear, the boundary-value problem is explained here for a homogeneous linear differential equation of second order

$$y'' + P(x)y' + Q(x)y = 0 \qquad (5.1.1)$$

The boundary conditions at $x = a$ and $x = b$ are, for instance,

$$y(a) = 0 \qquad y(b) = 0 \qquad (5.1.2)$$

These conditions are called homogeneous because if the differential equation and the boundary conditions are satisfied by a function $F(x)$, they are also satisfied by $cF(x)$, where c is an arbitrary constant. A general solution of (5.1.1) is

$$y = C_1 y_1(x) + C_2 y_2(x) \qquad (5.1.3)$$

where $y_1(x)$ and $y_2(x)$ are two independent solutions of (5.1.1) and C_1 and C_2 two integration constants. The boundary conditions (5.1.2) yield two linear algebraic

5.1

5.2 THEORY

equations for the two unknown constants

$$C_1 y_1(a) + C_2 y_2(a) = 0 \qquad (5.1.4a)$$

$$C_1 y_1(b) + C_2 y_2(b) = 0 \qquad (5.1.4b)$$

A nontrivial solution requires the determinant of the system (5.1.4a, b) to vanish

$$\begin{vmatrix} y_1(a) & y_2(a) \\ y_1(b) & y_2(b) \end{vmatrix} = 0 \qquad (5.1.4c)$$

This condition implies that Eqs. (5.1.4a, b) are not independent. Therefore one of the integration constants, now denoted C, can be chosen arbitrarily; the solution of the boundary-value problem (5.1.2) is either of the functions

$$y = C[y_2(a) y_1(x) - y_1(a) y_2(x)] \qquad (5.1.5a)$$

and

$$y = C[y_2(b) y_1(x) - y_1(b) y_2(x)] \qquad (5.1.5b)$$

If

$$y_1(a) = y_2(a) = 0 \qquad y_1(b) \neq 0 \qquad y_2(b) \neq 0 \qquad (5.1.6a)$$

Eq. (5.1.5b) is to be used because (5.1.5a) is a trivial identity. On the other hand, if

$$y_1(b) = y_2(b) = 0 \qquad y_1(a) \neq 0 \qquad y_2(a) \neq 0 \qquad (5.1.6b)$$

Eq. (5.1.5b) is a trivial identity and Eq. (5.1.5a) is to be used. If

$$y_1(a) = y_1(b) = y_2(a) = y_2(b) = 0 \qquad (5.1.6c)$$

each of the independent solutions $y_1(x)$, $y_2(x)$ satisfies the boundary conditions (5.1.2), and the function (5.1.3) with two arbitrary constants is the solution of the boundary-value problem. This may be the case when the differential equation contains a parameter λ whose value has been chosen such that both functions $y_1(x)$, $y_2(x)$ satisfy (5.1.6c). In other common cases one of the particular integrals in (5.1.3), say $y_2(x)$, is singular on the integration domain and must be excluded from the solution; the boundary conditions are now satisfied by a proper choice of the parameter λ.

Other homogeneous boundary conditions

$$y'(a) = 0 \qquad y'(b) = 0 \qquad (5.1.7)$$

or

$$y(a) = 0 \qquad y'(b) = 0 \qquad (5.1.8)$$

lead to similar conclusions.

The differential equation (5.1.1), in which

$$Q(x) = Q_1(x) + \lambda Q_2(x) \qquad (5.1.9)$$

and $P(x)$, $Q_1(x)$, and $Q_2(x)$ are given functions, can be reduced to a Sturm-Liouville differential equation

$$\frac{d}{dx} p(x) y' + [q(x) + \lambda w(x)] y = 0 \qquad (5.1.10a)$$

by the substitutions

$$p(x) = \exp\left[\int P(x)\, dx\right] \qquad (5.1.11a)$$

$$q(x) = Q_1(x)p(x) \qquad w(x) = Q_2(x)p(x) \qquad (5.1.11b)$$

Formula (5.1.11a) indicates that the function $p(x)$ is always positive. The weighting function $w(x)$ is assumed positive, $w(x) > 0$, on the whole integration domain, with possibly $w(x) = 0$ at isolated points. The function $q(x)$ is also a given continuous function; usually it is nonpositive within the integration interval (but it may be positive).

The solution of (5.1.10a) satisfies certain boundary conditions (specified later) only if the parameter λ takes on any of the discrete set of values $\lambda_1, \lambda_2, \ldots$, called *eigenvalues* (*characteristic values, proper values*) of the given boundary-value problem. The solutions $y_1, y_2, \ldots$ corresponding to $\lambda_1, \lambda_2, \ldots$ are the *eigenfunctions* (*characteristic functions, proper functions*).

If y_j, y_k are two eigenfunctions with the eigenvalues λ_j and λ_k, the following equation can be deduced from (5.1.10a):

$$(\lambda_k - \lambda_j) \int_a^b w(x) y_j(x) y_k(x)\, dx = [p(x)\{y_k(x) y_j'(x) - y_j(x) y_k'(x)\}]_a^b \qquad (5.1.12)$$

The right-hand side of this equation vanishes if all eigenfunctions y_j, $j = 1, 2, \ldots$, satisfy (1) the homogeneous Dirichlet conditions

$$y(a) = y(b) = 0 \qquad (5.1.13a)$$

or (2) the homogeneous Neumann conditions

$$y'(a) = y'(b) = 0 \qquad (5.1.13b)$$

or (3) a linear combination of conditions 1 and 2

$$y_j(a) + c_1 y_j'(a) = y_j(b) + c_2 y_j'(b) = 0 \qquad (5.1.13c)$$

in which c_1, c_2 are two arbitrary constants, or (4) one of the above conditions at $x = a$ and another at $x = b$, or (5) the periodicity conditions

$$y_j(a) = y_j(b) \qquad y_j'(a) = y_j'(b) \qquad p(a) = p(b) \qquad (5.1.13d)$$

or (6) one of the conditions (5.1.13a, b, c) at $x = a$ or $x = b$ and

$$p(b) = 0 \quad \text{or} \quad p(a) = 0 \quad \text{respectively} \qquad (5.1.13e)$$

or (7) the conditions

$$p(a) = p(b) = 0 \qquad (5.1.13f)$$

Conditions 6 and 7 imply the existence of singular points at $x = a$ or $x = b$ or at both boundaries in the differential equation (5.1.10a) which can be reduced to the form

$$p(x) y'' + p'(x) y' + [q(x) + \lambda w(x)] y = 0 \qquad (5.1.10b)$$

5.4 THEORY

In the presence of singular points additional conditions, e.g., continuity or finiteness of the solution or singularity of an order not higher than prescribed, take over the role of boundary conditions [20].

When any of these boundary conditions are satisfied, the right-hand side of (5.1.12) vanishes and the equation can be written in the form of the orthogonality relations

$$\int_a^b w(x) y_j(x) y_k(x) \, dx = \begin{cases} N_j^2 \neq 0 & \text{if } j = k \\ 0 & \text{if } \lambda_j \neq \lambda_k \end{cases} \quad (5.1.14)$$

The value N_j^2 of this integral for $j = k$ is called the *square of the norm* (sometimes simply the norm). If a multiplicative constant for eigenfunctions is so chosen that the norm equals 1, eigenfunctions are *normalized* (*orthonormal eigenfunctions*). In general, the condition $i \neq k$ does not imply $\lambda_j \neq \lambda_k$. If there are several, say n, eigenfunctions belonging to the same eigenvalue, the problem is n-fold degenerate. All the eigenfunctions of the Sturm-Liouville problem are simple except for the problems with periodicity conditions (5.1.13d) when two eigenfunctions belong to each eigenvalue [20].

Orthogonality conditions also exist for differential equations of the fourth order; the right-hand side of an equation corresponding to (5.1.12) is, of course, more complicated and consequently so are the boundary conditions.

The Sturm-Liouville differential equation (5.1.10a) can also be written in the form

$$Sy = -\lambda y \quad (5.1.10c)$$

where

$$S = \frac{1}{w(x)} \frac{d}{dx} \left[p(x) \frac{d}{dx} \right] + \frac{q(x)}{w(x)} \quad (5.1.15)$$

is the Sturm-Liouville linear differential operator. Equation (5.1.10c) states that the Sturm-Liouville operator acting upon a function $y(x)$ generates a function that is proportional to the function y, the eigenvalue $-\lambda$ being the coefficient of proportionality. As a consequence of (5.1.10c) all eigenvalues are real and discretely distributed. There are infinitely many eigenvalues; if the eigenfunction $q(x)$ is nonpositive everywhere on the integration domain they are all nonnegative; if $q(x)$ also takes on positive values, there may be at most a finite number of negative eigenvalues [20].

The eigenfunctions of Sturm-Liouville operator form a complete set of orthogonal eigenfunctions. Any piecewise continuous function with piecewise continuous first and second derivatives which satisfies the boundary conditions can be expanded into an absolutely and uniformly convergent series of eigenfunctions

$$f(x) = \sum_{j=0}^{\infty} A_j y_j(x) \quad (5.1.16a)$$

At points of discontinuity x_d this series converges to

$$\lim_{\varepsilon \to 0} \tfrac{1}{2} [f(x_d + \varepsilon) + f(x_d - \varepsilon)]$$

The coefficients A_j of the series (5.1.16a) are determined by a formula that follows from the orthogonality relations (5.1.14)

$$A_j = \frac{1}{N_j^2} \int_a^b f(x) w(x) y_j(x) \, dx \tag{5.1.16b}$$

If the upper limit of the sum in (5.1.16a) is finite, the coefficients of the series determined by (5.1.16b) reduce the value of the following integral to a minimum:

$$\int_a^b \left[f(x) - \sum_{j=0}^n A_j y_j(x) \right]^2 dx = \min \tag{5.1.17}$$

Since the eigenfunctions of a Sturm-Liouville differential equation form a complete set of orthogonal eigenfunctions, the value of the integral can be made arbitrarily small by increasing the upper limit of the sum; if the series in (5.1.17) is infinite, the value of the integral vanishes (this is the definition of completeness of a system of eigenfunctions). Thus the finite series (5.1.16a) with coefficients computed by the integral in (5.1.16b) is a least-squares approximation of the function $f(x)$, in precise analogy with the least-squares approximations of Secs. 2.8 to 2.11, where the coefficients of orthogonal functions were determined by a finite sum of terms at prescribed abscissas.

Once the homogeneous Sturm-Liouville differential equation has been solved, the integral of the corresponding nonhomogeneous differential equation

$$\frac{d}{dx}[p(x)y'] + [q(x) + \Lambda w(x)]y = h(x) \tag{5.1.18a}$$

with the same boundary conditions and with a given piecewise continuous and piecewise differentiable function $h(x)$ can be determined by assuming it to be in the form of a series of the eigenfunctions $y_j(x)$ satisfying the homogeneous differential equation

$$y = \sum_{j=0}^\infty C_j y_j(x) \tag{5.1.19a}$$

According to the homogeneous equation (5.1.10c) and the definition (5.1.15), the nonhomogeneous differential equation divided by $w(x)$ can thus be reduced with the help of the series (5.1.19a) to the form

$$\sum_{j=0}^\infty (\Lambda - \lambda_j) C_j y_j = \frac{h(x)}{w(x)} \tag{5.1.18b}$$

The order of differentiation and summation has been reversed because the series of orthogonal eigenfunctions (5.1.19a) is absolutely and uniformly convergent. The right-hand side of (5.1.18b) is now expanded into a series of orthogonal eigenfunctions $y_j(x)$

$$\frac{h(x)}{w(x)} = \sum_{j=0}^\infty A_j y_j(x) \tag{5.1.20a}$$

5.6 THEORY

with
$$A_j = \frac{1}{N_j^2} \int_a^b h(x) y_j(x) \, dx \tag{5.1.20b}$$

The substitution of (5.1.20a) into (5.1.18b) yields the formula for the unknown coefficients C_j in the series (5.1.19)

$$C_j = \frac{A_j}{\Lambda - \lambda_j} \tag{5.1.19b}$$

After Eqs. (5.3.1) and (5.1.20b) have been substituted into (5.1.19), the solution of the homogeneous boundary-value problem for the nonhomogeneous differential equation (5.1.18a) can also be written in the form

$$y = \int_a^b h(t) G(x, t) \, dt \tag{5.1.19c}$$

where
$$G(x, t) = G(t, x) = \frac{\sum_{j=0}^{\infty} \bar{y}_j(x) \bar{y}_j(t)}{\Lambda - \lambda_j} \tag{5.1.21}$$

is the Green function of the given boundary-value problem and

$$\bar{y}_j(x) = \frac{y_j(x)}{N_j} \tag{5.1.22}$$

is the normalized eigenfunction of the problem [N_j is defined by the orthogonality relations (5.1.14)].

It is important to realize that if $h(x) = 0$ in (5.1.18a), the homogeneous boundary-value problem of the homogeneous differential equation has a solution $y_j(x)$ only if $\Lambda = \lambda_j$; however, when $h(x) \neq 0$, the solution of the same boundary-value problem of the nonhomogeneous differential equation exists only if (1) $\Lambda \neq \lambda_j$ or (2) if the function $h(x)/w(x)$ is orthogonal to the eigenfunction $y_j(x)$ corresponding to the eigenvalue $\lambda_j = \Lambda$, in which case

$$\int_a^b \frac{h(x)}{w(x)} w(x) y_j(x) \, dx = 0$$

The corresponding coefficient C_j is now undetermined.

The boundary-value problems of nonhomogeneous differential equations of higher order are solved in the same way.

5.2 ORTHOGONAL EIGENFUNCTIONS DEFINED AS SOLUTIONS OF A BOUNDARY-VALUE PROBLEM

All the differential equations of this section are reduced to a differential equation of Sturm-Liouville type.

The differential equation of a harmonic oscillator

$$y'' + \lambda y = 0 \tag{5.2.1a}$$

with $\qquad p = 1 \qquad q = 0 \qquad w = 1 \tag{5.2.1b}$

and Dirichlet boundary conditions (5.1.13a) at

$$a = 0 \quad \text{and} \quad b = 2\pi \tag{5.2.1c}$$

defines a complete set of orthogonal eigenfunctions

$$y_j = \sin jx \tag{5.2.1d}$$

with the eigenvalues

$$\lambda = j^2 \qquad j = 1, 2, \ldots \tag{5.2.1e}$$

The square of the norm is

$$N_j^2 = \pi \qquad j = 1, 2, \ldots \tag{5.2.1f}$$

The differential equation (5.2.1a, b) with Neumann boundary conditions (5.1.13b) at the boundaries (5.2.1c) defines a complete system of orthogonal eigenfunctions

$$y_j = \cos jx \tag{5.2.2a}$$

with the eigenvalues

$$\lambda = j^2 \qquad j = 0, 1, 2, \ldots \tag{5.2.2b}$$

The square of the norm is

$$N_0^2 = 2\pi \qquad N_j^2 = \pi \qquad j = 1, 2, \ldots \tag{5.2.2c}$$

The differential equation (5.2.1a, b) with the periodicity boundary conditions (5.1.13d) at the boundaries (5.2.1c) defines a complete system of eigenfunctions with one eigenfunction

$$y_0 = 1 \tag{5.2.3a}$$

belonging to the eigenvalue

$$\lambda = 0 \tag{5.2.3b}$$

with the square of its norm

$$N_0^2 = 2\pi \tag{5.2.3c}$$

but two linearly independent eigenfunctions

$$y_j = \cos jx \qquad y_j = \sin jx \qquad j = 1, 2, \ldots \tag{5.2.3d}$$

belonging to one eigenvalue

$$\lambda = j^2 \qquad j = 1, 2, \ldots \tag{5.2.3e}$$

with the square of their norms

$$N_j^2 = \pi \qquad j = 1, 2, \ldots \tag{5.2.3f}$$

5.8 THEORY

The Bessel differential equation

$$xy'' + y' - \left(\frac{n^2}{x} - \lambda x\right)y = 0 \qquad (5.2.4a)$$

where n is a given constant and

$$p = x \qquad q = -\frac{n^2}{x} \qquad w = x \qquad (5.2.4b)$$

has a regular singular point at the boundary $x = a = 0$. The solution is required to be finite at this point; the other boundary condition at $x = b$ may be, for instance,

$$y(b) = 0 \qquad (5.2.4c)$$

The eigenfunctions are Bessel functions of the first kind

$$y_j = J_n\left(\frac{xc_{nj}}{b}\right) \quad j = 1, 2, \ldots \qquad (5.2.4d)$$

with the eigenvalues

$$\lambda = \left(\frac{c_{nj}}{b}\right)^2 \quad j = 1, 2, \ldots \qquad (5.2.4e)$$

The c_{nj} are the roots of the transcendental equation

$$J_n(c_{nj}) = 0 \quad j = 1, 2, \ldots \qquad (5.2.4f)$$

The square of the norm is

$$N_j^2 = \tfrac{1}{2}b^2[J_{n+1}(c_{nj})]^2 \quad j = 1, 2, \ldots \qquad (5.2.4g)$$

If boundary condition (5.2.4c) is replaced by

$$y'(b) = 0 \qquad (5.2.4h)$$

the coefficients c_{nj} become the roots of the transcendental equation

$$J_n'(c_{nj}) = 0 \quad j = 1, 2, \ldots \qquad (5.2.4i)$$

This is the only difference caused by boundary condition (5.2.4h).

The spherical Bessel differential equation

$$x^2 y'' + 2xy' - [n(n+1) - \lambda x^2]y = 0 \qquad (5.2.5a)$$

where n is a given integer constant and

$$p = x^2 \qquad q = -n(n+1) \qquad w = x^2 \qquad (5.2.5b)$$

has a regular singular point at the boundary $x = a = 0$. The solution is required to be finite at this point; the other boundary condition at $x = b$ may be, for instance,

$$y(b) = 0 \qquad (5.2.5c)$$

The eigenfunctions are the spherical Bessel functions of the first kind

$$y_j = j_n\left(\frac{xd_{nj}}{b}\right) \qquad j = 1, 2, \ldots \tag{5.2.5d}$$

with the eigenvalues

$$\lambda = \left(\frac{d_{nj}}{b}\right)^2 \qquad j = 1, 2, \ldots \tag{5.2.5e}$$

The d_{nj} are the roots of the transcendental equation

$$j_n(d_{nj}) = 0 \qquad j = 1, 2, \ldots \tag{5.2.5f}$$

The square of the norm is

$$N_j^2 = \tfrac{1}{2}b^3[j_{n+1}(d_{nj})]^2 \qquad j = 1, 2, \ldots \tag{5.2.5g}$$

Here too the change of boundary condition (5.2.5c) to

$$y'(b) = 0 \tag{5.2.5h}$$

results only in the change of the constants d_{nj}, which are now the roots of the transcendental equation

$$j_n'(d_{nj}) = 0 \qquad j = 1, 2, \ldots \tag{5.2.5i}$$

The Legendre differential equation

$$(1 - x^2)y'' - 2xy' + \lambda y = 0 \tag{5.2.6a}$$

with

$$p = 1 - x^2 \qquad q = 0 \qquad w = 1 \tag{5.2.6b}$$

has regular singular points at both boundaries

$$a = -1 \qquad b = 1 \tag{5.2.6c}$$

The solution is required to be finite here. The eigenfunctions are the Legendre polynomials

$$y_n = P_n(x) \qquad n = 0, 1, 2, \ldots \tag{5.2.6d}$$

with the eigenvalues

$$\lambda = n(n + 1) \qquad n = 0, 1, 2, \ldots \tag{5.2.6e}$$

The square of the norm is

$$N_n^2 = \frac{2}{1 + 2n} \qquad n = 0, 1, 2, \ldots \tag{5.2.6f}$$

The associated Legendre differential equation

$$(1 - x^2)y'' - 2xy' - \left(\frac{m^2}{1 - x^2} - \lambda\right)y = 0 \tag{5.2.7a}$$

5.10 THEORY

with
$$p = 1 - x^2 \qquad q = -\frac{m^2}{1 - x^2} \qquad w = 1 \qquad (5.2.7b)$$

and n, m integer constants, $0 \le m \le n$, has regular singular points at both boundaries

$$a = -1 \qquad b = 1 \qquad (5.2.7c)$$

The solution is required to be finite here. The eigenfunctions are the associated Legendre functions

$$y_{nm} = P_n^m(x) \qquad \begin{array}{l} n = 0, 1, 2, \ldots \\ m = 0, 1, \ldots, n \end{array} \qquad (5.2.7d)$$

with the eigenvalues

$$\lambda = n(n+1) \qquad n = 0, 1, 2, \ldots \qquad (5.2.7e)$$

The square of the norm is

$$N_{nm}^2 = \frac{2}{1 + 2n} \frac{(n+m)!}{(n-m)!} \qquad \begin{array}{l} n = 0, 1, 2, \ldots \\ m = 0, 1, \ldots, n \end{array} \qquad (5.2.7f)$$

The differential equation of Chebyshev polynomials of the first kind

$$(1 - x^2)^{1/2} y'' - x(1 - x^2)^{-1/2} y' + \lambda(1 - x^2)^{-1/2} y = 0 \qquad (5.2.8a)$$

with
$$p = (1 - x^2)^{1/2} \qquad q = 0 \qquad w = (1 - x^2)^{-1/2} \qquad (5.2.8b)$$

has regular singular points at both boundaries

$$a = -1 \qquad b = 1 \qquad (5.2.8c)$$

The solution is required to be finite here. The eigenfunctions are the Chebyshev polynomials of the first kind

$$y_j = T_j(x) \qquad j = 0, 1, 2, \ldots \qquad (5.2.8d)$$

with the eigenvalues

$$\lambda = j^2 \qquad j = 0, 1, 2, \ldots \qquad (5.2.8e)$$

The square of the norm is

$$N_0^2 = \pi \qquad N_j^2 = \tfrac{1}{2}\pi \qquad j = 1, 2, \ldots \qquad (5.2.8f)$$

The Laguerre differential equation

$$e^{-x} x y'' + e^{-x}(1 - x) y' + \lambda e^{-x} y = 0 \qquad (5.2.9a)$$

with
$$p = x e^{-x} \qquad q = 0 \qquad w = e^{-x} \qquad (5.2.9b)$$

has regular singular points at both boundaries

$$a = 0 \qquad b = \infty \qquad (5.2.9c)$$

The solution $L_j(x)$ is required to be finite at $x = a$ and to go to infinity as a positive power of x at $x = b$. The eigenfunctions are

$$y_j = e^{-x}L_j(x) \qquad j = 0, 1, 2, \ldots \qquad (5.2.9d)$$

where $L_j(x)$ are the Laguerre polynomials, in which $L_j(0) = 1$. The eigenvalues are

$$\lambda = j \qquad j = 0, 1, 2, \ldots \qquad (5.2.9e)$$

The square of the norm is

$$N_j^2 = j! \qquad j = 0, 1, 2, \ldots \qquad (5.2.9f)$$

The Hermite differential equation

$$e^{-x^2}y'' - 2xe^{-x^2}y' + \lambda e^{-x^2}y = 0 \qquad (5.2.10a)$$

with
$$p = e^{-x^2} \qquad q = 0 \qquad w = e^{-x^2} \qquad (5.2.10b)$$

has regular singular points at both boundaries

$$a = -\infty \qquad b = \infty \qquad (5.2.10c)$$

The solution $H_j(x)$ is required to go to infinity at both boundaries as a positive power of x. The eigenfunctions are

$$y_j = e^{-x^2}H_j(x) \qquad j = 0, 1, 2, \ldots \qquad (5.2.10d)$$

where $H_j(x)$ are the Hermite polynomials. The eigenvalues are

$$\lambda = 2j \qquad j = 0, 1, 2, \ldots \qquad (5.2.10e)$$

The square of the norm is

$$N_j^2 = 2^j \pi^{1/2} j! \qquad j = 0, 1, 2, \ldots \qquad (5.2.10f)$$

5.3 NUMERICAL APPROACH TO BOUNDARY-VALUE PROBLEMS

Each boundary-value problem should first be investigated thoroughly by analytical methods. Is the differential equation linear? If so, does it have singular points on the integration domain? If so, where does the eigenvalue appear in the integral? As a coefficient at the independent variable? Such questions should be asked and analytical methods exhausted before a numerical approach is attempted.

This section describes a method for the numerical solution of boundary-value problems of one second-order differential equation which may be linear or nonlinear.

The initial data of a second-order differential equation are the values of the

5.12 THEORY

function and of its derivative at the initial point. In a boundary-value problem only one of these values, say $y(a)$, is given at the initial point; the other value, $y'(a)$, must be estimated: $y'_1(a)$. The differential equation is then integrated numerically by any suitable method of Chap. 4. The integration yields a certain value $y_1(b)$, which differs from the given boundary value $y(b)$. Another guess $y'_2(a)$ must be made and the integration repeated to give the corresponding $y_2(b)$. A linear interpolation in which $y_1(b)$, $y_2(b)$ are the values of the independent variable and the estimates $y'_1(a)$, $y'_2(a)$ are the values of the dependent variable yields for the given $y(b)$ an approximation $y'_3(a)$ that is closer to the true $y'(a)$ than both estimates $y'_1(a)$ and $y'_2(a)$. The differential equation is integrated again with the initial data $y(a)$, $y'_3(a)$; the result is $y_3(b)$. The new point $(y_3(b), y'_3(a))$ is added to $(y_1(b), y'_1(a))$ and $(y_2(b), y'_2(a))$; a quadratic interpolation yields a further approximation $y'_4(a)$. Another integration with the initial data $y(a)$, $y'_4(a)$ gives the value $y_4(b)$. This point $(y_4(b), y'_4(a))$ is added to the first three; the cubic interpolation results in the approximation $y'_5(a)$. Henceforth each new point $(y_j(b), y'_j(a))$, $j = 5, 6, \ldots$, replaces the point with the subscript $j - 4$ [or $(j - 8)$ if $j > 8$] in the cubic interpolation.

If the initial guess $y'_1(a)$, $y'_2(a)$ has been made sufficiently close to the solution of the boundary-value problem, i.e., to the true value $y'(a)$ corresponding to the boundary condition $y(b)$, the sequence $y'_j(a)$ converges to the true $y'(a)$. Usually by quadratic interpolation $y'_4(a)$ is already very close to $y'(a)$.

If the boundary data correspond to Neumann or other boundary conditions in which one of the four values $y(a)$, $y'(a)$; $y(b)$, $y'(b)$ is given, the procedure is analogous to the one described above. A boundary condition like (5.1.13c), in which none of $y(a)$, $y'(a)$; $y(b)$, $y'(b)$ is known, or a boundary-value problem of a differential equation of higher order with two or more unknown initial values is much more complicated; it requires linear approximations of two functions with two independent variables, which is outside the scope of this handbook.

Program P501 is a Lagrange interpolation adapted to the present problem. The input data are used for a linear, quadratic, and repeatedly cubic interpolation.

NONLINEAR EQUATIONS

A nonlinear equation is either algebraic or transcendental. For an algebraic equation all roots, both real and complex, are to be determined, but for a transcendental equation the search is usually restricted to real roots. Direct methods exist, in general, only for quadratic, cubic, and biquadratic polynomials. All other equations are solved numerically by iterative methods. After a short survey of these methods, a new method with a global cubic convergence suitable for both algebraic and transcendental equations is described in detail. In the concluding section methods for the computation of real roots of a system of nonlinear equations are discussed.

6.1 DIRECT METHODS FOR ALGEBRAIC EQUATIONS

The fundamental theorem of algebra states that an algebraic equation of the nth degree has n real or complex roots if a k-fold root is counted as k roots. If all the coefficients of an algebraic equation are real, the roots are either real or complex conjugate.

After an important contribution in 1824 by Abel, Galois gave the definite proof in 1832 that it is impossible to solve a general equation of the fifth or higher degree by radicals. Formulas exist for solution of an algebraic equation of higher than fourth degree, but such equations must be of a special form.

The quadratic equation

$$Ax^2 + Bx + C = 0 \qquad A \neq 0 \qquad (6.1.1)$$

is transformed into

$$x^2 + 2px + q = 0 \qquad (6.1.2a)$$

where

$$p = \frac{B}{2A} \qquad q = \frac{C}{A} \qquad (6.1.2b)$$

to make the subsequent formulas simpler. The roots of (6.1.2a) are determined by

6.2 THEORY

the formula
$$x_{1,2} = -p \pm (p^2 - q)^{1/2} \tag{6.1.3}$$

The roots are complex conjugate if $p^2 < q$. In this case

$$\text{Re } x_{1,2} = -p \qquad \text{Im } x_{1,2} = \pm(q - p^2)^{1/2} \qquad \text{if } q > p^2 \tag{6.1.4}$$

If the roots are real, the accuracy of the root of smaller absolute value computed by (6.1.3) may be very poor because of roundoff errors. Therefore one root should be computed by the formula

$$x_1 = \begin{cases} -p - \text{sgn}(p)(p^2 - q)^{1/2} & \text{if } p^2 \geq q,\ p \neq 0 \tag{6.1.5a} \\ (-q)^{1/2} & q < 0,\ p = 0 \tag{6.1.5b} \end{cases}$$

and the other by

$$x_2 = \frac{q}{x_1} \tag{6.1.5c}$$

where

$$\text{sgn } p = \begin{cases} 1 & \text{if } p > 0 \\ -1 & \text{if } p < 0 \end{cases}$$

Program P601 computes the roots of the quadratic equation (6.1.1).

The cubic equation

$$Ax^3 + Bx^2 + Cx + D = 0 \qquad A \neq 0 \tag{6.1.6a}$$

is transformed by the substitution

$$x = y - \frac{B}{3A} \tag{6.1.6b}$$

into the reduced form

$$y^3 + 3py + 2q = 0 \tag{6.1.7a}$$

where

$$p = \frac{C}{3A} - \left(\frac{B}{3A}\right)^2$$

$$q = \left(\frac{B}{3A}\right)^3 - \frac{BC}{6A^2} + \frac{D}{2A} \tag{6.1.7b}$$

If

$$q^2 + p^3 < 0 \tag{6.1.8}$$

(the irreducible case), then all three roots are real

$$x_1 = s \cos \frac{\phi}{3} - \frac{B}{3A}$$

$$x_2 = s \cos \left(\frac{\phi}{3} + \frac{2\pi}{3}\right) - \frac{B}{3A} \tag{6.1.9}$$

$$x_3 = s \cos \left(\frac{\phi}{3} + \frac{4\pi}{3}\right) - \frac{B}{3A}$$

where $\qquad s = 2(-p)^{1/2}\qquad$ and $\qquad \phi = \arccos \dfrac{-q}{(-p^3)^{1/2}}\qquad$ (6.1.10)

If
$$q^2 + p^3 > 0 \qquad (6.1.11)$$

Cardano's formulas show that one root is real and two roots complex conjugate

$$x_1 = u + v - \frac{B}{3A}$$

$$\operatorname{Re} x_{2,3} = -\tfrac{1}{2}x_1 - \frac{B}{2A} \qquad (6.1.12)$$

$$\operatorname{Im} x_{2,3} = \pm(u - v)\frac{3^{1/2}}{2}$$

where
$$u = |-q + r|^{1/3} \operatorname{sgn}(-q + r)$$
$$v = |-q - r|^{1/3} \operatorname{sgn}(-q - r) \qquad (6.1.13)$$
$$r = (q^2 + p^3)^{1/2}$$

If
$$q^2 + p^2 = 0 \qquad (6.1.14)$$

all three roots are real, of which at least two are equal

$$x_1 = 2(-p)^{1/2} \operatorname{sgn}(-q) - \frac{B}{3A} \qquad (6.1.15a)$$

$$x_2 = x_3 = -\tfrac{1}{2}x_1 - \frac{B}{2A} \qquad (6.1.15b)$$

Program P602 computes the roots of the cubic equation (6.1.6a).
The biquadratic equation

$$Ax^4 + Bx^3 + Cx^2 + Dx + E = 0 \qquad A \neq 0 \qquad (6.1.16a)$$

is transformed by the substitution

$$x = y - \frac{B}{4A} \qquad (6.1.16b)$$

into the reduced form

$$y^4 + py^2 + qy + r = 0 \qquad (6.1.17a)$$

where
$$p = \frac{C}{A} - 6\left(\frac{B}{4A}\right)^2$$

$$q = \frac{D}{A} + 2\frac{B}{4A}\left[4\left(\frac{B}{4A}\right)^2 - \frac{C}{A}\right] \qquad (6.1.17b)$$

$$r = \frac{E}{A} + \frac{B}{4A}\left\{\frac{B}{4A}\left[\frac{C}{A} - 3\left(\frac{B}{4A}\right)^2\right] - \frac{D}{A}\right\}$$

6.4 THEORY

With the signs of roots chosen so that $z_1^{1/2} z_2^{1/2} z_3^{1/2} = -q$ its four roots

$$y_1 = \frac{z_1^{1/2} + z_2^{1/2} + z_3^{1/2}}{2} \qquad y_2 = \frac{z_1^{1/2} - z_2^{1/2} - z_3^{1/2}}{2}$$

$$y_3 = \frac{-z_1^{1/2} + z_2^{1/2} - z_3^{1/2}}{2} \qquad y_4 = \frac{-z_1^{1/2} - z_2^{1/2} + z_3^{1/2}}{2} \qquad (6.1.18)$$

are computed from the roots of the cubic resolvent equation

$$z^3 + 2pz^2 + (p^2 - 4r)z - q^2 = 0 \qquad (6.1.19)$$

If all three roots z_1, z_2, z_3 are real and positive, all four roots x_1, x_2, x_3, x_4 are real. If one of the roots z_1, z_2, z_3 is positive and the other two negative, the roots x_1, x_2, x_3, x_4 are pairwise complex conjugate. If one of the roots z_1, z_2, z_3 is real and the other two complex conjugate, two of the roots x_1, x_2, x_3, x_4 are real and the other two complex conjugate. The roots x_1, x_2, x_3, x_4 are computed from the roots y_1, y_2, y_3, y_4 by (6.1.16b).

The computation of the roots y_1, y_2, y_3, y_4 and x_1, x_2, x_3, x_4 is carried out in the arithmetic of complex numbers. For two complex numbers z_1 and z_2

$$z_1 = a_1 + ib_1 \qquad \text{and} \qquad z_2 = a_2 + ib_2 \qquad (6.1.20a)$$

1. Addition and subtraction

$$z_1 \pm z_2 = (a_1 \pm a_2) + i(b_1 \pm b_2) \qquad (6.1.20b)$$

2. Multiplication

$$z_1 z_2 = (a_1 a_2 - b_1 b_2) + i(a_2 b_1 + a_1 b_2) \qquad (6.1.20c)$$

3. Division

$$\frac{z_1}{z_2} = \frac{a_1 a_2 + b_1 b_2}{a_2^2 + b_2^2} + \frac{i(a_2 b_1 - a_1 b_2)}{a_2^2 + b_2^2} \qquad (6.1.20d)$$

The absolute value of a complex number is

$$|z| = (a^2 + b^2)^{1/2} \qquad (6.1.20e)$$

Two square roots of a complex number are of opposite sign. They are computed most efficiently from the formulas

$$z^{1/2} = \begin{cases} \pm \left(W + \dfrac{ib}{2W} \right) & \text{if } a \geq 0 \qquad (6.1.20f) \\ \pm \left(\dfrac{b}{2W} + iW \right) & \text{if } a < 0 \qquad (6.1.20g) \end{cases}$$

where

$$W = \{\tfrac{1}{2}[(a^2 + b^2)^{1/2} + |a|]\}^{1/2} \qquad (6.1.20h)$$

These formulas are also valid for $b = 0$, that is, z is a real number; in this case of course, it is faster to compute $z^{1/2}$ by the simpler formulas

$$z^{1/2} = \begin{cases} \pm |a|^{1/2} & \text{if } a \geq 0 & (6.1.20i) \\ \pm i|a|^{1/2} & \text{if } a < 0 & (6.1.20j) \end{cases}$$

Program P603 computes the roots of the biquadratic equation (6.1.16a).

The evaluation of the radicals in the formulas for cubic and biquadratic equations is the source of many roundoff errors that are not negligible. Even if the coefficients of these equations are integers chosen such that all the roots are also integers, it often happens that the formulas yield a root that should be real with a small imaginary part. Because the formulas for the roots of cubic and (especially) biquadratic equations are so complicated, computation by an iterative method, Programs P605 and P606, is usually faster and more accurate.

6.2 ITERATIVE METHODS

The roots of a nonlinear equation

$$f(x) = 0 \qquad (6.2.1)$$

are the intersections of the curve

$$y = f(x) \qquad (6.2.2)$$

with the x axis

$$y = 0 \qquad (6.2.3)$$

The real intersections are the real roots; the imaginary intersections are the complex roots.

For a preliminary orientation on the location of real roots of a nonlinear equation (6.2.1) within a given interval $[a, b]$ the function (6.2.2) can be evaluated at equally spaced abscissas

$$x_j = a + jh \qquad j = 0, 1, 2, \ldots, \qquad (6.2.4)$$

tabulated, and possibly plotted.

Instead of tabulation the function $f(x)$ and its derivative $f'(x)$ can be evaluated at equally spaced abscissas (6.2.4), but the data $x_j, f(x_j), f'(x_j)$ and $x_k, f(x_k), f'(x_k)$ are printed only if they satisfy either the condition

$$\operatorname{sgn} f(x_j) = -\operatorname{sgn} f(x_k) \qquad k = j + 1 \qquad (6.2.5)$$

or the conditions

$$\begin{aligned} \operatorname{sgn} f'(x_j) &= -\operatorname{sgn} f'(x_k) \qquad x_j < x_k \\ \operatorname{sgn} f'(x_k) &= \operatorname{sgn} f(x_k) \end{aligned} \qquad (6.2.6)$$

If condition (6.2.5) is satisfied, then Eq. (6.2.1) between x_j and x_k has one or an

6.6 THEORY

odd number of real roots. If both conditions (6.2.6) are simultaneously satisfied, then Eq. (6.2.1) between x_j and x_k has no or an even number of real roots (possibly multiple roots).

After the preliminary location of real roots the process can be repeated with a shorter interval h and the search restricted to the subintervals where the roots have been located.

Program P604 locates the real roots of a nonlinear equation (6.2.1) either by tabulation or by printing the data $x_j, f(x_j), f'(x_j)$ and $x_k, f(x_k), f'(x_k)$ when they satisfy either condition (6.2.5) or (6.2.6).

After the simple real roots have been located within certain subintervals, further computation can be accelerated by using the bisection method. The function (6.2.2) is evaluated at both ends of a subinterval $[a_0, b_0]$, where sgn $f(a_0) = -\text{sgn } f(b_0)$, and at the midpoint

$$m_k = \tfrac{1}{2}(a_{k-1} + b_{k-1}) \qquad k = 1, 2, \ldots \qquad (6.2.7a)$$

If $f(m_k) = 0$, the root has been found. If $f(m_k) \neq 0$ and

$$\text{sgn } f(a_{k-1}) = -\text{sgn } f(m_k) \qquad (6.2.7b)$$

the root is within the subinterval $[a_0, m_1]$. The midpoint is computed again by (6.2.7a), in which now $a_{k-1} = a_0$ and $b_{k-1} = m_1$. If condition (6.2.7b) is not satisfied, the root is within the subinterval $[m_1, b_0]$. In this case the midpoint is computed by (6.2.7a) with $a_{k-1} = m_1$ and $b_{k-1} = b_0$. After n such steps the root is contained in a subinterval of length $2^{-n}(b_0 - a_0)$. The bisection method always converges to a simple root, but it is very slow. It may miss some real roots if there is more than one in the starting interval $[a_0, b_0]$. It fails to determine a root of even multiplicity.

Therefore function (6.2.2) in the neighborhood $[x_0, b]$ of a real root is approximated by a straight line. The intersection of this line with the x axis indicates an improved value x_1 which is closer to the root than the initial estimate x_0.

In the regula falsi method the straight line is determined by two points on the function (6.2.2), one with the fixed abscissa b and the other with the varying abscissa x_j. The intersection with the x axis is

$$x_{j+1} = x_j - \frac{f(x_j)(x_j - b)}{f(x_j) - f(b)} \qquad j = 0, 1, 2, \ldots \qquad (6.2.8)$$

For any continuous function $f(x)$ the sequence $x_1, x_2, \ldots$ always converges to a real root but from one side. In the limit it yields the real root, simple or of an odd multiplicity

$$x = \lim_{j \to \infty} x_j \qquad (6.2.9)$$

The variants of this method are the modified regula falsi and the secant method. In the latter both points determining the straight line again lie on the curve (6.2.2), but their abscissas are the successive improved values x_{j-2}, x_{j-1}.

The intersection with the x axis is now

$$x_{j+1} = x_j - \frac{f(x_j)(x_j - x_{j-1})}{f(x_j) - f(x_{j-1})} \qquad j = 1, 2, \ldots \qquad (6.2.10)$$

In Newton's method the straight line approximating the function (6.2.2) is determined as a tangent at a point with abscissa x_j. Its intersection with the x axis is

$$x_{j+1} = x_j - \frac{f(x_j)}{f'(x_j)} \qquad j = 0, 1, 2, \ldots \qquad (6.2.11)$$

This formula differs from (6.2.8) and (6.2.10) only in the slope of the straight line, which here is the slope of the tangent $f'(x_j)$ whereas in (6.2.8) it is the slope of the secant

$$\frac{f(x_j) - f(b)}{x_j - b}$$

and in (6.2.10) the slope of another secant

$$\frac{f(x_j) - f(x_{j-1})}{x_j - x_{j-1}}$$

If point x_0 has been chosen sufficiently close to the root, the sequence $x_1, x_2, \ldots$ determined by (6.2.11) converges to the root as indicated in (6.2.9) but from the side opposite that for regula falsi.

When regula falsi gives an approximation sufficiently close to the root, it can be combined with Newton's method. Each method approaches the real root from different sides and can thus bracket it with any desired accuracy as long as $f'(x)$ at the root is not too close to zero.

The error in the jth iteration is defined by

$$e_j = x - x_j \qquad (6.2.12)$$

The rate of convergence of iteration methods is expressed by relating the errors of two consecutive iterations. Regula falsi has a linear convergence

$$e_{j+1} \approx e_j \qquad (6.2.13)$$

The secant method converges faster than linear convergence but not so fast as Newton's method, which has a quadratic convergence

$$e_{j+1} \approx e_j^2 \qquad (6.2.14)$$

Newton's method can also determine a multiple root, but in this case [or if the derivative $f'(x)$ at the root is small] the convergence is linear and the numerical errors in the computation of $f(x)$ during the iteration process are amplified because $f'(x) \to 0$. Therefore the iteration must stop before the errors in further steps destroy the achieved accuracy.

In order to improve the rate of convergence of an iteration method the function (6.2.2) must be approximated by a quadratic parabola.

6.8 THEORY

Muller's method [14] is an extension of the secant method; the parabola $p(x)$

$$p(x) = a_0 + a_1 x + a_2 x^2 \qquad (6.2.15a)$$

is given by three points with abscissas x_j, x_{j-1}, x_{j-2} on the curve (6.2.2). The coefficients a_2, a_1, a_0 are expressed by the divided differences of Sec. 2.3

$$\begin{aligned} a_2 &= f[x_j, x_{j-1}, x_{j-2}] \\ a_1 &= f[x_j, x_{j-1}] - a_2(x_j + x_{j-1}) \\ a_0 &= f(x_j) - x_j(f[x_j, x_{j-1}] - a_2 x_{j-1}) \end{aligned} \qquad (6.2.15b)$$

The next approximation x_{j+1} follows from the quadratic equation

$$p(x_{j+1}) = 0 \qquad (6.2.15c)$$

From the two roots of this equation Muller's method chooses the smaller one as the next approximation

$$x_{j+1} = \frac{a_0}{-a_1 - \operatorname{sgn}(a_1)(a_1^2 - 4a_0 a_2)^{1/2}} \qquad (6.2.15d)$$

In the author's experience this decision may be a source of serious difficulty. If the discriminant in (6.2.15d) is a real number, the author computes both roots

$$\bar{x} = -\frac{a_1}{2a_2} - \left(\operatorname{sgn}\frac{a_1}{2a_2}\right)\left[\left(\frac{a_1}{2a_2}\right)^2 - \frac{a_0}{a_2}\right]^{1/2} \qquad (6.2.15e)$$

and

$$\bar{\bar{x}} = \frac{a_0}{a_2 \bar{x}}$$

and the differences $|\bar{x} - x_j|$ and $|\bar{\bar{x}} - x_j|$ and takes the root for which the difference is smaller as the next approximation x_{j+1}. If the discriminant is a small number, it can be neglected; the sequence of iterations approaches a real root. If, the discriminant is an imaginary number comparable in size with $a_1/2a_2$, however, the arithmetic must be carried out in the domain of complex numbers [cf. Eqs. (6.1.20)] since the sequence is approaching a complex root.

Muller's method has a global convergence; i.e., it converges for any choice of three initial points. In each step it need evaluate the function (6.2.2) only once, and it also can determine complex roots, but Eqs. (6.2.15b) to (6.2.15e) are rather complicated, especially for complex numbers.

The Laguerre method for algebraic equations is an extension of Newton's method. The function (6.2.2) is approximated by a quadratic parabola which has two real intersections with the x axis when approaching a real root and is determined by a point $(x_j, f(x_j))$ on the curve (6.2.2) and by its first and second derivatives $f'(x_j), f''(x_j)$. The next approximation x_{j+1} of the iteration process is given by

$$x_{j+1} = x_j - \frac{nf(x_j)}{f'(x_j) + [\operatorname{sgn} f'(x_j)]\{[(n-1) f'(x_j)]^2 - n(n-1) f(x_j) f''(x_j)\}^{1/2}} \qquad (6.2.16)$$

where $n > 2$ is the degree of the algebraic equation. The method requires the calculation of $f(x_j)$, $f'(x_j)$, and $f''(x_j)$ at each step of iteration, but it has a global cubic convergence

$$e_{j+1} \approx e_j^3 \tag{6.2.17}$$

when it is approaching a simple real or complex root. The convergence becomes linear [Eq. (6.2.13)] when the iteration is close to a multiple root. The method also determines complex roots when the computation is carried out in complex numbers. In this case it may converge to a complex root even if the initial approximation is a real number. Empirical evidence shows that an absence of convergence is very unusual.

Other methods for the computation of all roots, real and complex, of an algebraic equation exist. The Sturm sequence determines the number of real roots and helps locate them and find their multiplicity. The Lehmer-Schur method, Bernoulli's method, and Graeffe's root-squaring method are always convergent if certain special cases are taken care of, but the first two converge slowly and a program using Graeffe's method is rather complicated.

The next section describes a slight modification of the Laguerre method which is very efficient in computing the real and complex roots of an algebraic equation and the real roots of a transcendental equation.

6.3 REAL ROOTS OF ALGEBRAIC AND TRANSCENDENTAL EQUATIONS

In the method developed by the author and described below the function (6.2.2) is approximated by a quadratic parabola which goes through a point $(x_j, f(x_j))$ on the curve (6.2.2); its first and second derivatives are equal to the derivatives $f'(x_j)$ and $f''(x_j)$. This parabola intersects the x axis at the point x_{j+1}

$$x_{j+1} = x_j - \frac{2f(x_j)}{f'(x_j) + [\operatorname{sgn} f'(x_j)]\{[f'(x_j)]^2 - 2f(x_j)f''(x_j)\}^{1/2}} \tag{6.3.1}$$

A comparison of this formula with (6.2.16) shows that for $n = 2$ the latter becomes identical with (6.3.1). If $f(x_j)$ and $f''(x_j)$ are of the same sign and $f'(x_j)$ is close to zero, both the determinant and the intersection become imaginary. When a real root is sought, the imaginary discriminant should be neglected no matter how large it is. The correction term is then

$$-\frac{2f(x_j)}{f'(x_j)}$$

which is twice the correction term in Newton's formula (6.2.11). The improved approximation x_{j+1} is thus shifted to the other side of the real root, where $f(x_j)$ has the opposite sign so that in the next iteration determining x_{j+2} the discriminant becomes positive. Henceforth the root is approached from the side where $f(x_j)$ and $f''(x_j)$ are of opposite signs.

6.10 THEORY

The method has a global cubic convergence when applied to the computation of simple real roots of an algebraic or transcendental equation. When the root is of double or higher multiplicity and the approximation is close to its true value, the method converges at least linearly.

The iteration stops when $|f(x)|$ becomes smaller than a chosen accuracy of computation or when the number of iterations equals a predetermined maximum of 10 to 15 steps. In the latter (very rare) case the program asks for another initial estimate and recomputes the root.

When a transcendental equation has many real roots, it is sometimes necessary to start the iteration sufficiently close to the root to be determined to avoid convergence to another root already computed. In this case Program P604 is very helpful.

An algebraic equation of the nth degree is assumed to have all real coefficients

$$f(x) = \sum_{k=0}^{n} a_k x^k \qquad (6.3.2)$$

The function and its first and second derivatives are computed by repeated synthetic division using the formulas

$$b_n = a_n \qquad b_k = b_{k+1} x_j + a_k \qquad k = n-1, n-2, \ldots, 0 \quad (6.3.3a)$$

$$c_{n-1} = b_n \qquad c_k = c_{k+1} x_j + b_{k+1} \qquad k = n-2, n-3, \ldots, 0 \quad (6.3.3b)$$

$$d_{n-2} = c_{n-1} \qquad d_k = d_{k+1} x_j + c_{k+1} \qquad k = n-3, n-4, \ldots, 0 \quad (6.3.3c)$$

The function and its first and second derivatives are now given by

$$f(x_j) = b_0 \qquad f'(x_j) = c_0 \qquad f''(x_j) = 2d_0 \qquad (6.3.3d)$$

To prevent convergence to an already known real root the function $f(x)$ given at the start of computation by Eq. (6.3.2) should be repeatedly deflated, i.e., its degree lowered by 1 as soon as a root has been computed. If the real root x_r that has just been determined is substituted into (6.3.3a), the synthetic division gives the result $b_0 = 0$. The coefficients $\bar{a}_k$ of the new polynomial of degree lower by 1 are thus given by

$$\bar{a}_n = a_n \qquad \bar{a}_k = \bar{a}_{k+1} x_r + a_k \qquad k = n-1, n-2, \ldots, 0 \qquad (6.3.4)$$

The last coefficient, $\bar{a}_{-1}$, vanishes if x_r is a real root of the polynomial with the coefficients a_k, $k = 0, 1, 2, \ldots, n$.

The deflation process is to be repeated: once a real root has been found, it should be factored out. The degree of the polynomial is decreasing step by step and the computation becomes faster and faster. Because the numerical value of the computed root contains a numerical error, the coefficients of the deflated polynomials are becoming less and less accurate. A closer theoretical examination shows [18] that these errors are negligible if the roots are computed in order of increasing absolute value and if the whole computation is carried out with the maximum accuracy possible. To satisfy the first condition the initial estimate should always be close to zero. Further, after all the real roots have been determined, each should be recomputed from the original polynomial, Eq. (6.3.2), with the initial estimate always equal to one of the computed roots. This whole recomputation is accomplished in a short time, since usually only one or two iterations are needed to redetermine a root with improved accuracy.

Program P605 computes the real roots of an algebraic or transcendental equation by iteration process using Eq. (6.3.1). For algebraic equations a subroutine can be applied for deflating the polynomial each time a root is found. The program can use Newton's method [Eq. (6.2.11)] as an option.

6.4 COMPLEX ROOTS OF ALGEBRAIC EQUATIONS

After all the roots of a polynomial of nth degree all of whose coefficients are real have been computed and the polynomial deflated to the mth degree, the complex roots are determined in complex-number arithmetic [Eqs. (6.1.20)] by the formula

$$z_{j+1} = z_j - \frac{2f(z_j)}{f'(z_j) \pm \{[f'(z_j)]^2 - 2f(z_j)f''(z_j)\}^{1/2}} \tag{6.4.1}$$

where z is a complex number,

$$z = a + ib \tag{6.4.2}$$

The function $f(z)$ and its derivatives $f'(z)$ and $f''(z)$ are also evaluated here by synthetic division [Eqs. (6.3.3)] in which x_j is now replaced by the complex z_j and all arithmetic operations are carried out in complex numbers [Eqs. (6.1.20)]. There are two signs standing before the discriminant in Eq. (6.4.1) because the two square roots of a complex number are of opposite sign; experience shows that it does not matter which sign is used.

The iteration stops when $|f(z)|$ becomes smaller than a chosen accuracy of computation or when the number of iterations equals a predetermined maximum of 10 to 15 steps. In the latter case the program asks for another initial estimate and recomputes the root.

To avoid convergence to an already known complex root the function $f(z)$ should be repeatedly deflated. Since there are always two complex-conjugate roots (all coefficients of the polynomial are assumed real), the degree of the polynomial is lowered in each deflation by 2. All computation in the deflation process is carried out in real numbers; the coefficients of the deflated polynomial are also real. Using the substitution

$$p = 2a_r \qquad q = -(a_r^2 + b_r^2) \tag{6.4.3a}$$

in which a_r and b_r are the real and imaginary part of a complex root z_r

$$z_r = a_r + ib_r \tag{6.4.3b}$$

the coefficients $\bar{a}_k$ of the new polynomial of the degree lower by 2 are given by

$$\bar{a}_n = a_n \qquad \bar{a}_{n-1} = p\bar{a}_n + a_{n-1} \qquad (6.4.3c)$$

$$\bar{a}_k = q\bar{a}_{k+2} + p\bar{a}_{k+1} + a_k \qquad k = n-2, n-3, \ldots, 0 \qquad (6.4.3d)$$

The last two coefficients, $\bar{a}_1$ and $\bar{a}_0$, vanish if z_r is one of a pair of complex-conjugate roots.

Once a complex root has been found, it should immediately be factored out together with its complex conjugate. Because of the repeated deflation the computation becomes faster and faster, but its accuracy is decreasing. To eliminate the errors generated by the repeated deflation all the complex roots should be recomputed from the original polynomial [Eq. (6.3.2)], as suggested in Sec. 6.3. The initial estimate always equals one of the computed complex roots. The iteration is now completed in one or two steps.

Program P606 computes the real and complex roots of a polynomial. It locates the real roots by the process used in P604 and computes them by iteration using Eq. (6.3.1). A subroutine can be applied for deflating the polynomial each time a real root has been found. Then the complex roots are computed in complex arithmetic by the iteration process using Eq. (6.4.1) with the positive sign before the discriminant. A subroutine can be applied for deflating the polynomial when a complex root has been determined.

All roots, real and complex, can be computed in P606 using only complex arithmetic throughout. A real initial approximation can converge to either a real or a complex root. In the former case the deflation for a real root must be used to lower the degree of the equation by 1; in the latter case the deflation for two complex-conjugate roots must be applied, lowering the degree of the polynomial by 2. When the computation results in a complex root with a negligibly small imaginary part apparently generated by numerical errors, that part should be suppressed and the deflation process for real roots used. The recomputation of all roots from the original polynomial [Eq. (6.3.2)] is always carried out in complex arithmetic. When the iteration converges to a root already recomputed, this recomputation is neglected, the sign in Eq. (6.4.1) automatically changed, and the root is recomputed once again.

The second alternative, using complex arithmetic for computation of both complex and real roots, requires more operations and longer computer time, but since the location process is now skipped, there is no great time difference between the alternatives.

6.5 ANALYSIS OF ERRORS IN ROOTS OF ALGEBRAIC EQUATIONS

In the method described in Secs. 6.3 and 6.4 and used in Programs P605 and P606 for the computation of real and complex roots of an algebraic equation or real roots of a transcendental equation each iteration is a new beginning; there is no propagation of numerical errors although such a propagation does exist in the deflation process. The only source of possibly serious numerical errors is the roundoff errors in the computation of the function $f(x)$ or $f(z)$ and their first and second derivatives. These errors become more relevant with the increasing degree of the algebraic equation and with the increasing magnitude of its roots. They set a limit upon the attainable accuracy of computation so that once this limit has been reached, no further iteration can improve the accuracy.

An algebraic equation with multiple roots or with nearly multiple roots, i.e., roots very close to each other, is in general ill-conditioned. The accuracy in the computation of roots is then limited, and a relatively small perturbation in the coefficients of the equation can generate a large change in the computed roots.

In an equation with real roots, $x_1 < x_2 < \cdots < x_n$, the product

$$s = \prod_{j=1}^{n-1} \frac{x_{j+1} - x_j}{|x_j|} \qquad x_j \neq 0 \tag{6.5.1}$$

can be used as a measure of separation. The smaller the value of s the more ill-conditioned the equation. If, for instance, $x_1 = 1$ and the difference between all n roots, $x_{j+1} - x_j$, equals 1, then $s = 1/(n-1)!$. This number clearly indicates how difficult it may be to obtain good accuracy when the degree of the polynomial is high. It is therefore essential to carry out the computation in double precision.

6.6 REAL ROOTS OF A SYSTEM OF NONLINEAR EQUATIONS

The solution of a system of nonlinear equations is a difficult problem because the convergence of iteration methods for such problems is not global. Hence the initial estimate must be quite close to the true roots of the equations. An attempt should therefore be made first to eliminate the unknowns by successive algebraic operations and then to reduce the system to one nonlinear equation with only one unknown. Numerical computation using Program P605 is then straightforward.

If this approach fails, the problem can be solved by an iteration method developed for a system of nonlinear equations. Three methods are discussed here; for simplicity the number of equations is restricted to two.

In *fixed-point iteration* the given two equations

$$f(x, y) = 0 \qquad g(x, y) = 0 \tag{6.6.1}$$

are reduced to a form suitable for this method

$$x_{j+1} = F(x_j, y_j) \qquad y_{j+1} = G(x_j, y_j) \tag{6.6.2}$$

The iteration functions F and G are derived from (6.6.1), so that any solution of (6.6.2) is also a solution of (6.6.1) and vice versa. A proper choice of the iteration functions requires mathematical skill and experience. The iteration converges under the following sufficient, but not necessary, conditions:

6.14 THEORY

1. F and G and their derivatives with respect to x and y are continuous in the neighborhood of the solution x, y.
2. The inequalities

$$\left|\frac{\partial F}{\partial x}\right| + \left|\frac{\partial F}{\partial y}\right| < 1 \qquad \left|\frac{\partial G}{\partial x}\right| + \left|\frac{\partial G}{\partial y}\right| < 1$$

are satisfied in the neighborhood of the solution where the iteration starts.

When the iteration converges, the convergence is linear.

Program P607 computes the roots of two nonlinear equations by fixed-point iteration.

In Newton's method the given equations (6.6.1) are expanded into a Taylor series for two independent variables

$$f(x, y) = f(x_0, y_0) + f_x(x_0, y_0)(x - x_0) + f_y(x_0, y_0)(y - y_0) + \cdots$$

$$g(x, y) = g(x_0, y_0) + g_x(x_0, y_0)(x - x_0) + g_y(x_0, y_0)(y - y_0) + \cdots$$

where $f_x(x_0, y_0)$, $f_y(x_0, y_0)$ and $g_x(x_0, y_0)$, $g_y(x_0, y_0)$ are the partial derivatives with respect to x and y, respectively, computed at the point (x_0, y_0). Thus the functions $f(x, y)$ and $g(x, y)$ are approximated by two straight lines whose intersection is an improved approximation to the initial estimates x_0, y_0. The two linear equations are solved by Cramer's rule for the unknowns $x - x_0$ and $y - y_0$. The iteration formulas are

$$x_{j+1} = x_j - \frac{fg_y - gf_y}{J} \qquad j = 0, 1, 2, \ldots$$

$$y_{j+1} = y_j - \frac{gf_x - fg_x}{J} \qquad j = 0, 1, 2, \ldots$$

(6.6.3a)

where J is the jacobian of the system

$$J = f_x g_y - f_y g_x \qquad (6.6.3b)$$

The iteration process converges if (1) the functions $f(x, y)$ and $g(x, y)$ and their first and second partial derivatives are continuous and bounded in the region close to the solution x,y, (2) if the jacobian does not vanish in this region, and (3) if the initial estimate is chosen sufficiently close to the solution. When the iteration converges, the convergence is quadratic.

In the modified Newton's method the iteration formulas are

$$x_{j+1} = x_j - \frac{f(x_j, y_j)}{f_x(x_j, y_j)} \qquad j = 0, 1, 2, \ldots \qquad (6.6.4a)$$

$$y_{j+1} = y_j - \frac{g(x_{j+1}, y_j)}{g_y(x_{j+1}, y_j)} \qquad j = 0, 1, 2, \ldots \qquad (6.6.4b)$$

The improved value x_{j+1} computed in (6.6.4a) is used immediately in the same iteration to improve y_{j+1} in (6.6.4b). The amount of computation is less than in Newton's method, but the modified method may diverge for any choice of initial estimate whereas Newton's method usually converges if the initial guess is sufficiently close to the roots.

Program P608 computes the roots of two nonlinear equations by Newton's method or the modified Newton's method.

LINEAR ALGEBRA II

This chapter deals with the algebraic eigenvalue problem, the numerical computation of eigenvalues of a real matrix, and the corresponding eigenvectors.

7.1 ALGEBRAIC EIGENVALUE PROBLEM

The matrix form (1.1.1b) of a system of n linear algebraic equations

$$AX = B$$

can also be interpreted as an equation determining a vector X which, after transformation by a linear algebraic operator, the square matrix A of nth order, equals a given vector B. In general, vectors X and B differ in both direction and magnitude. In many applications a vector X after linear transformation by the matrix A should equal a vector λX of the same direction but different magnitude

$$AX = \lambda X \qquad (7.1.1a)$$

Hence

$$(A - \lambda I)X = 0 \qquad (7.1.1b)$$

where I is the identity matrix. This system of n homogeneous linear algebraic equations has a nontrivial solution $X \neq 0$ only if the determinant of the system vanishes

$$\det(A - \lambda I) = 0 \qquad (7.1.2a)$$

Explicitly,

$$\begin{vmatrix} a_{11} - \lambda & a_{12} & \cdots & a_{1n} \\ a_{21} & a_{22} - \lambda & \cdots & a_{2n} \\ \cdots & \cdots & \cdots & \cdots \\ a_{n1} & a_{n2} & \cdots & a_{nn} - \lambda \end{vmatrix} = 0 \qquad (7.1.2b)$$

The evaluation of this determinant yields an algebraic equation of nth degree

$$f(\lambda) \equiv \lambda^n + \sum_{k=0}^{n-1} b_k \lambda^k = 0 \qquad (7.1.2c)$$

7.1

7.2 THEORY

Thus the algebraic eigenvalue problem (7.1.1a) has a nontrivial solution only if the coefficient λ equals a root of the secular (characteristic) equation (7.1.2c). The roots are called eigenvalues (characteristic values or proper values) and the corresponding vectors X are the eigenvectors (characteristic vectors or proper vectors). If there are only simple eigenvalues, all eigenvectors are orthogonal. If some eigenvalues are multiple roots, the eigenvectors corresponding to the same eigenvalue are no longer orthogonal but by a suitable linear combination they can always be made orthogonal (the Gram-Schmidt orthogonalization procedure).

There is a precise analogy between the algebraic eigenvalue problem and the boundary-value problem of a Sturm-Liouville differential equation. A third equivalent approach, sometimes more powerful, is represented in the theory of integral equations.

Some important theorems concerning the eigenvalues follow:

If the matrix A is real and symmetric or hermitian, all eigenvalues are real.

If P is a nonsingular matrix, the matrices A and PAP^{-1} have the same eigenvalues.

Cayley-Hamilton theorem: The secular equation (7.1.2c) is also satisfied if the powers λ^k of an eigenvalue are replaced by the powers A^k of the matrix A and the absolute term multiplied by the identity matrix I (for $A^0 = I$)

$$f(A) = 0 \tag{7.1.3}$$

The sum of eigenvalues equals the negative value of the trace of the matrix A

$$\sum_{j=1}^{n} \lambda_j = -\text{Tr } A \tag{7.1.4}$$

Schur theorem: The sum of squares of absolute values of all eigenvalues λ_j of a matrix A of nth order has an upper bound given by

$$\sum_{j=1}^{n} |\lambda_j|^2 \leq \sum_{j=1}^{n} \sum_{k=1}^{n} a_{jk}^2 \tag{7.1.5}$$

Gershgorin theorem: All eigenvalues of a matrix A lie within a domain D that is the union of all circles C_j

$$D = \bigcup_{j=1}^{n} C_j \tag{7.1.6a}$$

with centers at a_{jj} and radii r_j

$$r_j = \sum_{\substack{k=1 \\ k \neq j}}^{n} |a_{jk}| \tag{7.1.6b}$$

If the matrix A is strictly diagonally dominant, the radii of circles C_j may be sometimes so small that the intersection of any two circles is an empty set

$$C_j \cap C_k = \emptyset \qquad \begin{array}{l} j \neq k \\ j,k = 1, 2, \ldots, n \end{array} \qquad (7.1.6c)$$

All the eigenvalues lie within the interval

$$\lambda_{\min} \leq \lambda \leq \lambda_{\max} \qquad (7.1.7a)$$

The limits (not the minimum or maximum eigenvalue) are

$$\lambda_{\min} = \min_{1 \leq j \leq n} (a_{jj} - r_j) \qquad \lambda_{\max} = \max_{1 \leq j \leq n} (a_{jj} + r_j) \qquad (7.1.7b)$$

For a given matrix A Program P701 computes the centers a_{jj}, the corresponding radii r_j, and both limits $\lambda_{\min}$ and $\lambda_{\max}$ within which all the eigenvalues lie.

With the help of eigenvalues it is now possible to define the spectral radius $\rho(A)$ and spectral norm $\|A\|_S$ of a matrix A

$$\rho(A) = \max_{1 \leq j \leq n} |\lambda_j(A)| \qquad (7.1.8a)$$

$$\|A\|_S = \max_{1 \leq j \leq n} [\lambda_j(\tilde{A}A)]^{1/2} \qquad (7.1.9)$$

The matrix $\tilde{A}A$ is always symmetric and positive definite. The spectral radius and spectral norm can be determined when the eigenvalues are known. There exists, however, an upper bound for the spectral radius

$$\rho(A) \leq \max_{1 \leq j \leq n} \sum_{k=1}^{n} |a_{jk}| \qquad (7.1.8b)$$

This formula does not require previous computation of eigenvalues. The upper bound is identical with the maximum norm of the matrix (1.5.9).

7.2 NUMERICAL COMPUTATION OF THE EIGENVALUES OF A REAL MATRIX

The eigenvalues can be determined either from the secular polynomial (7.1.2c) or by an iteration process from Eq. (7.1.1a) or (7.1.1b).

The coefficients b_k in the secular equation can easily be computed from the elements of matrix A only if it is of second or third order. Second-order matrix:

$$\begin{aligned} b_0 &= \det A \\ b_1 &= -\operatorname{Tr} A \end{aligned} \qquad (7.2.1)$$

Third-order matrix:

$$b_0 = -\det A$$

$$b_1 = \begin{vmatrix} a_{11} & a_{12} \\ a_{21} & a_{22} \end{vmatrix} + \begin{vmatrix} a_{11} & a_{13} \\ a_{31} & a_{33} \end{vmatrix} + \begin{vmatrix} a_{22} & a_{23} \\ a_{32} & a_{33} \end{vmatrix} \qquad (7.2.2)$$

$$b_2 = -\operatorname{Tr} A$$

7.4 THEORY

If matrix A is of higher order, the coefficients b_k of the secular equation can be determined by the Krylov method, which is based on the Cayley-Hamilton theorem. Its explicit form [compare Eq. (7.1.3) with (7.1.2c)] is

$$A^n + \sum_{k=0}^{n-1} b_k A^k = 0$$

If this operator acts upon an arbitrary vector Y, it generates a nonhomogeneous system of n linear algebraic equations for n unknown $b_0, b_1, \ldots, b_{n-1}$,

$$\sum_{k=0}^{n-1} b_k A^k Y = -A^n Y \qquad (7.2.3)$$

For simplicity the arbitrary vector Y (a single-column matrix) can be chosen as

$$Y = \begin{bmatrix} 1 \\ 0 \\ \vdots \\ 0 \end{bmatrix} \qquad (7.2.4a)$$

and successively transformed by the matrix A

$$A^0 Y = IY = Y \qquad (7.2.4b)$$

Further $\qquad A^k Y = A(A^{k-1} Y) \qquad k = 1, 2, \ldots, n \qquad (7.2.4c)$

These vectors (single-column matrices) form the square matrix in (7.2.3) and the vector on the right-hand side of that equation. The system is solved by the Doolittle method [Eqs. (1.4.4a) to (1.4.6b)]. The coefficients b_k determined in this way are substituted into the secular equation (7.1.2c), which is solved by an iteration method for algebraic equations.

The choice of (7.2.4a) makes the computation of the coefficients b_k in the secular equation easier but sometimes produces a singular determinant of the system of n nonhomogeneous linear algebraic equations (7.2.3). In such a case the vector Y is replaced by a new one in which the first two, or more, components equal suitably chosen constants. For certain matrices, however, the Krylov method generates incorrect coefficients in the secular equation or it breaks down. This may happen for matrices with multiple eigenvalues.

Program P702 computes the coefficients of the secular equation by the Krylov method. It contains the Doolittle method, Program P102, for solving a system of linear algebraic equations as a subroutine.

Program P703 computes the eigenvalues from the secular equation. It contains as subroutines Programs P604, for the location of real eigenvalues, and P606, for the computation of real and complex eigenvalues.

Krylov and Doolittle methods used for the computation of the coefficients in the secular equation are direct methods. The only errors they generate are the roundoff errors. Since each of n elements of the single-column matrix on the right-hand side of Eq. (7.2.3) is the result of approximately n^2 multiplications, where n is the number of eigenvalues, and since there are $n^3/3$ multiplications in the Doolittle method, the total of $\frac{4}{3}n^3$ multiplications may be the source of a considerable number of errors, which are further amplified by roundoff errors in solving the secular equation (Program P606). These errors must not be underestimated; Programs P702 and P703 must therefore be run in double precision to keep them as small as possible. On the other hand, no symmetry restriction is imposed upon matrix A; all its eigenvalues are computed, real as well as complex.

After all real and simple eigenvalues have been determined, they are substituted into the determinant (7.1.2a), which is evaluated in Program P704 by the Doolittle method to test the accuracy of the whole procedure of the computation of eigenvalues. As long as the order of matrix A is not too high, say $n \leq 10$, the results are usually quite satisfactory. For sparse matrices the results may be excellent even for $n = 21$.

Iteration methods for eigenvalues, such as the power method, the methods based on similarity transformations (Jacobi's and Given's methods, Householder's method for symmetric matrices, Lanczos's method, supertriangularization and deflation, and the QR algorithm) are very complicated and therefore demand a detailed description. Some methods determine only a few eigenvectors, others require the same number of algebraic operations as the direct method expounded above, and all approach the true eigenvalue by iteration. For all these reasons they are not discussed here.

7.3 EIGENVECTORS

In most eigenvalue problems only the eigenvalues are to be determined. Therefore the computation of eigenvectors and their properties is discussed here briefly without omitting anything essential.

The components of the eigenvector belonging to a given eigenvalue λ_j are determined by the homogeneous system of n linear algebraic equations (7.1.1b). If the eigenvalue λ_j is of multiplicity m_j, the system has rank $\leq n - m_j$; that is, m_j or more equations can be expressed as a linear combination of the remaining equations and are satisfied by their solution, in which m_j or more components of the eigenvector are to be chosen freely. If all eigenvalues are simple, $m_j = 1$, all the eigenvectors are orthogonal to each other. If some of the eigenvalues are of multiplicity $m_j > 1$, then m_j components in each of m_j eigenvectors belonging to the same eigenvalues are to be chosen so that they make these eigenvectors orthogonal to each other.

The orthogonal eigenvectors can be normalized by multiplying each of them by a constant which reduces their magnitude to 1. The eigenvectors are then called *orthonormal eigenvectors*.

When the matrix A is real and symmetric or hermitian (orthogonal and unitary matrices are special cases of real and symmetric or hermitian matrices, respectively), all the eigenvalues are real. It is then possible to diagonalize matrix A by a similarity transformation $S^{-1}AS$. The matrix S is formed by n columns, each corresponding to one orthonormal eigenvector. For a real symmetric matrix

7.6 THEORY

A the matrix S is orthogonal; for a hermitian matrix A the matrix S is unitary. Their inverses S^{-1} are the transposed matrix $\tilde{S}$ and adjoint matrix $S\dagger$, respectively. The result of such a similarity transformation is a diagonal matrix with n elements equal to n eigenvalues.

The actual problem in the computation of eigenvectors is simply to find the basis of the singular matrix $A - \lambda I$, that is, a nonsingular matrix of order $\leq n - m_j$ formed by eliminating m_j or more rows and columns. Once this problem has been solved for each eigenvalue, the remaining computation is straightforward.

Program P704 is restricted to the computation of eigenvectors for simple real eigenvalues. First it tests the accuracy of computation of eigenvalues by substituting each into the singular matrix $A - \lambda I$ and and evaluating its determinant. Then it eliminates the last column and last row of this singular matrix and tests whether the new matrix of $(n - 1)$th order is nonsingular. If it is still singular, the program automatically eliminates another column and another row and tests again. It repeats this procedure until it finds a nonsingular matrix of $(n - 1)$th order. After this matrix has been found, the program solves the nonhomogeneous system of $n - 1$ linear algebraic equations whose right-hand side has been formed by transferring there the eliminated column, and finally it normalizes the orthogonal eigenvectors. If the matrix of $(n - 1)$th order is singular for a certain eigenvalue, the program cannot compute the corresponding eigenvector. It continues computing the eigenvector for the next eigenvalue. The program contains the Doolittle method (Program P102) as a subroutine.

SPECIAL FUNCTIONS

In response to the demand for accurate mathematical tables in a volume of reasonable size the editors of such compilations had no other choice than to tabulate the functions to many significant figures (10, 15, or more decimal places) and with relatively large intervals of the argument. To achieve the same accuracy the users must apply an interpolation of higher order which is not only time-consuming but always attended by the risk of human error when several data with many figures are keyed in as an input to an interpolation program. The programs of this chapter yield the value of a function, accurate in many digits, which corresponds to an argument given by the user. Since the input consists just of a single argument and perhaps some integer parameters, the chances of human error are reduced to a minimum.

A function can be defined by a series, either finite or infinite, or a recurrence formula; by an ordinary differential equation; as a boundary-value problem; or by a definite integral or contour integral. Whenever the definition requires a finite number of algebraic operations (a finite series, a recurrence formula), this definition is applied in the subsequent programs. Since the evaluation of definite integrals by gaussian or Gauss-Laguerre integration is very accurate and relatively fast, the integral representation has been chosen in computing most of the functions in this chapter. A truncated infinite series is used only when necessary.

When a function should repeatedly be used within a certain range, with equally spaced arguments, or at a large number of given values of the argument, a tabulation of the function may be quite useful. In this case the numerical integration of the ordinary differential equation defining the function may prove to be the most efficient method of evaluation.

8.1 THE POLYNOMIAL FUNCTION

The polynomial function

$$y = \sum_{k=0}^{n} a_k x^k \qquad (8.1.1a)$$

8.2 THEORY

and its first and second derivatives

$$\frac{dy}{dx} = \sum_{k=1}^{n} k a_k x^{k-1} \qquad \frac{d^2 y}{dx^2} = \sum_{k=2}^{n} k(k-1) a_k x^{k-2} \qquad (8.1.1b)$$

are evaluated by repeated synthetic division with the help of the formulas

$$\begin{array}{lll} b_n = a_n & b_k = b_{k+1} x + a_k & k = n-1, n-2, \ldots, 0 \\ c_{n-1} = b_n & c_k = c_{k+1} x + b_{k+1} & k = n-2, n-3, \ldots, 0 \qquad (8.1.2) \\ d_{n-2} = c_{n-1} & d_k = d_{k+1} x + c_{k+1} & k = n-3, n-4, \ldots, 0 \end{array}$$

The function and its derivatives are now given by

$$y(x) = b_0 \qquad \frac{dy}{dx} = c_0 \qquad \frac{d^2 y}{dx^2} = 2 d_0 \qquad (8.1.3)$$

Program P801 computes the polynomial function and its first and second derivatives.

8.2 ORTHOGONAL POLYNOMIALS

Orthogonal polynomials are defined in Sec. 5.2 as the solutions of boundary-value problems. Here they are defined by the recurrence relation

$$y_n = (Ax + B) y_{n-1} - C y_{n-2} \qquad (8.2.1a)$$

By differentiation their first and second derivatives are

$$y'_n = A y_{n-1} + (Ax + B) y'_{n-1} - C y'_{n-2}$$

and

$$y''_n = 2 A y'_{n-1} + (Ax + B) y''_{n-1} - C y''_{n-2} \qquad (8.2.1b)$$

The coefficients A, B, C are given in Table 8.1, along with initial polynomials necessary for starting the recurrence relations. Chebyshev polynomials are also

TABLE 8.1 Coefficients for orthogonal polynomials

Polynomial	Coefficients			Initial polynomials for starting recurrence relations	
	A	B	C		
Legendre	$2 - \dfrac{1}{n}$	0	$1 - \dfrac{1}{n}$	$P_0 = 1$	$P_1 = x$
Laguerre	$-\dfrac{1}{n}$	$2 - \dfrac{1}{n}$	$1 - \dfrac{1}{n}$	$L_0 = 1$	$L_1 = 1 - x$
Hermite	2	0	$2n - 2$	$H_0 = 1$	$H_1 = 2x$
Chebyshev:					
First kind	2	0	1	$T_0 = 1$	$T_1 = x$
Second kind	2	0	1	$U_0 = 0$	$U_1 = 1$

defined by

$$T_n = \cos(n \arccos x) \quad (8.2.3a)$$

and

$$U_n = \frac{\sin(n \arccos x)}{(1-x^2)^{1/2}} \quad (8.2.3b)$$

Hence

$$T_n^2 + U_n^2(1-x^2) = 1 \quad (8.2.3c)$$

The explicit formulas for $T_n(x)$, $n = 0, 1, \ldots, 12$, are given in Sec. 2.12. Two linearly independent solutions of the differential equation [cf. Eq. (4.2.10a)]

$$(1-x^2)y'' - xy' + n^2 y = 0 \quad (8.2.4)$$

are the functions $T_n(x)$ and $(1-x^2)^{1/2}U_n(x)$. Chebyshev polynomials of the second kind are also defined as a polynomial solution of the differential equation [cf. Eq. (4.2.10e)]

$$(1-x^2)y'' - 3xy' + (n^2 - 1)y = 0 \quad (8.2.5)$$

The other linearly independent solution is of the form $(1-x^2)^{-1/2}f_n(x)$, where $f_n(x)$ is a polynomial of nth degree. The polynomial solutions are sometimes defined by a slightly different differential equation

$$(1-x^2)y'' - 3xy' + n(n+2)y = 0$$

If this definition is accepted, the second independent solution of (8.2.4) is not $(1-x^2)^{1/2}U_n$ but $(1-x^2)^{1/2}\bar{U}_{n-1}$. The functions U_n and $\bar{U}_n$ are related to each other by the equation

$$U_{n+1} = \bar{U}_n$$

It is evident from (8.2.3a) that the zeros of Chebyshev polynomials of first kind are

$$x_k = \pm \cos\left[\left(\frac{1}{2} - \frac{k}{n}\right)\pi\right] \quad (8.2.6a)$$

where

$$k = \begin{cases} 0, 1, \ldots, \dfrac{n-1}{2} & \text{if } n \text{ is odd} \\ 1, 2, \ldots, \dfrac{n}{2} & \text{if } n \text{ is even} \end{cases} \quad (8.2.6b)$$

Program P802 computes the orthogonal polynomials of Legendre, Laguerre, Hermite, Chebyshev (first and second kind), their first and second derivatives, and zeros of Chebyshev polynomials of the first kind.

8.3 HYPERGEOMETRIC SERIES AND CONFLUENT HYPERGEOMETRIC FUNCTIONS

The hypergeometric series $F(a, b; c; x)$ is defined in Sec. 4.2 by the differential equation (4.2.7a) and by the series (4.2.7b). The confluent hypergeometric function

8.4 THEORY

$M(a, c; x)$ is defined by the differential equation (4.2.9a) and by the series (4.2.9b). Both functions can be expressed by the series

$$y = \sum_{k=0}^{\infty} m_k \tag{8.3.1a}$$

In hypergeometric series the terms m_k are given by

$$m_0 = 1 \qquad m_1 = \frac{ab}{c} \qquad m_{k+1} = m_k \frac{x}{k+1} \frac{(a+k)(b+k)}{c+k} \tag{8.3.1b}$$

In confluent hypergeometric functions the terms m_k are

$$m_0 = 1 \qquad m_1 = \frac{a}{c} \qquad m_{k+1} = m_k \frac{x}{k+1} \frac{a+k}{c+k} \tag{8.3.1c}$$

The first and second derivatives are

$$\frac{d}{dx} F(a, b; c; x) = \frac{ab}{c} F(a+1, b+1; c+1; x)$$

$$\frac{d^2}{dx^2} F(a, b; c; x) = \frac{ab}{c} \frac{(a+1)(b+1)}{c+1} F(a+2, b+2; c+2; x) \tag{8.3.2}$$

$$\frac{d}{dx} M(a, c; x) = \frac{a}{c} M(a+1, c+1; x)$$

$$\frac{d^2}{dx^2} M(a, c; x) = \frac{a}{c} \frac{a+1}{c+1} M(a+2, c+2; x) \tag{8.3.3}$$

In both functions the parameter c must be other than zero. The hypergeometric series terminates if a or b is a negative integer and $|c| > |a|$ or $|c| > |b|$ if c is a negative integer. The infinite series converges if c is not a negative integer for $|x| < 1$, at $x = 1$ if $c > a + b$ and at $x = -1$ if $c > a + b - 1$. The confluent hypergeometric series terminates if a is a negative integer and $|c| > |a|$ if c is a negative integer. The infinite series converges for all values of x, a, and c, provided c is not a negative integer.

Program P803 computes the hypergeometric series and confluent hypergeometric functions and their first and second derivatives. In both series the summation stops when $|m_{k+1} - m_k|$ is smaller than a certain chosen constant or when the number of terms in the series exceeds a given limit.

Many elementary and special functions can be represented in terms of hypergeometric series or confluent hypergeometric functions. For instance,

$$F(n, b; b; x) = (1 + x)^n \qquad b \text{ arbitrary} \tag{8.3.4a}$$

$$F(1, 1, 2; x) = \frac{\ln(1 + x)}{x} \tag{8.3.4b}$$

Legendre polynomials:

$$F(-n, n+1; 1; x) = P_n(1 - 2x) \tag{8.3.4c}$$

Chebyshev polynomials of the first kind:

$$F(-n, n; \tfrac{1}{2}; x) = T_n(1 - 2x) \qquad (8.3.4d)$$

Jacobi polynomials:

$$F(-n, \alpha + 1 + \beta + n; \alpha + 1; x) = \frac{n!\Gamma(\alpha + 1)}{\Gamma(\alpha + 1 + n)} P_n^{(\alpha, \beta)}(1 - 2x) \qquad (8.3.4e)$$

Gegenbauer's ultraspherical polynomials:

$$F(-n, n + 2\alpha; \alpha + \tfrac{1}{2}; x) = \frac{n!\Gamma(2\alpha)}{\Gamma(2\alpha + n)} C_n^{(\alpha)}(1 - 2x) \qquad (8.3.4f)$$

Complete elliptic integral of the first kind:

$$F(\tfrac{1}{2}, \tfrac{1}{2}; 1; k^2) = \frac{2}{\pi} K(k) \qquad (8.3.4g)$$

Complete elliptic integral of the second kind:

$$F(-\tfrac{1}{2}, \tfrac{1}{2}; 1; k^2) = \frac{2}{\pi} E(k) \qquad (8.3.4h)$$

Laguerre polynomials:

$$M(-n, 1; x) = L_n(x) \qquad (8.3.5a)$$

$$M(-n, \alpha + 1; x) = \frac{n!\Gamma(\alpha + 1)}{\Gamma(\alpha + 1 + n)} L_n^{(\alpha)}(x) \qquad (8.3.5b)$$

Incomplete gamma function:

$$M(a, a + 1; -x) = ax^{-a}\gamma(a, x) \qquad (8.3.5c)$$

$$M(1, a + 1; x) = ax^{-a}e^x\gamma(a, x) \qquad (8.3.5d)$$

8.4 INCOMPLETE ELLIPTIC INTEGRALS OF THE FIRST, SECOND, AND THIRD KIND

The definition of the incomplete elliptic integral of the first kind is

$$F(\phi, k) = \int_0^\phi (1 - k^2 \sin^2 x)^{-1/2} \, dx \qquad (8.4.1)$$

of the second kind is

$$E(\phi, k) = \int_0^\phi (1 - k^2 \sin^2 x)^{1/2} \, dx \qquad (8.4.2)$$

and of the third kind is

$$\Pi(\phi, n, k) = \int_0^\phi (1 - n \sin^2 x)^{-1}(1 - k^2 \sin^2 x)^{-1/2} \, dx \qquad (8.4.3)$$

If the upper limit is $\pi/2$, incomplete elliptic integrals of the first and second kind become complete elliptic integrals

$$K(k) = F\left(\frac{\pi}{2}, k\right) \tag{8.4.4}$$

$$E(k) = E\left(\frac{\pi}{2}, k\right) \tag{8.4.5}$$

Programs P804 and P805 evaluate the incomplete elliptic integrals of the first, second, and third kind by gaussian integration in single and double precision, respectively.

8.5 THE BESSEL FUNCTION $J_n(x)$ AND MODIFIED BESSEL FUNCTION $I_n(x)$ OF INTEGER ORDER

The Bessel function $J_n(x)$ and modified Bessel function $I_n(x)$ are defined by the differential equations (4.2.1a) and (4.2.2). If they are of integer order, their integral representation is

$$J_n(x) = \frac{1}{\pi} \int_0^\pi \cos(x \sin t - nt) \, dt \tag{8.5.1}$$

$$I_n(x) = \frac{1}{\pi} \int_0^\pi e^{x \cos t} \cos nt \, dt \tag{8.5.2}$$

Programs P804 and P805 evaluate the Bessel function $J_n(x)$ and modified Bessel function $I_n(x)$ of integer order by gaussian integration in single and double precision, respectively.

8.6 THE BESSEL FUNCTION $J_\nu(x)$ OF ORDER $\nu > -\frac{1}{2}$

The Bessel function $J_\nu(x)$ is defined by the differential equation (4.2.1a). If its order is $\nu > -\frac{1}{2}$, its representation by a Poisson integral is

$$J_\nu(x) = \frac{2(x/2)^\nu \pi^{-1/2}}{\Gamma(\nu + \frac{1}{2})} \int_0^{\pi/2} \cos(x \cos t) \sin^{2\nu} t \, dt \tag{8.6.1}$$

Programs P806 and P807 evaluate the Bessel function $J_\nu(x)$ of order $\nu > -\frac{1}{2}$ by gaussian integration in single and double precision, respectively.

8.7 THE BESSEL FUNCTION $Y_v(x)$ OF ORDER $v > -\frac{1}{2}$

The Bessel function of the second kind $Y_v(x)$ is defined by the differential equation (4.2.1a). If its order is $v > -\frac{1}{2}$, its representation by a Poisson integral is

$$Y_v(x) = \frac{2(x/2)^v \pi^{-1/2}}{\Gamma(v + \frac{1}{2})} \left[\int_0^{\pi/2} \sin(x \sin t) \cos^{2v} t \, dt \right.$$

$$\left. - \int_0^\infty \exp(2vt - x \sinh t) \left(\frac{1 + e^{-2t}}{2} \right)^{2v} dt \right] \quad (8.7.1)$$

Programs P806 and P807 evaluate the Bessel function of the second kind $Y_v(x)$ of order $v > -\frac{1}{2}$ by gaussian integration (the first integral) and by Gauss-Laguerre integration (the second integral) in single and double precision, respectively.

8.8 THE MODIFIED BESSEL FUNCTION $K_v(x)$

The modified Bessel function $K_v(x)$ is defined by the differential equation (4.2.2). Its integral representation is

$$K_v(x) = \frac{1}{2} \int_0^\infty \exp(vt - x \cosh t)(1 + e^{-2vt}) \, dt \quad (8.8.1)$$

Programs P806 and P807 evaluate the modified Bessel function $K_v(x)$ by Gauss-Laguerre integration in single and double precision, respectively.

8.9 THE SPHERICAL BESSEL FUNCTIONS $j_n(x)$ AND $y_n(x)$

The spherical Bessel functions $j_n(x)$ and $y_n(x)$ are related to the Bessel functions $J_{n+1/2}(x)$ and $Y_{n+1/2}(x)$ by

$$j_n(x) = \left(\frac{\pi}{2x}\right)^{1/2} J_{n+1/2}(x) \quad (8.9.1a)$$

$$y_n(x) = \left(\frac{\pi}{2x}\right)^{1/2} Y_{n+1/2}(x) \quad (8.9.2a)$$

They can be computed from the expressions

$$j_n(x) = \frac{1}{x} [zi_n(x) \sin x - zr_n(x) \cos x] \quad (8.9.1b)$$

$$y_n(x) = -\frac{1}{x} [zi_n(x) \cos x + zr_n(x) \sin x] \quad (8.9.2b)$$

where $zi_n(x)$ and $zr_n(x)$ are polynomials in negative powers of x which can be

8.8 THEORY

determined with the help of the formula

$$j_n(x) = x^n \left(-\frac{1}{x} \frac{d}{dx} \right)^n \frac{\sin x}{x} \qquad (8.9.3)$$

Explicit formulas for $zi_n(x)$ and $zr_n(x)$, $n = 0, 1, \ldots, 10$, can be found in Ref. 26.

Program P809 computes the spherical Bessel functions $j_n(x)$ and $y_n(x)$ of order $n = 0, 1, \ldots, 9$. If the order n is large and the argument x small, the program can give erroneous results because of roundoff errors. In this case the spherical Bessel functions should be evaluated by Program P806 in single precision and by P807 in double precision.

8.10 THE GAMMA FUNCTION OF A REAL ARGUMENT

The gamma function $\Gamma(x)$ of a real positive argument is defined by Euler's integral

$$\Gamma(x) = \int_0^\infty t^{x-1} e^{-t} \, dt \qquad (8.10.1)$$

If x equals a positive integer n, the gamma function can be reduced to factorial function

$$\Gamma(n) = (n-1)! \qquad (8.10.2)$$

For $x > 0$ the gamma function is computed in Program P810 by the recurrence formula

$$\Gamma(x) = \Gamma(y) \prod_{k=1}^m (x-k) \quad \text{where} \quad \begin{matrix} m = \text{Int}(x) \\ y = x - m \end{matrix} \qquad (8.10.3)$$

For $x < 0$ but $x \neq -n$, it is determined in P810 by

$$\Gamma(x) = \Gamma(y) \prod_{k=1}^p (y-p) \quad \text{where} \quad \begin{matrix} p = 1 + \text{Int}(|x|) \\ y = x + p \end{matrix} \qquad (8.10.4)$$

Int (x) denotes the integer part of x.

Equations (8.10.3) and (8.10.4) express $\Gamma(x)$ in terms of $\Gamma(y)$, where $0 < y < 1$. The gamma function $\Gamma(y)$ is now computed with the help of a power series

$$\Gamma(y) = \frac{1}{\sum_{k=1}^\infty c_k y^k} \qquad (8.10.5)$$

Although this series converges for any finite argument, to increase its accuracy and speed up its convergence the argument has been reduced with the help of the functional equation

$$\Gamma(x+1) = x\Gamma(x) \qquad (8.10.6)$$

to a number between 0 and 1. The coefficients of the series (8.10.5) are given in Ref. 36.

Program P809 computes the gamma function for any real argument with exception of zero and negative integers, for which the gamma function is singular.

8.11 THE GAMMA FUNCTION $\Gamma(\frac{1}{2} + n)$ FOR $n = 0, \pm 1, \pm 2, \ldots$

The gamma function for the argument $\frac{1}{2} + n$, where $n = 0, \pm 1, \pm 2, \ldots$, is computed with the help of the functional equation (8.10.6), which reduces the argument to $\Gamma(\frac{1}{2}) = \pi^{1/2}$. It is given by

$$\Gamma(\tfrac{1}{2} + n) = \begin{cases} \pi^{1/2} \prod_{k=0}^{n-1}(\tfrac{1}{2} + k) & n > 0 \quad (8.11.1) \\ \dfrac{\pi^{1/2}(-1)^{|n|}}{\prod_{k=0}^{n-1}(\tfrac{1}{2} + k)} & n < 0 \quad (8.11.2) \end{cases}$$

Program P810 computes the gamma function $\Gamma(\frac{1}{2} + n)$ for $n = 0, \pm 1, \pm 2, \ldots$.

8.12 THE INCOMPLETE GAMMA FUNCTION

The incomplete gamma function $\gamma(a, x)$ is defined by the intergal

$$\gamma(a, x) = \int_0^x t^{a-1} e^{-t} \, dt \qquad a > 0 \qquad (8.12.1)$$

A comparison with Euler's formula (8.10.1) shows that the argument x in Euler's integral is now denoted by a and the upper limit of the integral reduced to x.

Other definitions of the incomplete gamma function can be expressed with the help of $\gamma(a, x)$ by

$$\Gamma(a, x) = \Gamma(a) - \gamma(a, x) \qquad (8.12.2)$$

and

$$P(a, x) = \frac{\gamma(a, x)}{\Gamma(a)} \qquad a > 0 \qquad (8.12.3)$$

The incomplete gamma function is closely related to the chi-square distribution of mathematical statistics (cf. Sec. 9.8).

Programs P804 and P805 evaluate the incomplete gamma function by gaussian integration of the integral in (8.12.1) in single and double precision, respectively.

8.13 THE BETA FUNCTION

The beta function $B(x, y)$ is defined by

$$B(x, y) = B(y, x) = \frac{\Gamma(x)\Gamma(y)}{\Gamma(x + y)} = \int_0^1 t^{x-1}(1 - t)^{y-1} \, dt \qquad (8.13.1)$$

where x and y are positive numbers.

8.10 THEORY

Under the assumption that $x \leq y$ the constants m_1, m_2, m_3 and the variables u_1, u_2, u_3 are computed first

$$\begin{aligned} m_1 &= \text{Int}(x) & u_1 &= x - m_1 \\ m_2 &= \text{Int}(y) & u_2 &= y - m_2 \\ m_3 &= \text{Int}(x+y) & u_3 &= x + y - m_3 \end{aligned} \qquad (8.13.2)$$

If $m_1 < m_2$, the beta function is determined by

$$B(x, y) = \frac{\Gamma(u_1)\Gamma(u_2)}{\Gamma(u_3)} \frac{\left[\prod_{k=1}^{m_1} \frac{(x-k)(y-k)}{x+y-k}\right]\left[\prod_{k=m_1+1}^{m_2} \frac{y-k}{x+y-k}\right]}{\prod_{k=m_1+1}^{m_3}(x+y-k)} \qquad (8.13.3a)$$

If $m_1 = m_2$, then

$$B(x, y) = \frac{\Gamma(u_1)\Gamma(u_2)}{\Gamma(u_3)} \frac{\prod_{k=1}^{m_1} \frac{(x-k)(y-k)}{x+y-k}}{\prod_{k=m_1+1}^{m_3}(x+y-k)} \qquad (8.13.3b)$$

The gamma functions $\Gamma(u_1)$, $\Gamma(u_2)$, and $\Gamma(u_3)$ are evaluated by Eq. (8.10.5).

If x and y are positive integers m, n, the beta function can be reduced to an expression involving the binomial coefficient

$$B(m, n) = \frac{1}{n\binom{m+n-1}{n}} \qquad (8.13.4)$$

Program P811 computes the beta function. If m and n are positive integers, the beta function is computed by (8.13.4). The binomial coefficient is evaluated by formula (9.1.7).

8.14 THE ERROR FUNCTION

The error function is defined by the integral

$$\text{erf } x = \frac{2}{\pi^{1/2}} \int_0^x e^{-t^2} dt \qquad (8.14.1)$$

Its complement erfc x is related to erf x by the formula

$$\text{erfc } x = \frac{2}{\pi^{1/2}} \int_x^\infty e^{-t^2} dt = 1 - \text{erf } x \qquad (8.14.2)$$

The $(n+1)$th derivative of the error function erf x is given by

$$\left(\frac{d}{dx}\right)^{n+1} \text{erf } x = (-1)^n f(x) H_n(x) \qquad (8.14.3)$$

where $H_n(x)$ is the Hermite polynomial of nth degree (Sec. 8.2) and $f(x)$ stands for the integrand in (8.14.1).

$$f(x) = \frac{2}{\pi^{1/2}} e^{-x^2} \tag{8.14.4}$$

Programs P804 and P805 compute the error function erf x by gaussian integration in formula (8.14.1) in single and double precision, respectively.

8.15 THE FRESNEL INTEGRALS $C(x)$ AND $S(x)$

The Fresnel integrals $C(x)$ and $S(x)$ are defined by

$$C(x) = \int_0^x \cos\frac{\pi t^2}{2} dt \tag{8.15.1a}$$

$$S(x) = \int_0^x \sin\frac{\pi t^2}{2} dt \tag{8.15.2a}$$

Other definitions in use, namely,

$$C_1(x) = \left(\frac{2}{\pi}\right)^{1/2} \int_0^x \cos t^2 \, dt \qquad C_2(x) = (2\pi)^{-1/2} \int_0^x (\cos t) t^{-1/2} \, dt \tag{8.15.1b}$$

$$S_1(x) = \left(\frac{2}{\pi}\right)^{1/2} \int_0^x \sin t^2 \, dt \qquad S_2(x) = (2\pi)^{-1/2} \int_0^x (\sin t) t^{-1/2} \, dt \tag{8.15.2b}$$

are interrelated by

$$C(x) = C_1\left(x\left(\frac{\pi}{2}\right)^{1/2}\right) = C_2\left(\frac{x^2\pi}{2}\right) \tag{8.15.1c}$$

$$S(x) = S_1\left(x\left(\frac{\pi}{2}\right)^{1/2}\right) = S_2\left(\frac{x^2\pi}{2}\right) \tag{8.15.2c}$$

Programs P804 and P805 compute the Fresnel integrals $C(x)$ and $S(x)$ by gaussian integration in (8.15.1a) and (8.15.2a) in single and double precision, respectively.

8.16 THE SINE INTEGRAL AND COSINE INTEGRAL

The sine and cosine integrals are defined by

$$\operatorname{Si} x = \int_0^x \sin t \, \frac{dt}{t} \tag{8.16.1a}$$

8.12 THEORY

and
$$\text{Ci } x = \gamma + \ln x + \int_0^x (\cos t - 1) \frac{dt}{t} \quad (8.16.2a)$$

where γ is Euler's constant

$$\gamma = \lim_{m \to \infty} \left(\sum_{k=1}^{m} \frac{1}{k} - \ln m \right) \quad (8.16.3a)$$

Its numerical value is

$$\gamma = 0.57721\ 56649\ 01532\ 9 \quad (8.16.3b)$$

The cosine integral has a logarithmic singularity at $x = 0$. Other notations also in use are

$$\text{si } x = \text{Si } x - \frac{\pi}{2} \quad (8.16.1b)$$

and
$$\text{ci } x = \text{Ci } x \quad (8.16.2b)$$

Programs P804 and P805 compute the sine and cosine integrals by gaussian integration in (8.16.1a) and (8.16.2a) in single and double precision, respectively.

8.17 THE EXPONENTIAL INTEGRAL AND LOGARITHMIC INTEGRAL

The exponential integral Ei x is an infinitely many-valued function of a complex variable z. For a real argument it is defined by the series

$$\text{Ei } x = \gamma + \ln x + \sum_{k=1}^{\infty} \frac{x^k}{kk!} \quad (8.17.1a)$$

where γ is Euler's constant given by (8.16.3a, b). A slightly different form of (8.17.1a) is

$$\text{Ei } x = \gamma + \ln x + \sum_{k=1}^{\infty} e_k \quad (8.17.1b)$$

where
$$e_1 = x \qquad e_{k+1} = e_k \frac{xk}{(k+1)^2} \quad (8.17.1c)$$

The exponential integral has a logarithmic singularity at $x = 0$.
The logarithmic integral, defined by

$$\text{li } x = \int_0^x \frac{dt}{\ln t} \quad (8.17.2)$$

is related to the exponential integral Ei x through the relations

$$\text{li } x = \text{Ei } (\ln x) \qquad \text{Ei } x = \text{li } e^x \quad (8.17.3)$$

Program P803 computes the exponential integral by Eqs. (8.17.1b, c). The summation of the series in (8.17.1b) stops when $|e_{k+1} - e_k|$ is smaller than a certain chosen constant or the number of terms in the series exceeds a given limit. The logarithmic integral is reduced to the exponential integral by Eqs. (8.17.3) and also computed in P803.

8.18 THE GUDERMANNIAN AND ITS INVERSE

The gudermannian (the hyperbolic amplitude) relates the circular and hyperbolic functions without using functions of imaginary argument. The gudermannian gd x and its inverse $gd^{-1}\, x$ are defined by

$$\text{gd } x = 2 \arctan e^x - \frac{\pi}{2} \tag{8.18.1a}$$

and

$$gd^{-1}\, x + \ln \tan\left(\frac{x}{2} + \frac{\pi}{4}\right) \tag{8.18.2a}$$

They are also represented by the definite integrals

$$\text{gd } x = \int_0^x \frac{dt}{\cosh t} \tag{8.18.1b}$$

and

$$gd^{-1}\, x = \int_0^x \frac{dt}{\cos t} \tag{8.18.2b}$$

Some important relations follow:

$$\sinh x = \tan (gd\, x) \tag{8.18.3}$$

$$\cosh x = \frac{1}{\cos (gd\, x)} \tag{8.18.4}$$

$$\tanh x = \sin (gd\, x) \tag{8.18.5}$$

$$\coth x = \frac{1}{\sin (gd\, x)} \tag{8.18.6}$$

$$\tanh \frac{x}{2} = \tan (\tfrac{1}{2}\, gd\, x) \tag{8.18.7}$$

$$\coth \frac{x}{2} = \frac{1}{\tan (\tfrac{1}{2}\, gd\, x)} \tag{8.18.8}$$

Program P812 computes the gudermannian and its inverse by Eqs. (8.18.1a) and (8.18.2a) and the hyperbolic functions $\sinh x$, $\cosh x$, $\tanh x$, $\coth x$, $\tanh (x/2)$, and $\cosh (x/2)$ by Eqs. (8.18.3) to (8.18.8).

SELECTED PROBLEMS OF MATHEMATICAL STATISTICS

Four problems of mathematical statistics are discussed in this chapter: elementary questions of combinatorial analysis occurring in probability considerations; basic concepts of mathematical statistics, namely, mean value, deviation, regression, and correlation; curve fitting; and the definitions and evaluations of probability distributions.

9.1 ELEMENTS OF COMBINATORIAL ANALYSIS

In probability considerations it is sometimes necessary to know in how many ways a given number of elements can be arranged under certain conditions. These questions are answered in combinatorial analysis.

The number of different arrangements of n different elements is called a *permutation* P_n. There exist $n!$ different sequences of elements

$$P_n = n! \tag{9.1.1}$$

For example, $n = 3$; $P_3 = 3! = 6$. The different sequences are abc, acb, bac, bca, cab, and cba.

If element a occurs p_1 times, element b occurs p_2 times, and so on, the number of permutations with repetitions is

$$P_{Rn} = \frac{n!}{\prod_{k=1}^{m} p_k!} \qquad \sum_{k=1}^{m} p_k = n \tag{9.1.2}$$

For example, $p_1 = 2$, $p_2 = 1$; $n = 3$, $P_{Rn} = 3$. The sequences are: aab, aba, and baa.

Any arrangement of k elements selected from n elements is called a *combination*. A combination with regard to the order of selected elements is called a *variation*. If all k selected elements in such sequences are different, the number of variations without repetition is

$$V_n^k = \frac{n!}{(n-k)!} = \prod_{j=n-k+1}^{n} j \qquad (9.1.3)$$

For example, $n = 3$, $k = 2$; $V_3^2 = 6$. The sequences are: ab, ba, ac, ca, bc, and cb.

If the selected elements in a sequence are allowed to repeat, the number of variations with repetitions is

$$V_{Rn}^k = n^k \qquad (9.1.4)$$

For example, $n = 3$, $k = 2$; $V_{R3}^2 = 9$. The sequences are aa, ab, ba, bb, bc, cb, ca, ac, and cc.

A combination without regard to the order of selected elements is simply called a combination. If all k selected elements in such groups are different, the number of combinations without repetition is

$$C_n^k = \frac{n!}{k!(n-k)!} = \binom{n}{k} \qquad (9.1.5)$$

For example, $n = 3$, $k = 2$; $C_3^2 = 3$. The sequences are ab, ac, bc.

If each element may occur in each group of selected k elements several times, the number of combinations with repetition is

$$C_{Rn}^k = \binom{n+k-1}{k} \qquad (9.1.6)$$

For example, $n = 3$, $k = 2$; $C_{R3}^2 = 6$. The sequences are ab, ac, bc, aa, bb, and cc.

In order to prevent an overflow of the computer when $n > 33$, the binomial coefficients occurring in (9.1.5) and (9.1.6) should be evaluated by the formula

$$\binom{n}{k} = \prod_{j=1}^{n-k} \frac{k+j}{j} = \prod_{j=1}^{k} \frac{n-k+j}{j} \qquad (9.1.7)$$

The first formula is used if $k > n/2$ and the second when $k \leq n/2$. Also the variations without repetition should be computed in (9.1.3) by multiplication.

Program P901 computes the permutations, variations, and combinations with or without repetition.

9.2 BASIC CONCEPTS OF MATHEMATICAL STATISTICS

The arithmetic mean value $\bar{x}$ (arithmetic average) of a sample of n values x_j, $j = 1, 2, \ldots, n$, is given by

$$\bar{x} = \frac{1}{n} \sum_{j=1}^{n} x_j \qquad (9.2.1)$$

The geometric mean value $\bar{x}_G$ is defined by

$$\bar{x}_G = \left(\prod_{j=1}^{n} x_j\right)^{1/n} \tag{9.2.2}$$

The harmonic mean value $\bar{x}_H$ is

$$\bar{x}_H = \frac{n}{\sum_{j=1}^{n} x_j^{-1}} \tag{9.2.3}$$

The sample variance is based upon the arithmetic mean value. It is given by either

$$S^2 = \frac{\sum_{j=1}^{n}(x_j - \bar{x})^2}{n} \tag{9.2.4a}$$

or

$$s^2 = \frac{\sum_{j=1}^{n}(x_j - \bar{x})^2}{n - 1} \tag{9.2.5a}$$

These quantities are the measures of dispersion of the sample distribution. The positive square root of the sample variance S or s is called the (sample) *standard deviation* or (sample) *dispersion*. Since s^2 is an unbiased, consistent estimate of the sample variance, the standard deviation s is often more useful than S.

In order to lower the roundoff errors in the computation of S^2 and s^2 the following slightly different but equivalent formulas are used

$$S^2 = \frac{\sum_{j=1}^{n} x_j^2 - n\bar{x}^2}{n} \tag{9.2.4b}$$

$$s^2 = \frac{\sum_{j=1}^{n} x_j^2 - n\bar{x}^2}{n - 1} \tag{9.2.5b}$$

There are two definitions of the standard error, one based on the standard deviation S

$$E = \pm \frac{S}{n^{1/2}} \tag{9.2.6}$$

and the other on the standard deviation s

$$e = \pm \frac{s}{n^{1/2}} \tag{9.2.7}$$

If the values $x_j, j = 1, 2, \ldots, n$, occur with frequencies f_j (grouped data), the arithmetic mean value, both standard deviations and both standard errors are

9.4 THEORY

computed as follows

$$\bar{x} = \frac{\sum_{j=1}^{n} f_j x_j}{\sum_{j=1}^{n} f_j} \tag{9.2.8}$$

$$S = \left(\frac{\sum_{j=1}^{n} f_j x_j^2 - \bar{x}^2 \sum_{j=1}^{n} f_j}{\sum_{j=1}^{n} f_j} \right)^{1/2} \tag{9.2.9}$$

$$s = \left(\frac{\sum_{j=1}^{n} f_j x_j^2 - \bar{x}^2 \sum_{j=1}^{n} f_j}{\sum_{j=1}^{n} f_j - 1} \right)^{1/2} \tag{9.2.10}$$

$$E = \pm \frac{S}{\left(\sum_{j=1}^{n} f_j \right)^{1/2}} \tag{9.2.11}$$

$$e = \pm \frac{s}{\left(\sum_{j=1}^{n} f_j \right)^{1/2}} \tag{9.2.12}$$

Program P902 computes the arithmetic, geometric, and harmonic mean value, both variances, both standard deviations, and both standard errors for a sample consisting of n values x_j. Program P903 computes the arithmetic mean value, both variances, both standard deviations, and both standard errors for a sample of n values x_j occurring with frequencies f_j (grouped data).

In regression analysis the independent variable x is considered to be measured without a significant error, while the other variable y is a random variable. The problem is to find a curve, the regression curve y on x, for which the sum of squares of deviations of y_j from the regression curve y is a minimum (the least-squares technique)

$$\sum_{j=1}^{n} (y_j - y)^2 = \min \tag{9.2.13}$$

The coefficients of the regression curve are called *regression coefficients*. The coefficient of determination r^2, $0 \leq r^2 \leq 1$, measures the quality of the fit; if $r^2 = 1$, the fit is perfect. Regression curves are discussed in detail in Sec. 9.3.

In correlation analysis both x and y are random variables. Their arithmetic mean values $\bar{x}$, $\bar{y}$ and standard deviations s_x, s_y or S_x, S_y are computed by (9.2.1), (9.2.4b), and (9.2.5b); when $\bar{y}$, s_y, and S_y are to be determined, the x_j, $\bar{x}$ are replaced by y_j, $\bar{y}$. The formula for the covariance S_{xy} or s_{xy} is an obvious generalization of Eqs. (9.2.4) and (9.2.5) to two variables:

$$S_{xy} = \sum_{j=1}^{n} \frac{(x_j - \bar{x})(y_j - \bar{y})}{n} = \frac{\sum_{j=1}^{n} x_j y_j - n\bar{x}\bar{y}}{n} \qquad (9.2.14)$$

$$s_{xy} = \sum_{j=1}^{n} \frac{(x_j - \bar{x})(y_j - \bar{y})}{n-1} = \frac{\sum_{j=1}^{n} x_j y_j - n\bar{x}\bar{y}}{n-1} \qquad (9.2.15)$$

The covariance takes positive as well as negative values. The coefficients of correlation is the normalized covariance

$$r_{xy} = \frac{S_{xy}}{S_x S_y} = \frac{s_{xy}}{s_x s_y} \qquad (9.2.16)$$

It is independent of the units used in the standard deviations and covariance and varies within the interval $-1 \leq r_{xy} \leq 1$. If both random variables tend to rise or fall together, the variables are positively correlated. If the value of one variable tends to fall when the other rises, the variables are negatively correlated. If $r_{xy} = \pm 1$, there is a direct or indirect linear dependence of one variable upon the other. If $r_{xy} = 0$, there is no interdependence, no correlation, of the variables; they change in a completely random way.

It is important to realize that the correlation analysis presumes a linear dependence of variables. If they fail to show a linear regression, the coefficient of correlation (9.2.16) underestimates the degree of relationship.

9.3 CURVE FITTING

In linear regression a straight line

$$y = a + bx \qquad (9.3.1a)$$

should fit a sample of n pairs of data (x_j, y_j), $j = 1, 2, \ldots, n$, to satisfy the least-squares condition (9.2.9). The regression coefficients are

$$b = \frac{\sum_{j=1}^{n} x_j y_y - n\bar{x}\bar{y}}{\sum_{j=1}^{n} x_j^2 - n\bar{x}^2} \qquad (9.3.1b)$$

and

$$a = \bar{y} - b\bar{x} \qquad (9.3.1c)$$

where

$$\bar{x} = \frac{1}{n}\sum_{j=1}^{n} x_j \quad \bar{y} = \frac{1}{n}\sum_{j=1}^{n} y_j \qquad (9.3.1d)$$

9.6 THEORY

The coefficient of determination r^2 is given by

$$r^2 = \frac{b^2 \left(\sum_{j=1}^{n} x_j^2 - n\bar{x}^2 \right)^2}{\left(\sum_{j=1}^{n} y_j^2 - n\bar{y}^2 \right)^2} \tag{9.3.1e}$$

In this case of linear regression it equals the square of the coefficient of correlation. The sum S of squares of residuals, i.e., of deviations of y_j from the regression line, is given by

$$S = \sum_{j=1}^{n} (a + bx_j - y_j)^2 \tag{9.3.1f}$$

In the power-curve fit

$$y = ax^b \tag{9.3.2a}$$

the sample must consist of n pairs (x_j, y_j) with only positive data, $x_j > 0$, $y_j > 0$. The regression coefficients a, b and the coefficient of determination r^2 now are

$$b = \frac{\sum_{j=1}^{n} \ln x_j \ln y_j - n\bar{x}\bar{y}}{\sum_{j=1}^{n} (\ln x_j)^2 - n\bar{x}^2} \tag{9.3.2b}$$

and

$$a = \exp(\bar{y} - b\bar{x}) \tag{9.3.2c}$$

where

$$\bar{x} = \frac{1}{n} \sum_{j=1}^{n} \ln x_j \qquad \bar{y} = \frac{1}{n} \sum_{j=1}^{n} \ln y_j \tag{9.3.2d}$$

and

$$r^2 = \frac{b^2 \left[\sum_{j=1}^{n} (\ln x_j)^2 - n\bar{x}^2 \right]^2}{\left[\sum_{j=1}^{n} (\ln y_j)^2 - n\bar{y}^2 \right]^2} \tag{9.3.2e}$$

The sum of squares of residuals is here given by

$$S = \sum_{j=1}^{n} (ax_j^b - y_j)^2 \tag{9.3.2f}$$

In the exponential-curve fit

$$y = ae^{bx} \tag{9.3.3a}$$

the sample must consist of n pairs (x_j, y_j) with $y_j > 0$. The regression coefficients

a, b and the coefficient of determination r^2 are

$$b = \frac{\sum_{j=1}^{n} x_j \ln y_j - n\bar{x}\bar{y}}{\sum_{j=1}^{n} x_j^2 - n\bar{x}^2} \qquad (9.3.3b)$$

and

$$a = \exp(\bar{y} - b\bar{x}) \qquad (9.3.3c)$$

where

$$\bar{x} = \frac{1}{n}\sum_{j=1}^{n} x_j \qquad \bar{y} = \frac{1}{n}\sum_{j=1}^{n} \ln y_j \qquad (9.3.3d)$$

and

$$r^2 = \frac{b^2\left(\sum_{j=1}^{n} x_j^2 - n\bar{x}^2\right)^2}{\left[\sum_{j=1}^{n} (\ln y_j)^2 - n\bar{y}^2\right]^2} \qquad (9.3.3e)$$

The sum of squares of residuals in this case is

$$S = \sum_{j=1}^{n} (ae^{bx_j} - y_j)^2 \qquad (9.3.3f)$$

In the logarithmic-curve fit

$$y = a + b \ln x \qquad (9.3.4a)$$

the sample must consist of n pairs (x_j, y_j) with $x_j > 0$. The regression coefficients a, b and the coefficient of determination r^2 are

$$b = \frac{\sum_{j=1}^{n} y_j \ln x_j - n\bar{x}\bar{y}}{\sum_{j=1}^{n} (\ln x_j)^2 - n\bar{x}^2} \qquad (9.3.4b)$$

$$a = \bar{y} - b\bar{x} \qquad (9.3.4c)$$

where

$$\bar{x} = \frac{1}{n}\sum_{j=1}^{n} \ln x_j \qquad \bar{y} = \frac{1}{n}\sum_{j=1}^{n} y_j \qquad (9.3.4d)$$

and

$$r^2 = \frac{b^2\left[\sum_{j=1}^{n} (\ln x_j)^2 - n\bar{x}^2\right]^2}{\left(\sum_{j=1}^{n} y_j^2 - n\bar{y}^2\right)^2} \qquad (9.3.4e)$$

The sum of squares of residuals here is

$$S = \sum_{j=1}^{n} (a + b \ln x_j - y_j)^2 \qquad (9.3.4f)$$

9.8 THEORY

Program P904 computes the regression coefficients a, b, the coefficient of determination r^2, the sum of squares of residuals S, and the value y on the regression curve for a given x for the linear, power, exponential, and logarithmic curve fit.

In the binomial-power curve fit

$$y = ax^c + bx^d \qquad c,d \text{ given constants} \qquad (9.3.5a)$$

for a sample consisting of n pairs of data (x_j, y_j) the regression coefficients a, b are

$$b = \frac{\left(\sum_{j=1}^{n} x_j^d y_j\right)\left(\sum_{j=1}^{n} x_j^{2c}\right) - \left(\sum_{j=1}^{n} x_j^c y_j\right)\left(\sum_{j=1}^{n} x_j^{c+d}\right)}{\left(\sum_{j=1}^{n} x_j^{2c}\right)\left(\sum_{j=1}^{n} x_j^{2d}\right) - \left(\sum_{j=1}^{n} x_j^{c+d}\right)^2} \qquad (9.3.5b)$$

$$a = \frac{\sum_{j=1}^{n} x_j^c y_j - b \sum_{j=1}^{n} x_j^{c+d}}{\sum_{j=1}^{n} x_j^{2c}} \qquad (9.3.5c)$$

The quality of fit here is measured by the sum of squares of residuals

$$S = \sum_{j=1}^{n} (ax_j^c + bx_j^d - y_j)^2 \qquad (9.3.5d)$$

Program P905 computes the regression coefficients a and b, the sum of squares of residuals S, and the value of y on the regression curve for the binomial-power curve fit.

In the parabolic curve fit

$$y = a + bx + cx^2 \qquad (9.3.6a)$$

for a sample consisting of n pairs of data (x_j, y_j) the regression coefficients are

$$c = \frac{d - e}{\left[f\left(n \sum_{j=1}^{n} x_j^4\right) - \left(\sum_{j=1}^{n} x_j^2\right)^2\right] - g^2}$$

$$b = \frac{n \sum_{j=1}^{n} x_j y_j - \left(\sum_{j=1}^{n} x_j\right)\left(\sum_{j=1}^{n} y_j\right) - cg}{f} \qquad (9.3.6b)$$

$$a = \frac{\sum_{j=1}^{n} y_j - b \sum_{j=1}^{n} x_j - c \sum_{j=1}^{n} x_j^2}{n}$$

where

$$f = n \sum_{j=1}^{n} x_j^2 - \left(\sum_{j=1}^{n} x_j\right)^2$$

$$g = n \sum_{j=1}^{n} x_j^3 - \left(\sum_{j=1}^{n} x_j\right)\left(\sum_{j=1}^{n} x_j^2\right)$$

$$d = f\left[n \sum_{j=1}^{n} x_j^2 y_j - \left(\sum_{j=1}^{n} x_j^2\right)\left(\sum_{j=1}^{n} y_j\right)\right] \quad (9.3.6c)$$

$$e = g\left[n \sum_{j=1}^{n} x_j y_j - \left(\sum_{j=1}^{n} x_j\right)\left(\sum_{j=1}^{n} y_j\right)\right]$$

The quality of fit here is also measured by the sum of squares of residuals

$$S = \sum_{j=1}^{n} (a + bx_j + cx_j^2 - y_j)^2 \quad (9.3.6d)$$

Program P906 computes the regression coefficients a, b, c, the sum of squares of residuals S, and the value of y on the regression curve for the parabolic curve fit.

9.4 BINOMIAL AND NEGATIVE BINOMIAL DISTRIBUTION

If p is the probability for the occurrence of an event A in a single trial, the probability that the event A occurs in n independent trials exactly x times, $0 \le x \le n$, is determined by the *binomial distribution* (*Bernoulli distribution*)

$$f(x) = \binom{n}{x} p^x (1-p)^{n-x} \quad x = 0, 1, 2, \ldots, n \quad (9.4.1)$$

The mean value m and variance s^2 are given by

$$m = np \qquad s^2 = m(1-p) \quad (9.4.2)$$

The cumulative binomial distribution is defined by

$$P(x) = \sum_{k=0}^{x} f(k) \quad (9.4.3)$$

This distribution can be computed by the recurrence relation

$$f(k+1) = \frac{p}{1-p} \frac{n-k}{k+1} f(k) \quad (9.4.4a)$$

with the initial term

$$f(0) = (1-p)^n \quad (9.4.4b)$$

Program P907 computes the binomial distribution and the cumulative binomial distribution. It uses formulas (9.1.7) for computation of the binomial coefficient.

The probability that the same event A occurs n times in $x + n$ trials is determined by the negative binomial distribution

$$f(x) = \binom{x + n - 1}{n - 1} p^n (1 - p)^x \qquad x = 0, 1, 2, \ldots \qquad (9.4.5)$$

The mean value m and the variance s^2 are given by

$$m = \frac{n(1 - p)}{p} \qquad s^2 = \frac{m}{p} \qquad (9.4.6)$$

The cumulative negative binomial distribution is defined by Eq. (9.4.3), where now $f(k)$ is given by (9.4.5). It is computed by the recurrence relation

$$f(k + 1) = (1 - p) \frac{n + k}{1 + k} f(k) \qquad (9.4.7a)$$

with the initial term

$$f(0) = p^n \qquad (9.4.7b)$$

Program P908 computes the negative binomial distribution and the cumulative negative binomial distribution. It uses formulas (9.1.7) for computation of the binomial coefficient.

9.5 HYPERGEOMETRIC DISTRIBUTION

In a sample of N objects there are M objects with a certain property A. The probability that there will be precisely x objects with the property A in drawing a sample of n objects is determined by the hypergeometric distribution (sampling without replacement)

$$f(x) = \frac{\binom{M}{x}\binom{N - M}{n - x}}{\binom{N}{n}} \qquad x = 0, 1, 2, \ldots, n \qquad (9.5.1)$$

The mean value m and the variance s^2 are given by

$$m = \frac{nM}{N} \qquad s^2 = \frac{nM(N - M)(N - n)}{N^2(N - 1)} \qquad (9.5.2)$$

The cumulative hypergeometric distribution is defined by

$$P(x) = \sum_{k=0}^{x} f(k) \qquad (9.5.3)$$

and computed by the recurrence relation

$$f(k + 1) = \frac{(k - M)(k - n)}{(k + 1)(N - M + 1 - n + k)} f(k) \qquad (9.5.4a)$$

with the initial term

$$f(0) = \frac{\binom{N-M}{n}}{\binom{N}{n}} \qquad (9.5.4b)$$

If N, M, and $N - M$ are large in comparison with n, the hypergeometric distribution can be approximated by the binomial distribution with $p = M/N$.

Program P909 computes the hypergeometric distribution and the cumulative hypergeometric distribution. It uses formulas (9.1.7) for computation of the binomial coefficients.

9.6 POISSON DISTRIBUTION

The Poisson distribution is a limiting case of the binomial distribution when $p \to 0$ and $n \to \infty$ in such a way that the mean value $m = pn$ is approaching a finite value. If the mean value of the probability for the occurrence of an event A is m, the probability that the event occurs exactly x times is determined by the Poisson distribution

$$f(x) = \frac{m^x e^{-m}}{x!} \qquad x = 0, 1, 2, \ldots \qquad (9.6.1)$$

The cumulative Poisson distribution is defined by

$$P(x) = \sum_{k=0}^{x} f(k) \qquad (9.6.2)$$

It is computed by the recurrence relation

$$f(k+1) = \frac{m}{k+1} f(k) \qquad (9.6.3a)$$

with the initial term

$$f(0) = e^{-m} \qquad (9.6.3b)$$

Program P910 computes the Poisson distribution and the cumulative Poisson distribution.

9.7 NORMAL DISTRIBUTION AND INVERSE NORMAL DISTRIBUTION

The normal distribution (*gaussian distribution*) is a limiting case of the binomial distribution when n, $n - x$, and x are all very large numbers. While the binomial, negative binomial, hypergeometric, and Poisson distributions are discrete distributions in which the independent variable x indicating the number of occurrences of the event under consideration equals zero or a positive integer, the normal distribution is a continuous distribution in which the random variable x takes all values between $-\infty$ and $+\infty$. The standard deviation from the mean value m is denoted by s. The normal distribution is defined by the function

$$f(x) = (2\pi)^{-1/2} s^{-1} \exp \frac{-\tfrac{1}{2}(x-m)^2}{s^2} \tag{9.7.1}$$

The probability that the variable lies within the infinitesimal interval between x and $x + dx$ is proportional to its width $f(x)\,dx$. The probability that the variable has any value between $-\infty$ and $+\infty$ equals 1, that is, certainty, for

$$\int_{-\infty}^{+\infty} f(x)\,dx = 1 \tag{9.7.2}$$

The probability $P(a, b)$ that the variable lies between a and b is expressed by the relation

$$P(a, b) = (2\pi)^{-1/2} s^{-1} \int_a^b \exp\left[\frac{-\tfrac{1}{2}(t-m)^2}{s^2}\right] dt = \pi^{-1/2} \int_A^B e^{-t^2}\,dt$$

$$= \tfrac{1}{2}(\operatorname{erf} B - \operatorname{erf} A) \tag{9.7.3a}$$

where the limits are

$$A = \frac{a-m}{2^{1/2} s} \qquad B = \frac{b-m}{2^{1/2} s} \tag{9.7.3b}$$

The probability that the random variable lies between $-\infty$ and a or between b and $+\infty$ is denoted $Q(a, b)$ and given by

$$Q(a, b) = 1 - P(a, b) \tag{9.7.3c}$$

Instead of the probability $P(a, b)$ the probability integral $\Pr\{X \leq x\}$ may be used; it indicates the probability that the random variable X is less than or equal to an upper limit x and is related to the probability $P(a, b)$ by

$$\Pr\{X \leq x\} = (2\pi)^{-1/2} s^{-1} \int_{-\infty}^x \exp\left[\frac{-\tfrac{1}{2}(t-m)^2}{s^2}\right] dt$$

$$= \tfrac{1}{2}(1 + \operatorname{erf} B) = P(-\infty, x) \tag{9.7.4a}$$

where

$$B = \frac{x-m}{2^{1/2} s} \tag{9.7.4b}$$

The meaning of the standard deviation is elucidated by computing the numerical value of the probability that the variable x lies within the interval $m - s$ and $m + s$:

$$P(m - s, m + s) = 0.683 \tag{9.7.5a}$$

Similarly

$$P(m - 2s, m + 2s) = 0.954 \tag{9.7.5b}$$

$$P(m - 3s, m + 3s) = 0.997 \tag{9.7.5c}$$

Program P911 computes for given mean value m and standard deviation s of the normal distribution the probability $P(a, b)$ that the independent random variable lies within the interval $[a, b]$ and the probability $Q(a, b)$ that the variable lies outside the interval $[a, b]$.

The inverse normal distribution is the function

$$x = x(Q) \tag{9.7.6a}$$

which determines the lower limit of the integral Q,

$$Q = (2\pi)^{-1/2} s^{-1} \int_x^\infty \exp\left[\frac{-\frac{1}{2}(t-m)^2}{s^2}\right] dt = \pi^{-1/2} \int_A^\infty e^{-t^2} dt \tag{9.7.6b}$$

where

$$A = \frac{x - m}{2^{1/2} s} \tag{9.7.6c}$$

The integral Q expresses the probability that the random variable is greater than or equal to its lower limit A. It is approximated by the rational function

$$A = t - \frac{c_0 + c_1 t + c_2 t^2}{1 + d_1 t + d_2 t^2 + d_3 t^3} \tag{9.7.6d}$$

in which c_j, d_j are constants (Ref. 36) and

$$t = (-2 \ln Q)^{1/2} \qquad 0 \le Q \le \tfrac{1}{2} \tag{9.7.6e}$$

The lower limit x is now given by the formula that follows from (9.7.6c)

$$x = 2^{1/2} s A + m \tag{9.7.6f}$$

Program P912 computes for given mean value m and standard deviation s of the inverse normal distribution the lower limit of the random variable for which the probability is Q that the variable is greater than or equal to this limit.

9.8 CHI-SQUARE DISTRIBUTION

If $X_1, X_2, \ldots, X_v$ are independent random variables with normal distribution, mean value $m = 0$, and variance $s^2 = 1$, then

$$X^2 = \sum_{j=1}^v X_j^2$$

follows the chi-square distribution with v degrees of freedom

$$f(x) = \frac{x^{(v/2)-1} e^{-x/2}}{2^{v/2} \Gamma(\tfrac{1}{2} v)} \tag{9.8.1}$$

The probability that $X^2 \le \chi^2$ is given by the integral

$$P(\chi^2 \mid v) = \int_0^{\chi^2} f(x)\, dx \qquad 0 \le \chi^2 < \infty \tag{9.8.2}$$

9.14 THEORY

A comparison of this definition with Eq. (8.12.3) shows that the chi-square distribution can be expressed in terms of the incomplete gamma function

$$P(\chi^2 \mid v) = P(\tfrac{1}{2}v, \tfrac{1}{2}\chi^2) = \frac{\gamma(\tfrac{1}{2}v, \tfrac{1}{2}\chi^2)}{\Gamma(\tfrac{1}{2}v)} \tag{9.8.3}$$

The probability that $X^2 \geq \chi^2$ is determined by the formula

$$Q(\chi^2 \mid v) = 1 - P(\chi^2 \mid v) \qquad 0 \leq \chi^2 < \infty \tag{9.8.4}$$

Program P911 computes the probabilities $P(\chi^2 \mid v)$ and $Q(\chi^2 \mid v)$.

9.9 t DISTRIBUTION

If X is a random variable with normal distribution, mean value $m = 0$, and variance $s^2 = 1$ and χ^2 is a random variable following an independent chi-square distribution with v degrees of freedom, then the distribution of $X/(\chi^2/v)^{1/2}$ is called the t distribution with v degrees of freedom. The probability that the absolute value of $X/(\chi^2/v)^{1/2}$ is less than or equal to a given constant $t > 0$ is

$$A(t \mid v) = \Pr\left\{ \left| \frac{X}{(\chi^2/v)^{1/2}} \right| \leq t \right\}$$

$$= \frac{\Gamma(\tfrac{1}{2} + \tfrac{1}{2}v)}{(\pi v)^{1/2} \Gamma(\tfrac{1}{2}v)} \int_{-t}^{+t} \left(1 + \frac{x^2}{v}\right)^{-(1+v)/2} dx \tag{9.9.1}$$

In another definition of t distribution the lower limit of the integral in (9.9.1) is $-\infty$. This distribution, denoted $A'(t \mid v)$, is related to $A(t \mid v)$ by

$$A' = \tfrac{1}{2}(1 + A) \tag{9.9.2}$$

Program P911 computes the probabilities $A(t \mid v)$ and $A'(t \mid v)$.

9.10 F DISTRIBUTION

Let X_1^2 and X_2^2 be independent random variables following the chi-square distribution with v_1 and v_2 degrees of freedom, respectively. Then the distribution of

$$F = \frac{X_1^2/v_1}{X_2^2/v_2}$$

is given by the F distribution with v_1 and v_2 degrees of freedom and with the distribution function

$$P(F \mid v_1, v_2) = \frac{v_1^{v_1/2} v_2^{v_2/2} \Gamma(\tfrac{1}{2}v_1 + \tfrac{1}{2}v_2)}{\Gamma(\tfrac{1}{2}v_1)\Gamma(\tfrac{1}{2}v_2)} \int_0^F t^{(v_1/2)-1}(v_1 t + v_2)^{-(v_1+v_2)/2} dt \tag{9.10.1}$$

The complementary integral with the lower limit equal to F and upper limit equal to ∞ is denoted by $Q(F\,|\,v_1, v_2)$ and given by

$$Q(F\,|\,v_1, v_2) = 1 - P(F\,|\,v_1, v_2) \qquad F > 0 \qquad (9.10.1b)$$

In the special case $v_1 = 1$ the F distribution is related to the t distribution by

$$P(F\,|\,1, v_2) = A(t\,|\,v_2) \qquad t = F^{1/2} \qquad (9.10.2a)$$

and

$$Q(F\,|\,1, v_2) = 1 - A(t\,|\,v_2) \qquad t = F^{1/2} \qquad (9.10.2b)$$

When $v_1 \to \infty$, the limit of the F distribution is the chi-square distribution

$$\lim_{v_1 \to \infty} Q(F\,|\,v_1, v_2) = P(\chi^2\,|\,v_2) \qquad \chi^2 = \frac{v_2}{F} \qquad (9.10.3)$$

When $v_2 \to \infty$, the limit of the F distribution is another chi-square distribution

$$\lim_{v_2 \to \infty} Q(F\,|\,v_1, v_2) = Q(\chi^2\,|\,v_1) \qquad \chi^2 = v_1 F \qquad (9.10.4)$$

Program P911 computes the integrals $P(F\,|\,v_1, v_2)$ and $Q(F\,|\,v_1, v_2)$. Since the integrand becomes singular at $t = 0$ when $v_1 = 1$, the integral $P(F\,|\,1, v_2)$ is reduced to the t distribution and evaluated by the formula (9.10.2a).

REFERENCES

There are many textbooks on numerical analysis, programming, and advanced analysis. The books most used by the author are listed.

Textbooks and Papers on Numerical Analysis and Programming

1. Ahlberg, J. H., E. N. Nilson, and J. L. Walsh: "The Theory of Splines and Their Application," Academic, New York, 1967.
2. Collatz, L.: "Numerical Treatment of Differential Equations," 3d ed., Springer, New York, 1966.
3. Conte, S. D., and C. de Boor: "Elementary Numerical Analysis," 2d ed., McGraw-Hill, New York, 1972.
4. Dahlquist, G., and Å. Björck: "Numerical Methods," trans. N. Anderson, Prentice-Hall, Englewood Cliffs, N.J., 1974.
5. Gill, S.: *Proc. Camb. Phil. Soc.*, **47**: 96 (1951).
6. Gottfried, B. S.: "Programming with Basic," Schaum's Outline Series, McGraw-Hill, New York, 1975.
7. Hamming, R. W.: "Numerical Methods for Engineers and Scientists," 2d ed., McGraw-Hill, New York, 1973.
8. Hildebrand, F. B.: "Introduction to Numerical Analysis," 2d ed., McGraw-Hill, New York, 1974.
9. Kahan, W.: *Can. Math. Bull.*, **9**: 757 (1966).
10. Lapidus, L., and J. H. Seinfeld: "Numerical Solution of Ordinary Differential Equations," Academic, New York, 1971.
11. McCracken, D. D. and W. S. Dorn: "Numerical Methods and Fortran Programming," Wiley, New York, 1964.
12. Milne, W. E.: "Numerical Calculus," Princeton University Press, Princeton, N.J., 1949.
13. Milne, W. E.: "Numerical Solutions of Differential Equations," Wiley, New York, 1953.
14. Muller, D. E.: *Math. Tables Aids Comput.*, **10**: 208 (1956).
15. Pachner, J.: in A. R. Prasanna, J. V. Narlikar, and C. V. Vishveshwara (eds.), "Gravitation, Quanta and the Universe," *Proc. Einstein Centenary Symp.*, Wiley Eastern, New Delhi, 1980.
16. Peterson, W. W.: "Introduction to Programming Languages," Prentice-Hall, Englewood Cliffs, N.J., 1974.
17. Ralston, A.: "A First Course in Numerical Analysis," McGraw-Hill, New York, 1965.
18. Wilkinson, J. H.: "Rounding Errors in Algebraic Processes," H. M. Stationery Office, London, 1963.

R.2 THEORY

Textbooks and Papers on Advanced Analysis

19. Arfken, G.: "Mathematical Methods for Physicists," 2d ed., Academic, New York, 1970.
20. Courant, R., and D. Hilbert: "Methods of Mathematical Physics," vol. I, Wiley, New York, 1953.
21. Hildebrand, F. B.: "Advanced Calculus for Applications," 2d ed., Prentice-Hall, Englewood Cliffs, N.J., 1976.
22. Jones, L. M.: "An Introduction to Mathematical Methods of Physics," Benjamin-Cummings, Menlo Park, Calif., 1979.
23. Kraut, E. A.: "Fundamentals of Mathematical Physics," McGraw-Hill, New York, 1967.
24. Kreyszig, E.: "Advanced Engineering Mathematics," 4th ed., Wiley, New York, 1979.
25. Morse, P. M., and H. Feshbach: "Methods of Theoretical Physics," 2 vols., McGraw-Hill, New York, 1953.
26. Pachner, J.: *J. Acoust. Soc. Am.*, **23**: 185 (1951).
27. Spiegel, M. S.: "Advanced Mathematics for Engineers and Scientists," Schaum's Outline Series, McGraw-Hill, New York, 1971.
28. Watson, G. N.: "A Treatise on the Theory of Bessel Functions," 2d ed., Cambridge University Press, Cambridge, 1944.
29. Whittaker, E. T., and G. N. Watson: "A Course of Modern Analysis," 4th ed., Cambridge University Press, Cambridge, 1962.

Handbooks

30. Bronshtein, I. N., and K. A. Semedyayev: "A Guide-Book to Mathematics," trans. J. Jaworoski and M. N. Bleicher, Deutsch, Frankfurt, 1971; distributed by Springer, New York.
31. Condon, E. U., and H. Odishaw (eds.): "Handbook of Physics," 2d ed., McGraw-Hill, New York, 1967.
32. Gradshteyn, I. S., and I. M. Ryzhik: "Table of Integrals, Series, and Products," 4th ed. prepared by Yu. V. Geronimus and M. Yu. Tseytlin, trans. ed. by A. Jeffrey, Academic, New York, 1965.
33. Kamke, E.: "Differentialgleichungen, Lösungsmethoden und Lösungen," Bd. I, 9th ed., Teubner, Stuttgart, 1977.
34. Korn, G. A., and T. M. Korn: "Mathematical Handbook for Scientists and Engineers," 2d ed., McGraw-Hill, New York, 1968.
35. Magnus, W., F. Oberhettinger, and R. P. Soni: "Formulas and Theorems for the Special Functions of Mathematical Physics," 3d ed., Springer, New York, 1966.

Tables

36. Abramowitz, M., and I. A. Stegun: "Handbook of Mathematical Functions," Dover, New York, 1965.

37. Jahnke-Emde-Lösch: "Tables of Higher Functions," 6th ed., McGraw-Hill, New York, 1960.
38. Selby, S. S. (ed.): "Standard Mathematical Tables," 19th ed., Chemical Rubber, Cleveland, 1971.

part two
Programs in BASIC

USING THE PROGRAMS

Part Two contains 85 programs plus 3 additional programs in the Appendix, each with a set of examples illustrating the applicability of the numerical method applied in the program. A rudimentary knowledge of programming in BASIC is necessary to enter these programs into a microcomputer's memory. Gottfried [6], among others, gives that information.

The programs were developed on a TRS-80 48K Model III microcomputer, but the statements were carefully chosen to ensure compatibility with other versions of BASIC. For instance, the assignment statements throughout are written with LET though the Model III does not require it. For the same reason the printout is never formatted with the help of the convenient statement PRINT USING. The author believes that the required modification of the programs to other computers has thus been reduced to a minimum. If the programs are run without any modification, the computer itself will indicate whether and where a modification should take place.

All the programs can be stored on two double-density 40-track, $5\frac{1}{4}$-inch diskettes. The programs of Chaps. 1 to 3 and the three programs in the Appendix can be placed on one diskette (15 granules free). The second diskette can hold the programs for Chaps. 4 to 9 (7 granules free).

Once the programs have been typed, tested, and saved on the diskettes, they should be duplicated by an appropriate backup routine. Thereafter, the only system commands required to execute the programs are those associated with loading and running the programs.

The input to each program consists of a set of integer parameters which enable the user to choose between several available options for running the program, plus either a set of numerical data or one or more analytical expressions to be edited on specified lines.

The advantage of using only integer variables as input parameters is obvious: integer variables can take more than two values (more than Yes and No) and, what is also very important, all these data can be typed just by using the right hand at the 12-key section of the keyboard for convenient numeric entry. For instance, the parameter O1 always determines the precision of computation. If O1 = 1, a single-precision computation is used; if O1 = 2, double precision is applied. The chosen value, either 1 or 2 in this case, is to be entered after O1? appears on the screen. In this way the user knows the meaning of the variable O1 when inspecting the program. Only in a few instances does the name of an input variable, integer or noninteger, differ from its name in the program, but this is evident from the INPUT statement. In certain programs some parameters can have more than two values. For example, if the printer is not used, O2 = 0; when it is used but only the final results are to be printed, O2 = 1; if the results are to be printed at each iteration, enter any value greater than 1 (cf. Program P106).

When numerical data are entered, a message appearing on the screen defines the name of a variable and its character (scalar, vector, or matrix). For instance, in program P102 N denotes the number of linear algebraic equations (a scalar), B(I) represents the right-hand side of each equation (a vector), and A(I,J) represents the coefficients on the left-hand side of these equations (a matrix). The question mark in N? indicates that a value for N should be entered. Similarly, when I = 3 J = 4 appears on the screen followed by A(I,J)? on the next line, the question mark asks for the A(3,4) component of a matrix.

The chance of making an error when typing a large number of data with many digits is high. Therefore after the input data have been entered, the program automatically asks whether there is to be a revision. Each datum then appears on the screen as previously typed to be checked out. If the value is correct, enter 0; if it is incorrect enter any number greater than 0 and then the correct value, which should be carefully checked before entering. The printed output of the input data is the last chance for discovering errors in the input.

When a program is to be run using single precision, the input data can include up to 7 significant digits; the computation is also carried out in this precision, but some computers print all single-precision data (not only the output of computation but also the input data) rounded to 6 digits. If the program is then rerun with the 6-digit input data taken from the printout of the previous run, the final results of the computation in some cases may be very different, e.g., for second derivatives computed by the programs of Chap. 2. To prevent such discrepancies the input must either always consist of no more than 6 digits or the printed input data rounded to 6 digits must be corrected manually to represent the true, unrounded input.

In some programs the input data are automatically saved in a file (stored on the diskette with the master program for BASIC) to be used again in the subsequent runs (this save routine will need to be modified slightly for use on microcomputers other than the TRS-80 Model III). For instance, when a system of linear algebraic equations is to be solved by an iterative method (Program P106), the coefficients on the left-hand side are saved in a file. The system can then be solved by another iterative method with the input taken from the file. When the program is to be run in double precision, special precautions must be taken in some microcomputers to guarantee that the data will really be stored in the file in double precision. This situation is discussed in Example 4 to Program P101.

Instead of a set of numerical data the input may be represented by one or more analytical expressions that must be edited on the lines specified by a message on the screen. This is the case when a definite integral with the integrand defined analytically is to be evaluated (Programs P305 to P312), when one or more differential equations are to be integrated (Programs P401 to P407), or when the real roots of a transcendental equation (Programs P604 and P605) or of two nonlinear equations (Programs P607 and P608) are to be found. When one of these programs is run, a message on the screen asks for the button BREAK to be pressed. Then the analytical expressions must be edited on specified lines and the program run again. When the number of differential equations in the set is greater than the number of specified lines (Programs P401 to P407), the line numbers may increase not by 10 but by 5 or 2 (or if necessary by 1). If the program is to be run more than once, it may be convenient to save it with the programmed lines either on the diskette with the master program for BASIC (if space is still available) or on a special diskette replacing the diskette with the programs. *Never save such a special program on the original diskette with programs of Part Two.*

After a program has been entered, it should be tested against possible typing errors. For this purpose run the examples following each program. The first example usually yields an exact solution. For instance, when Simpson's rule is applied for the evaluation of a definite integral (Program P301), the numerical input data lie on a cubic parabola for which Simspon's rule is exact. The other examples should also be run to test the typed program, but they also illustrate the applicability of the numerical method used in the program. These examples simplify the choice between different numerical methods that may be applied for solving the problem under investigation. Remember that there are certain limits in the accuracy of numerical computation which are determined by the method applied in the program. A definite integral with an integrand which is at least twice differentiable and with both derivatives nonsingular over a finite interval of integration can easily be evaluated with any desired accuracy. On the other hand, numerical differentiation is usually a difficult problem. The choice of the numerical method depends on the quality of the input data, and the accuracy of derivatives depends on both the method and the quality of data.

Appendix Programs PA1 and PA2 compute the derived elementary functions using single precision and all elementary functions using double precision. These functions can occur again and again in research problems and should be included in the standard library of every microcomputer capable of solving scientific problems. If they are not there, Programs PA1 and PA2 must be typed, tested, saved on the diskette, and used when needed in the research problem. The third Appendix Program, PA3, should also be carefully typed and stored on the diskette. As described in the Notes following the listing of PA3, one or more sections of PA3 must be used in Programs P305 to P310, P401 to P407, and P605 whenever they are run in double precision and one or more elementary functions appear in the integrand, differential equation, or transcendental equation, respectively. This kind of programming is used in Programs P805 and P807.

CHAPTER 1: LINEAR ALGEBRA I

P101   Condition Indicator for a System of Linear Algebraic Equations

```
10 CLS:PRINT"P101 COMPUTES INDICATORS AND THE CONDITION INDICATOR"
20 PRINT"FOR A SYSTEM OF LINEAR ALGEBRAIC EQUATIONS AND TESTS"
30 PRINT"THE ACCURACY OF SOLUTION IN P102. TO TEST THE IMPACT"
40 PRINT"OF ROUNDOFF ERRORS IN P102 A VECTOR X0 (A SOLUTION OF"
50 PRINT"THE SYSTEM) IS CHOSEN; P101 COMPUTES VECTOR B0 (THE"
60 PRINT"RIGHT-HAND SIDES OF EQUATIONS). IF P102 YIELDS THEN"
70 PRINT"A SOLUTION CLOSE TO CHOSEN X0, THE NUMERICAL ERRORS IN"
80 PRINT"P102 ARE ACCEPTABLE; P102 MAY THEN BE RERUN WITH"
90 PRINT"A GIVEN B(I) TO COMPUTE THE SOLUTION OF THE SYSTEM."
100 PRINT:PRINT"THE INPUT:"
110 PRINT"IF SINGLE PRECISION O1=1, IF DOUBLE O1=2."
120 DEFINT I-O:INPUT"O1";O1:CLS
130 PRINT"IF PRINTER NOT USED O2=0, IF USED O2>0."
140 INPUT"O2";O2:CLS
150 IF O1=2 THEN DEFDBL A-G,P-Z
160 PRINT"IF DATA A(I,J) FROM INPUT O3>0, IF FROM FILE P101A"
170 PRINT"WRITTEN IN THE PRECEDENT RUN OF P101 O3=0."
180 INPUT"O3";O3:CLS
190 PRINT"IF ONLY INDICATORS COMPUTED O4=0, IF ONLY VECTOR"
200 PRINT"B0(I) COMPUTED O4>1, IF INDICATORS AND VECTOR B0(I)"
210 PRINT"COMPUTED O4=1."
220 INPUT"O4";O4: CLS
230 PRINT"NUMBER OF EQUATIONS IN THE SYSTEM: N"
240 INPUT"N";N:CLS
250 PRINT"COEFFICIENTS AT LEFT-HAND SIDES OF EQUATIONS: A(I,J)"
260 PRINT"A CHOSEN SOLUTION OF THE SYSTEM (O4>0): X0(I)"
270 PRINT"P101 WRITES A FILE P101A TO STORE THE COEFFICIENTS"
280 PRINT"A(I,J) AND, OPTIONALLY, THE VECTOR B0(I).":PRINT
290 IF O2=0 THEN GOTO 310
300 LPRINT "P101: O1=";O1;"O2=";O2;"O3=";O3;"O4=";O4;"N=";N
310 DIM A(N,N),B(N),D(N),X(N)
320 PRINT "THE INPUT A(I,J):"
330 IF O2>0 THEN LPRINT "THE INPUT A(I,J):"
340 IF O3=0 THEN GOTO 670
350 REM INPUT A(I,J)
360 FOR I=1 TO N
370 FOR J=1 TO N
380 PRINT "I=";I;"J=";J
390 INPUT"A(I,J)";A(I,J)
400 NEXT J:PRINT
410 NEXT I
420 CLS:PRINT "CHECK A(I,J):
430 FOR I=1 TO N
440 FOR.J=1 TO N
450 PRINT "I=";I;"J=";J;"A(I,J)=";A(I,J)
460 PRINT"IF NO ERROR INPUT 0"
470 PRINT"IF ERROR INPUT ANY NUMBER>0 AND THEN THE CORRECT A(I,J)"
480 INPUT"ERROR";K
490 IF K=0 THEN GOTO 510
500 INPUT"A(I,J)";A(I,J)
510 CLS:NEXT J
520 NEXT I
530 PRINT"COMPUTING"
540 REM WRITING INPUT ON A FILE
550 OPEN "O",1,"P101A"
560 FOR I=1 TO N
```

PROGRAM P101

```
570 FOR J=1 TO N
580 PRINT #1, A(I,J)
590 IF O2>0 THEN LPRINT "I=";I;"J=";J;"A(I,J)=";A(I,J)
600 NEXT J
610 IF O2>0 THEN LPRINT
620 NEXT I
630 CLOSE 1
640 IF O2>0 THEN LPRINT
650 GOTO 780
660 REM READING INPUT FROM A FILE
670 OPEN "I",1,"P101A"
680 FOR I=1 TO N
690 FOR J=1 TO N
700 INPUT #1, A(I,J)
710 PRINT "I=";I;"J=";J;"A(I,J)=";A(I,J)
720 IF O2>0 THEN LPRINT "I=";I;"J=";J;"A(I,J)=";A(I,J)
730 NEXT J
740 IF O2>0 THEN LPRINT
750 NEXT I
760 CLOSE 1
770 IF O2>0 THEN LPRINT
780 IF O4>1 THEN GOTO 1100
790 REM COMPUTATION OF INDICATORS
800 FOR I=1 TO N
810 LET D(I)=0
820 FOR J=1 TO N
830 LET D(I)=D(I)+A(I,J)*A(I,J)
840 NEXT J
850 LET D(I)=SQR(D(I))
860 NEXT I
870 LET H0=1
880 LET M=N-1
890 FOR I=1 TO M
900 LET K=I+1
910 FOR J=K TO N
920 LET H1=0
930 FOR L=1 TO N
940 LET H1=H1+A(I,L)*A(J,L)
950 NEXT L
960 LET H1=1-ABS(H1)/(D(I)*D(J))
970 PRINT "INDICATOR=";H1;" FOR I,J=";I;J
980 IF O2>0 THEN LPRINT "INDICATOR=";H1;" FOR I,J=";I;J
990 IF H1<H0 THEN H0=H1
1000 NEXT J
1010 IF O2>0 THEN LPRINT
1020 NEXT I
1030 LET H2=SQR(2*H0)
1040 PRINT:PRINT "CONDITION INDICATOR=";H0;" ANGLE=";H2
1050 IF O2=0 THEN GOTO 1080
1060 LPRINT "CONDITION INDICATOR=";H0;" ANGLE=";H2
1070 LPRINT
1080 IF O4=0 THEN GOTO 1350
1090 REM TESTING BY CHOOSING VECTOR X0
1100 PRINT:PRINT "CHOSEN VECTOR X0:"
1110 IF O2>0 THEN LPRINT "CHOSEN VECTOR X0:"
1120 FOR I=1 TO N
1130 PRINT "I=";I
1140 INPUT"X(I)";X(I)
1150 IF O2>0 THEN LPRINT "I=";I;" X0(I)=";X(I)
1160 NEXT I
1170 IF O2>0 THEN LPRINT
1180 REM VECTOR B0 STORED AS VECTOR B
1190 PRINT:PRINT "COMPUTED VECTOR B0:"
1200 IF O2>0 THEN LPRINT "COMPUTED VECTOR B0:"
```

```
1210 FOR I=1 TO N
1220 LET B(I)=0
1230 FOR J=1 TO N
1240 LET B(I)=B(I)+A(I,J)*X(J)
1250 NEXT J
1260 PRINT "I=";I;" B0(I)=";B(I)
1270 IF O2>0 THEN LPRINT "I=";I;" B(I)=";B(I)
1280 NEXT I
1290 REM WRITING B0 ON A FILE
1300 OPEN "E",1,"P101A"
1310 FOR I=1 TO N
1320 PRINT #1, B(I)
1330 NEXT I
1340 CLOSE 1
1350 END
```

Examples

1. A system of 4 linear algebraic equations

$$4x_1+3x_2+2x_3+x_4=17$$
$$4x_1+5x_2+3x_3+2x_4=23$$
$$3x_1+5x_2+7x_3+x_4=31$$
$$x_1+2x_2+7x_3+9x_4=79$$

has the condition indicator 0.0310375 and the angle 0.249148. The chosen vector x_0 has the components 1, 2, 3, 7. The components of the corresponding computed vector b_0 are 23, 37, 41, 89.

2. A system of 4 linear algebraic equations

$$x_1+2x_2+3x_3+4x_4=31$$
$$5x_1+4x_2+3x_3+2x_4=25$$
$$7x_1+3x_2+3x_3+x_4=21$$
$$x_1+3x_2+7x_3+9x_4=64$$

has the condition indicator 0.012459 and the angle 0.157854. The chosen vector x_0 has the components 1, 2, 1, 5. The components of the corresponding computed vector b_0 are 28, 26, 25, 59.

3. A system of 4 linear algebraic equations

$$0.4721x_1+0.2352x_2-0.2613x_3+0.8421x_4=-0.2317$$
$$0.2352x_1+0.7411x_2-0.0463x_3+0.1569x_4=0.3219$$
$$-0.2613x_1-0.0463x_2+0.8955x_3+0.1748x_4=0.6217$$
$$0.8421x_1+0.1569x_2+0.1748x_3+0.9841x_4=0.9835$$

has the condition indicator 0.0997845 and the angle 0.446732. The chosen vector x_0 has the components 2, 0.1, 1.5, −1. The components of the corresponding vector b_0 are computed in double precision. Their values are

$\quad$ −0.26633, 0.31816, 0.6412200000000001, 0.97799.

4. The input of the coefficients on the left-hand side of the system of 4 algebraic equations of example 3 is changed. Each coefficient now has 16 decimal places, the first 4 digits are followed by 11 zeros and by 1 in the last sixteenth place. For instance, the input of 0.4721 is changed to

$\quad$ 0.4721000000000001

to guarantee that the file P101A will be written in double precision. Similarly the input of the chosen vector x_0 is now

$\quad$ 2.000000000000001, 0.1, 1.5, −1.

The components of the vector b_0 are

$\quad$ −0.2663299999999995, 0.3181600000000002,

$\quad$ 0.6412199999999996, 0.9779900000000012.

They are then written in double precision on the file P101A. There is no change of the condition indicator and no change of the corresponding angle.

P102 Solution of a System of Linear Algebraic Equations by the Doolittle Method with Partial Pivoting and/or Computation of the Determinant

```
10 CLS:PRINT"P102 SOLVES A SYSTEM OF LINEAR ALGEBRAIC EQUATIONS"
20 PRINT"BY DOOLITTLE METHOD WITH PARTIAL PIVOTING AND/OR"
30 PRINT"COMPUTES THE DETERMINANT OF THE SYSTEM."
40 PRINT:PRINT"THE INPUT:"
50 PRINT"IF SINGLE PRECISION O1=1; IF DOUBLE O1=2."
60 DEFINT I-P:INPUT"O1";O1:CLS
70 PRINT"IF PRINTER NOT USED O2=0, IF USED O2>0."
80 INPUT"O2";O2:CLS
90 IF O1=2 THEN DEFDBL A-H,Q-Z
100 PRINT"COEFFICIENTS AT LEFT-HAND SIDES OF EQUATIONS: A(I,J)"
110 PRINT"RIGHT-HAND SIDES OF EQUATIONS: B(I)":PRINT
120 PRINT"IF DATA A(I,J) AND B(I) FROM INPUT O3>1, IF ONLY"
130 PRINT"B(I) FROM INPUT BUT A(I,J) FROM FILE P101A O3=1,"
140 PRINT"IF A(I,J) AND B(I) FROM FILE P101A O3=0.":PRINT
150 PRINT"IF LU-DECOMPOSITION COMPUTED O4>0, IF NOT O4=0."
160 PRINT"THE LU-MATRICES COMPUTED IN THE FIRST RUN ARE WRITTEN"
170 PRINT"ON FILE P102A. IF THE SYSTEM IS REPEATEDLY SOLVED FOR"
180 PRINT"DIFFERENT B(I), THE LU-MATRICES TAKEN FROM FILE P102A"
190 PRINT"REPLACE NOW A(I,J); CHOOSE O3=1, O4=0.":PRINT
200 INPUT"O3";O3:INPUT"O4";O4:CLS
210 PRINT"IF ONLY THE SYSTEM SOLVED O5>1, IF ONLY THE DETERMINANT"
220 PRINT"COMPUTED O5=0, IF BOTH COMPUTED O5=1."
230 INPUT"O5";O5:CLS
240 IF O4=0 THEN GOTO 270
250 PRINT"IF PARTIAL PIVOTING APPLIED O6>0, IF NOT O6=0.":PRINT
260 INPUT"O6";O6:CLS
270 PRINT"NUMBER OF EQUATIONS IN THE SYSTEM: N"
280 INPUT"N";N: CLS
290 IF O2=0 THEN GOTO 330
300 LPRINT "P102: O1=";O1;"O2=";O2;"O3=";O3
310 IF O4>0 THEN LPRINT "O4=";O4;"O5=";O5;"O6=";O6;"N=";N
320 IF O4=0 THEN LPRINT "O4=";O4;"O5=";O5;"N=";N
330 DIM A(N,N),B(N),C0(N),P(N),X(N)
340 IF O4=0 THEN GOTO 1420
350 REM INPUT A(I,J)
360 PRINT "THE INPUT A(I,J):"
370 IF O2>0 THEN LPRINT "THE INPUT A(I,J):"
380 IF O3<2 THEN GOTO 650
390 FOR I=1 TO N
400 FOR J=1 TO N
410 PRINT "I=";I;"J=";J
420 INPUT"A(I,J)";A(I,J)
430 NEXT J:PRINT
440 NEXT I
450 CLS:PRINT "CHECK A(I,J):"
460 FOR I=1 TO N
470 FOR J=1 TO N
480 PRINT "I=";I;"J=";J;"A(I,J)=";A(I,J)
490 PRINT"IF NO ERROR INPUT 0"
500 PRINT"IF ERROR INPUT ANY NUMBER>0 AND THEN THE CORRECT A(I,J)"
510 INPUT"ERROR";K
520 IF K>0 THEN INPUT "A(I,J)";A(I,J)
530 CLS:NEXT J
540 NEXT I
550 PRINT"COMPUTING"
560 IF O2=0 THEN GOTO 870
570 FOR I=1 TO N
580 FOR J=1 TO N
590 LPRINT "I=";I;"J=";J;"A(I,J)=";A(I,J)
```

P. 11

```
600 NEXT J
610 LPRINT
620 NEXT I
630 GOTO 870
640 REM READING A(I,J) AND, OPTIONALLY, B(I) FROM A FILE
650 OPEN "I", 1, "P101A"
660 FOR I=1 TO N
670 FOR J=1 TO N
680 INPUT #1, A(I,J)
690 PRINT "I=";I"J=";J;"A(I,J)=";A(I,J)
700 IF O2>0 LPRINT "I=";I;"J=";J;"A(I,J)=";A(I,J)
710 NEXT J
720 IF O2>0 THEN LPRINT
730 NEXT I
740 IF O3>0 THEN GOTO 830
750 IF O2>0 THEN LPRINT
760 PRINT "THE INPUT B(I):"
770 IF O2>0 THEN LPRINT "THE INPUT B(I):"
780 FOR I=1 TO N
790 INPUT #1, B(I)
800 PRINT "I=";I;"B(I)=";B(I)
810 IF O2>0 THEN LPRINT "I=";I;"B(I)=";B(I)
820 NEXT I
830 CLOSE 1
840 IF O2>0 THEN LPRINT
850 REM ** LU-DECOMPOSITION **
860 REM LU-MATRICES ARE STORED AS MATRIX A
870 LET P(1)=1
880 IF O6=0 THEN GOTO 940
890 LET C0(1)=A(1,1)
900 FOR J=2 TO N
910 LET P(J)=0
920 LET C0(J)=A(J,1)
930 NEXT J
940 LET I=1
950 IF O6>0 THEN GOSUB 1530
960 REM COMPUTATION OF L(I,1), I=2,...N
970 FOR J=2 TO N
980 LET A(J,1)=A(J,1)/A(1,1)
990 NEXT J
1000 LET I=I+1
1010 IF I=N THEN GOTO 1120
1020 IF O6=0 THEN GOTO 1120
1030 REM PREPARATION FOR PIVOTING
1040 FOR J=I TO N
1050 LET C0(J)=A(J,I)
1060 FOR L=1 TO I-1
1070 LET C0(J)=C0(J)-A(J,L)*A(L,I)
1080 NEXT L
1090 NEXT J
1100 GOSUB 1530
1110 REM COMPUTATION OF U(I,J), J=I,...N
1120 LET K=I-1
1130 FOR J=I TO N
1140 FOR L=1 TO K
1150 LET A(I,J)=A(I,J)-A(I,L)*A(L,J)
1160 NEXT L
1170 IF A(I,I)<>0 THEN GOTO 1200
1180 PRINT "SINGULAR MATRIX: I=";I
1190 STOP
1200 NEXT J
1210 IF I=N THEN GOTO 1310
1220 REM COMPUTATION OF L(J,I), J=I+1,...N
1230 FOR J=I+1 TO N
```

```
1240 FOR L=1 TO K
1250 LET A(J,I)=A(J,I)-A(J,L)*A(L,I)
1260 NEXT L
1270 LET A(J,I)=A(J,I)/A(I,I)
1280 NEXT J
1290 GOTO 1000
1300 REM WRITING LU-MATRICES AND P-VECTOR ON A FILE
1310 OPEN "O", 1, "P102A"
1320 PRINT #1, O6
1330 FOR I=1 TO N
1340 FOR J=1 TO N
1350 PRINT #1, A(I,J)
1360 NEXT J
1370 PRINT #1, P(I)
1380 NEXT I
1390 CLOSE 1
1400 GOTO 1710
1410 REM READING LU-MATRICES AND P-VECTOR FROM A FILE
1420 OPEN "I", 1, "P102A"
1430 INPUT #1, O6
1440 FOR I=1 TO N
1450 FOR J=1 TO N
1460 INPUT #1, A(I,J)
1470 NEXT J
1480 INPUT #1, P(I)
1490 NEXT I
1500 CLOSE 1
1510 GOTO 1710
1520 REM SUBROUTINE FOR PARTIAL PIVOTING
1530 LET K=I+1
1540 LET M=0
1550 LET C=ABS(C0(I))
1560 FOR L=K TO N
1570 IF ABS(C0(L))<=C THEN GOTO 1620
1580 LET C=ABS(C0(L))
1590 LET P(1)=-P(1)
1600 LET P(K)=L
1610 LET M=1
1620 NEXT L
1630 IF M=0 THEN RETURN
1640 LET L=P(K)
1650 FOR J=1 TO N
1660 LET C=A(I,J)
1670 LET A(I,J)=A(L,J)
1680 LET A(L,J)=C
1690 NEXT J
1700 RETURN
1710 IF O5>1 THEN GOTO 1830
1720 REM COMPUTATION OF THE DETERMINANT
1730 LET D=A(1,1)
1740 FOR I=2 TO N
1750 LET D=D*A(I,I)
1760 NEXT I
1770 LET D=D*P(1)
1780 PRINT:PRINT "DETERMINANT=";D
1790 IF O2=0 THEN GOTO 1820
1800 LPRINT "DETERMINANT=";D
1810 LPRINT
1820 IF O5=0 THEN GOTO 2400
1830 IF O3=0 THEN GOTO 2060
1840 REM INPUT B(I)
1850 PRINT:PRINT "THE INPUT B(I):"
1860 IF O2>0 THEN LPRINT "THE INPUT B(I):"
1870 FOR I=1 TO N
```

P. 14   PROGRAM P102

```
1880 PRINT "I=";I
1890 INPUT"B(I)";B(I)
1900 NEXT I
1910 CLS:PRINT "CHECK B(I):"
1920 FOR I=1 TO N
1930 PRINT "I=";I;"B(I)=";B(I)
1940 PRINT "IF NO ERROR INPUT 0"
1950 PRINT"IF ERROR INPUT ANY NUMBER>0 AND THEN THE CORRECT B(I)"
1960 INPUT"ERROR";K
1970 IF K>0 THEN INPUT "B(I)";B(I)
1980 CLS:NEXT I
1990 PRINT"COMPUTING"
2000 IF O2=0 THEN GOTO 2060
2010 FOR I=1 TO N
2020 LPRINT "I=";I;"B(I)=";B(I)
2030 NEXT I
2040 LPRINT
2050 REM ADJUSTMENT OF B(I) TO PIVOTING
2060 IF O6=0 THEN GOTO 2170
2070 FOR I=2 TO N
2080 IF P(I)=0 THEN GOTO 2140
2090 LET J=P(I)
2100 LET K=I-1
2110 LET C=B(K)
2120 LET B(K)=B(J)
2130 LET B(J)=C
2140 NEXT I
2150 REM SOLUTION OF THE FIRST TRIANGULAR SYSTEM
2160 REM VECTOR Y IS STORED AS VECTOR B
2170 FOR I=2 TO N
2180 LET K=I-1
2190 FOR J=1 TO K
2200 LET B(I)=B(I)-A(I,J)*B(J)
2210 NEXT J
2220 NEXT I
2230 REM SOLUTION OF THE SECOND TRIANGULAR SYSTEM
2240 LET X(N)=B(N)/A(N,N)
2250 LET M=N-1
2260 FOR I=M TO 1 STEP -1
2270 LET K=I+1
2280 LET X(I)=B(I)
2290 FOR J=K TO N
2300 LET X(I)=X(I)-A(I,J)*X(J)
2310 NEXT J
2320 LET X(I)=X(I)/A(I,I)
2330 NEXT I
2340 PRINT:PRINT "THE SOLUTION:"
2350 IF O2>0 THEN LPRINT "THE SOLUTION:"
2360 FOR I=1 TO N
2370 PRINT "I=";I;"X(I)=";X(I)
2380 IF O2>0 THEN LPRINT "I=";I;"X(I)=";X(I)
2390 NEXT I
2400 END
```

Examples

1. The system of 4 linear algebraic equations of example 1 to program P101 is solved in single precision first for the computed vector b_0 with the result $x_1=1$, $x_2=2$, $x_3=3$, $x_4=7$, identical with the chosen vector x_0. This indicates that the roundoff errors in P102 do not seriously affect the accuracy of solution. Then the system is solved with the right-hand side given in example 1 to program P101. The solution is $x_1=2.22741$, $x_2=-1.92523$, $x_3=4.01558$, $x_4=5.83489$. The value of the determinant is 321.

2. The system of 4 linear algebraic equations of example 2 to program P101 is solved in single precision for the computed vector b_0 with the result $x_1=x_2=1.5$, $x_3=0.5$, $x_4=5.5$ differing substantially from the chosen vector x_0. The value of the determinant is $5.72205*10^{-6}$.

The system is solved once again in double precision for the computed vector b_0. The unknowns now have an order of magnitude of 10^{16}, and the determinant is $-2.22187*10^{-31}$.

The values of the determinant are of the order of magnitude of roundoff errors of computation in single and double precision. Together with the difference between the chosen vector x_0 and the solution of the system in single and double precision this is a clear indication that the true value of the determinant must be 0 (it is easy to prove analytically that the matrix is singular). Vanishing of the condition indicator is thus a sufficient but not necessary condition for the singularity of a matrix.

3. The system of 4 linear algebraic equations of example 3 to program P101 is solved in double precision for the computed vector b_0. The inputs of A(I,J) and B(I) in P102 read from the file P101A differ from the inputs in P101. The same is true for the chosen vector x_0 in P101 and the corresponding solution in P102. Since the components of the matrix and vectors are given in only 4 or 5 digits, they are written on the file P101A as single-precision quantities and the solution for the vector b_0 is identical with the chosen x_0 when it is rounded to 6 decimal places.

4. The input of the coefficients of the system of 4 linear algebraic equations of example 3 to program P101 is modified as described in example 4. The coefficients and the chosen vector b_0 are written on file P101A and read from it in double precision.

The solution rounded to 15 decimal places is identical with the chosen vector x_0. Thus the roundoff errors of Gaussian elimination do not compromise the accuracy of the solution.

The system with the coefficients $A(I,J)$ read from the file P101 and the input $b_1=-0.2317$, $b_2=0.3219$, $b_3=.6217$, $b_4=0.9835$ (cf. the system of example 3 to program P101) has the solution

$x_1=1.955386499887141$, $x_2=0.1056828793875768$,

$x_3=1.455569628474086$, $x_4=-0.9492390878855224$.

The value of the determinant is -0.2716100252102514. All these digits can be taken into further consideration only if the input data are exact to 16 decimal places.

When the four-digit data of example 3 to program P101 are used as the input in P102 (not read from file P101A), the solution rounded to 15 decimal places is identical with that of the preceding paragraph.

NOTE: It may happen in some very rare cases that pivoting generates a matrix in which a diagonal element vanishes. Then the program prints "A SINGULAR MATRIX" and stops. In such cases try to run the program once again without pivoting (O6=0). This situation occurs when an eigenvector in example 4 to P704 is computed (in P704 the switch to rerun without pivoting is automatic).

P103 Solution of a Tridiagonal System of Linear Algebraic Equations

```
10 CLS:PRINT"P103 SOLVES A TRIDIAGONAL SYSTEM OF LINEAR ALGEBRAIC"
20 PRINT"EQUATIONS BY LU-DECOMPOSITION."
30 PRINT:PRINT"THE INPUT:"
40 PRINT"IF SINGLE PRECISION O1=1, IF DOUBLE O1=2"
50 DEFINT I-O:INPUT"O1";O1:CLS
60 PRINT"IF PRINTER NOT USED O2=0, IF USED O2>0"
70 INPUT"O2";O2:CLS
80 IF O1=2 THEN DEFDBL A-H,P-Z
90 PRINT"COEFFICIENTS AT LEFT-HAND SIDES OF EQUATIONS: A(I,J)"
100 PRINT"THERE ARE ONLY TWO OR THREE NONZERO A(I,J) IN A ROW;"
110 PRINT"A(I,1) LIES BELOW THE DIAGONAL, A(I,2) ON THE"
120 PRINT"DIAGONAL, A(I,3) ABOVE THE DIAGONAL."
130 PRINT"RIGHT-HAND SIDES OF EQUATIONS: B(I)"
140 PRINT"IF DATA A(I,J) AND B(I) FROM INPUT O3>0, IF ONLY"
150 PRINT"B(I) FROM INPUT O3=0.":PRINT
160 PRINT"THE LU-MATRICES COMPUTED IN THE FIRST RUN ARE WRITTEN"
170 PRINT"ON FILE P103A. IF THE SYSTEM IS REPEATEDLY SOLVED FOR"
180 PRINT"DIFFERENT B(I), THE LU-MATRICES TAKEN FROM FILE P103A"
190 PRINT"REPLACE NOW A(I,J); CHOOSE O3=0.":PRINT
200 PRINT"IF ONLY THE SYSTEM SOLVED O5>1, IF ONLY THE"
210 PRINT"DETERMINANT COMPUTED O5=0, IF BOTH COMPUTED O5=1"
220 INPUT"O3";O3:INPUT"O5";O5:CLS
230 PRINT"WARNING: BECAUSE PIVOTING CANNOT BE APPLIED THE PROGRAM"
240 PRINT"MAY STOP (DIVISION BY 0) FOR SOME SYSTEMS. IN THIS CASE"
250 PRINT"SOLVE THE SYSTEM IN P102.":PRINT
260 PRINT"NUMBER OF EQUATIONS IN THE SYSTEM: N"
270 INPUT"N";N:CLS
280 IF O2=0 THEN GOTO 300
290 LPRINT "P103: O1=";O1;"O2=";O2;"O3=";O3;"O5=";O5;"N=";N
300 LET M=N-1
310 DIM A(N,3),B(N),X(N)
320 IF O3=0 THEN GOTO 880
330 PRINT "INPUT A(I,J):"
340 INPUT"A(1,2)";A(1,2):INPUT"A(1,3)";A(1,3):PRINT
350 FOR I=2 TO M
360 PRINT "I=";I
370 INPUT"A(I,1)";A(I,1):INPUT"A(I,2)";A(I,2):INPUT"A(I,3)";A(I,3)
380 PRINT:NEXT I
390 PRINT"I=N=";N
400 INPUT"A(I,1)";A(N,1):INPUT"A(I,2)";A(N,2)
410 CLS:PRINT "CHECK A(I,J):"
420 FOR I=1 TO N
430 IF I=1 THEN GOTO 470
440 IF I=N THEN GOTO 490
450 FOR J=1 TO 3
460 GOTO 500
470 FOR J=2 TO 3
480 GOTO 500
490 FOR J=1 TO 2
500 PRINT "I=";I;"J=";J;"A(I,J)=";A(I,J)
510 PRINT"IF NO ERROR INPUT 0"
520 PRINT"IF ERROR INPUT ANY NUMBER>0 AND THEN THE CORRECT A(I,J)"
530 INPUT"ERROR";K
540 IF K>0 THEN INPUT"A(I,J)";A(I,J)
550 CLS:NEXT J
560 NEXT I
570 PRINT"COMPUTING"
580 IF O2=0 THEN GOTO 730
590 LPRINT "THE INPUT A(I,J):"
600 LPRINT "A(1,2)=";A(1,2)
```

P. 17

```
610 LPRINT "A(1,3)=";A(1,3)
620 FOR I=2 TO M
630 LPRINT "I=";I;"A(I,1)=";A(I,1)
640 LPRINT "I=";I;"A(I,2)=";A(I,2)
650 LPRINT "I=";I;"A(I,3)=";A(I,3)
660 LPRINT
670 NEXT I
680 LPRINT "A(N,1)=";A(N,1)
690 LPRINT"A(N,2)=";A(N,2)
700 LPRINT
710 REM LU-DECOMPOSITION
720 REM LU-MATRICES STORED AS MATRIX A
730 FOR I=2 TO N
740 LET I1=I-1
750 LET A(I,1)=A(I,1)/A(I1,2)
760 LET A(I,2)=A(I,2)-A(I,1)*A(I1,3)
770 NEXT I
780 REM PRINTING LU-MATRICES INTO A FILE
790 OPEN "O",1,"P103A"
800 FOR I=1 TO N
810 FOR J=1 TO 3
820 PRINT #1, A(I,J)
830 NEXT J
840 NEXT I
850 CLOSE 1
860 GOTO 960
870 REM WRITING LU-MATRICES FROM A FILE
880 OPEN "I",1,"P103A"
890 FOR I=1 TO N
900 FOR J=1 TO 3
910 INPUT #1, A(I,J)
920 NEXT J
930 NEXT I
940 CLOSE 1
950 REM COMPUTATION OF THE DETERMINANT
960 IF O5>1 THEN GOTO 1070
970 LET D=A(1,2)
980 FOR I=2 TO N
990 LET D=D*A(I,2)
1000 NEXT I
1010 PRINT:PRINT "DETERMINANT=";D
1020 IF O2=0 THEN GOTO 1050
1030 LPRINT "DETERMINANT=";D
1040 LPRINT
1050 IF O5=0 THEN GOTO 1440
1060 PRINT:PRINT "THE INPUT B(I):"
1070 FOR I=1 TO N
1080 PRINT "I=";I
1090 INPUT"B(I)";B(I)
1100 NEXT I
1110 CLS:PRINT "CHECK B(I):"
1120 FOR I=1 TO N
1130 PRINT "I=";I;"B(I)=";B(I)
1140 PRINT"IF NO ERROR INPUT 0"
1150 PRINT"IF ERROR INPUT ANY NUMBER>0 AND THEN THE CORRECT B(I)"
1160 INPUT"ERROR";K
1170 IF K>0 THEN INPUT"B(I)";B(I)
1180 CLS:NEXT I
1190 PRINT"COMPUTING"
1200 IF O2=0 THEN GOTO 1280
1210 LPRINT "THE INPUT B(I):"
1220 FOR I=1 TO N
1230 LPRINT "I=";I;"B(I)=";B(I)
1240 NEXT I
```

```
1250 LPRINT
1260 REM SOLUTION OF THE FIRST TRIANGULAR SYSTEM
1270 REM VECTOR Y STORED AS VECTOR B
1280 FOR I=2 TO N
1290 LET I1=I-1
1300 LET B(I)=B(I)-A(I,1)*B(I1)
1310 NEXT I
1320 REM SOLUTION OF THE SECOND TRIANGULAR SYSTEM
1330 X(N)=B(N)/A(N,2)
1340 FOR I=M TO 1 STEP -1
1350 LET I1=I+1
1360 LET X(I)=(B(I)-A(I,3)*X(I1))/A(I,2)
1370 NEXT I
1380 PRINT:PRINT "THE SOLUTION:"
1390 IF O2>0 THEN LPRINT "THE SOLUTION:"
1400 FOR I=1 TO N
1410 PRINT "I=";I;"X(I)=";X(I)
1420 IF O2>0 THEN LPRINT "I=";I;"X(I)=";X(I)
1430 NEXT I
1440 END
```

Examples

1. A tridiagonal system of 6 algebraic equations

$$2x_1+x_2=4$$
$$x_1+2x_2+x_3=8$$
$$x_2+2x_3+x_4=12$$
$$x_3+2x_4+x_5=16$$
$$x_4+2x_5+x_6=20$$
$$x_5+2x_6=17$$

is solved in single precision. The determinant equals 7 and the unknowns are

$$x_1=1, \quad x_2=2, \quad x_3=3, \quad x_4=4, \quad x_5=5, \quad x_6=6.$$

2. The components of the matrix on the left-hand side are read from the file P103A. The terms on the right-hand sides of all 6 equations of the preceding example are now changed to 1. The solution is

$$x_1=0.428572, \quad x_2=0.142857, \quad x_3=0.285715,$$
$$x_4=0.285714, \quad x_5=0.142858, \quad x_6=0.428570.$$

3. The problem of the preceding example 2 is now solved in double precision. When all data are taken from the input, the

determinant, rounded to 15 digits, equals to 7 and the unknowns are

$x_1=0.4285714285714285$, $x_2=0.1428571428571424$,
$x_3=0.2857142857142858$, $x_4=0.2857142857142853$,
$x_5=0.1428571428571428$, $x_6=0.4285714285714282$.

4. When the components of the matrix on the left-hand side are read from file P103A, the solution is identical with the results of example 3 only in the first 6 decimal places. To achieve a solution identical in all but the last digits with the results of example 3 the input of the components of the matrix that are stored in file P103A must have 16 digits; that is, the integers are followed by 14 zeros and 1 in the last place.

P104 Solution of a Pentadiagonal System of Linear Algebraic Equations

```
10 CLS:PRINT"P104 SOLVES A FIVE-DIAGONAL SYSTEM OF LINEAR ALGEBRAIC"
20 PRINT"EQUATIONS BY LU-DECOMPOSITION."
30 PRINT:PRINT"THE INPUT:"
40 PRINT"IF SINGLE PRECISION O1=1, IF DOUBLE O1=2."
50 DEFINT I-O:INPUT"O1";O1:CLS
60 PRINT"IF PRINTER NOT USED O2=0, IF USED O2>0."
70 INPUT"O2";O2:CLS
80 IF O1=2 THEN DEFDBL A-H,P-Z
90 PRINT"COEFFICIENTS AT LEFT-HAND SIDES OF EQUATIONS: A(I,J)"
100 PRINT"THERE ARE ONLY THREE TO FIVE NONZERO A(I,J) IN A ROW;"
110 PRINT"A(I,1),A(I,2) LIE BELOW THE DIAGONAL, A(I,3) ON THE"
120 PRINT"DIAGONAL, A(I,4),A(I,5) ABOVE THE DIAGONAL."
130 PRINT"RIGHT-HAND SIDES OF EQUATIONS: B(I)"
140 PRINT"IF DATA A(I,J) AND B(I) FROM INPUT O3>0, IF ONLY"
150 PRINT"B(I) FROM INPUT O3=0.":PRINT
160 PRINT"THE LU-MATRICES COMPUTED IN THE FIRST RUN ARE WRITTEN"
170 PRINT"ON FILE P104A. IF THE SYSTEM IS REPEATEDLY SOLVED FOR"
180 PRINT"DIFFERENT B(I), THE LU-MATRICES TAKEN FROM FILE P104A"
190 PRINT"REPLACE NOW A(I,J); CHOOSE O3=0.":PRINT
200 PRINT"IF ONLY THE SYSTEM SOLVED O5>1, IF ONLY THE"
210 PRINT"DETERMINANT COMPUTED O5=0, IF BOTH COMPUTED O5=1."
220 INPUT"O3";O3:INPUT"O5";O5:CLS
230 PRINT"WARNING: BECAUSE PIVOTING CANNOT BE APPLIED THE PROGRAM"
240 PRINT"MAY STOP (DIVISION BY 0) FOR SOME SYSTEMS. IN THIS CASE"
250 PRINT"SOLVE THE SYSTEM IN P102.":PRINT
260 PRINT"NUMBER OF EQUATIONS IN THE SYSTEM: N>4"
270 INPUT"N";N:CLS
280 IF O2=0 THEN GOTO 300
290 LPRINT "P104: O1=";O1;"O2=";O2;"O3=";O3;"O5=";O5;"N=";N
300 LET M=N-1
310 LET M2=N-2
320 DIM A(N,5),B(N),X(N)
330 IF O3=0 THEN GOTO 1220
340 PRINT "INPUT A(I,J):"
350 INPUT"A(1,3)";A(1,3):INPUT"A(1,4)";A(1,4):INPUT"A(1,5)";A(1,5)
360 PRINT:INPUT"A(2,2)";A(2,2):INPUT"A(2,3)";A(2,3)
370 INPUT"A(2,4)";A(2,4):INPUT"A(2,5)";A(2,5):PRINT
380 FOR I=3 TO M2
390 PRINT "I=";I
400 INPUT"A(I,1)";A(I,1):INPUT"A(I,2)";A(I,2):INPUT"A(I,3)";A(I,3)
410 INPUT"A(I,4)";A(I,4):INPUT"A(I,5)";A(I,5):PRINT
420 NEXT I
430 PRINT "I=N-1=";M
440 INPUT"A(I,1)";A(M,1):INPUT"A(I,2)=";A(M,2)
450 INPUT"A(I,3)=";A(M,3):INPUT"A(I,4)=";A(M,4):PRINT
460 PRINT "I=N=";N
470 INPUT"A(N,1)";A(N,1):INPUT"A(N,2)";A(N,2):INPUT"A(N,3)";A(N,3)
480 CLS:PRINT "CHECK A(I,J):"
490 FOR I=1 TO N
500 IF I=1 THEN GOTO 560
510 IF I=2 THEN GOTO 580
520 IF I=M THEN GOTO 600
530 IF I=N THEN GOTO 620
540 FOR J=1 TO 5
550 GOTO 630
560 FOR J=3 TO 5
570 GOTO 630
580 FOR J=2 TO 5
590 GOTO 630
600 FOR J=1 TO 4
```

P. 21

```
610 GOTO 630
620 FOR J=1 TO 3
630 PRINT "I=";I;"J=";J;"A(I,J)=";A(I,J)
640 PRINT"IF NO ERROR INPUT 0"
650 PRINT"IF ERROR INPUT ANY NUMBER>0 AN THEN THE CORRECT A(I,J)"
660 INPUT"ERROR";K
670 IF K>0 THEN INPUT "A(I,J)";A(I,J)
680 CLS:NEXT J
690 NEXT I
700 PRINT"COMPUTING"
710 IF O2=0 THEN GOTO 1000
720 LPRINT "THE INPUT A(I,J):"
730 LPRINT "A(1,3)=";A(1,3)
740 LPRINT "A(1,4)=";A(1,4)
750 LPRINT "A(1,5)=";A(1,5)
760 LPRINT
770 LPRINT "A(2,2)=";A(2,2)
780 LPRINT "A(2,3)=";A(2,3)
790 LPRINT "A(2,4)=";A(2,4)
800 LPRINT "A(2,5)=";A(2,5)
810 FOR I=3 TO M2
820 LPRINT "I=";I;"A(I,1)=";A(I,1)
830 LPRINT "I=";I;"A(I,2)=";A(I,2)
840 LPRINT "I=";I;"A(I,3)=";A(I,3)
850 LPRINT "I=";I;"A(I,4)=";A(I,4)
860 LPRINT "I=";I;"A(I,5)=";A(I,5)
870 LPRINT
880 NEXT I
890 LPRINT "A(N-1,1)=";A(M,1)
900 LPRINT "A(N-1,2)=";A(M,2)
910 LPRINT "A(N-1,3)=";A(M,3)
920 LPRINT "A(N-1,4)=";A(M,4)
930 LPRINT
940 LPRINT "A(N,1)=";A(N,1)
950 LPRINT "A(N,2)=";A(N,2)
960 LPRINT "A(N,3)=";A(N,3)
970 LPRINT
980 REM LU-DECOMPOSITION
990 REM LU-MATRICES STORED AS MATRIX A
1000 LET A(2,2)=A(2,2)/A(1,3)
1010 LET A(2,3)=A(2,3)-A(2,2)*A(1,4)
1020 LET A(2,4)=A(2,4)-A(2,2)*A(1,5)
1030 FOR I=3 TO N
1040 LET I1=I-1
1050 LET I2=I-2
1060 LET A(I,1)=A(I,1)/A(I2,3)
1070 LET A(I,2)=(A(I,2)-A(I,1)*A(I2,4))/A(I1,3)
1080 LET A(I,3)=A(I,3)-A(I,2)*A(I1,4)-A(I,1)*A(I2,5)
1090 IF I=N THEN GOTO 1110
1100 LET A(I,4)=A(I,4)-A(I,2)*A(I1,5)
1110 NEXT I
1120 REM PRINTING LU-MATRICES INTO A FILE
1130 OPEN "O",1,"P104A"
1140 FOR I=1 TO N
1150 FOR J=1 TO 5
1160 PRINT #1, A(I,J)
1170 NEXT J
1180 NEXT I
1190 CLOSE 1
1200 GOTO 1300
1210 REM WRITING LU-MATRICES FROM A FILE
1220 OPEN "I",1,"P104A"
1230 FOR I=1 TO N
1240 FOR J=1 TO 5
```

```
1250 INPUT #1, A(I,J)
1260 NEXT J
1270 NEXT I
1280 CLOSE 1
1290 REM COMPUTATION OF THE DETERMINANT
1300 IF O5>1 THEN GOTO 1410
1310 LET D=A(1,3)
1320 FOR I=2 TO N
1330 LET D=D*A(I,3)
1340 NEXT I
1350 PRINT:PRINT "DETERMINANT=";D
1360 IF O2=0 THEN GOTO 1390
1370 LPRINT "DETERMINANT=";D
1380 LPRINT
1390 IF O5=0 THEN GOTO 1820
1400 PRINT:PRINT"THE INPUT B(I):"
1410 FOR I=1 TO N
1420 PRINT "I=";I
1430 INPUT"B(I)";B(I)
1440 NEXT I
1450 CLS:PRINT "CHECK B(I)"
1460 FOR I=1 TO N
1470 PRINT "I=";I;"B(I)=";B(I)
1480 PRINT"IF NO ERROR INPUT 0"
1490 PRINT"IF ERROR INPUT ANY NUMBER>0 AND THEN THE CORRECT B(I)"
1500 INPUT"ERROR";K
1510 IF K>0 THEN INPUT"B(I)";B(I)
1520 CLS:NEXT I
1530 PRINT"COMPUTING"
1540 IF O2=0 THEN GOTO 1620
1550 LPRINT "THE INPUT B(I):"
1560 FOR I=1 TO N
1570 LPRINT "I=";I;"B(I)=";B(I)
1580 NEXT I
1590 LPRINT
1600 REM SOLUTION OF THE FIRST TRIANGULAR SYSTEM
1610 REM VECTOR Y STORED AS VECTOR B
1620 B(2)=B(2)-A(2,2)*B(1)
1630 FOR I=3 TO N
1640 LET I1=I-1
1650 LET I2=I-2
1660 LET B(I)=B(I)-A(I,2)*B(I1)-A(I,1)*B(I2)
1670 NEXT I
1680 REM SOLUTION OF THE SECOND TRIANGULAR SYSTEM
1690 X(N)=B(N)/A(N,3)
1700 X(M)=(B(M)-A(M,4)*X(N))/A(M,3)
1710 FOR I=M2 TO 1 STEP -1
1720 LET I1=I+1
1730 LET I2=I+2
1740 LET X(I)=(B(I)-A(I,4)*X(I1)-A(I,5)*X(I2))/A(I,3)
1750 NEXT I
1760 IF O2>0 THEN LPRINT "THE SOLUTION:"
1770 FOR I=1 TO N
1780 PRINT "I=";I;"X(I)=";X(I)
1790 IF O2>0 THEN LPRINT "I=";I;"X(I)=";X(I)
1800 NEXT I
1810 END
```

Examples

1. A five-diagonal system of 6 algebraic equations

$$2x_1+x_2+x_3=4$$
$$2x_1+3x_2+2x_3+2x_4=9$$
$$2x_1+3x_2+4x_3+3x_4+2x_5=14$$
$$3x_2+4x_3+5x_4+4x_5+3x_6=19$$
$$4x_3+5x_4+6x_5+5x_6=20$$
$$5x_4+6x_5+7x_6=18$$

is solved in single precision. The determinant equals 144 and all 6 unknowns equal 1.

2. The components of the matrix on the left-hand side are read from file P104A. The terms on the right-hand sides of all 6 equations of the preceding example are now changed to 1. The solution is

$x_1=0.25$, $x_2=0.277778$, $x_3=0.222224$,
$x_4=-0.388889$, $x_5=-0.0277788$, $x_6=0.444445$.

3. The problem of example 2 is now solved in double precision. When all data are taken from the input, the determinant equals 144 and the unknowns are

$x_1=0.25$, $x_2=0.2777777777777778$,
$x_3=0.2222222222222222$, $x_4=-0.3888888888888889$,
$x_5=-0.02777777777777778$, $x_6=0.4444444444444444$.

4. When the components of the matrix on the left-hand side are read from file P104A, the solution is identical with the results of example 3 in all but the last one or two digits because in this special case the components of the LU matrices expressed in hexadecimal system are exact in single precision. In general, however, the input and output of these components to and from file P104A must have 16 digits to achieve double-precision accuracy of the solution; that is, zeros and 1 in the last (sixteenth) digit must be added to the given values.

P105 Reduction of an Overdetermined System of Linear Algebraic Equations to a Determined System of Normal Equations

```
10 CLS:PRINT"P105 REDUCES AN OVERDETERMINED SYSTEM"
20 PRINT"OF LINEAR ALGEBRAIC EQUATIONS"
30 PRINT"TO A DETERMINED SYSTEM OF NORMAL EQUATIONS."
40 PRINT:PRINT"THE INPUT:"
50 PRINT"IF SINGLE PRECISION O1=1, IF DOUBLE O1=2."
60 DEFINT I-O:INPUT"O1";O1:CLS
70 PRINT"IF PRINTER NOT USED O2=0, IF USED O2>0."
80 INPUT"O2";O2:CLS
90 IF O1=2 THEN DEFDBL A-H,Q-Z
100 PRINT"COEFFICIENTS AT LEFT-HAND SIDES OF EQUATIONS: A(I,J)"
110 PRINT"RIGHT-HAND SIDES OF EQUATIONS: B(I)":PRINT
120 PRINT"NUMBER OF EQUATIONS IN THE SYSTEM: M"
130 PRINT"NUMBER OF UNKNOWNS: N<M"
140 INPUT"M";M:INPUT"N";N:CLS
150 IF O2=0 THEN GOTO 170
160 LPRINT "P105: O1=";O1;"O2=";O2;"M=";M;"N=";N
170 DIM A(M,N),B(M),C(N,N),D(N)
180 REM INPUT A(I,J)
190 PRINT "THE INPUT A(I,J):"
200 IF O2>0 THEN LPRINT "THE INPUT A(I,J):"
210 FOR I=1 TO M
220 FOR J=1 TO N
230 PRINT "I=";I;"J=";J
240 INPUT"A(I,J)";A(I,J)
250 NEXT J:PRINT
260 NEXT I
270 CLS:PRINT "CHECK A(I,J):"
280 FOR I=1 TO M
290 FOR J=1 TO N
300 PRINT "I=";I;"J=";J;"A(I,J)=";A(I,J)
310 PRINT"IF NO ERROR INPUT 0"
320 PRINT"IF ERROR INPUT ANY NUMBER>0 AND THEN THE CORRECT A(I,J)"
330 INPUT"ERROR";K
340 IF K>0 THEN INPUT"A(I,J)";A(I,J)
350 CLS:NEXT J
360 NEXT I
370 PRINT"COMPUTING"
380 IF O2=0 THEN GOTO 450
390 FOR I=1 TO M
400 FOR J=1 TO N
410 LPRINT "I=";I;"J=";J;"A(I,J)=";A(I,J)
420 NEXT J
430 LPRINT
440 NEXT I
450 PRINT:PRINT "THE INPUT B(I):"
460 IF O2>0 THEN LPRINT "THE INPUT B(I):"
470 FOR I=1 TO M
480 PRINT "I=";I
490 INPUT"B(I)";B(I)
500 NEXT I:PRINT
510 CLS:PRINT "CHECK B(I):"
520 FOR I=1 TO M
530 PRINT "I=";I;"B(I)=";B(I)
540 PRINT"IF NO ERROR INPUT 0"
550 PRINT"IF ERROR INPUT ANY NUMBER>0 AND THEN THE CORRECT B(I)"
560 INPUT"ERROR";K
570 IF K>0 THEN INPUT "B(I)";B(I)
580 CLS:NEXT I
590 PRINT"COMPUTING"
```

```
600 IF O2=0 THEN GOTO 650
610 FOR I=1 TO M
620 LPRINT "I=";I;"B(I)=";B(I)
630 NEXT I
640 LPRINT
650 FOR I=1 TO N
660 LET D(I)=0
670 FOR J=1 TO N
680 LET C(I,J)=0
690 FOR L=1 TO M
700 LET C(I,J)=C(I,J)+A(L,I)*A(L,J)
710 NEXT L
720 NEXT J
730 FOR L=1 TO M
740 LET D(I)=D(I)+A(L,I)*B(L)
750 NEXT L
760 NEXT I
770 REM WRITING MATRIX C(I,J) AND VECTOR D(I) ON A FILE
780 OPEN "O",1,"P101A"
790 FOR I=1 TO N
800 FOR J=1 TO N
810 PRINT #1, C(I,J)
820 NEXT J
830 NEXT I
840 FOR I=1 TO N
850 PRINT #1, D(I)
860 NEXT I
870 CLOSE 1
880 PRINT "THE SYSTEM OF NORMAL EQUATIONS IS NOW TO BE"
890 PRINT "SOLVED IN P102 WITH O3=0 AND O6=0. THE SAME"
900 PRINT "PRECISION MUST BE USED IN P105 AND P102."
910 END
```

Example

An overdetermined system of 5 algebraic equations in 2 unknowns

$$x_1 - 3x_2 = -2.1$$
$$x_1 - x_2 = -0.9$$
$$x_1 = -0.6$$
$$x_1 + 2x_2 = 0.6$$
$$x_1 + 3x_2 = 0.9$$

is reduced to a determined system of two normal equations which are solved in P102.

The normal equations taken from P102 are

$$5x_1 + x_1 = -2.1$$
$$x_1 + 23x_2 = 11.1$$

Their solution is

$$x_1 = -0.521053, \quad x_2 = 0.505263.$$

P106 Iterative Methods for a System of Linear Algebraic Equations: Jacobi, Gauss-Seidel, and Successive-Overrelaxation Methods

```
10 CLS:PRINT"P106 SOLVES A SYSTEM OF LINEAR ALGEBRAIC EQUATIONS"
20 PRINT"BY AN ITERATIVE METHOD."
30 PRINT:PRINT"THE INPUT:"
40 PRINT"IF SINGLE PRECISION O1=1, IF DOUBLE O1=2."
50 DEFINT I-O:INPUT"O1";O1:CLS
60 PRINT"IF PRINTER NOT USED O2=0, IF ONLY THE SOLUTION PRINTED"
70 PRINT"O2=1, IF ALSO EACH ITERATION PRINTED O2>1."
80 INPUT"O2";O2:CLS
90 IF O1=2 THEN DEFDBL A-H,T-Z
100 PRINT"COEFFICIENTS AT LEFT-HAND SIDES OF EQUATIONS: A(I,J)"
110 PRINT"RIGHT-HAND SIDES OF EQUATIONS: B(I)":PRINT
120 PRINT"IF DATA A(I,J) AND B(I) FROM INPUT O3>1, IF ONLY"
130 PRINT"B(I) FROM INPUT BUT A(I,J) FROM FILE P101A O3=1,"
140 PRINT"IF A(I,J) AND B(I) FROM FILE P101A O3=0."
150 PRINT"THE A(I,J) AND B(I) ARE WRITTEN ON FILE P101A"
160 PRINT"IN THE FIRST RUN TO SIMPLIFY SUCCESSIVE RUNS FOR"
170 PRINT"OTHER ITERATION METHODS OR WITH DIFFERENT B(I)."
180 INPUT"O3";O3:CLS
190 PRINT"IF JACOBI METHOD O4=1, IF GAUSS-SEIDEL METHOD O4=2,"
200 PRINT"IF SOR METHOD O4=3.":INPUT"O4";O4:CLS
210 LET S=1
220 IF O4<3 THEN GOTO 250
230 PRINT"RELAXATION PARAMETER: S"
240 INPUT"S";S:CLS
250 PRINT"MAXIMUM RELATIVE ERROR OF SOLUTION: R"
260 INPUT"R";R:CLS
270 PRINT"NUMBER OF EQUATIONS IN THE SYSTEM: N"
280 PRINT"MAXIMUM NUMBER OF ITERATIONS: M"
290 INPUT"N";N:INPUT"M";M:CLS
300 IF O2=0 THEN GOTO 330
310 LPRINT "P106: O1=";O1;"O2=";O2;"O3=";O3;"O4=";O4
320 LPRINT "S=";S;"R=";R;"N=";N;"M=";M
330 DIM A(N,N),B(N),X(N),X0(N)
340 IF O3=0 THEN GOTO 830
350 IF O3=1 THEN GOTO 620
360 REM INPUT A(I,J)
370 PRINT "THE INPUT A(I,J):"
380 IF O2>0 THEN LPRINT "THE INPUT A(I,J):"
390 FOR I=1 TO N
400 FOR J=1 TO N
410 PRINT "I=";I;"J=";J
420 INPUT"A(I,J)";A(I,J)
430 NEXT J:PRINT
440 NEXT I
450 CLS:PRINT "CHECK A(I,J):"
460 FOR I=1 TO N
470 FOR J=1 TO N
480 PRINT "I=";I;"J=";J;"A(I,J)=";A(I,J)
490 PRINT"IF NO ERROR INPUT 0"
500 PRINT"IF ERROR INPUT ANY NUMBER>0 AND THEN THE CORRECT A(I,J)"
510 INPUT"ERROR";K
520 IF K>0 THEN INPUT"A(I,J)";A(I,J)
530 CLS:NEXT J
540 NEXT I
550 IF O2=0 THEN GOTO 620
560 FOR I=1 TO N
570 FOR J=1 TO N
580 LPRINT "I=";I;"J=";J;"A(I,J)=";A(I,J)
590 NEXT J
```

```
600 LPRINT
610 NEXT I
620 PRINT:PRINT "THE INPUT B(I):"
630 IF O2>0 THEN LPRINT "THE INPUT B(I):"
640 FOR I=1 TO N
650 PRINT "I=";I
660 INPUT"B(I)";B(I)
670 NEXT I
680 CLS:PRINT "CHECK B(I):"
690 FOR I=1 TO N
700 PRINT "I=";I;"B(I)=";B(I)
710 PRINT"IF NO ERROR INPUT 0"
720 PRINT"IF ERROR INPUT ANY NUMBER>0 AND THEN THE CORRECT B(I)"
730 INPUT"ERROR";K
740 IF K>0 THEN INPUT"B(I)";B(I)
750 CLS:NEXT I
760 PRINT"COMPUTING"
770 IF O2=0 THEN GOTO 820
780 FOR I=1 TO N
790 LPRINT "I=";I;"B(I)=";B(I)
800 NEXT I
810 LPRINT
820 IF O3>1 THEN GOTO 1050
830 IF O2=0 THEN GOTO 860
840 LPRINT "INPUT TAKEN FROM A FILE"
850 LPRINT
860 OPEN "I",1,"P101A"
870 IF O2>0 THEN LPRINT "THE INPUT A(I,J):"
880 FOR I=1 TO N
890 FOR J=1 TO N
900 INPUT #1, A(I,J)
910 IF O2>0 THEN LPRINT "I=";I;"J=";J;"A(I,J)=";A(I,J)
920 NEXT J
930 IF O2>0 THEN LPRINT
940 NEXT I
950 IF O3=1 THEN GOTO 1010
960 IF O2>0 THEN LPRINT "THE INPUT B(I):"
970 FOR I=1 TO N
980 INPUT #1, B(I)
990 IF O2>0 THEN LPRINT "I=";I;"B(I)=";B(I)
1000 NEXT I
1010 CLOSE 1
1020 IF O2>0 THEN LPRINT
1030 REM WRITING A(I,J) AND B(I) ON A FILE
1040 IF O3=0 THEN GOTO 1220
1050 OPEN "O",1,"P101A"
1060 FOR I=1 TO N
1070 FOR J=1 TO N
1080 PRINT #1, A(I,J)
1090 NEXT J
1100 NEXT I
1110 FOR I=1 TO N
1120 PRINT #1, B(I)
1130 NEXT I
1140 CLOSE 1
1150 REM INPUT OF THE INITIAL ESTIMATE X0(I)
1160 PRINT "INPUT OF THE INITIAL ESTIMATE X0(I):"
1170 FOR I=1 TO N
1180 PRINT "I=";I
1190 INPUT"X0(I)";X0(I)
1200 NEXT I
1210 REM PREPARATIONS FOR THE ITERATION PROCESS
1220 IF O2=0 THEN GOTO 1270
1230 IF O4=1 THEN LPRINT "JACOBI METHOD"
```

```
1240 IF O4=2 THEN LPRINT "GAUSS-SEIDEL METHOD"
1250 IF O4=3 THEN LPRINT "SOR METHOD"
1260 LPRINT
1270 IF O4<3 THEN GOTO 1320
1280 FOR I=1 TO N
1290 LET A(I,I)=A(I,I)/S
1300 NEXT I
1310 LET S=1-S
1320 FOR I=1 TO N
1330 FOR J=1 TO N
1340 IF J=I THEN GOTO 1360
1350 LET A(I,J)=A(I,J)/A(I,I)
1360 NEXT J
1370 LET B(I)=B(I)/A(I,I)
1380 NEXT I
1390 LET K=0
1400 REM ITERATION STARTS
1410 LET R0=0
1420 FOR I=1 TO N
1430 IF O4=3 THEN X(I)=S*X(I)+B(I)
1440 IF O4<3 THEN X(I)=B(I)
1450 FOR J=1 TO N
1460 IF J=I THEN GOTO 1560
1470 IF O4>1 THEN GOTO 1520
1480 REM JACOBI METHOD
1490 LET X(I)=X(I)-A(I,J)*X0(J)
1500 GOTO 1560
1510 REM GAUSS-SEIDEL AND SOR METHODS
1520 IF J>I THEN GOTO 1550
1530 LET X(I)=X(I)-A(I,J)*X(J)
1540 GOTO 1560
1550 LET X(I)=X(I)-A(I,J)*X0(J)
1560 NEXT J
1570 IF X(I)=0 THEN GOTO 1600
1580 LET R1=ABS((X(I)-X0(I))/X(I))
1590 IF R1>R0 THEN R0=R1
1600 NEXT I
1610 LET K=K+1
1620 PRINT "NUMBER OF ITERATIONS=";K
1630 IF O2>1 THEN LPRINT "NUMBER OF ITERATIONS=";K
1640 FOR I=1 TO N
1650 LET X0(I)=X(I)
1660 PRINT "I=";I;"X(I)=";X(I)
1670 IF O2>1 THEN LPRINT "I=";I;"X(I)=";X(I)
1680 NEXT I
1690 IF O2>1 THEN LPRINT
1700 IF K=M THEN GOTO 1740
1710 IF R0<R THEN GOTO 1740
1720 GOTO 1410
1730 REM ITERATION ENDS
1740 PRINT:PRINT "THE SOLUTION:"
1750 PRINT "NUMBER OF ITERATIONS=";K
1760 FOR I=1 TO N
1770 PRINT "I=";I;"X(I)=";X(I)
1780 NEXT I
1790 IF O2=0 THEN GOTO 1850
1800 LPRINT "THE SOLUTION:"
1810 LPRINT "NUMBER OF ITERATIONS=";K
1820 FOR I=1 TO N
1830 LPRINT "I=";I;"X(I)=";X(I)
1840 NEXT I
1850 END
```

PROGRAM P106

Examples

1. A system of 4 algebraic equations
$$4x_1 - x_2 - x_3 = 1$$
$$-x_1 + 4x_2 - x_4 = 2$$
$$-x_1 + 4x_3 - x_4 = 0$$
$$-x_2 - x_3 + 4x_4 = 1$$
is solved by the Jacobi iterative method. The initial estimate is 0 for all 4 unknowns, maximum relative error of solution $1*10^{-6}$. After 21 iterations the exact solution is reached
$$x_1 = 0.5, \quad x_2 = 0.75, \quad x_3 = 0.25, \quad x_4 = 0.5.$$

2. The same system of 4 algebraic equations is solved by the Gauss-Seidel iterative method with the same initial estimate and the same maximum relative error of solution as in example 1. After 12 iterations the same exact solution is reached.

3. The same system of 4 algebraic equations is solved by the SOR iterative method with the relaxation parameter 1.0718, with the same initial estimate and the same maximum relative error of solution as in example 1. After 8 iterations the same exact solution is reached.

CHAPTER 2: INTERPOLATION, APPROXIMATION, AND NUMERICAL DIFFERENTIATION

P201 Lagrange Interpolation

```
10 CLS:PRINT"P201 COMPUTES Y(X) OF A FUNCTION DEFINED BY A SET"
20 PRINT"OF N POINTS BY LAGRANGE INTERPOLATION."
30 PRINT:PRINT"THE INPUT:"
40 PRINT"IF SINGLE PRECISION O1=1, IF DOUBLE O1=2."
50 DEFINT I-O:INPUT"O1";O1:CLS
60 PRINT"IF PRINTER NOT USED O2=0, IF USED O2>0."
70 INPUT"O2";O2:CLS
80 IF O1=2 THEN DEFDBL C,F,X,Y
90 PRINT"ABSCISSAS OF POINTS DEFINING THE FUNCTION: X0(I)"
100 PRINT"ORDINATES OF POINTS DEFINING THE FUNCTION: F(I)":PRINT
110 PRINT"IF THE ABSCISSAS OF POINTS DEFINING THE FUNCTION ARE"
120 PRINT"CHEBYSHEV ABSCISSAS O5>0, IF THE ABSCISSAS ARE"
130 PRINT"GIVEN BY THE INPUT O5=0."
140 INPUT"O5";O5:CLS
150 PRINT"NUMBER OF POINTS DEFINING THE FUNCTION: N"
160 INPUT"N";N:CLS
170 IF O2=0 THEN GOTO 190
180 LPRINT "P201: O1=";O1;"O2=";O2;"O5=";O5;"N=";N
190 LET O3=2
200 DIM X0(N),F(N),C(N),T(N)
210 IF O5=0 THEN GOTO 380
220 PRINT "CHEBYSHEV ABSCISSAS COMPUTED ON THE INTERVAL (A,B)"
230 IF O2>0 THEN LPRINT "CHEBYSHEV ABSCISSAS ON (A,B)"
240 INPUT"A";A:INPUT"B";B:CLS
250 IF O2>0 THEN LPRINT "A=";A;"B=";B
260 LET P=(B+A)/2
270 LET Q=(B-A)/2
280 LET S=1.570796/N
290 LET M=1+2*N
300 FOR I=1 TO N
310 LET T(I)=P+Q*COS(S*(M-2*I))
320 PRINT "I=";I;"X0(I)=";T(I)
330 IF O2>0 THEN LPRINT "I=";I;"X0(I)=";T(I)
340 NEXT I
350 IF O2>0 THEN LPRINT
360 PRINT "ROUNDOFF X0(I) AND FIND THE F(I) IN THE TABLE"
370 GOTO 410
380 IF O3=1 THEN GOTO 520
390 IF O3=-1 THEN GOTO 630
400 PRINT "INPUT OF ALL POINTS:"
410 IF O2>0 THEN LPRINT "THE INPUT:"
420 FOR I=1 TO N
430 PRINT "I=";I
440 INPUT"X0(I)";X0(I):INPUT"F(I)";F(I)
450 IF O2>0 THEN LPRINT "I=";I;"X0(I)=";X0(I);"F(I)=";F(I)
460 NEXT I
470 IF O2=0 THEN GOTO 740
480 LPRINT
490 LPRINT "THE INTERPOLATION:"
500 GOTO 740
510 REM INPUT OF ONE POINT ON THE RIGHT
520 FOR I=2 TO N
530 LET J=I-1
540 LET X0(J)=X0(I)
550 LET F(J)=F(I)
560 NEXT I
570 INPUT"X0(N)";X0(N):INPUT"F(N)";F(N)
580 IF O2=0 THEN GOTO 740
```

PROGRAM P201

```
590 LPRINT "X0(N)=";X0(N);"F(N)=";F(N)
600 LPRINT
610 GOTO 740
620 REM INPUT OF ONE POINT ON THE LEFT
630 LET K=N-1
640 FOR I=K TO 1 STEP -1
650 LET J=I+1
660 LET X0(J)=X0(I)
670 LET F(J)=F(I)
680 NEXT I
690 INPUT"X0(1)";X0(1):INPUT"F(1)";F(1)
700 IF O2=0 THEN GOTO 740
710 LPRINT
720 LPRINT "X0(1)=";X0(1);"F(1)=";F(1)
730 REM COMPUTATION OF C(I)
740 FOR J=1 TO N
750 LET C0=1
760 FOR I=1 TO N
770 IF I=J THEN GOTO 790
780 LET C0=C0*(X0(J)-X0(I))
790 NEXT I
800 LET C(J)=F(J)/C0
810 NEXT J
820 CLS:PRINT "THE INTERPOLATION:"
830 INPUT"X";X
840 LET Y0=1
850 FOR I=1 TO N
860 LET X1=X-X0(I)
870 IF X1=0 THEN GOTO 890
880 LET Y0=Y0*X1
890 NEXT I
900 LET K=0
910 LET Y=0
920 FOR I=1 TO N
930 LET X1=X-X0(I)
940 IF X1=0 THEN GOTO 970
950 LET Y=Y+C(I)/X1
960 GOTO 990
970 LET Y1=C(I)
980 LET K=1
990 NEXT I
1000 IF K=1 THEN Y=Y1
1010 LET Y=Y*Y0
1020 CLS:PRINT "X=";X;"Y(X)=";Y
1030 IF O2>0 THEN LPRINT "X=";X;"Y(X)=";Y
1040 IF O2>0 THEN LPRINT
1050 PRINT"TO STOP PRESS BREAK"
1060 IF O5>0 THEN GOTO 830
1070 PRINT"IF NO NEW POINTS DEFINING THE FUNCTION O3=0"
1080 PRINT"IF ONE NEW POINT ON THE RIGHT O3=1"
1090 PRINT"IF ONE NEW POINT ON THE LEFT O3=-1"
1100 PRINT"IF ALL POINTS ARE NEW O3>1"
1110 INPUT"O3";O3
1120 IF O3=0 THEN GOTO 820
1130 GOTO 380
1140 END
```

Examples

1. The function $y=x^3$ is given by 6 points
    (1,1), (1.5,3.375), (2,8),
    (3.5,42.875), (5.5,166.375), (6,216).
Some results of interpolation in single precision:
    x=2.5,   y=15.625.
    x=4.5,   y=91.125.

2. The function $y=e^x$ is given by 6 points
    (0.95,2.58571), (0.96,2.61170), (0.965,2.62479),
    (0.985,2.67781), (0.99,2.69123), (1,2.71828).
Some results of interpolation in single precision:
    x=0.955, y=2.59868.
    x=0.975, y=2.65117.
    x=0.990, y=2.69123.

3. The function $y=e^x$ is given by 6 points with Chebyshev abscissas on the interval $0.95 \leq x \leq 1$:
    (0.950852,2.58791), (0.957322,2.60471), (0.968530,2.63407),
    (0.981471,2.66838), (0.992678,2.69845), (0.999148,2.71597).
Some results of interpolation in single precision:
    x=0.955, y=2.59867.
    x=0.975, y=2.65117.
    x=0.990, y=2.69123.

4. The function $y=e^x$ is given by 6 points
    (0.95,2.585709659315846), (0.96,2.611696473423118),
    (0.965,2.624787656474575), (0.985,2.677811884421050),
    (0.99,2.691234472349262), (1,2.718281828459045).
Some results of interpolation in double precision:
    x=0.955, y=2.598670582919564.
    x=0.975, y=2.651167210982659.
    x=0.990, y=2.691234472349262.

P202 Lagrange Interpolation with Equally Spaced Abscissas

```
10 CLS:PRINT"P202 COMPUTES Y(X) OF A FUNCTION DEFINED BY A SET"
20 PRINT"OF N EQUALLY SPACED POINTS BY LAGRANGE INTERPOLATION."
30 PRINT:PRINT"THE INPUT:"
40 PRINT"IF SINGLE PRECISION O1=1, IF DOUBLE O1=2."
50 DEFINT I-O:INPUT"O1";O1:CLS
60 PRINT"IF PRINTER NOT USED O2=0, IF USED O2>0."
70 INPUT"O2";O2:CLS
80 IF O1=2 THEN DEFDBL A-H,X,Y
90 PRINT"NUMBER OF POINTS DEFINING THE FUNCTION: N"
100 INPUT"N";N:CLS
110 PRINT"INTERVAL COVERED BY N EQUALLY SPACED POINTS: (A,B)"
120 INPUT"A";A:INPUT"B";B:CLS
130 IF O2=0 THEN GOTO 160
140 LPRINT "P202: O1=";O1;"O2=";O2;"N=";N
150 LPRINT "A=";A;"B=";B
160 LET O3=2
170 DIM F(N),C(N)
180 IF O3=0 THEN GOTO 650
190 IF O3=1 THEN GOTO 350
200 IF O3=-1 THEN GOTO 460
210 PRINT"ORDINATES OF POINTS DEFINING THE FUNCTION: F(I)"
220 PRINT "INPUT OF ALL POINTS:"
230 LET H=(B-A)/(N-1)
240 IF O2>0 THEN LPRINT "THE INPUT:"
250 FOR I=1 TO N
260 PRINT "I=";I
270 INPUT"F(I)";F(I)
280 IF O2>0 THEN LPRINT "I=";I;"F(I)=";F(I)
290 NEXT I
300 IF O2=0 THEN GOTO 570
310 LPRINT
320 LPRINT "THE INTERPOLATION:"
330 GOTO 570
340 REM INPUT OF ONE POINT ON THE RIGHT
350 FOR I=2 TO N
360 LET J=I-1
370 LET F(J)=F(I)
380 NEXT I
390 INPUT"F(N)";F(N)
400 LET A=A+H
410 IF O2=0 THEN GOTO 570
420 LPRINT "F(N)=";F(N)
430 LPRINT
440 GOTO 570
450 REM INPUT OF ONE POINT ON THE LEFT
460 LET K=N-1
470 FOR I=K TO 1 STEP -1
480 LET J=I+1
490 LET F(J)=F(I)
500 NEXT I
510 INPUT"F(1)";F(1)
520 LET A=A-H
530 IF O2=0 THEN GOTO 570
540 LPRINT "F(1)=";F(1)
550 LPRINT
560 REM COMPUTATION OF C(I)
570 FOR J=1 TO N
580 LET J0=1
590 FOR I=1 TO N
600 IF I=J THEN GOTO 620
610 LET J0=J0*(J-I)
620 NEXT I
```

```
630 LET C(J)=F(J)/J0
640 NEXT J
650 CLS:PRINT "THE INTERPOLATION:"
660 INPUT"X";X
670 IF O2>0 THEN LPRINT "X=";X
680 LET X=1+(X-A)/H
690 LET Y0=1
700 FOR I=1 TO N
710 LET X1=X-I
720 IF X1=0 THEN GOTO 740
730 LET Y0=Y0*X1
740 NEXT I
750 LET K=0
760 LET Y=0
770 FOR I=1 TO N
780 LET X1=X-I
790 IF X1=0 THEN GOTO 820
800 LET Y=Y+C(I)/X1
810 GOTO 840
820 LET Y1=C(I)
830 LET K=1
840 NEXT I
850 IF K=1 THEN Y=Y1
860 LET Y=Y*Y0
870 PRINT "Y(X)=";Y
880 IF O2>0 THEN LPRINT "Y(X)=";Y
890 IF O2>0 THEN LPRINT
900 PRINT "TO STOP PRESS BREAK"
910 PRINT "IF NO NEW POINTS DEFINING THE FUNCTION O3=0"
920 PRINT "IF ONE NEW POINT ON THE RIGHT O3=1"
930 PRINT "IF ONE NEW POINT ON THE LEFT O3=-1"
940 PRINT "IF ALL POINTS ARE NEW O3>1"
950 INPUT"O3";O3
960 GOTO 180
970 END
```

Examples

1. The function $y=x^3$ is given by 6 points
   (1,1), (2,8), (3,27), (4,64), (5,125), (6,216).
Some results of interpolation in single precision:
   x=2.5,   y=15.625.
   x=4.5,   y=91.125.

2. The function $y=e^x$ is given by 6 points
   (0.95,2.58571), (0.96,2.61170), (0.97,2.63794),
   (0.98,2.66446), (0.99,2.69123), (1,2.71828).

Some results of interpolation in single precision:
        x=0.955,   y=2.59868.
        x=0.975,   y=2.65117.
        x=0.990,   y=2.69123.

   3. The function $y=e^x$ is given by 6 points
        (0.95,2.585709659315846), (0.96,2.611696473423118),
        (0.97,2.637944459354153), (0.98,2.664456241929417),
        (0.99,2.691234472349262), (1,2.718281828459045).
Some results of interpolation in double precision:
        x=0.955,   y=2.598670582919576.
        x=0.975,   y=2.651167210982628.
        x=0.990,   y=2.691234472349262.

P203 Newton Interpolation for the Function and Its First and Second Derivatives

```
10 CLS:PRINT"P203 COMPUTES Y(X) AND THE FIRST AND SECOND DERIVATIVE"
20 PRINT"Y1,Y2 OF A FUNCTION DEFINED BY A SET OF N POINTS"
30 PRINT"BY NEWTON INTERPOLATION."
40 PRINT:PRINT"THE INPUT:"
50 PRINT"IF SINGLE PRECISION O1=1, IF DOUBLE O1=2."
60 DEFINT I-O:INPUT"O1";O1:CLS
70 PRINT"IF PRINTER NOT USED O2=0, IF USED O2>0."
80 PRINT"USE DOUBLE PRECISION IF SECOND DERIVATIVE IS COMPUTED"
90 INPUT"O2";O2:CLS
100 IF O1=2 THEN DEFDBL C-H,U-Z
110 PRINT"IF ONLY FUNCTION COMPUTED O4=0, IF FUNCTION AND ITS"
120 PRINT"FIRST DERIVATIVE COMPUTED O4=1, IF FUNCTION AND ITS"
130 PRINT"FIRST AND SECOND DERIVATIVES COMPUTED O4=2."
140 INPUT"O4";O4:CLS
150 PRINT"ABSCISSAS OF POINTS DEFINING THE FUNCTION: X0(I)"
160 PRINT"ORDINATES OF POINTS DEFINING THE FUNCTION: F(I)":PRINT
170 PRINT"IF THE ABSCISSAS OF POINTS DEFINING THE FUNCTION ARE"
180 PRINT"CHEBYSHEV ABSCISSAS O5>0, IF THE ABSISSAS ARE"
190 PRINT"GIVEN BY THE INPUT O5=0."
200 INPUT"O5";O5:CLS
210 PRINT"NUMBER OF POINTS DEFINING THE FUNCTION: N<7"
220 INPUT"N";N:CLS
230 IF O2=0 THEN GOTO 260
240 LPRINT "P203: O1=";O1;"O2=";O2;"O4=";O4
250 LPRINT "O5=";O5;"N=";N
260 DIM F(6),F0(5),G(6),U(5),X0(6)
270 IF O5=0 THEN GOTO 470
280 PRINT "CHEBYSHEV ABSCISSAS COMPUTED ON THE INTERVAL (A,B)"
290 IF O2>0 THEN LPRINT "CHEBYSHEV ABSCISSAS ON (A,B)"
300 INPUT"A";A:INPUT"B";B:CLS
310 IF O2>0 THEN LPRINT "A=";A;"B=";B
320 LET P=(B+A)/2
330 LET Q=(B-A)/2
340 LET S=1.570796/N
350 LET M=1+2*N
360 FOR I=1 TO N
370 LET T(I)=P+Q*COS(S*(M-2*I))
380 PRINT "I=";I;"X0(I)=";T(I)
390 IF O2>0 THEN LPRINT "I=";I;"X0(I)=";T(I)
400 NEXT I
410 IF O2>0 THEN LPRINT
420 PRINT "ROUNDOFF X0(I) AND FIND THE F(I) IN THE TABLE"
430 GOTO 470
440 IF O3=1 THEN GOTO 580
450 IF O3=-1 THEN GOTO 690
460 PRINT "INPUT OF ALL POINTS:"
470 IF O2>0 THEN LPRINT "THE INPUT:"
480 FOR I=1 TO N
490 PRINT "I=";I
500 INPUT"X0(I)";X0(I):INPUT"F(I)";F(I)
510 IF O2>0 THEN LPRINT "I=";I;"X0(I)=";X0(I);"F(I)=";F(I)
520 NEXT I
530 IF O2=0 THEN GOTO 800
540 LPRINT
550 LPRINT "THE INTERPOLATION:"
560 GOTO 800
570 REM INPUT OF ONE POINT ON THE RIGHT
580 FOR I=2 TO N
590 LET J=I-1
600 LET X0(J)=X0(I)
610 LET F(J)=F(I)
```

PROGRAM P203

```
620 NEXT I
630 INPUT"X0(N)";X0(N):INPUT"F(N)";F(N)
640 IF O2=0 THEN GOTO 800
650 LPRINT "X0(N)=";X0(N);"F(N)=";F(N)
660 LPRINT
670 GOTO 800
680 REM INPUT OF ONE POINT ON THE LEFT
690 LET K=N-1
700 FOR I=K TO 1 STEP -1
710 LET J=I+1
720 LET X0(J)=X0(I)
730 LET F(J)=F(I)
740 NEXT I
750 INPUT"X0(1)";X0(1):INPUT"F(1)";F(1)
760 IF O2=0 THEN GOTO 800
770 LPRINT "X0(1)=";X0(1);"F(1)=";F(1)
780 LPRINT
790 REM COMPUTATION OF DIVIDED DIFFERENCES
800 LET M=N
810 LET K=1
820 LET L0=1
830 FOR I=1 TO N
840 LET G(I)=F(I)
850 NEXT I
860 LET M=M-1
870 FOR I=1 TO M
880 LET J=I+1
890 LET L=I+L0
900 LET G(I)=(G(J)-G(I))/(X0(L)-X0(I))
910 NEXT I
920 LET F0(K)=G(1)
930 LET K=K+1
940 LET L0=L0+1
950 IF M>1 THEN GOTO 860
960 CLS:PRINT "THE INTERPOLATION:"
970 INPUT"X";X
980 CLS:PRINT "X=";X
990 IF O2>0 THEN LPRINT "X =";X
1000 LET M=N-1
1010 FOR I=1 TO M
1020 LET U(I)=X-X0(I)
1030 NEXT I
1040 LET Y=F0(M)*U(M)
1050 LET M=M-1
1060 LET Y=(Y+F0(M))*U(M)
1070 IF M>1 THEN GOTO 1050
1080 LET Y=Y+F(1)
1090 PRINT "Y =";Y
1100 IF O2>0 THEN LPRINT "Y =";Y
1110 IF O4=0 THEN GOTO 1390
1120 REM COMPUTATION OF THE FIRST DERIVATIVE
1130 LET M=N-1
1140 LET Y1=0
1150 LET V1=U(1)+U(2)
1160 LET V2=U(1)*U(2)
1170 IF N<5 THEN GOTO 1200
1180 LET V3=U(3)+U(4)
1190 LET V4=U(3)*U(4)
1200 ON M GOTO 1250,1240,1230,1220
1210 LET Y1=F0(5)*(U(5)*V1*V4+V2*(U(5)*V3+V4))
1220 LET Y1=Y1+F0(4)*(V1*V4+V2*V3)
1230 LET Y1=Y1+F0(3)*(U(3)*V1+V2)
1240 LET Y1=Y1+F0(2)*V1
1250 LET Y1=Y1+F0(1)
```

```
1260 PRINT "Y1=";Y1
1270 IF O2>0 THEN LPRINT "Y1=";Y1
1280 IF O4=1 THEN GOTO 1390
1290 REM COMPUTATION OF THE SECOND DERIVATIVE
1300 LET Y2=0
1310 ON M GOTO 1360,1350,1340,1330
1320 LET Y2=F0(5)*(V1*V4+V2*V3+U(5)*(V1*V3+V2+V3))
1330 LET Y2=Y2+F0(4)*(V1*V3+V2+V4)
1340 LET Y2=Y2+F0(3)*(V1+U(3))
1350 LET Y2=Y2+F0(2)
1360 LET Y2=2*Y2
1370 PRINT "Y2=";Y2
1380 IF O2>0 THEN LPRINT "Y2=";Y2
1390 IF O2>0 THEN LPRINT
1400 PRINT "TO STOP PRESS BREAK"
1410 IF O5>0 THEN GOTO 970
1420 PRINT "IF NO NEW POINTS DEFINING THE FUNCTION O3=0"
1430 PRINT "IF ONE NEW POINT ON THE RIGHT O3=1"
1440 PRINT "IF ONE NEW POINT ON THE LEFT O3=-1"
1450 PRINT "IF ALL POINTS ARE NEW O3>1"
1460 INPUT"O3";O3
1470 IF O3=0 THEN GOTO 960
1480 GOTO 440
1490 END
```

Examples

1. The function $y=x^3$ is given by 6 points

    (1,1), (1.5,3.375), (2,8),

    (3.5,42.875), (5.5,166.375), (6,216).

Some results of interpolation in single precision:

   $x=2.5$, $y=15.625$, $y_1=18.75$, $y_2=15$.

   $x=4.5$, $y=91.125$, $y_1=60.75$, $y_2=27$.

Since the function is a parabola of third degree determined by 6 points, the interpolation of the function and both its derivatives is exact.

2. The function $y=e^x$ is given by 6 points

    (0.95,2.58571), (0.96,2.61170), (0.965,2.62479),

    (0.985,2.67781), (0.99,2.69123), (1,2.71828).

Some results of interpolation in single precision:

   $x=0.955$, $y=2.59867$, $y_1=2.59884$, $y_2=7.74303$.

   $x=0.975$, $y=2.65117$, $y_1=2.65105$, $y_2=2.62134$.

   $x=0.990$, $y=2.69123$, $y_1=2.69071$, $y_2=2.73699$.

3. The function $y=e^x$ is given by 6 points with Chebyshev abscissas on the interval $0.95 \leq x \leq 1$:

(0.950852, 2.58791), (0.957322, 2.60471), (0.968530, 2.63407),
(0.981471, 2.66838), (0.992678, 2.69845), (0.999148, 2.71597).

Some results of interpolation in single precision:

$x=0.955$, $y=2.59867$, $y_1=2.59890$, $y_2=6.01753$.
$x=0.975$, $y=2.65117$, $y_1=2.65128$, $y_2=2.61638$.
$x=0.990$, $y=2.69123$, $y_1=2.69102$, $y_2=2.55821$.

4. The function $y=e^x$ is given by 6 points

(0.95, 2.585709659315846), (0.96, 2.611696473423118),
(0.965, 2.624787656474575), (0.985, 2.677811884421050),
(0.99, 2.691234472349262), (1, 2.718281828459045).

Some results of interpolation in double precision:

$x=0.955$, $y=2.598670582919566$, $y_1=2.59867058291152$,
$y_2=2.5987322908695873$.
$x=0.975$, $y=2.651167210982658$, $y_1=2.6511672109827$,
$y_2=2.6511671430390B1$.
$x=0.990$, $y=2.691234472349262$, $y_1=2.691234472354647$,
$y_2=2.691234474541773$.

Examples 2 to 4 show that the interpolation of the function is exact in single precision and exact to 13 decimal places in double precision; the values of the first derivatives are satisfactory, but the second derivatives may be used only if computed in double precision.

P204 Newton Interpolation with Equally Spaced Abscissas for the Function and Its First and Second Derivatives

```
10 CLS:PRINT"P204 COMPUTES Y(X) AND THE FIRST AND SECOND DERIVATIVE"
20 PRINT"Y1,Y2 OF A FUNCTION DEFINED BY A SET OF N EQUALLY"
30 PRINT"SPACED POINTS BY NEWTON INTERPOLATION."
40 PRINT:PRINT"THE INPUT:"
50 PRINT"IF SINGLE PRECISION O1=1, IF DOUBLE O1=2."
60 PRINT"USE DOUBLE PRECISION IF SECOND DERIVATIVE IS COMPUTED."
70 DEFINT I-O:INPUT"O1";O1:CLS
80 PRINT"IF PRINTER NOT USED O2=0, IF USED O2>0."
90 INPUT"O2";O2:CLS
100 IF O1=2 THEN DEFDBL A-H,P-Z
110 PRINT"IF ONLY FUNCTION COMPUTED O4=0, IF FUNCTION AND ITS"
120 PRINT"FIRST DERIVATIVE COMPUTED O4=1, IF FUNCTION AND ITS"
130 PRINT"FIRST AND SECOND DERIVATIVES COMPUTED O4=2."
140 INPUT"O4";O4:CLS
150 PRINT"NUMBER OF POINTS DEFINING THE FUNCTION: N<7"
160 INPUT"N";N:CLS
170 PRINT"INTERVAL COVERED BY N EQUALLY SPACED POINTS: (A,B)"
180 INPUT"A";A:INPUT"B";B:CLS
190 IF O2=0 THEN GOTO 220
200 LPRINT "P204: O1=";O1;"O2=";O2;"O4=";O4;"N=";N
210 LPRINT "A=";A;"B=";B
220 DIM F(6),F0(5),G(6),U(5)
230 GOTO 290
240 IF O3=0 THEN GOTO 780
250 IF O3=1 THEN GOTO 410
260 IF O3=-1 THEN GOTO 520
270 PRINT"ORDINATES OF POINTS DEFINING THE FUNCTION: F(I)"
280 PRINT "INPUT OF ALL POINTS:"
290 IF O2>0 THEN LPRINT "THE INPUT:"
300 LET H=(B-A)/(N-1)
310 FOR I=1 TO N
320 PRINT "I=";I
330 INPUT"F(I)";F(I)
340 IF O2>0 THEN LPRINT "I=";I;"F(I)=";F(I)
350 NEXT I
360 IF O2=0 THEN GOTO 630
370 LPRINT
380 LPRINT "THE INTERPOLATION:"
390 GOTO 630
400 REM INPUT OF ONE POINT ON THE RIGHT
410 FOR I=2 TO N
420 LET J=I-1
430 LET F(J)=F(I)
440 NEXT I
450 INPUT"F(N)";F(N)
460 LET A=A+H
470 IF O2=0 THEN GOTO 630
480 LPRINT "F(N)=";F(N)
490 LPRINT
500 GOTO 630
510 REM INPUT OF ONE POINT ON THE LEFT
520 LET K=N-1
530 FOR I=K TO 1 STEP -1
540 LET J=I+1
550 LET F(J)=F(I)
560 NEXT I
570 INPUT"F(1)";F(1)
580 LET A=A-H
590 IF O2=0 THEN GOTO 630
600 LPRINT "F(1)=";F(1)
610 LPRINT
```

```
620 REM COMPUTATION OF DIVIDED DIFFERENCES
630 LET M=N
640 LET K=1
650 LET L=1
660 FOR I=1 TO N
670 LET G(I)=F(I)
680 NEXT I
690 LET M=M-1
700 FOR I=1 TO M
710 LET J=I+1
720 LET G(I)=(G(J)-G(I))/L
730 NEXT I
740 LET F0(K)=G(1)
750 LET K=K+1
760 LET L=L+1
770 IF M>1 THEN GOTO 690
780 CLS:PRINT "THE INTERPOLATION:"
790 INPUT"X";X
800 IF O2>0 THEN LPRINT "X =";X
810 LET X=1+(X-A)/H
820 LET M=N-1
830 FOR I=1 TO M
840 LET U(I)=X-I
850 NEXT I
860 LET Y=F0(M)*U(M)
870 LET M=M-1
880 LET Y=(Y+F0(M))*U(M)
890 IF M>1 THEN GOTO 870
900 LET Y=Y+F(1)
910 PRINT "Y =";Y
920 IF O2>0 THEN LPRINT "Y =";Y
930 IF O4=0 THEN GOTO 1210
940 REM COMPUTATION OF THE FIRST DERIVATIVE
950 LET M=N-1
960 LET Y1=0
970 LET V1=U(1)+U(2)
980 LET V2=U(1)*U(2)
990 IF N<5 THEN GOTO 1020
1000 LET V3=U(3)+U(4)
1010 LET V4=U(3)*U(4)
1020 ON M GOTO 1070,1060,1050,1040
1030 LET Y1=F0(5)*(U(5)*V1*V4+V2*(U(5)*V3+V4))
1040 LET Y1=Y1+F0(4)*(V1*V4+V2*V3)
1050 LET Y1=Y1+F0(3)*(U(3)*V1+V2)
1060 LET Y1=Y1+F0(2)*V1
1070 LET Y1=(Y1+F0(1))/H
1080 PRINT "Y1=";Y1
1090 IF O2>0 THEN LPRINT "Y1=";Y1
1100 IF O4=1 THEN GOTO 1210
1110 REM COMPUTATION OF THE SECOND DERIVATIVE
1120 LET Y2=0
1130 ON M GOTO 1180,1170,1160,1150
1140 LET Y2=F0(5)*(V1*V4+V2*V3+U(5)*(V1*V3+V2+V3))
1150 LET Y2=Y2+F0(4)*(V1*V3+V2+V4)
1160 LET Y2=Y2+F0(3)*(V1+U(3))
1170 LET Y2=Y2+F0(2)
1180 LET Y2=2*Y2/(H*H)
1190 PRINT "Y2=";Y2
1200 IF O2>0 THEN LPRINT "Y2=";Y2
1210 IF O2>0 THEN LPRINT
1220 PRINT "IF NO NEW POINTS DEFINING THE FUNCTION O3=0"
1230 PRINT "IF ONE NEW POINT ON THE RIGHT O3=1"
1240 PRINT "IF ONE NEW POINT ON THE LEFT O3=-1"
1250 PRINT "IF ALL POINTS ARE NEW O3>1"
1260 PRINT "TO STOP PRESS BREAK"
```

```
1270 INPUT"O3";O3
1280 IF O3=0 THEN GOTO 780
1290 GOTO 250
1300 END
```

Examples

1. The function $y=x^3$ is given by 6 points
   $(1,1)$, $(2,8)$, $(3,27)$, $(4,64)$, $(5,125)$, $(6,216)$.

Some results of interpolation in single precision:

   $x=2.5$, $y=15.625$, $y_1=18.75$, $y_2=15$.
   $x=4.5$, $y=91.125$, $y_1=60.75$, $y_2=27$.

2. The function $y=e^x$ is given by 6 points
   $(0.95, 2.58571)$, $(0.96, 2.61170)$, $(0.97, 2.63794)$,
   $(0.98, 2.66446)$, $(0.99, 2.69123)$, $(1, 2.71828)$.

Some results of interpolation in single precision:

   $x=0.955$, $y=2.59868$, $y_1=2.59820$, $y_2=2.49369$.
   $x=0.975$, $y=2.65117$, $y_1=2.65215$, $y_2=2.64240$.
   $x=0.990$, $y=2.69123$, $y_1=2.68944$, $y_2=2.65379$.

This interpolation may be considered as satisfactory only for the function and its first derivative, not for the second derivative (to improve it the input must be given in more than 6 digits).

3. The function $y=e^x$ is given by 6 points
   $(0.95, 2.585709659315846)$, $(0.96, 2.611696473423118)$,
   $(0.97, 2.637944459354153)$, $(0.98, 2.664456241929417)$,
   $(0.99, 2.691234472349262)$, $(1, 2.718281828459045)$.

Some results of interpolation in double precision:

   $x=0.955$, $y=2.598670582919576$, $y_1=2.598670582919576$,
      $y_2=2.598670582911064$.
   $x=0.975$, $y=2.651167210982620$, $y_1=2.651167210982533$,
      $y_2=2.651167193218523$.
   $x=0.990$, $y=2.691234472349262$, $y_1=2.691234472358145$,
      $y_2=2.691234474278359$.

This interpolation is very satisfactory for the function and its first derivative; for the second derivative it is accurate in 7 to 8 decimal places.

P205 Hermite Interpolation for the Function and Its First and Second Derivatives

```
10 CLS:PRINT"P205 COMPUTES Y(X) AND THE FIRST AND SECOND DERIVATIVE"
20 PRINT"Y1,Y2 OF A FUNCTION DEFINED AT N POINTS BY FUNCTION"
30 PRINT"AND ITS FIRST DERIVATIVE BY HERMITE INTERPOLATION."
40 PRINT:PRINT"THE INPUT:"
50 PRINT"IF SINGLE PRECISION O1=1, IF DOUBLE O1=2."
60 PRINT"USE DOUBLE PRECISION IF SECOND DERIVATIVE IS COMPUTED."
70 DEFINT I-O:INPUT"O1";O1:CLS
80 PRINT"IF PRINTER NOT USED O2=0, IF USED O2>0."
90 INPUT"O2";O2:CLS
100 IF O1=2 THEN DEFDBL C-H,U-Z
110 PRINT"IF ONLY FUNCTION COMPUTED O4=0, IF FUNCTION AND ITS"
120 PRINT"FIRST DERIVATIVE COMPUTED O4=1, IF FUNCTION AND ITS"
130 PRINT"FIRST AND SECOND DERIVATIVES COMPUTED O4=2."
140 INPUT"O4";O4:CLS
150 PRINT"NUMBER OF POINTS DEFINING THE FUNCTION: N<4"
160 INPUT"N";N:CLS
170 IF O2=0 THEN GOTO 190
180 LPRINT "P205: O1=";O1;"O2=";O2;"O4=";O4;"N=";N
190 DIM F(3),F1(3),F0(5),G(5),U(5),X0(3),X1(6)
200 GOTO 230
210 IF O3=1 THEN GOTO 390
220 IF O3=-1 THEN GOTO 510
230 PRINT"ABSCISSAS OF POINTS DEFINING THE FUNCTION: X0(I)"
240 PRINT"ORDINATES OF POINTS DEFINING THE FUNCTION: F(I)"
250 PRINT"FIRST DERIVATIVES DEFINING THE FUNCTION: F1(I)"
260 PRINT "INPUT OF ALL POINTS:"
270 IF O2>0 THEN LPRINT "THE INPUT:"
280 FOR I=1 TO N
290 PRINT "I=";I
300 INPUT"X0(I)";X0(I):INPUT"F(I)";F(I):INPUT"F1(I)";F1(I)
310 IF O2=0 THEN GOTO 330
320 LPRINT "I=";I;"X0(I)=";X0(I);"F(I)=";F(I);"F1(I)=";F1(I)
330 NEXT I
340 IF O2=0 THEN GOTO 630
350 LPRINT
360 LPRINT "THE INTERPOLATION:"
370 GOTO 630
380 REM INPUT OF ONE POINT ON THE RIGHT
390 FOR I=2 TO N
400 LET J=I-1
410 LET X0(J)=X0(I)
420 LET F(J)=F(I)
430 LET F1(J)=F1(I)
440 NEXT I
450 INPUT"X0(N)";X0(N):INPUT"F(N)";F(N):INPUT"F1(N)";F1(N)
460 IF O2=0 THEN GOTO 630
470 LPRINT "X0(N)=";X0(N);"F(N)=";F(N);"F1(N)=";F1(N)
480 LPRINT
490 GOTO 630
500 REM INPUT OF ONE POINT ON THE LEFT
510 LET K=N-1
520 FOR I=K TO 1 STEP -1
530 LET J=I+1
540 LET X0(J)=X0(I)
550 LET F(J)=F(I)
560 LET F1(J)=F1(I)
570 NEXT I
580 INPUT"X0(1)";X0(1):INPUT"F(1)";F(1):INPUT"F1(1)";F1(1)
590 IF O2=0 THEN GOTO 630
600 LPRINT "X0(1)=";X0(1);"F(1)=";F(1);"F1(1)=";F1(1)
610 LPRINT
```

```
620 REM COMPUTATION OF DIVIDED DIFFERENCES
630 FOR I=1 TO N
640 LET J=2*I-1
650 LET X1(J)=X0(I)
660 LET G(J)=F1(I)
670 LET J=2*I
680 LET K=I+1
690 LET X1(J)=X0(I)
700 IF I=N THEN GOTO 720
710 LET G(J)=(F(K)-F(I))/(X0(K)-X0(I))
720 NEXT I
730 LET F0(1)=F1(1)
740 LET M=2*N-1
750 LET K=2
760 LET L0=2
770 LET M=M-1
780 FOR I=1 TO M
790 LET J=I+1
800 LET L=I+L0
810 LET G(I)=(G(J)-G(I))/(X1(L)-X1(I))
820 NEXT I
830 LET F0(K)=G(1)
840 LET K=K+1
850 LET L0=L0+1
860 IF M>1 THEN GOTO 770
870 CLS:PRINT "THE INTERPOLATION:"
880 INPUT"X";X
890 IF O2>0 THEN LPRINT "X =";X
900 LET M=2*N-1
910 FOR I=1 TO M
920 LET U(I)=X-X1(I)
930 NEXT I
940 LET Y=F0(M)*U(M)
950 LET M=M-1
960 LET Y=(Y+F0(M))*U(M)
970 IF M>1 THEN GOTO 950
980 LET Y=Y+F(1)
990 PRINT "Y =";Y
1000 IF O2>0 THEN LPRINT "Y =";Y
1010 IF O4=0 THEN GOTO 1290
1020 REM COMPUTATION OF THE FIRST DERIVATIVE
1030 LET M=2*N-1
1040 LET Y1=0
1050 LET V1=U(1)+U(2)
1060 LET V2=U(1)*U(2)
1070 IF N<5 THEN GOTO 1100
1080 LET V3=U(3)+U(4)
1090 LET V4=U(3)*U(4)
1100 ON M GOTO 1150,1140,1130,1120
1110 LET Y1=F0(5)*(U(5)*V1*V4+V2*(U(5)*V3+V4))
1120 LET Y1=Y1+F0(4)*(V1*V4+V2*V3)
1130 LET Y1=Y1+F0(3)*(U(3)*V1+V2)
1140 LET Y1=Y1+F0(2)*V1
1150 LET Y1=Y1+F0(1)
1160 PRINT "Y1=";Y1
1170 IF O2>0 THEN LPRINT "Y1=";Y1
1180 IF O4=1 THEN GOTO 1290
1190 REM COMPUTATION OF THE SECOND DERIVATIVE
1200 LET Y2=0
1210 ON M GOTO 1260,1250,1240,1230
1220 LET Y2=F0(5)*(V1*V4+V2*V3+U(5)*(V1*V3+V2+V3))
1230 LET Y2=Y2+F0(4)*(V1*V3+V2+V4)
1240 LET Y2=Y2+F0(3)*(V1+U(3))
1250 LET Y2=Y2+F0(2)
```

```
1260 LET Y2=2*Y2
1270 PRINT "Y2=";Y2
1280 IF O2>0 THEN LPRINT "Y2=";Y2
1290 IF O2>0 THEN LPRINT
1300 PRINT "IF NO NEW POINTS DEFINING THE FUNCTION O3=0"
1310 PRINT "IF ONE NEW POINT ON THE RIGHT O3=1"
1320 PRINT "IF ONE NEW POINT ON THE LEFT O3=-1"
1330 PRINT "IF ALL POINTS ARE NEW O3>1"
1340 PRINT "TO STOP PRESS BREAK"
1350 INPUT"O3";O3
1360 IF O3=0 THEN GOTO 870
1370 GOTO 210
1380 END
```

Examples

1. The function $y=x^3$ is given at 3 points by the function F and its derivative F1

$x_1=1$, $F_1=1$, $F1_1=3$,
$x_2=3$, $F_2=27$, $F1_2=27$,
$x_3=6$, $F_3=216$, $F1_3=108$.

Some results of interpolation in single precision:

$x=2.5$, $y=15.625$, $y_1=18.75$, $y_2=15$.
$x=4.5$, $y=91.125$, $y_1=60.75$, $y_2=27$.

2. The function $y=e^x$ is given at 3 points by the function F and its first derivative F1

$x_1=0.95$, $F_1=F1_1=2.58571$,
$x_2=0.985$, $F_2=F1_2=2.67781$,
$x_3=1$, $F_3=F1_3=2.71828$.

Some results of interpolation in single precision:

$x=0.955$, $y=2.59867$, $y_1=2.59863$, $y_2=2.59171$.
$x=0.975$, $y=2.65117$, $y_1=2.65110$, $y_2=2.65053$.
$x=0.990$, $y=2.69123$, $y_1=2.69128$, $y_2=2.69350$.

The values of the function are exact in 6 digits, the loss of accuracy in the first and second derivatives is acceptable.

3. The function $y=e^x$ is given at 3 points by the function F and its first derivative F1

$x_1=0.95$, $F_1=F1_1=2.585709659315846$,
$x_2=0.985$, $F_2=F1_2=2.67781188442105$,
$x_3=1$, $F_3=F1_3=2.718281828459045$.

Some results of interpolation in double precision:

$x=0.955$, $y=2.598670582919354$, $y_1=2.598669762163584$,
   $y_2=2.598604633657784$.
$x=0.975$, $y=2.651167210982462$, $y_1=2.651168033159792$,
   $y_2=2.651365189652281$.
$x=0.990$, $y=2.691234472349248$, $y_1=2.691232488992467$,
   $y_2=2.691051948799385$.

P206 Hermite Interpolation with Equally Spaced Abscissas for the Function and Its First and Second Derivatives

```
10 CLS:PRINT"P206 COMPUTES Y(X) AND THE FIRST AND SECOND"
20 PRINT"DERIVATIVE Y1,Y2 OF A FUNCTION DEFINED AT N EQUALLY"
30 PRINT"SPACED POINTS BY THE FUNCTION AND ITS FIRST DERIVATIVE"
40 PRINT"BY HERMITE INTERPOLATION."
50 PRINT:PRINT"THE INPUT:"
60 PRINT"IF SINGLE PRECISION O1=1, IF DOUBLE O1=2."
70 PRINT"USE DOUBLE PRECISION IF SECOND DERIVATIVE IS COMPUTED."
80 DEFINT I-O:INPUT"O1";O1:CLS
90 PRINT"IF PRINTER NOT USED O2=0, IF USED O2>0."
100 INPUT"O2";O2:CLS
110 IF O1=2 THEN DEFDBL A-H,U-Z
120 PRINT"IF ONLY FUNCTION COMPUTED O4=0, IF FUNCTION AND ITS"
130 PRINT"FIRST DERIVATIVE COMPUTED O4=1, IF FUNCTION AND ITS"
140 PRINT"FIRST AND SECOND DERIVATIVES COMPUTED O4=2."
150 INPUT"O4";O4:CLS
160 PRINT"NUMBER OF POINTS DEFINING THE FUNCTION: N<4"
170 INPUT"N";N:CLS
180 PRINT"INTERVAL COVERED BY N EQUALLY SPACED POINTS: (A,B)"
190 INPUT"A";A:INPUT"B";B:CLS
200 IF O2=0 THEN GOTO 230
210 LPRINT "P206: O1=";O1;"O2=";O2;"O4=";O4
220 LPRINT "N=";N;"A=";A;"B=";B
230 DIM F(3),F1(3),F0(5),G(5),U(5),X0(3),X1(6)
240 GOTO 270
250 IF O3=1 THEN GOTO 460
260 IF O3=-1 THEN GOTO 600
270 PRINT"ORDINATES OF POINTS DEFINING THE FUNCTION: F(I)"
280 PRINT"FIRST DERIVATIVES DEFINING THE FUNCTION: F1(I)"
290 PRINT "INPUT OF ALL POINTS:"
300 IF O2>0 THEN LPRINT "THE INPUT:"
310 LET H=(B-A)/(N-1)
320 FOR I=1 TO N
330 LET X0(I)=A+H*(I-1)
340 NEXT I
350 FOR I=1 TO N
360 PRINT "I=";I
370 INPUT"F(I)";F(I):INPUT"F1(I)";F1(I)
380 IF O2=0 THEN GOTO 400
390 LPRINT "I=";I;"F(I)=";F(I);"F1(I)=";F1(I)
400 NEXT I
410 IF O2=0 THEN GOTO 730
420 LPRINT
430 LPRINT "THE INTERPOLATION:"
440 GOTO 730
450 REM INPUT OF ONE POINT ON THE RIGHT
460 FOR I=2 TO N
470 LET J=I-1
480 LET X0(J)=X0(I)
490 LET F(J)=F(I)
500 LET F1(J)=F1(I)
510 NEXT I
520 LET M=N-1
530 LET X0(N)=X0(M)+H
540 INPUT"F(N)";F(N):INPUT"F1(N)";F1(N)
550 IF O2=0 THEN GOTO 730
560 LPRINT "F(N)=";F(N);"F1(N)=";F1(N)
570 LPRINT
580 GOTO 730
590 REM INPUT OF ONE POINT ON THE LEFT
600 LET K=N-1
610 FOR I=K TO 1 STEP -1
```

```
620 LET J=I+1
630 LET X0(J)=X0(I)
640 LET F(J)=F(I)
650 LET F1(J)=F1(I)
660 NEXT I
670 LET X0(1)=X0(2)-H
680 INPUT"F(1)";F(1):INPUT"F1(1)";F1(1)
690 IF O2=0 THEN GOTO 730
700 LPRINT "F(1)=";F(1);"F1(1)=";F1(1)
710 LPRINT
720 REM COMPUTATION OF DIVIDED DIFFERENCES
730 FOR I=1 TO N
740 LET J=2*I-1
750 LET X1(J)=X0(I)
760 LET G(J)=F1(I)
770 LET J=2*I
780 LET K=I+1
790 LET X1(J)=X0(I)
800 IF I=N THEN GOTO 820
810 LET G(J)=(F(K)-F(I))/(X0(K)-X0(I))
820 NEXT I
830 LET F0(1)=F1(1)
840 LET M=2*N-1
850 LET K=2
860 LET L0=2
870 LET M=M-1
880 FOR I=1 TO M
890 LET J=I+1
900 LET L=I+L0
910 LET G(I)=(G(J)-G(I))/(X1(L)-X1(I))
920 NEXT I
930 LET F0(K)=G(1)
940 LET K=K+1
950 LET L0=L0+1
960 IF M>1 THEN GOTO 870
970 CLS:PRINT "THE INTERPOLATION:"
980 INPUT"X";X
990 IF O2>0 THEN LPRINT "X =";X
1000 LET M=2*N-1
1010 FOR I=1 TO M
1020 LET U(I)=X-X1(I)
1030 NEXT I
1040 LET Y=F0(M)*U(M)
1050 LET M=M-1
1060 LET Y=(Y+F0(M))*U(M)
1070 IF M>1 THEN GOTO 1050
1080 LET Y=Y+F(1)
1090 PRINT "Y =";Y
1100 IF O2>0 THEN LPRINT "Y =";Y
1110 IF O4=0 THEN GOTO 1390
1120 REM COMPUTATION OF THE FIRST DERIVATIVE
1130 LET M=2*N-1
1140 LET Y1=0
1150 LET V1=U(1)+U(2)
1160 LET V2=U(1)*U(2)
1170 IF N<5 THEN GOTO 1200
1180 LET V3=U(3)+U(4)
1190 LET V4=U(3)*U(4)
1200 ON M GOTO 1250,1240,1230,1220
1210 LET Y1=F0(5)*(U(5)*V1*V4+V2*(U(5)*V3+V4))
1220 LET Y1=Y1+F0(4)*(V1*V4+V2*V3)
1230 LET Y1=Y1+F0(3)*(U(3)*V1+V2)
1240 LET Y1=Y1+F0(2)*V1
1250 LET Y1=Y1+F0(1)
```

P. 50 PROGRAM P206

```
1260 PRINT "Y1=";Y1
1270 IF O2>0 THEN LPRINT "Y1=";Y1
1280 IF O4=1 THEN GOTO 1390
1290 REM COMPUTATION OF THE SECOND DERIVATIVE
1300 LET Y2=0
1310 ON M GOTO 1360,1350,1340,1330
1320 LET Y2=F0(5)*(V1*V4+V2*V3+U(5)*(V1*V3+V2+V3))
1330 LET Y2=Y2+F0(4)*(V1*V3+V2+V4)
1340 LET Y2=Y2+F0(3)*(V1+U(3))
1350 LET Y2=Y2+F0(2)
1360 LET Y2=2*Y2
1370 PRINT "Y2=";Y2
1380 IF O2>0 THEN LPRINT "Y2=";Y2
1390 IF O2>0 THEN LPRINT
1400 PRINT "IF NO NEW POINTS DEFINING THE FUNCTION O3=0"
1410 PRINT "IF ONE NEW POINT ON THE RIGHT O3=1"
1420 PRINT "IF ONE NEW POINT ON THE LEFT O3=-1"
1430 PRINT "IF ALL POINTS ARE NEW O3>1"
1440 PRINT "TO STOP PRESS BREAK"
1450 INPUT"O3";O3
1460 IF O3=0 THEN GOTO 970
1470 GOTO 250
1480 END
```

Examples

1. The function $y=x^3$ is given at 3 equally spaced points by the function F and its derivative F1

$x_1=1$, $F_1=1$, $F1_1=3$,
$x_2=3.5$, $F_2=42.875$, $F1_2=36.75$,
$x_3=6$, $F_3=216$, $F1_3=108$.

Some results of interpolation in single precision:

$x=2.5$, $y=15.625$, $y_1=18.75$, $y_2=15$.
$x=4.5$, $y=91.125$, $y_1=60.75$, $y_2=27$.

2. The function $y=e^x$ is given at 3 equally spaced points by the function F and its first derivative F1

$x_1=0.95$, $F_1=F1_1=2.58571$,
$x_2=0.975$, $F_2=F1_2=2.65117$,
$x_3=1$, $F_3=F1_3=2.71828$.

Some results of interpolation in single precision:

$x=0.955$, $y=2.59867$, $y_1=2.59877$, $y_2=2.61320$.
$x=0.975$, $y=2.65117$, $y_1=2.65117$, $y_2=2.62865$.
$x=0.990$, $y=2.69124$, $y_1=2.69065$, $y_2=2.64898$.

The values of the function are exact in 6 digits, the loss of accuracy in the first and second derivatives is acceptable.

3. The function $y=e^x$ is given at 3 equally spaced points by the function F and its first derivative F1

$x_1=0.95$, $F_1=F1_1=2.585709659315846$,
$x_2=0.975$, $F_2=F1_2=2.651167210982607$,
$x_3=1$, $F_3=F1_3=2.718281828459045$.

Some results of interpolation in double precision:

$x=0.955$, $y=2.598670582919447$, $y_1=2.598670255909560$,
   $y_2=2.598670451891907$.
$x=0.975$, $y=2.651167210982607$, $y_1=2.651167210982607$,
   $y_2=2.651167208102486$.
$x=0.990$, $y=2.691234472349129$, $y_1=2.691227224420492$,
   $y_2=2.690655652668159$.

P207 Hermite Interpolation of a Function Defined at Two Points by the Function and Its First and Second Derivatives for the Function and Its First and Second Derivatives

```
10 CLS:PRINT"P207 COMPUTES Y(X) AND THE FIRST AND SECOND DERIVATIVE"
20 PRINT"Y1,Y2 OF A FUNCTION DEFINED AT TWO POINTS BY FUNCTION"
30 PRINT"AND ITS FIRST AND SECOND DERIVATIVES BY HERMITE"
40 PRINT"INTERPOLATION."
50 PRINT:PRINT"THE INPUT:"
60 PRINT"IF SINGLE PRECISION O1=1, IF DOUBLE O1=2."
70 PRINT"USE DOUBLE PRECISION IF SECOND DERIVATIVE IS COMPUTED."
80 DEFINT I-O:INPUT"O1";O1:CLS
90 PRINT"IF PRINTER NOT USED O2=0, IF USED O2>0."
100 INPUT"O2";O2:CLS
110 IF O1=2 THEN DEFDBL A-H,U-Z
120 PRINT"IF ONLY FUNCTION COMPUTED O4=0, IF FUNCTION AND ITS"
130 PRINT"FIRST DERIVATIVE COMPUTED O4=1, IF FUNCTION AND ITS"
140 PRINT"FIRST AND SECOND DERIVATIVES COMPUTED O4=2."
150 INPUT"O4";O4:CLS
160 IF O2=0 THEN GOTO 210
170 LPRINT "P207: O1=";O1;"O2=";O2;"O4=";O4
180 GOTO 210
190 IF O3=1 THEN GOTO 360
200 IF O3=-1 THEN GOTO 460
210 PRINT"ABSCISSAS OF THE TWO GIVEN POINTS: X1,X2"
220 PRINT"FUNCTION AND ITS TWO DERIVATIVES AT X1: F,F1,F2"
230 PRINT"FUNCTION AND ITS TWO DERIVATOVES AT X2: G,G1,G2"
240 PRINT "INPUT OF BOTH POINTS:"
250 IF O2>0 THEN LPRINT "THE INPUT:"
260 INPUT"X1";X1:INPUT"F";F:INPUT"F1";F1:INPUT"F2";F2
270 INPUT"X2";X2:INPUT"G";G:INPUT"G1";G1:INPUT"G2";G2
280 IF O2=0 THEN GOTO 550
290 LPRINT "X1=";X1;"F=";F;"F1=";F1;"F2=";F2
300 LPRINT "X2=";X2;"G=";G;"G1=";G1;"G2=";G2
310 IF O2=0 THEN GOTO 550
320 LPRINT
330 LPRINT "THE INTERPOLATION:"
340 GOTO 550
350 REM INPUT OF ONE POINT ON THE RIGHT
360 LET X1=X2
370 LET F=G
380 LET F1=G1
390 LET F2=G2
400 INPUT"X2";X2:INPUT"G";G:INPUT"G1";G1:INPUT"G2";G2
410 IF O2=0 THEN GOTO 550
420 LPRINT "X2=";X2;"G=";G;"G1=";G1;"G2=";G2
430 LPRINT
440 GOTO 550
450 REM INPUT OF ONE POINT ON THE LEFT
460 LET X2=X1
470 LET G=F
480 LET G1=F1
490 LET G2=F2
500 INPUT"X1";X1:INPUT"F";F:INPUT"F1";F1:INPUT"F2";F2
510 IF O2=0 THEN GOTO 550
520 LPRINT "X1=";X1;"F=";F;"F1=";F1;"F2=";F2
530 LPRINT
540 REM COMPUTATION OF COEFFICIENTS A3,A4,A5
550 LET H=X2-X1
560 LET B1=(((G-F)/H-F1)/H-F2/2)/H
570 LET B2=((G1-F1)/H-F2)/H
580 LET B3=(G2-F2)/H
590 LET A3=10*B1-4*B2+B3/2
```

```
600 LET A4=(7*B2-15*B1-B3)/H
610 LET A5=(6*B1-3*B2+B3/2)/(H*H)
620 CLS:PRINT "THE INTERPOLATION:"
630 INPUT"X";X
640 IF O2>0 THEN LPRINT "X =";X
650 LET H=X-X1
660 LET Y=F+H*(F1+H*(F2/2+H*(A3+H*(A4+H*A5))))
670 PRINT "Y =";Y
680 IF O2>0 THEN LPRINT "Y =";Y
690 IF O4=0 THEN GOTO 790
700 REM COMPUTATION OF THE FIRST DERIVATIVE
710 LET Y1=F1+H*(F2+H*(3*A3+H*(4*A4+5*H*A5)))
720 PRINT "Y1=";Y1
730 IF O2>0 THEN LPRINT "Y1=";Y1
740 IF O4=1 THEN GOTO 790
750 REM COMPUTATION OF THE SECOND DERIVATIVE
760 LET Y2=F2+H*(6*A3+H*(12*A4+20*H*A5))
770 PRINT "Y2=";Y2
780 IF O2>0 THEN LPRINT "Y2=";Y2
790 IF O2>0 THEN LPRINT
800 PRINT "IF NO NEW POINTS DEFINING THE FUNCTION O3=0"
810 PRINT "IF ONE NEW POINT ON THE RIGHT O3=1"
820 PRINT "IF ONE NEW POINT ON THE LEFT O3=-1"
830 PRINT "IF BOTH POINTS ARE NEW O3>1"
840 PRINT "TO STOP PRESS BREAK"
850 INPUT"O3";O3
860 IF O3=0 THEN GOTO 620
870 GOTO 190
880 END
```

Examples

1. The function $y=x^3$ is given at 2 points by the function F, G and its first and second derivatives F1, F2, G1, G2

$x_1=1$,   F=1,   F1=3,   F2=6,

$x_2=6$,   G=216,   G1=108,   G2=36.

Some results of interpolation in single precision:

x=2.5,   y=15.625, y_1=18.75, y_2=15.

x=4.5,   y=91.125, y_1=60.75, y_2=27.

2. The function $y=e^x$ is given at 2 points by the function F, G and its first and second derivatives F1, F2, G1, G2

x_1=0.95,   F=F1=F2=2.58571,

x_2=1,   G=G1=G2=2.71828.

Some results of interpolation in single precision:

x=0.955,   y=2.59867, y_1=2.59866, y_2=2.59537.

x=0.975,   y=2.65117, y_1=2.65110, y_2=2.65111.

x=0.990,   y=2.69123, y_1=2.69120, y_2=2.69557.

P. 54   PROGRAM P207

The values of the function are exact in 6 digits, the loss of accuracy in the first and second derivatives is acceptable.

3. The function $y=e^x$ is given at 2 points by the function F, G and its first and second derivatives F1, F2, G1, G2

$x_1=0.95$,   F=F1=F2=2.585709659315846,
$x_2=1$,   G=G1=G2=2.718281828459045.

Some results of interpolation in double precision:

x=0.955,   y=2.598670582919563, y_1=2.598670582941833,
y_2=2.598670589741525.

x=0.975,   y=2.651167210983506, y_1=2.651167210982732,
y_2=2.651167202352246.

x=0.990,   y=2.691234472349498, y_1=2.691234472296156,
y_2=2.691234476762503.

The accuracy of interpolation is here excellent.

P208 Cubic Spline for the Function and Its First and Second Derivatives

```
10 CLS:PRINT"P208 COMPUTES Y(X) AND THE FIRST AND SECOND DERIVATIVE"
20 PRINT"Y1,Y2 BY CUBIC SPLINE."
30 PRINT:PRINT"THE INPUT:"
40 PRINT"IF SINGLE PRECISION O1=1, IF DOUBLE O1=2."
50 DEFINT I-O:INPUT"O1";O1:CLS
60 PRINT"IF PRINTER NOT USED O2=0, IF USED O2>0."
70 INPUT"O2";O2:CLS
80 IF O1=2 THEN DEFDBL A-H,P-Z
90 PRINT"ABSCISSAS OF POINTS DEFINING THE FUNCTION: X0(I)"
100 PRINT"ORDINATES OF POINTS DEFINING THE FUNCTION: F(I)":PRINT
110 PRINT"IF X0(I) AND F(I) FROM INPUT O3>1,"
120 PRINT"IF X0(I) FROM FILE P208A AND F(I) FROM INPUT O3=1,"
130 PRINT"IF DATA FROM FILES P208A AND P208B O3=0."
140 PRINT"FILES P208A AND P208B ARE WRITTEN IN THE FIRST RUN."
150 INPUT"O3";O3:CLS
160 PRINT"IF ONLY FUNCTION COMPUTED O4=0, IF FUNCTION AND ITS"
170 PRINT"FIRST DERIVATIVE COMPUTED O4=1, IF FUNCTION AND ITS"
180 PRINT"FIRST AND SECOND DERIVATIVES COMPUTED O4=2."
190 INPUT"O4";O4:CLS:IF O3=0 THEN GOTO 350
200 PRINT"MODE OF BOUNDARY CONDITIONS:"
210 PRINT"IF NO DERIVATIVES AT BOUNDARIES GIVEN O9=1"
220 PRINT"IF SECOND DERIVATIVE D1 AT X0(1) GIVEN O9=2"
230 PRINT"IF SECOND DERIVATIVE D2 AT X0(N) GIVEN O9=3"
240 PRINT"IF SECOND DERIVATIVES D1,D2 GIVEN O9=4"
250 PRINT"IF FIRST DERIVATIVE F1(1) AT X0(1) GIVEN O9=5"
260 PRINT"IF FIRST DERIVATIVE F1(N) AT X0(N) GIVEN O9=6"
270 PRINT"IF FIRST DERIVATIVES F1(1),F1(N) GIVEN O9=7"
280 PRINT"IF FIRST DERIVATIVE F1(1) AT X0(1) AND SECOND"
290 PRINT"DERIVATIVE D2 AT X0(N) GIVEN O9=8"
300 PRINT"IF FIRST DERIVATIVE F1(N) AT X0(N) AND SECOND"
310 PRINT"DERIVATIVE D1 AT X0(1) GIVEN O9=9"
320 INPUT"O9";O9:CLS:IF O9<1 THEN GOTO 340
330 IF O9<10 THEN GOTO 350
340 PRINT"ERROR IN O9":GOTO 200
350 PRINT"NUMBER OF POINTS DEFINING THE FUNCTION: N"
360 INPUT"N";N:CLS
370 IF O2=0 THEN GOTO 390
380 LPRINT "P208: O1=";O1;"O2=";O2;"O3=";O3;"O4=";O4;"N=";N
390 DIM A(N,3),B(N),F(N),F1(N),X0(N)
400 IF O3<2 THEN GOTO 630
410 PRINT "INPUT OF ABSCISSAS:"
420 FOR I=1 TO N
430 PRINT "I=";I
440 INPUT"X0(I)";X0(I)
450 NEXT I
460 CLS:PRINT "CHECK THE ABSCISSAS:"
470 FOR I=1 TO N
480 PRINT "I=";I;"X0(I)=";X0(I)
490 PRINT"IF NO ERROR INPUT 0"
500 PRINT"IF ERROR INPUT ANY NUMBER>0 AND THEN THE CORRECT X0(I)"
510 INPUT"ERROR";K
520 IF K>0 THEN INPUT"X0(I)";X0(I)
530 CLS:NEXT I
540 PRINT"COMPUTING"
550 REM WRITING VECTOR X0(I) ON A FILE
560 OPEN "O",1,"P208A"
570 FOR I=1 TO N
580 PRINT #1, X0(I)
590 NEXT I
600 CLOSE 1
610 GOTO 680
```

PROGRAM P208

```
620 REM READING VECTOR X0(I) FROM A FILE
630 OPEN "I",1,"P208A"
640 FOR I=1 TO N
650 INPUT #1, X0(I)
660 NEXT I
670 CLOSE 1
680 IF O3=0 THEN GOTO 1930
690 PRINT "INPUT OF ORDINATES:"
700 FOR I=1 TO N
710 PRINT "I=";I
720 INPUT"F(I)";F(I)
730 NEXT I
740 CLS:PRINT "CHECK THE ORDINATES:"
750 FOR I=1 TO N
760 PRINT "I=";I;"F(I)=";F(I)
770 PRINT"IF NO ERROR INPUT 0"
780 PRINT"IF ERROR INPUT ANY NUMBER>0 AND THEN THE CORRECT F(I)"
790 INPUT"ERROR";K
800 IF K>0 THEN INPUT"F(I)";F(I)
810 CLS:NEXT I
820 PRINT"COMPUTING"
830 IF O2=0 THEN GOTO 900
840 LPRINT "THE INPUT I, X0(I), F(I)"
850 FOR I=1 TO N
860 LPRINT "I=";I;"X0(I)=";X0(I);"F(I)=";F(I)
870 NEXT I
880 LPRINT
890 REM COMPUTATION OF MATRIX A AND VECTOR B
900 LET M=N-1
910 FOR I=2 TO M
920 LET A(I,2)=2
930 LET J=I-1
940 LET K=I+1
950 LET S=X0(K)-X0(J)
960 LET S1=X0(K)-X0(I)
970 LET S2=X0(I)-X0(J)
980 LET A(I,1)=S1/S
990 LET A(I,3)=S2/S
1000 LET B(I)=3*(A(I,1)*(F(I)-F(J))/S2+A(I,3)*(F(K)-F(I))/S1)
1010 NEXT I
1020 REM BOUNDARY CONDITIONS
1030 LET A(1,2)=2
1040 LET A(1,3)=2/A(2,1)
1050 LET B(1)=(2+A(2,3))*(F(2)-F(1))/(X0(2)-X0(1))
1060 LET C=(A(2,3)*A(2,3)/A(2,1))*(F(3)-F(2))/(X0(3)-X0(2))
1070 LET B(1)=2*(B(1)+C)
1080 LET A(N,2)=2
1090 LET A(N,1)=2/A(M,3)
1100 LET M2=N-2
1110 LET C=(A(M,1)*A(M,1)/A(M,3))*(F(M)-F(M2))/(X0(M)-X0(M2))
1120 LET B(N)=(2+A(M,1))*(F(N)-F(M))/(X0(N)-X0(M))
1130 LET B(N)=2*(C+B(N))
1140 LET I0=1
1150 LET I2=N
1160 ON O9 GOTO 1650,1210,1240,1270,1310,1340,1370,1410,1170
1170 INPUT"D1";D1:INPUT"F1(N)";F1(N)
1180 GOSUB 1460
1190 GOSUB 1590
1200 GOTO 1650
1210 INPUT"D1";D1
1220 GOSUB 1460
1230 GOTO 1650
1240 INPUT"D2";D2
1250 GOSUB 1500
```

```
1260 GOTO 1650
1270 INPUT"D1";D1:INPUT"D2";D2
1280 GOSUB 1460
1290 GOSUB 1500
1300 GOTO 1650
1310 INPUT"F1(1)";F1(1)
1320 GOSUB 1540
1330 GOTO 1650
1340 INPUT"F1(N)";F1(N)
1350 GOSUB 1590
1360 GOTO 1650
1370 INPUT"F1(1)";F1(1):INPUT"F1(N)";F1(N)
1380 GOSUB 1540
1390 GOSUB 1590
1400 GOTO 1650
1410 INPUT"F1(1)";F1(1):INPUT"D2";D2
1420 GOSUB 1500
1430 GOSUB 1540
1440 GOTO 1650
1450 REM SUBROUTINE FOR BOUNDARY CONDITION D1
1460 LET A(1,3)=1
1470 LET B(1)=3*(F(2)-F(1))/(X0(2)-X0(1))-(X0(2)-X0(1))*D1/2
1480 RETURN
1490 REM SUBROUTINE FOR BOUNDARY CONDITION D2
1500 LET A(N,1)=1
1510 LET B(N)=3*(F(N)-F(M))/(X0(N)-X0(M))+(X0(N)-X0(M))*D2/2
1520 RETURN
1530 REM SUBROUTINE FOR BOUNDARY CONDITION F1(1)
1540 LET B(2)=A(2,1)*(3*(F(2)-F(1))/(X0(2)-X0(1))-F1(1))
1550 LET B(2)=B(2)+3*A(2,3)*(F(3)-F(2))/(X0(3)-X0(2))
1560 LET I0=2
1570 RETURN
1580 REM SUBROUTINE FOR BOUNDARY CONDITION F1(N)
1590 LET B(M)=A(M,3)*(3*(F(N)-F(M))/(X0(N)-X0(M))-F1(N))
1600 LET B(M)=B(M)+3*A(M,1)*(F(M)-F(M2))/(X0(M)-X0(M2))
1610 LET I2=M
1620 RETURN
1630 REM LU-DECOMPOSITION
1640 REM LU-MATRICES STORED AS MATRIX A
1650 LET I1=I0+1
1660 FOR I=I1 TO I2
1670 LET J=I-1
1680 LET A(I,1)=A(I,1)/A(J,2)
1690 LET A(I,2)=A(I,2)-A(I,1)*A(J,3)
1700 NEXT I
1710 REM SOLUTION OF THE FIRST TRIANGULAR SYSTEM
1720 REM VECTOR Y STORED AS VECTOR B
1730 FOR I=I1 TO I2
1740 LET J=I-1
1750 LET B(I)=B(I)-A(I,1)*B(J)
1760 NEXT I
1770 REM SOLUTION OF THE SECOND TRIANGULAR SYSTEM
1780 F1(I2)=B(I2)/A(I2,2)
1790 LET I3=I2-1
1800 FOR I=I3 TO I0 STEP -1
1810 LET J=I+1
1820 LET F1(I)=(B(I)-A(I,3)*F1(J))/A(I,2)
1830 NEXT I
1840 REM WRITING VECTORS F(I),F1(I) ON A FILE
1850 OPEN "O",1,"P208B"
1860 PRINT #1, 09
1870 FOR I=1 TO N
1880 PRINT #1, F(I),F1(I)
1890 NEXT I
```

P. 58 PROGRAM P208

```
1900 CLOSE 1
1910 GOTO 1990
1920 REM READING VECTORS F(I),F1(I) FROM A FILE
1930 OPEN "I",1,"P208B"
1940 INPUT #1, O9
1950 FOR I=1 TO N
1960 INPUT #1, F(I),F1(I)
1970 NEXT I
1980 CLOSE 1
1990 PRINT "MODE OF BOUNDARY CONDITIONS:";O9
2000 PRINT "THE INTERPOLATION:"
2010 IF O2=0 THEN GOTO 2050
2020 LPRINT "MODE OF BOUNDARY CONDITIONS:";O9
2030 LPRINT
2040 LPRINT "THE INTERPOLATION:"
2050 INPUT"X";X
2060 IF O2>0 THEN LPRINT "X =";X
2070 LET J=2
2080 IF X<=X0(J) THEN GOTO 2110
2090 LET J=J+1
2100 GOTO 2080
2110 LET I=J-1
2120 LET H=X0(J)-X0(I)
2130 LET H2=H*H
2140 LET X1=X-X0(I)
2150 LET X2=X0(J)-X
2160 LET Y=F(I)*X2*X2*(2*X1+H)+F(J)*X1*X1*(2*X2+H)
2170 LET Y=Y/H+(F1(I)*X2-F1(J)*X1)*X1*X2
2180 LET Y=Y/H2
2190 PRINT "Y =";Y
2200 IF O2>0 THEN LPRINT "Y =";Y
2210 IF O4=0 THEN GOTO 2350
2220 REM COMPUTATION OF THE FIRST DERIVATIVE
2230 LET X3=X2-2*X1
2240 LET X4=2*X2-X1
2250 LET Y1=6*X1*X2*(F(J)-F(I))/H+F1(I)*X2*X3-F1(J)*X1*X4
2260 LET Y1=Y1/H2
2270 PRINT "Y1=";Y1
2280 IF O2>0 THEN LPRINT "Y1=";Y1
2290 IF O4=1 THEN GOTO 2350
2300 REM COMPUTATION OF THE SECOND DERIVATIVE
2310 LET Y2=3*(X2-X1)*(F(J)-F(I))/H-F1(I)*X4-F1(J)*X3
2320 LET Y2=2*Y2/H2
2330 PRINT "Y2=";Y2
2340 IF O2>0 THEN LPRINT "Y2=";Y2
2350 IF O2>0 THEN LPRINT
2360 PRINT "TO STOP PRESS BREAK":PRINT
2370 GOTO 2050
2380 END
```

Examples

1. The function $y=x^3$ is given by 6 points
    (1,1), (1.5,3.375), (2,8),
    (3.5,42.875), (5.5,166.375), (6,216).

Mode of boundary conditions: 1. Since the function is a parabola of third degree, the cubic spline interpolates the function y and its first and second derivatives y_1, y_2 exactly. Some results of interpolation in single precision:

    x=2.5,   y=15.625, y_1=18.75, y_2=15.
    x=4.5,   y=91.125, y_1=60.75, y_2=27.

2. The function $y=e^x$ is given by 6 points
    (0.95,2.58571), (0.96,2.61170), (0.965,2.62479),
    (0.985,2.67781), (0.99,2.69123), (1,2.71828).

Mode of boundary conditions: 7. Some results of interpolation in single precision:

    x=0.955,   y=2.59867, y_1=2.59912, y_2=2.60412.
    x=0.975,   y=2.65117, y_1=2.65107, y_2=2.63172.
    x=0.990,   y=2.69123, y_1=2.69102, y_2=2.94216.

The values of the function are exact in 6 digits, but note the loss of accuracy in the first and second derivatives.

3. The function $y=e^x$ is given by 6 points
    (0.95,2.585709659315846), (0.96,2.611696473423118),
    (0.965,2.624787656474575), (0.985,2.677811884421050),
    (0.99,2.691234472349262), (1,2.718281828459045).

Mode of boundary conditions: 7. Some results of interpolation in double precision:

    x=0.955,   y=2.5986705828, y_1=2.5986705699, y_2=2.5986865974.
    x=0.975,   y=2.6511672093, y_1=2.6511672106, y_2=2.6512229244.
    x=0.990,   y=2.6912344723, y_1=2.6912344197, y_2=2.6912329966.

The values of the function are exact in 9 to 11 digits, the values of the first derivative in 8 to 10 digits, and the values of the second derivative in 4 to 6 digits because of the improved accuracy of the input of ordinates (exact in 16 digits while in example 2 they are exact only in 6 digits).

NOTE: The boundary conditions have no effect on the interpolation only if the function is a parabola of third or lower degree. The choice of the mode of boundary conditions is determined by the available information. Whenever possible use mode 7 (first derivatives on both ends of interval are given).

P209 Cubic Spline with Equally Spaced Abscissas for the Function and Its First and Second Derivatives

```
10 CLS:PRINT"P209 COMPUTES Y(X) AND THE FIRST AND SECOND DERIVATIVE"
20 PRINT"Y1,Y2 BY CUBIC SPLINE WITH EQUALLY SPACED ABSCISSAS."
30 PRINT:PRINT"THE INPUT:"
40 PRINT"IF SINGLE PRECISION O1=1, IF DOUBLE O1=2."
50 DEFINT I-O:INPUT"O1";O1:CLS
60 PRINT"IF PRINTER NOT USED O2=0, IF USED O2>0."
70 INPUT"O2";O2:CLS
80 IF O1=2 THEN DEFDBL A-H,P-Z
90 PRINT"ORDINATES OF POINTS DEFINING THE FUNCTION: F(I)":PRINT
100 PRINT"IF F(I) FROM INPUT O3>0, IF DATA FROM FILE P209A O3=0."
110 PRINT"FILE P209A WRITTEN IN THE FIRST RUN."
120 INPUT"O3";O3:CLS
130 PRINT"IF ONLY FUNCTION COMPUTED O4=0, IF FUNCTION AND ITS"
140 PRINT"FIRST DERIVATIVE COMPUTED O4=1, IF FUNCTION AND ITS"
150 PRINT"FIRST AND SECOND DETIVATIVES COMPUTED O4=2."
160 INPUT"O4";O4:CLS:IF O3=0 THEN GOTO 320
170 PRINT"MODE OF BOUNDARY CONDITIONS:"
180 PRINT"IF NO DERIVATIVES AT BOUNDARIES GIVEN O9=1"
190 PRINT"IF SECOND DERIVATIVE D1 AT P GIVEN O9=2"
200 PRINT"IF SECOND DERIVATIVE D2 AT Q GIVEN O9=3"
210 PRINT"IF SECOND DERIVATIVES D1,D2 GIVEN O9=4"
220 PRINT"IF FIRST DERIVATIVE F1(1) AT P GIVEN O9=5"
230 PRINT"IF FIRST DERIVATIVE F1(N) AT Q GIVEN O9=6"
240 PRINT"IF FIRST DERIVATIVES F1(1),F1(N) GIVEN O9=7"
250 PRINT"IF FIRST DERIVATIVE F1(1) AT P AND SECOND"
260 PRINT"DERIVATIVE D2 AT Q GIVEN O9=8"
270 PRINT"IF FIRST DERIVATIVE F1(N) AT Q AND SECOND"
280 PRINT"DERIVATIVE D1 AT P GIVEN O9=9"
290 INPUT"O9";O9:CLS:IF O9<1 THEN GOTO 310
300 IF O9<10 THEN GOTO 320
310 PRINT"ERROR IN O9":GOTO 170
320 PRINT"NUMBER OF POINTS DEFINING THE FUNCTION: N"
330 INPUT"N";N:CLS
340 PRINT"INTERVAL COVERED BY N EQUALLY SPACED POINTS: (P,Q)"
350 INPUT"P";P:INPUT"Q";Q:CLS
360 IF O2=0 THEN GOTO 380
370 LPRINT "P209: O1=";O1;"O2=";O2;"O3=";O3;"O4=";O4;"N=";N
380 DIM A(N,3),B(N),F(N),F1(N)
390 IF O2>0 THEN LPRINT "INTERVAL P=";P;"Q=";Q
400 LET H0=(Q-P)/(N-1)
410 IF O3=0 THEN GOTO 1570
420 PRINT "INPUT OF ORDINATES:"
430 FOR I=1 TO N
440 PRINT "I=";I
450 INPUT"F(I)";F(I)
460 NEXT I
470 CLS:PRINT "CHECK THE ORDINATES:"
480 FOR I=1 TO N
490 PRINT "I=";I;"F(I)=";F(I)
500 PRINT"IF NO ERROR INPUT 0"
510 PRINT"IF ERROR INPUT ANY NUMBER>0 AND THEN THE CORRECT F(I)"
520 INPUT"ERROR";K
530 IF K>0 THEN INPUT"F(I)";F(I)
540 CLS:NEXT I
550 PRINT"COMPUTING"
560 IF O2=0 THEN GOTO 630
570 LPRINT "THE INPUT I, F(I)"
580 FOR I=1 TO N
590 LPRINT "I=";I;"F(I)=";F(I)
600 NEXT I
610 LPRINT
```

```
620 REM COMPUTATION OF MATRIX A AND VECTOR B
630 LET M=N-1
640 FOR I=2 TO M
650 LET A(I,2)=2
660 LET J=I-1
670 LET K=I+1
680 LET A(I,1)=.5
690 LET A(I,3)=.5
700 LET B(I)=1.5*(F(K)-F(J))/H0
710 NEXT I
720 REM BOUNDARY CONDITIONS
730 LET A(1,2)=2
740 LET A(1,3)=4
750 LET B(1)=(F(3)+4*F(2)-5*F(1))/H0
760 LET A(N,2)=2
770 LET A(N,1)=4
780 LET M2=N-2
790 LET B(N)=(5*F(N)-4*F(M)-F(M2))/H0
800 LET I0=1
810 LET I2=N
820 ON O9 GOTO 1290,870,900,930,970,1000,1030,1070
830 INPUT"D1";D1:INPUT"F1(N)";F1(N)
840 GOSUB 1120
850 GOSUB 1240
860 GOTO 1290
870 INPUT"D1";D1
880 GOSUB 1120
890 GOTO 1290
900 INPUT"D2";D2
910 GOSUB 1160
920 GOTO 1290
930 INPUT"D1";D1:INPUT"D2";D2
940 GOSUB 1120
950 GOSUB 1160
960 GOTO 1290
970 INPUT"F1(1)";F1(1)
980 GOSUB 1200
990 GOTO 1290
1000 INPUT"F1(N)";F1(N)
1010 GOSUB 1240
1020 GOTO 1290
1030 INPUT"F1(1)";F1(1):INPUT"F1(N)";F1(N)
1040 GOSUB 1200
1050 GOSUB 1240
1060 GOTO 1290
1070 INPUT"F1(1)";F1(1):INPUT"D2";D2
1080 GOSUB 1160
1090 GOSUB 1200
1100 GOTO 1290
1110 REM SUBROUTINE FOR BOUNDARY CONDITION D1
1120 LET A(1,3)=1
1130 LET B(1)=3*(F(2)-F(1))/H0-H0*D1/2
1140 RETURN
1150 REM SUBROUTINE FOR BOUNDARY CONDITION D2
1160 LET A(N,1)=1
1170 LET B(N)=3*(F(N)-F(M))/H0+H0*D2/2
1180 RETURN
1190 REM SUBROUTINE FOR BOUNDARY CONDITION F1(1)
1200 LET B(2)=1.5*(F(3)-F(1))/H0-F1(1)/2
1210 LET I0=2
1220 RETURN
1230 REM SUBROUTINE FOR BOUNDARY CONDITION F1(N)
1240 LET B(M)=1.5*(F(N)-F(M2))/H0-F1(N)/2
1250 LET I2=M
```

PROGRAM P209

```
1260 RETURN
1270 REM LU-DECOMPOSITION
1280 REM LU-MATRICES STORED AS MATRIX A
1290 LET I1=I0+1
1300 FOR I=I1 TO I2
1310 LET J=I-1
1320 LET A(I,1)=A(I,1)/A(J,2)
1330 LET A(I,2)=A(I,2)-A(I,1)*A(J,3)
1340 NEXT I
1350 REM SOLUTION OF THE FIRST TRIANGULAR SYSTEM
1360 REM VECTOR Y STORED AS VECTOR B
1370 FOR I=I1 TO I2
1380 LET J=I-1
1390 LET B(I)=B(I)-A(I,1)*B(J)
1400 NEXT I
1410 REM SOLUTION OF THE SECOND TRIANGULAR SYSTEM
1420 F1(I2)=B(I2)/A(I2,2)
1430 LET I3=I2-1
1440 FOR I=I3 TO I0 STEP -1
1450 LET J=I+1
1460 LET F1(I)=(B(I)-A(I,3)*F1(J))/A(I,2)
1470 NEXT I
1480 REM WRITING VECTORS F(I),F1(I) ON A FILE
1490 OPEN "O",1,"P209A"
1500 PRINT #1, O9,P,H0
1510 FOR I=1 TO N
1520 PRINT #1, F(I),F1(I)
1530 NEXT I
1540 CLOSE 1
1550 GOTO 1670
1560 REM READING VECTORS F(I),F1(I) FROM A FILE
1570 OPEN "I",1,"P209A"
1580 INPUT #1, O9,Z1,Z2
1590 LET Z2=Z1+Z2*(N-1)
1600 PRINT "P=";Z1;"Q=";Z2;"ARE THESE VALUES EQUAL TO INPUT?"
1610 PRINT "IF NOT PRESS BREAK, CORRECT INPUT AND RERUN"
1620 FOR I=1 TO N
1630 INPUT #1, F(I),F1(I)
1640 NEXT I
1650 CLOSE 1
1660 PRINT "P=";P;"H0=";H0
1670 PRINT "MODE OF BOUNDARY CONDITIONS:";O9
1680 PRINT "THE INTERPOLATION:"
1690 IF O2=0 THEN GOTO 1750
1700 LPRINT "P=";P;"H0=";H0
1710 LPRINT "MODE OF BOUNDARY CONDITIONS:";O9
1720 LPRINT
1730 REM INTERPOLATION OF THE FUNCTION
1740 LPRINT "THE INTERPOLATION:"
1750 INPUT "X";X
1760 IF O2>0 THEN LPRINT "X =";X
1770 LET I=INT((X-P)/H0)+1
1780 IF I=N THEN LET I=I-1
1790 LET J=I+1
1800 LET H2=H0*H0
1810 LET X1=X-P-H0*(I-1)
1820 LET X2=H0-X1
1830 LET Y=F(I)*X2*X2*(2*X1+H0)+F(J)*X1*X1*(2*X2+H0)
1840 LET Y=Y/H0+(F1(I)*X2-F1(J)*X1)*X1*X2
1850 LET Y=Y/H2
1860 PRINT "Y =";Y
1870 IF O2>0 THEN LPRINT "Y =";Y
1880 IF O4=0 THEN GOTO 2020
1890 REM COMPUTATION OF THE FIRST DERIVATIVE
```

```
1900 LET X3=X2-2*X1
1910 LET X4=2*X2-X1
1920 LET Y1=6*X1*X2*(F(J)-F(I))/H0+F1(I)*X2*X3-F1(J)*X1*X4
1930 LET Y1=Y1/H2
1940 PRINT "Y1=";Y1
1950 IF O2>0 THEN LPRINT "Y1=";Y1
1960 IF O4=1 THEN GOTO 2020
1970 REM COMPUTATION OF THE SECOND DERIVATIVE
1980 LET Y2=3*(X2-X1)*(F(J)-F(I))/H0-F1(I)*X4-F1(J)*X3
1990 LET Y2=2*Y2/H2
2000 PRINT "Y2=";Y2
2010 IF O2>0 THEN LPRINT "Y2=";Y2
2020 IF O2>0 THEN LPRINT
2030 PRINT "TO STOP PRESS BREAK":PRINT
2040 GOTO 1750
2050 END
```

Examples

1. The function $y=x^3$ is given by 6 equally spaced points on the interval $1 \le x \le 6$. The ordinates are: 1, 8, 27, 64, 125, 216. Mode of boundary conditions: 1. Since the function is a parabola of third degree, the cubic spline interpolates the function y and its first and second derivatives y_1, y_2 exactly. Some results in single precision:

   $x=2.5$, $y=15.625$, $y_1=18.75$, $y_2=15$.
   $x=4.5$, $y=91.125$, $y_1=60.75$, $y_2=27$.

2. The function $y=e^x$ is given by 6 equally spaced points on the interval $0.95 \le x \le 1.00$. The ordinates are exact in 6 digits:
   2.58571, 2.61170, 2.63794, 2.66446, 2.69123, 2.71828.
Mode of boundary conditions: 7. Some results in single precision:

   $x=0.955$, $y=2.59867$, $y_1=2.59924$, $y_2=2.56140$.
   $x=0.975$, $y=2.65117$, $y_1=2.65233$, $y_2=2.64998$.
   $x=0.990$, $y=2.69123$, $y_1=2.69082$, $y_2=3.02896$.

The values of the function are exact in 6 digits, but note the loss of accuracy in the first and second derivatives.

3. The function $y=e^x$ is given by 6 equally spaced points on the interval $0.95 \le x \le 1.00$. The ordinates are exact in 16 digits:
   2.585709659315846,   2.611696473423118,   2.637944459354153,
   2.664456241929417,   2.691234472349262,   2.718281828459045.

Mode of boundary conditions: 7. Some results computed in double precision and rounded to 11 digits:

   $x=0.955$, $y=2.5986705829$, $y_1=2.5986705830$, $y_2=2.5986813924$.
   $x=0.975$, $y=2.6511672109$, $y_1=2.6511672110$, $y_2=2.6511782574$.
   $x=0.990$, $y=2.6912344723$, $y_1=2.6912344721$, $y_2=2.6912120312$.

The values of the function are exact in 11 digits, the values of the first derivative in 10 digits, and the values of the second derivative in 5 or 6 digits because of the improved accuracy of the input of ordinates (exact in 16 digits while in example 2 they are exact only in 6 digits).

NOTE: The boundary conditions have no effect on the interpolation only if the function is a parabola of third or lower degree. The choice of the mode of boundary conditions is determined by the available information. Whenever possible use mode 7 (first derivatives on both ends of interval are given).

P210 Trigonometric Interpolation

```
10 CLS:PRINT"P210 COMPUTES Y(T) BY TRIGONOMETRIC INTERPOLATION."
20 PRINT:PRINT"THE INPUT:"
30 PRINT"SINGLE PRECISION ONLY."
40 PRINT"IF PRINTER NOT USED O2=0, IF USED O2>0."
50 DEFINT I-O
60 INPUT"O2";O2:CLS
70 PRINT"IF THE FUNCTION IS ODD O3=1, IF IT IS EVEN O3=2,"
80 PRINT"IF IT HAS NO SYMMETRY O3>2"
90 INPUT"O3";O3:CLS
100 PRINT"NUMBER OF POINTS DEFINING THE FUNCTION: N":PRINT
110 PRINT"INTERVAL COVERED BY N EQUALLY SPACED POINTS: (A,B)"
120 PRINT"IF THE FUNCTION HAS NO SYMMETRY THE N POINTS ARE"
130 PRINT"DISTRIBUTED ON THE INTERVAL (A,B)."
140 PRINT"IF THE FUNCTION IS ODD OR EVEN WITH RESPECT TO THE"
150 PRINT"POINT (A+B)/2 AND N IS ODD IT IS DEFINED BY (1+N)/2"
160 PRINT"POINTS DISTRIBUTED ON THE INTERVAL (A,(A+B)/2)."
170 PRINT:INPUT"N";N:INPUT"A";A:INPUT"B";B:CLS
180 PRINT"ABSCISSAS OF POINTS DEFINING THE FUNCTION: T0(I)"
190 PRINT"ORDINATES OF POINTS DEFINING THE FUNCTION: F(I)"
200 IF O2=0 THEN GOTO 230
210 LPRINT "P210: O2=";O2;"O3=";O3;"N=";N
220 LPRINT "A=";A;"B=";B
230 DIM F(N),T0(N),X0(N)
240 LET O4=1
250 IF ((N/2)-INT(N/2))=0 THEN LET O4=2
260 PRINT "INPUT OF ORDINATES:"
270 IF O2>0 THEN LPRINT "THE INPUT:"
280 LET Q=(B-A)/(N-1)
290 LET Q1=6.283185/(N-1)
300 LET Q2=6.283185/(B-A)
310 LET M=(3+N)/2
320 FOR I=1 TO N
330 LET T0(I)=Q*(I-1)+A
340 LET X0(I)=Q1*(I-1)
350 IF O4=2 THEN GOTO 420
360 IF O3>2 THEN GOTO 420
370 IF I<M THEN GOTO 420
380 LET J=1+N-I
390 LET F(I)=F(J)
400 IF O3=1 THEN F(I)=-F(I)
410 GOTO 460
420 PRINT "I=";I;"T0(I)=";T0(I)
430 INPUT"F(I)";F(I)
440 IF O2>0 THEN LPRINT "I=";I;"T0(I)=";T0(I);"F(I)=";F(I)
450 LET F(I)=F(I)/(N-1)
460 NEXT I
470 LET F(1)=(F(1)+F(N))/2
480 LET N=N-1
490 IF O2>0 THEN LPRINT
500 IF O2>0 THEN LPRINT "THE INTERPOLATION:"
510 CLS:PRINT "THE VARIABLE T LIES WITHIN THE INTERVAL A<=T<B"
520 PRINT "THE INTERPOLATION:"
530 INPUT"T";T
540 LET X=Q2*(T-A)
550 LET Y=0
560 FOR I=1 TO N
570 LET X1=(X0(I)-X)/2
580 LET Y0=SIN(X1)
590 IF Y0=0 THEN GOTO 640
600 LET Y1=F(I)*SIN(N*X1)/Y0
610 IF O4=2 THEN GOTO 650
620 LET Y1=Y1*COS(X1)
```

```
630 GOTO 650
640 LET Y1=N*F(I)
650 LET Y=Y+Y1
660 NEXT I
670 PRINT "Y(T)=";Y
680 IF O2=0 THEN GOTO 710
690 LPRINT
700 LPRINT "T=";T;"Y(T)=";Y
710 PRINT "TO STOP PRESS BREAK":PRINT
720 GOTO 530
730 END
```

Examples

1. An odd periodic function
$$y=\sin(x)+[\sin(2x)]/5+[\sin(3x)]/10$$
in which
$$x=2(PI)t/(B-A), \quad N=9, \quad A=1, \quad B=9,$$
is interpolated. The input data are
 0, 0.977817, 0.9, 0.577817, 0.
Some results of interpolation:
 t=1.5, y=0.616492, t=3.5, y=0.744190, t=7.25, y=-0.774177.

2. An even periodic function
$$y=\cos(x)+(2/5)*\cos(2x)+[\cos(3x)]/7$$
in which
$$x=2(PI)t/(B-A), \quad N=9, \quad A=1, \quad B=9,$$
is interpolated. The input data are
 1.54286, 0.606092, -0.4, -0.606092, -0.742857.
Some results of interpolation:
 t=2.5, y=-0.0321423, t=5.77, y=-0.646812,
 t=8.1257, y=0.784691.

3. A periodic function
$$y=\sin(x)+[\sin(2x)]/5+[\sin(3x)]/10+$$
$$+\cos(x)+(2/5)*\cos(2x)+[\cos(3x)]/7$$
in which
$$x=2(PI)t/(B-A), \quad N=9, \quad A=1, \quad B=9,$$
is interpolated. The 9 input data are
 1.54286, 1.58391, 0.5, -0.028274, -0.742857,
 -1.18391, -1.3, -0.371726, 1.54286.

Some results of interpolation:
    t=2.784,   y=0.686913,   t=6.501, y=-1.27789,
    t=8.999,   y=1.54079.
    4. A periodic function of example 3 with A=1, B=8 is interpolated. The 8 input data are
    1.54286, 1.42598, 0.316130, -0.308351,
    -1.05837, -1.30381, -0.614434, 1.54286.
Some results of interpolation:
    t=1.25,   y=1.81866,   t=5.787,   y=-1.27321.

P211 Least-Squares Approximation by Chebyshev Polynomials for the Function and Its First and Second Derivatives

```
10 CLS:PRINT"P211 APPROXIMATES Y(T) AND ITS FIRST AND SECOND"
20 PRINT"DERIVATIVE Y1,Y2 BY CHEBYSHEV POLYNOMIALS."
30 PRINT:PRINT"THE INPUT:"
40 PRINT"SINGLE PRECISION ONLY"
50 PRINT"IF PRINTER NOT USED O2=0, IF USED O2>0"
60 DEFINT I-O
70 INPUT"O2";O2:CLS
80 PRINT"ABSCISSAS OF POINTS DEFINING THE FUNCTION: X1(I)"
90 PRINT"ORDINATES OF POINTS DEFINING THE FUNCTION: F(I)":PRINT
100 PRINT"IF ALL DATA ARE NEW O3>1, IF ONLY ORDINATES F(I) ARE"
110 PRINT"NEW O3=1, IF ALL DATA ARE TAKEN FROM FILES P211A AND"
120 PRINT"P211B WRITTEN IN THE FIRST RUN O3=0."
130 INPUT"O3";O3:CLS
140 PRINT"IF ONLY FUNCTION COMPUTED O4=0, IF FUNCTION AND ITS"
150 PRINT"FIRST DERIVATIVE COMPUTED O4=1, IF FUNCTION AND ITS"
160 PRINT"FIRST AND SECOND DERIVATIVES COMPUTED O4=2."
170 INPUT"O4";O4:CLS:IF O3=0 THEN GOTO 200
180 PRINT"IF SEQUENCE OF VARIANCES COMPUTED O5>0, IF NOT O5=0."
190 INPUT"O5";O5:CLS
200 PRINT"NUMBER OF POINTS DEFINING THE FUNCTION: M>2"
210 PRINT"NUMBER OF TERMS IN THE SERIES: N<M, N>2"
220 PRINT"DEGREE OF THE HIGHEST POLYNOMIAL: N-1"
230 PRINT"IF N=M THE APPROXIMATION BECOMES INTERPOLATION"
240 INPUT"M";M:INPUT"N";N:CLS
250 PRINT"INTERVAL COVERED BY M POINTS: (P,Q)"
260 INPUT"P";P:INPUT"Q";Q:CLS
270 IF O2=0 THEN GOTO 310
280 LPRINT "P211: O2=";O2;"O3=";O3;"O4=";O4
290 IF O3>0 THEN LPRINT "O5=";O5
300 LPRINT "M=";M;"N=";N
310 DIM A(N),F(M),S(N)
320 DIM T(N,M),T0(N),T1(N),T2(N),X0(M),X1(M)
330 IF O3=0 THEN GOTO 1260
340 IF O3=1 THEN GOTO 530
350 REM COMPUTATION OF CHEBYSHEV ABSCISSAS ON INTERVAL (-1,1)
360 REM WRITING VECTOR X0(I) AND MATRIX T(J,I) ON A FILE
370 LET C=1.570796/M
380 LET M0=1+2*M
390 OPEN "O",1,"P211A"
400 FOR I=1 TO M
410 LET X0(I)=COS(C*(M0-2*I))
420 PRINT #1, X0(I)
430 LET X=X0(I)
440 GOSUB 1850
450 FOR J=1 TO N
460 LET T(J,I)=T0(J)
470 PRINT #1, T(J,I)
480 NEXT J
490 NEXT I
500 CLOSE 1
510 GOTO 620
520 REM READING VECTOR X0(I) AND MATRIX T(J,I) FROM A FILE
530 OPEN "I",1,"P211A"
540 FOR I=1 TO M
550 INPUT #1, X0(I)
560 FOR J=1 TO N
570 INPUT #1, T(J,I)
580 NEXT J
590 NEXT I
600 CLOSE 1
610 REM CHEBYSHEV ABSCISSAS ON INTERVAL (P,Q)
```

```
620 LET P0=(Q+P)/2
630 LET Q0=(Q-P)/2
640 PRINT "INPUT OF ORDINATES:"
650 IF O2>0 THEN LPRINT "THE INPUT:"
660 FOR I=1 TO M
670 LET X1(I)=P0+Q0*X0(I)
680 PRINT "I=";I;"X1(I)=";X1(I)
690 INPUT"F(I)";F(I)
700 NEXT I
710 CLS:PRINT "CHECK THE ORDINATES F(I):"
720 FOR I=1 TO M
730 PRINT "I=";I;"X1(I)=";X1(I);"F(I)=";F(I)
740 PRINT"IF NO ERROR INPUT 0"
750 PRINT"IF ERROR INPUT ANY NUMBER>0 AND THEN THE CORRECT F(I)"
760 INPUT"ERROR";K
770 IF K>0 THEN INPUT"F(I)";F(I)
780 CLS:NEXT I
790 PRINT"COMPUTING"
800 IF O2=0 THEN GOTO 860
810 LPRINT "I,X1(I),F(I):"
820 FOR I=1 TO M
830 LPRINT I;X1(I),F(I)
840 NEXT I
850 LPRINT
860 PRINT "COEFFICIENTS OF THE SERIES:"
870 FOR J=1 TO N
880 LET A(J)=0
890 FOR I=1 TO M
900 LET A(J)=A(J)+T(J,I)*F(I)
910 NEXT I
920 IF J=1 THEN GOTO 940
930 LET A(J)=2*A(J)
940 LET A(J)=A(J)/M
950 PRINT J;A(J)
960 NEXT J
970 IF O5=0 THEN GOTO 1150
980 PRINT "THE VARIANCES:"
990 FOR L=2 TO N
1000 LET S(L)=0
1010 FOR I=1 TO M
1020 LET B=0
1030 FOR J=1 TO L
1040 LET B=B+A(J)*T(J,I)
1050 NEXT J
1060 LET B=B-F(I)
1070 LET B=B*B
1080 LET S(L)=S(L)+B
1090 NEXT I
1100 IF L=M THEN GOTO 1120
1110 LET S(L)=S(L)/(M-L)
1120 PRINT "L=";L;"S(L)=";S(L)
1130 NEXT L
1140 REM WRITING COEFFICIENTS A(J) AND VARIANCES S(J) ON A FILE
1150 OPEN "O",1,"P211B"
1160 PRINT #1, O5,P,Q,P0,Q0,A(1)
1170 FOR J=2 TO N
1180 IF O5=0 THEN GOTO 1210
1190 PRINT #1, A(J),S(J)
1200 GOTO 1220
1210 PRINT #1, A(J)
1220 NEXT J
1230 CLOSE 1
1240 GOTO 1350
1250 REM READING COEFFICIENTS A AND VARIANCES S FROM A FILE
```

```
1260 OPEN "I",1,"P211B"
1270 INPUT #1, O5,P,Q,P0,Q0,A(1)
1280 FOR J=2 TO N
1290 IF O5=0 THEN GOTO 1320
1300 INPUT #1, A(J),S(J)
1310 GOTO 1330
1320 INPUT #1, A(J)
1330 NEXT J
1340 CLOSE 1
1350 PRINT "P=";P;"Q=";Q
1360 IF O2=0 THEN GOTO 1490
1370 LPRINT "P=";P;"Q=";Q
1380 LPRINT
1390 LPRINT "COEFFICIENTS A(J) AND VARIANCES S(J):"
1400 LPRINT "J=1    A(1)=";A(1)
1410 FOR J=2 TO N
1420 IF O5=0 THEN GOTO 1450
1430 LPRINT J;A(J),S(J)
1440 GOTO 1460
1450 LPRINT J;A(J)
1460 NEXT J
1470 LPRINT
1480 LPRINT "THE APPROXIMATION:"
1490 PRINT:PRINT "THE APPROXIMATION:"
1500 INPUT"T";X
1510 IF O2>0 THEN LPRINT "T =";X;"N=";N
1520 LET X=(X-P0)/Q0
1530 GOSUB 1850
1540 LET Y=0
1550 FOR J=1 TO N
1560 LET Y=Y+A(J)*T0(J)
1570 NEXT J
1580 PRINT "Y =";Y
1590 IF O2>0 THEN LPRINT "Y =";Y
1600 IF O4=0 THEN GOTO 1800
1610 REM COMPUTATION OF THE FIRST DERIVATIVE
1620 GOSUB 1940
1630 LET Y1=0
1640 FOR J=2 TO N
1650 LET Y1=Y1+A(J)*T1(J)
1660 NEXT J
1670 LET Y1=Y1/Q0
1680 PRINT "Y1=";Y1
1690 IF O2>0 THEN LPRINT "Y1=";Y1
1700 IF O4=1 THEN GOTO 1800
1710 REM COMPUTATION OF THE SECOND DERIVATIVE
1720 GOSUB 2030
1730 LET Y2=0
1740 FOR J=3 TO N
1750 LET Y2=Y2+A(J)*T2(J)
1760 NEXT J
1770 LET Y2=Y2/(Q0*Q0)
1780 PRINT "Y2=";Y2
1790 IF O2>0 THEN LPRINT "Y2=";Y2
1800 IF O2>0 THEN LPRINT
1810 PRINT "TO STOP PRESS BREAK":PRINT
1820 INPUT"N";N
1830 GOTO 1500
1840 REM SUBROUTINE FOR CHEBYSHEV POLYNOMIALS
1850 LET T0(1)=1
1860 LET T0(2)=X
1870 FOR L=3 TO N
1880 LET L1=L-1
1890 LET L2=L-2
```

```
1900 LET T0(L)=2*X*T0(L1)-T0(L2)
1910 NEXT L
1920 RETURN
1930 REM SUBROUTINE FOR THE FIRST DERIVATIVES OF CH. P.
1940 LET T1(1)=0
1950 LET T1(2)=1
1960 FOR L=3 TO N
1970 LET L1=L-1
1980 LET L2=L-2
1990 LET T1(L)=2*T0(L1)+2*X*T1(L1)-T1(L2)
2000 NEXT L
2010 RETURN
2020 REM SUBROUTINE FOR THE SECOND DERIVATIVES OF CH. P.
2030 LET T2(1)=0
2040 LET T2(2)=0
2050 FOR L=3 TO N
2060 LET L1=L-1
2070 LET L2=L-2
2080 LET T2(L)=4*T1(L1)+2*X*T2(L1)-T2(L2)
2090 NEXT L
2100 RETURN
2110 END
```

Examples

1. The function $y=x^3$ is given by 6 points with Chebyshev abscissas on the interval $1 \leq x \leq 6$:

(1.08519,1.27796), (1.73223,5.19777), (2.85295,23.2211),

(4.14705,71.3211), (5.26777,146.177), (5.91481,206,929).

The coefficients of the series are

75.6873, 103.593, 32.8122, 3.90601, $-1.09355*10^{-4}$,

$-5.72205*10^{-5}$,

and the variances

818.924, 15.257, $1.30642*10^{-8}$, (increasing values).

The minimum of variances indicates that the series should be restricted to 4 terms. Some results of approximation rounded to 5 digits:

x=2.5,   y=15.625, y_1=18.75, y_2=15,   N=4.
x=4.5,   y=91.125, y_1=60.750, y_2=27,   N=4.

2. The function $y=e^x$ is given by 6 points with Chebyshev abscissas on the interval $0.95 \leq x \leq 1$:

(0.950852,2.58791), (0.957322,2.60471), (0.968530,2.63407),

(0.981471,2.66838), (0.992678,2.69845), (0.999148,2.71597).

PROGRAM P211

The coefficients of the series are
 2.65158, 0.0662887, $4.11193*10^{-4}$, $3.616602*10^{-6}$,
 $-5.96046*10^{-7}$, $9.73543*10^{-7}$,
and the variances
 $1.27736*10^{-7}$, $1.34719*10^{-11}$, (increasing values).

The minimum of variances indicates that the series should be restricted to 3 or 4 terms. Some results of approximation:

x=0.955,	y=2.59866,	y_1=2.59892,	y_2=2.63163,	N=3.
x=0.955,	y=2.59867,	y_1=2.59959,	y_2=2.52055,	N=4.
x=0.975,	y=2.65117,	y_1=2.65155,	y_2=2.65163,	N=3.
x=0.975,	y=2.65117,	y_1=2.65111,	y_2=2.63163,	N=4.
x=0.990,	y=2.69124,	y_1=2.69102,	y_2=2.63163,	N=3.
x=0.990,	y=2.69124,	y_1=2.29121,	y_2=2.71495,	N=4.

P212 Least-Squares Approximation by Orthogonal Polynomials with Arbitrarily Spaced Abscissas and a Given Weight Function for the Function and Its First and Second Derivatives

```
10 CLS:PRINT"P212 APPROXIMATES Y(X) AND ITS FIRST AND SECOND"
20 PRINT"DERIVATIVE Y1,Y2 BY ORTHOGONAL POLYNOMIALS WITH"
30 PRINT"ARBITRARILY SPACED ABSCISSAS AND GIVEN WEIGHT"
40 PRINT"FUNCTION."
50 PRINT:PRINT"THE INPUT:"
60 PRINT"IF SINGLE PRECISION O1=1, IF DOUBLE O1=2."
70 DEFINT I-O:INPUT"O1";O1:CLS
80 PRINT"IF PRINTER NOT USED O2=0, IF USED O2>0."
90 INPUT "O2";O2:CLS
100 IF O1=2 THEN DEFDBL A-H,P-Z
110 PRINT"ABSCISSAS OF POINTS DEFINING THE FUNCTION: X0(I)"
120 PRINT"WEIGHT FUNCTION: W(I)"
130 PRINT"ORDINATES OF POINTS DEFINING THE FUNCTION: F(I)":PRINT
140 PRINT"IF ALL DATA NEW O3>1, IF ONLY ORDINATES F(I) ARE"
150 PRINT"NEW O3=1, IF ALL DATA ARE TAKEN FROM FILES P212A AND"
160 PRINT"P212B WRITTEN IN THE FIRST RUN O3=0."
170 INPUT"O3";O3:CLS
180 PRINT"IF ONLY FUNCTION COMPUTED O4=0, IF FUNCTION AND ITS"
190 PRINT"FIRST DERIVATIVE COMPUTED O4=1, IF FUNCTION AND ITS"
200 PRINT"FIRST AND SECOND DERIVATIVES COMPUTED O4=2."
210 INPUT"O4";O4:CLS:IF O3=0 THEN GOTO 240
220 PRINT"IF SEQUENCE OF VARIANCES COMPUTED O5>0, IF NOT O5=0."
230 INPUT"O5";O5:CLS
240 PRINT"NUMBER OF POINTS DEFINING THE FUNCTION: M>2"
250 PRINT"NUMBER OF TERMS IN THE SERIES: N<M, N>2"
260 PRINT"DEGREE OF THE HIGHEST POLYNOMIAL: N-1"
270 PRINT"IF N=M THE APPROXIMATION BECOMES INTERPOLATION"
280 INPUT"M";M:INPUT"N";N:CLS
290 REM VECTORS N(j),a(j),b(j) ARE STORED IN MATRIX C(J,L),
300 REM L=1,2,3. POLYNOMIALS p(j,(xi)) ARE STORED
310 REM IN MATRIX P(J,I)
320 IF O2=0 THEN GOTO 360
330 LPRINT "P212: O1=";O1;"O2=";O2;"O3=";O3;"O4=";O4
340 IF O3>0 THEN LPRINT "O5=";O5
350 LPRINT "M=";M;"N=";N
360 DIM A(N),C(N,3),S(N),W(M)
370 DIM P(N,M),P0(N),P1(N),P2(N),X0(M)
380 IF O3=0 THEN GOTO 1110
390 IF O2>0 THEN LPRINT "THE INPUT:"
400 IF O3=1 THEN GOTO 1110
410 PRINT "INPUT OF ABSCISSAS AND WEIGHTS:"
420 FOR I=1 TO M
430 PRINT "I=";I
440 INPUT"X0(I)";X0(I):INPUT"W(I)";W(I)
450 NEXT I
460 CLS:PRINT "CHECK THE ABSCISSAS AND WEIGHTS:"
470 FOR I=1 TO M
480 PRINT "I=";I;"X0(I)=";X0(I);"W(I)=";W(I)
490 PRINT"IF NO ERROR IN X0(I) INPUT 0"
500 PRINT"IF ERROR INPUT ANY NUMBER>0 AND THEN THE CORRECT X0(I)"
510 INPUT"ERROR";K
520 IF K>0 THEN INPUT"X0(I)";X0(I)
530 PRINT:PRINT"IF NO ERROR IN W(I) INPUT 0"
540 PRINT"IF ERROR INPUT ANY NUMBER>0 AND THEN THE CORRECT W(I)"
550 INPUT"ERROR";K
560 IF K>0 THEN INPUT"W(I)";W(I)
570 CLS:NEXT I
580 PRINT"COMPUTING"
```

PROGRAM P212

```
590 REM COMPUTATION OF MATRICES C(J,L) AND P(J,I)
600 LET C(1,1)=0
610 FOR I=1 TO M
620 LET C(1,1)=C(1,1)+W(I)
630 LET P(1,I)=1
640 NEXT I
650 LET C(2,2)=0
660 FOR I=1 TO M
670 LET C(2,2)=C(2,2)+W(I)*X0(I)
680 NEXT I
690 LET C(2,2)=C(2,2)/C(1,1)
700 FOR I=1 TO M
710 LET P(2,I)=X0(I)-C(2,2)
720 NEXT I
730 LET N1=N+1
740 FOR J=3 TO N1
750 LET J1=J-1
760 LET J2=J-2
770 LET C(J1,1)=0
780 IF J=N1 THEN GOTO 800
790 LET C(J,2)=0
800 FOR I=1 TO M
810 LET B=W(I)*P(J1,I)*P(J1,I)
820 LET C(J1,1)=C(J1,1)+B
830 IF J=N1 THEN GOTO 850
840 LET C(J,2)=C(J,2)+B*X0(I)
850 NEXT I
860 IF J=N1 THEN GOTO 920
870 LET C(J,2)=C(J,2)/C(J1,1)
880 LET C(J1,3)=C(J1,1)/C(J2,1)
890 FOR I=1 TO M
900 LET P(J,I)=(X0(I)-C(J,2))*P(J1,I)-C(J1,3)*P(J2,I)
910 NEXT I
920 NEXT J
930 REM WRITING VECTORS X0(I),W(I) AND MATRICES C(J,L),...
940 REM ...P(J,I) ON A FILE
950 OPEN "O",1,"P212A"
960 FOR I=1 TO M
970 PRINT #1, X0(I),W(I)
980 NEXT I
990 FOR J=1 TO N
1000 FOR I=1 TO M
1010 PRINT #1, P(J,I)
1020 NEXT I
1030 FOR L=1 TO 3
1040 PRINT #1, C(J,L)
1050 NEXT L
1060 NEXT J
1070 CLOSE 1
1080 GOTO 1250
1090 REM READING VECTORS X0(I),W(I) AND MATRICES C(J,L),...
1100 REM ...P(J,I) FROM A FILE
1110 OPEN "I",1,"P212A"
1120 FOR I=1 TO M
1130 INPUT #1, X0(I),W(I)
1140 NEXT I
1150 FOR J=1 TO N
1160 FOR I=1 TO M
1170 INPUT #1, P(J,I)
1180 NEXT I
1190 FOR L=1 TO 3
1200 INPUT #1, C(J,L)
1210 NEXT L
1220 NEXT J
```

```
1230 CLOSE 1
1240 IF O3=0 THEN GOTO 1820
1250 PRINT "INPUT OF ORDINATES:"
1260 FOR I=1 TO M
1270 PRINT "I=";I;"X0(I)=";X0(I)
1280 INPUT"F(I)";F(I)
1290 NEXT I
1300 CLS:PRINT "CHECK THE ORDINATES F(I):"
1310 FOR I=1 TO M
1320 PRINT "I=";I;"X0(I)=";X0(I);"F(I)=";F(I)
1330 PRINT"IF NO ERROR INPUT 0"
1340 PRINT"IF ERROR INPUT ANY NUMBER>0 AND THEN THE CORRECT F(I)"
1350 INPUT"ERROR";K
1360 IF K>0 THEN INPUT"F(I)";F(I)
1370 CLS:NEXT I
1380 PRINT"COMPUTING"
1390 IF O2=0 THEN GOTO 1450
1400 LPRINT "I,X0(I),W(I),F(I):"
1410 FOR I=1 TO M
1420 LPRINT I;X0(I),W(I),F(I)
1430 NEXT I
1440 LPRINT
1450 PRINT "COEFFICIENTS OF THE SERIES:"
1460 FOR J=1 TO N
1470 LET A(J)=0
1480 FOR I=1 TO M
1490 LET A(J)=A(J)+P(J,I)*W(I)*F(I)
1500 NEXT I
1510 LET A(J)=A(J)/C(J,1)
1520 PRINT J;A(J)
1530 NEXT J
1540 IF O5=0 THEN GOTO 1720
1550 PRINT "THE VARIANCES:"
1560 FOR L=2 TO N
1570 LET S(L)=0
1580 FOR I=1 TO M
1590 LET B=0
1600 FOR J=1 TO L
1610 LET B=B+A(J)*P(J,I)
1620 NEXT J
1630 LET B=B-F(I)
1640 LET B=B*B*W(I)
1650 LET S(L)=S(L)+B
1660 NEXT I
1670 IF L=M THEN GOTO 1690
1680 LET S(L)=S(L)/(M-L)
1690 PRINT "L=";L;"S(L)=";S(L)
1700 NEXT L
1710 REM WRITING COEFFICIENTS A(J) AND VARIANCES S(J) ON A FILE
1720 OPEN "O",1,"P212B"
1730 PRINT #1, O5
1740 FOR J=1 TO N
1750 PRINT #1, A(J)
1760 IF J=1 THEN GOTO 1780
1770 IF O5>0 THEN PRINT #1, S(J)
1780 NEXT J
1790 CLOSE 1
1800 GOTO 1900
1810 REM READING COEFFICIENTS A AND VARIANCES S FROM A FILE
1820 OPEN "I",1,"P212B"
1830 INPUT #1, O5
1840 FOR J=1 TO N
1850 INPUT #1, A(J)
1860 IF J=1 THEN GOTO 1880
```

PROGRAM P212

```
1870 IF O5>0 THEN INPUT #1, S(J)
1880 NEXT J
1890 CLOSE 1
1900 IF O2=0 THEN GOTO 2010
1910 LPRINT "COEFFICIENTS A(J) AND VARIANCES S(J):"
1920 LPRINT "J=1   A(1)=";A(1)
1930 FOR J=2 TO N
1940 IF O5=0 THEN GOTO 1970
1950 LPRINT J;A(J),S(J)
1960 GOTO 1980
1970 LPRINT J;A(J)
1980 NEXT J
1990 LPRINT
2000 LPRINT "THE APPROXIMATION:"
2010 PRINT:PRINT "THE APPROXIMATION:"
2020 INPUT"X";X
2030 IF O2>0 THEN LPRINT "X =";X;"N=";N
2040 GOSUB 2340
2050 LET Y=0
2060 FOR J=1 TO N
2070 LET Y=Y+A(J)*P0(J)
2080 NEXT J
2090 PRINT "Y =";Y
2100 IF O2>0 THEN LPRINT "Y =";Y
2110 IF O4=0 THEN GOTO 2290
2120 REM COMPUTATION OF THE FIRST DERIVATIVE
2130 GOSUB 2430
2140 LET Y1=0
2150 FOR J=2 TO N
2160 LET Y1=Y1+A(J)*P1(J)
2170 NEXT J
2180 PRINT "Y1=";Y1
2190 IF O2>0 THEN LPRINT "Y1=";Y1
2200 IF O4=1 THEN GOTO 2290
2210 REM COMPUTATION OF THE SECOND DERIVATIVE
2220 GOSUB 2520
2230 LET Y2=0
2240 FOR J=3 TO N
2250 LET Y2=Y2+A(J)*P2(J)
2260 NEXT J
2270 PRINT "Y2=";Y2
2280 IF O2>0 THEN LPRINT "Y2=";Y2
2290 IF O2>0 THEN LPRINT
2300 PRINT "TO STOP PRESS BREAK":PRINT
2310 INPUT"N";N
2320 GOTO 2020
2330 REM SUBROUTINE FOR ORTHOGONAL POLYNOMIALS
2340 LET P0(1)=1
2350 LET P0(2)=X-C(2,2)
2360 FOR J=3 TO N
2370 LET J1=J-1
2380 LET J2=J-2
2390 LET P0(J)=(X-C(J,2))*P0(J1)-C(J1,3)*P0(J2)
2400 NEXT J
2410 RETURN
2420 REM SUBROUTINE FOR THE FIRST DERIVATIVES OF O. P.
2430 LET P1(1)=0
2440 LET P1(2)=1
2450 FOR J=3 TO N
2460 LET J1=J-1
2470 LET J2=J-2
2480 LET P1(J)=P0(J1)+(X-C(J,2))*P1(J1)-C(J1,3)*P1(J2)
2490 NEXT J
2500 RETURN
```

```
2510 REM SUBROUTINE FOR THE SECOND DERIVATIVES OF O. P.
2520 LET P2(1)=0
2530 LET P2(2)=0
2540 FOR J=3 TO N
2550 LET J1=J-1
2560 LET J2=J-2
2570 LET P2(J)=2*P1(J1)+(X-C(J,2))*P2(J1)-C(J1,3)*P2(J2)
2580 NEXT J
2590 RETURN
2600 END
```

Examples

1. The function $y=x^3$ is given by 6 points
    (1,1), (1.5,3.375), (2,8),
    (3.5,42.875), (5.5,166.375), (6,216).
The weights at all 6 points are W=1. The coefficients of the series are
    72.9375, 42.9455, 10.2694, 0.999989, $-6.1589*10^{-6}$,
    $-1.2669*10^{-6}$,
and the variances
    766.349, 13.3045, $2.72519*10^{-9}$, (increasing values).
The minimum of variances indicates that the series should be restricted to 4 terms. Some results of approximation:
    x=2.5,   y=15.625, y_1=18.75, y_2=15.
    x=4.5,   y=91.125, y_1=60.75, y_2=27.

2. The function $y=e^x$ is given by 6 points
    (0.95,2.58571), (0.96,2.61170), (0.965,2.62479),
    (0.985,2.67781), (0.99,2.69123), (1,2.71828).
The weights at all 6 points W=1. The coefficients of the series are
    2.65159, 2.65151, 1.32307, 3.55712, 15.7048, -188217,
and the variances
    $1.31915*10^{-7}$, $5.92119*10^{-11}$, (increasing values).
The minimum of variances and the sharp increase of coefficients indicate that the series should be restricted to 3 terms (that is, to a quadratic parabola with a constant value of its second derivative). Some results of approximation in single precision:

```
x=0.955,    y=2.59867,    y₁=2.59859,    y₂=2.64614.
x=0.975,    y=2.65117,    y₁=2.65151,    y₂=2.64614.
x=0.990,    y=2.69124,    y₁=2.69120,    y₂=2.64614.
```
The second derivative is correct only in 2 digits, which is nevertheless a success compared with the results of interpolation or appoximation by other methods.

3. The function $y=e^x$ is given by 6 points
(0.95,2.585709659315846), (0.96,2.611696473423118),
(0.965,2.624787656474575), (0.985,2.677811884421050),
(0.99,2.691234472349262), (1,2.718281828459045).

The weights at all 6 points are W=1. The coefficients of the series are

2.651586995740483, 2.651377101019293, 1.325667060618282,
0.441878693393626, 0.1104687768539612, 0.02209024652667949,

and the variances steadily decrease from 10^{-7} to 10^{-28}. This indicates that the series should contain all 6 terms. Some results of approximation in double precision:

```
x=0.955,    y=2.598670582919566,    y₁=2.598670582912549,
    y₂=2.598670580396540.
x=0.975,    y=2.651167210982659,    y₁=2.651167210981691,
    y₂=2.6511672093409.
x=0.990,    y=2.691234472349259,    y₁=2.691234472355450,
    y₂=2.691234474605408.
```

The examples show how important it is to truncate the series after the term with the minimum variance. The results of approximations are satisfactory even in single precision and excellent in double precision.

P213 Least-Squares Approximation by Orthogonal Polynomials with Arbitrarily Spaced Abscissas and a Weight Function Equal to 1 for the Function and Its First and Second Derivatives

```
10 CLS:PRINT"P213 APPROXIMATES Y(X) AND ITS FIRST AND SECOND"
20 PRINT"DERIVATIVE Y1,Y2 BY ORTHOGONAL POLYNOMIALS WITH"
30 PRINT"ARBITRARILY SPACED ABSCISSAS AND WEIGHT FUNCTION=1."
40 PRINT:PRINT"THE INPUT:"
50 PRINT"IF SINGLE PRECISION O1=1, IF DOUBLE O1=2."
60 DEFINT I-O:INPUT"O1";O1:CLS
70 PRINT"IF PRINTER NOT USED O2=0, IF USED O2>0."
80 INPUT"O2";O2:CLS
90 IF O1=2 THEN DEFDBL A-H,P-Z
100 PRINT"ABSCISSAS OF POINTS DEFINING THE FUNCTION: X0(I)"
110 PRINT"ORDINATES OF POINTS DEFINING THE FUNCTION: F(I)":PRINT
120 PRINT"IF ALL DATA NEW O3>1, IF ONLY ORDINATES F(I) ARE"
130 PRINT"NEW O3=1, IF ALL DATA ARE TAKEN FROM FILES P213A"
140 PRINT"AND P213B WRITTEN IN THE FIRST RUN O3=0."
150 INPUT"O3";O3:CLS
160 PRINT"IF ONLY FUNCTION COMPUTED O4=0, IF FUNCTION AND ITS"
170 PRINT"FIRST DERIVATIVE COMPUTED O4=1, IF FUNCTION AND ITS"
180 PRINT"FIRST AND SECOND DERIVATIVES COMPUTED O4=2."
190 INPUT"O4";O4:CLS:IF O3=0 THEN GOTO 220
200 PRINT"IF SEQUENCE OF VARIANCES COMPUTED O5>0, IF NOT O5=0."
210 INPUT"O5";O5:CLS
220 PRINT"NUMBER OF POINTS DEFINING THE FUNCTION: M>2"
230 PRINT"NUMBER OF TERMS IN THE SERIES: N<M, N>2"
240 PRINT"DEGREE OF THE HIGHEST POLYNOMIAL: N-1"
250 PRINT"IF N=M THE APPROXIMATION BECOMES INTERPOLATION."
260 INPUT"M";M:INPUT"N";N:CLS
270 REM VECTORS N(j),a(j),b(j) ARE STORED IN MATRIX C(J,L),
280 REM L=1,2,3. POLYNOMIALS p(j,(xi)) ARE STORED
290 REM IN MATRIX P(J,I)
300 IF O2=0 THEN GOTO 340
310 LPRINT "P213: O1=";O1;"O2=";O2;"O3=";O3;"O4=";O4
320 IF O3>0 THEN LPRINT "O5=";O5
330 LPRINT "M=";M;"N=";N
340 DIM A(N),C(N,3),S(N)
350 DIM P(N,M),P0(N),P1(N),P2(N),X0(M)
360 IF O3=0 THEN GOTO 1040
370 IF O2>0 THEN LPRINT "THE INPUT:"
380 IF O3=1 THEN GOTO 1040
390 PRINT "INPUT OF ABSCISSAS:"
400 FOR I=1 TO M
410 PRINT "I=";I
420 INPUT"X0(I)";X0(I)
430 NEXT I
440 CLS:PRINT "CHECK THE ABSCISSAS:"
450 FOR I=1 TO M
460 PRINT "I=";I;"X0(I)=";X0(I)
470 PRINT"IF NO ERROR INPUT 0"
480 PRINT"IF ERROR INPUT ANY NUMBER>0 AND THEN THE CORRECT X0(I)"
490 INPUT"ERROR";K
500 IF K>0 THEN INPUT"X0(I)";X0(I)
510 CLS:NEXT I
520 PRINT"COMPUTING"
530 REM COMPUTATION OF MATRICES C(J,L) AND P(J,I)
540 LET C(1,1)=M
550 FOR I=1 TO M
560 LET P(1,I)=1
570 NEXT I
```

PROGRAM P213

```
580 LET C(2,2)=0
590 FOR I=1 TO M
600 LET C(2,2)=C(2,2)+X0(I)
610 NEXT I
620 LET C(2,2)=C(2,2)/C(1,1)
630 FOR I=1 TO M
640 LET P(2,I)=X0(I)-C(2,2)
650 NEXT I
660 LET N1=N+1
670 FOR J=3 TO N1
680 LET J1=J-1
690 LET J2=J-2
700 LET C(J1,1)=0
710 IF J=N1 THEN GOTO 730
720 LET C(J,2)=0
730 FOR I=1 TO M
740 LET B=P(J1,I)*P(J1,I)
750 LET C(J1,1)=C(J1,1)+B
760 IF J=N1 THEN GOTO 780
770 LET C(J,2)=C(J,2)+B*X0(I)
780 NEXT I
790 IF J=N1 THEN GOTO 850
800 LET C(J,2)=C(J,2)/C(J1,1)
810 LET C(J1,3)=C(J1,1)/C(J2,1)
820 FOR I=1 TO M
830 LET P(J,I)=(X0(I)-C(J,2))*P(J1,I)-C(J1,3)*P(J2,I)
840 NEXT I
850 NEXT J
860 REM WRITING VECTOR X0(I) AND MATRICES C(J,L), P(J,I)...
870 REM ...ON A FILE
880 OPEN "O",1,"P213A"
890 FOR I=1 TO M
900 PRINT #1, X0(I)
910 NEXT I
920 FOR J=1 TO N
930 FOR I=1 TO M
940 PRINT #1, P(J,I)
950 NEXT I
960 FOR L=1 TO 3
970 PRINT #1, C(J,L)
980 NEXT L
990 NEXT J
1000 CLOSE 1
1010 GOTO 1180
1020 REM READING VECTOR X0(I) AND MATRICES C(J,L), P(J,I)...
1030 REM ...FROM A FILE
1040 OPEN "I",1,"P213A"
1050 FOR I=1 TO M
1060 INPUT #1, X0(I)
1070 NEXT I
1080 FOR J=1 TO N
1090 FOR I=1 TO M
1100 INPUT #1, P(J,I)
1110 NEXT I
1120 FOR L=1 TO 3
1130 INPUT #1, C(J,L)
1140 NEXT L
1150 NEXT J
1160 CLOSE 1
1170 IF O3=0 THEN GOTO 1760
1180 PRINT "INPUT OF ORDINATES:"
1190 FOR I=1 TO M
1200 PRINT "I=";I;"X0(I)=";X0(I)
```

```
1210 INPUT"F(I)";F(I)
1220 NEXT I
1230 CLS:PRINT "CHECK THE ORDINATES F(I):"
1240 FOR I=1 TO M
1250 PRINT "I=";I;"X0(I)=";X0(I);"F(I)=";F(I)
1260 PRINT"IF NO ERROR INPUT 0"
1270 PRINT"IF ERROR INPUT ANY NUMBER>0 AND THEN THE CORRECT F(I)"
1280 INPUT"ERROR";K
1290 IF K>0 THEN INPUT"F(I)";F(I)
1300 CLS:NEXT I
1310 PRINT"COMPUTING"
1320 IF O2=0 THEN GOTO 1380
1330 LPRINT "I,X0(I),F(I):"
1340 FOR I=1 TO M
1350 LPRINT I;X0(I),F(I)
1360 NEXT I
1370 LPRINT
1380 PRINT "COEFFICIENTS OF THE SERIES:"
1390 FOR J=1 TO N
1400 LET A(J)=0
1410 FOR I=1 TO M
1420 LET A(J)=A(J)+P(J,I)*F(I)
1430 NEXT I
1440 LET A(J)=A(J)/C(J,1)
1450 PRINT J;A(J)
1460 NEXT J
1470 IF O5=0 THEN GOTO 1650
1480 PRINT "THE VARIANCES:"
1490 FOR L=2 TO N
1500 LET S(L)=0
1510 FOR I=1 TO M
1520 LET B=0
1530 FOR J=1 TO L
1540 LET B=B+A(J)*P(J,I)
1550 NEXT J
1560 LET B=B-F(I)
1570 LET B=B*B
1580 LET S(L)=S(L)+B
1590 NEXT I
1600 IF L=M THEN GOTO 1620
1610 LET S(L)=S(L)/(M-L)
1620 PRINT "L=";L;"S(L)=";S(L)
1630 NEXT L
1640 REM WRITING COEFFICIENTS A(J) AND VARIANCES S(J) ON A FILE
1650 OPEN "O",1,"P213B"
1660 PRINT #1, O5
1670 FOR J=1 TO N
1680 PRINT #1, A(J)
1690 IF J=1 THEN GOTO 1710
1700 IF O5>0 THEN PRINT #1, S(J)
1710 NEXT J
1720 CLOSE 1
1730 GOTO 1840
1740 REM READING COEFFICIENTS A(J) AND VARIANCES S(J)...
1750 REM ...FROM A FILE
1760 OPEN "I",1,"P213B"
1770 INPUT #1, O5
1780 FOR J=1 TO N
1790 INPUT #1, A(J)
1800 IF J=1 THEN GOTO 1820
1810 IF O5>0 THEN INPUT #1, S(J)
1820 NEXT J
1830 CLOSE 1
```

PROGRAM P213

```
1840 IF O2=0 THEN GOTO 1950
1850 LPRINT "COEFFICIENTS A(J) AND VARIANCES S(J):"
1860 LPRINT "J=1   A(1)=";A(1)
1870 FOR J=2 TO N
1880 IF O5=0 THEN GOTO 1910
1890 LPRINT J;A(J),S(J)
1900 GOTO 1920
1910 LPRINT J;A(J)
1920 NEXT J
1930 LPRINT
1940 LPRINT "THE APPROXIMATION:"
1950 PRINT:PRINT "THE APPROXIMATION:"
1960 INPUT"X";X
1970 IF O2>0 THEN LPRINT "X =";X;"N=";N
1980 GOSUB 2280
1990 LET Y=0
2000 FOR J=1 TO N
2010 LET Y=Y+A(J)*P0(J)
2020 NEXT J
2030 PRINT "Y =";Y
2040 IF O2>0 THEN LPRINT "Y =";Y
2050 IF O4=0 THEN GOTO 2230
2060 REM COMPUTATION OF THE FIRST DERIVATIVE
2070 GOSUB 2370
2080 LET Y1=0
2090 FOR J=2 TO N
2100 LET Y1=Y1+A(J)*P1(J)
2110 NEXT J
2120 PRINT "Y1=";Y1
2130 IF O2>0 THEN LPRINT "Y1=";Y1
2140 IF O4=1 THEN GOTO 2230
2150 REM COMPUTATION OF THE SECOND DERIVATIVE
2160 GOSUB 2460
2170 LET Y2=0
2180 FOR J=3 TO N
2190 LET Y2=Y2+A(J)*P2(J)
2200 NEXT J
2210 PRINT "Y2=";Y2
2220 IF O2>0 THEN LPRINT "Y2=";Y2
2230 IF O2>0 THEN LPRINT
2240 PRINT "TO STOP PRESS BREAK":PRINT
2250 INPUT"N";N
2260 GOTO 1960
2270 REM SUBROUTINE FOR ORTHOGONAL POLYNOMIALS
2280 LET P0(1)=1
2290 LET P0(2)=X-C(2,2)
2300 FOR J=3 TO N
2310 LET J1=J-1
2320 LET J2=J-2
2330 LET P0(J)=(X-C(J,2))*P0(J1)-C(J1,3)*P0(J2)
2340 NEXT J
2350 RETURN
2360 REM SUBROUTINE FOR THE FIRST DERIVATIVES OF O. P.
2370 LET P1(1)=0
2380 LET P1(2)=1
2390 FOR J=3 TO N
2400 LET J1=J-1
2410 LET J2=J-2
2420 LET P1(J)=P0(J1)+(X-C(J,2))*P1(J1)-C(J1,3)*P1(J2)
2430 NEXT J
2440 RETURN
2450 REM SUBROUTINE FOR THE SECOND DERIVATIVES OF O. P.
2460 LET P2(1)=0
```

```
2470 LET P2(2)=0
2480 FOR J=3 TO N
2490 LET J1=J-1
2500 LET J2=J-2
2510 LET P2(J)=2*P1(J1)+(X-C(J,2))*P2(J1)-C(J1,3)*P2(J2)
2520 NEXT J
2530 RETURN
2540 END
```

Examples

Since the weight function in the examples to program P212 is chosen as 1, those examples serve this program as well.

P214 Least-Squares Approximation by Orthogonal Polynomials with Equally Spaced Abscissas and a Given Weight Function for the Function and Its First and Second Derivatives

```
10 CLS:PRINT"P214 APPROXIMATES Y(X) AND ITS FIRST AND SECOND"
20 PRINT"DERIVATIVE Y1,Y2 BY ORTHOGONAL POLYNOMIALS WITH"
30 PRINT"EQUALLY SPACED ABSCISSAS AND GIVEN WEIGHT FUNCTION."
40 PRINT:PRINT"THE INPUT:"
50 PRINT"IF SINGLE PRECISION O1=1, IF DOUBLE O1=2."
60 DEFINT I-O:INPUT"O1";O1:CLS
70 PRINT"IF PRINTER NOT USED O2=0, IF USED O2>0."
80 INPUT"O2";O2:CLS
90 IF O1=2 THEN DEFDBL A-H,P-Z
100 PRINT"ABSCISSAS OF POINTS DEFINING THE FUNCTION: X0(I)"
110 PRINT"WEIGHT FUNCTION: W(I)"
120 PRINT"ORDINATES OF POINTS DEFINING THE FUNCTION: F(I)":PRINT
130 PRINT"IF ALL DATA NEW O3>1, IF ONLY ORDINATES F(I) ARE"
140 PRINT"NEW O3=1, IF ALL DATA ARE TAKEN FROM FILES P214A"
150 PRINT"AND P214B WRITTEN IN THE FIRST RUN O3=0."
160 INPUT"O3";O3:CLS
170 PRINT"IF ONLY FUNCTION COMPUTED O4=0, IF FUNCTION AND ITS"
180 PRINT"FIRST DERIVATIVE COMPUTED O4=1, IF FUNCTION AND ITS"
190 PRINT"FIRST AND SECOND DERIVATIVES COMPUTED O4=2."
200 INPUT"O4";O4:CLS:IF O3=0 THEN GOTO 230
210 PRINT"IF SEQUENCE OF VARIANCES COMPUTED O5>0, IF NOT O5=0."
220 INPUT"O5";O5:CLS
230 PRINT"NUMBER OF POINTS DEFINING THE FUNCTION: M>2"
240 PRINT"NUMBER OF TERMS IN THE SERIES: N<M, N>2"
250 PRINT"DEGREE OF THE HIGHEST POLYNOMIAL: N-1"
260 PRINT"IF N=M THE APPROXIMATION BECOMES INTERPOLATION."
270 INPUT"M";M:INPUT"N";N:CLS
280 PRINT "INTERVAL COVERED BY N POINTS: (U,V)"
290 INPUT"U";U:INPUT"V";V:CLS
300 REM VECTORS N(j),a(j),b(j) ARE STORED IN MATRIX C(J,L),
310 REM L=1,2,3. POLYNOMIALS p(j,(xi)) ARE STORED
320 REM IN MATRIX P(J,I)
330 IF O2=0 THEN GOTO 370
340 LPRINT "P214: O1=";O1;"O2=";O2;"O3=";O3;"O4=";O4
350 IF O3>0 THEN LPRINT "O5=";O5
360 LPRINT "M=";M;"N=";N
370 DIM A(N),C(N,3),S(N),W(M)
380 DIM P(N,M),P0(N),P1(N),P2(N),X0(M)
390 IF O3<2 THEN GOTO 1150
400 REM INPUT OF INTERVAL
410 IF O2>0 THEN LPRINT "INTERVAL U=";U;"V=";V
420 LET H=(V-U)/(M-1)
430 LET X0(1)=U
440 FOR I=2 TO M
450 LET I1=I-1
460 LET X0(I)=X0(I1)+H
470 NEXT I
480 IF O2>0 THEN LPRINT "THE INPUT:"
490 PRINT "INPUT OF WEIGHTS:"
500 FOR I=1 TO M
510 PRINT "I=";I
520 INPUT"W(I)";W(I)
530 NEXT I
540 CLS:PRINT "CHECK THE WEIGHTS:"
550 FOR I=1 TO M
560 PRINT "I=";I;"W(I)=";W(I)
570 PRINT"IF NO ERROR INPUT 0"
580 PRINT"IF ERROR INPUT ANY NUMBER>0 AND THEN THEN CORRECT W(I)"
```

```
590 INPUT"ERROR";K
600 IF K>0 THEN INPUT"W(I)";W(I)
610 CLS:NEXT I
620 PRINT"COMPUTING"
630 REM COMPUTATION OF MATRICES C(J,L) AND P(J,I)
640 LET C(1,1)=0
650 FOR I=1 TO M
660 LET C(1,1)=C(1,1)+W(I)
670 LET P(1,I)=1
680 NEXT I
690 LET C(2,2)=0
700 FOR I=1 TO M
710 LET C(2,2)=C(2,2)+W(I)*X0(I)
720 NEXT I
730 LET C(2,2)=C(2,2)/C(1,1)
740 FOR I=1 TO M
750 LET P(2,I)=X0(I)-C(2,2)
760 NEXT I
770 LET N1=N+1
780 FOR J=3 TO N1
790 LET J1=J-1
800 LET J2=J-2
810 LET C(J1,1)=0
820 IF J=N1 THEN GOTO 840
830 LET C(J,2)=0
840 FOR I=1 TO M
850 LET B=W(I)*P(J1,I)*P(J1,I)
860 LET C(J1,1)=C(J1,1)+B
870 IF J=N1 THEN GOTO 890
880 LET C(J,2)=C(J,2)+B*X0(I)
890 NEXT I
900 IF J=N1 THEN GOTO 960
910 LET C(J,2)=C(J,2)/C(J1,1)
920 LET C(J1,3)=C(J1,1)/C(J2,1)
930 FOR I=1 TO M
940 LET P(J,I)=(X0(I)-C(J,2))*P(J1,I)-C(J1,3)*P(J2,I)
950 NEXT I
960 NEXT J
970 REM WRITING VECTORS X0(I),W(I) AND MATRICES C(J,L),...
980 REM ...P(J,I) ON A FILE
990 OPEN "O",1,"P214A"
1000 FOR I=1 TO M
1010 PRINT #1, X0(I),W(I)
1020 NEXT I
1030 FOR J=1 TO N
1040 FOR I=1 TO M
1050 PRINT #1, P(J,I)
1060 NEXT I
1070 FOR L=1 TO 3
1080 PRINT #1, C(J,L)
1090 NEXT L
1100 NEXT J
1110 CLOSE 1
1120 GOTO 1290
1130 REM READING VECTORS X0(I),W(I) AND MATRICES C(J,L),...
1140 REM ...P(J,I) FROM A FILE
1150 OPEN "I",1,"P214A"
1160 FOR I=1 TO M
1170 INPUT #1, X0(I),W(I)
1180 NEXT I
1190 FOR J=1 TO N
1200 FOR I=1 TO M
1210 INPUT #1, P(J,I)
1220 NEXT I
```

PROGRAM P214

```
1230 FOR L=1 TO 3
1240 INPUT #1, C(J,L)
1250 NEXT L
1260  NEXT J
1270 CLOSE 1
1280 IF O3=0 THEN GOTO 1870
1290 PRINT "INPUT OF ORDINATES:"
1300 FOR I=1 TO M
1310 PRINT "I=";I;"X0(I)=";X0(I)
1320 INPUT"F(I)";F(I)
1330 NEXT I
1340 CLS:PRINT "CHECK THE ORDINATES F(I):"
1350 FOR I=1 TO M
1360 PRINT "I=";I;"X0(I)=";X0(I);"F(I)=";F(I)
1370 PRINT"IF NO ERROR INPUT 0"
1380 PRINT"IF ERROR INPUT ANY NUMBER>0 AND THEN THE CORRECT F(I)"
1390 INPUT"ERROR";K
1400 IF K>0 THEN INPUT"F(I)";F(I)
1410 CLS:NEXT I
1420 PRINT"COMPUTING"
1430 IF O2=0 THEN GOTO 1490
1440 LPRINT "I,X0(I),W(I),F(I):"
1450 FOR I=1 TO M
1460 LPRINT I;X0(I),W(I),F(I)
1470 NEXT I
1480 LPRINT
1490 PRINT "COEFFICIENTS OF THE SERIES:"
1500 FOR J=1 TO N
1510 LET A(J)=0
1520 FOR I=1 TO M
1530 LET A(J)=A(J)+P(J,I)*W(I)*F(I)
1540 NEXT I
1550 LET A(J)=A(J)/C(J,1)
1560 PRINT J;A(J)
1570 NEXT J
1580 IF O5=0 THEN GOTO 1760
1590 PRINT "THE VARIANCES:"
1600 FOR L=2 TO N
1610 LET S(L)=0
1620 FOR I=1 TO M
1630 LET B=0
1640 FOR J=1 TO L
1650 LET B=B+A(J)*P(J,I)
1660 NEXT J
1670 LET B=B-F(I)
1680 LET B=B*B*W(I)
1690 LET S(L)=S(L)+B
1700 NEXT I
1710 IF L=M THEN GOTO 1730
1720 LET S(L)=S(L)/(M-L)
1730 PRINT "L=";L;"S(L)=";S(L)
1740 NEXT L
1750 REM WRITING COEFFICIENTS A(J) AND VARIANCES S(J) ON A FILE
1760 OPEN "O",1,"P214B"
1770 PRINT #1, O5
1780 FOR J=1 TO N
1790 PRINT #1, A(J)
1800 IF J=1 THEN GOTO 1820
1810 IF O5>0 THEN PRINT #1, S(J)
1820 NEXT J
1830 CLOSE 1
1840 GOTO 1950
1850 REM READING COEFFICIENTS A(J) AND VARIANCES S(J)...
1860 REM ...FROM A FILE
```

```
1870 OPEN "I",1,"P214B"
1880 INPUT #1, O5
1890 FOR J=1 TO N
1900 INPUT #1, A(J)
1910 IF J=1 THEN GOTO 1930
1920 IF O5>0 THEN INPUT #1, S(J)
1930 NEXT J
1940 CLOSE 1
1950 IF O2=0 THEN GOTO 2060
1960 LPRINT "COEFFICIENTS A(J) AND VARIANCES S(J):"
1970 LPRINT "J=1   A(1)=";A(1)
1980 FOR J=2 TO N
1990 IF O5=0 THEN GOTO 2020
2000 LPRINT J;A(J),S(J)
2010 GOTO 2030
2020 LPRINT J;A(J)
2030 NEXT J
2040 LPRINT
2050 LPRINT "THE APPROXIMATION:"
2060 PRINT:PRINT "THE APPROXIMATION:"
2070 INPUT"X";X
2080 IF O2>0 THEN LPRINT "X =";X;"N=";N
2090 GOSUB 2390
2100 LET Y=0
2110 FOR J=1 TO N
2120 LET Y=Y+A(J)*P0(J)
2130 NEXT J
2140 PRINT "Y =";Y
2150 IF O2>0 THEN LPRINT "Y =";Y
2160 IF O4=0 THEN GOTO 2340
2170 REM COMPUTATION OF THE FIRST DERIVATIVE
2180 GOSUB 2480
2190 LET Y1=0
2200 FOR J=2 TO N
2210 LET Y1=Y1+A(J)*P1(J)
2220 NEXT J
2230 PRINT "Y1=";Y1
2240 IF O2>0 THEN LPRINT "Y1=";Y1
2250 IF O4=1 THEN GOTO 2340
2260 REM COMPUTATION OF THE SECOND DERIVATIVE
2270 GOSUB 2570
2280 LET Y2=0
2290 FOR J=3 TO N
2300 LET Y2=Y2+A(J)*P2(J)
2310 NEXT J
2320 PRINT "Y2=";Y2
2330 IF O2>0 THEN LPRINT "Y2=";Y2
2340 IF O2>0 THEN LPRINT
2350 PRINT "TO STOP PRESS BREAK":PRINT
2360 INPUT"N";N
2370 GOTO 2070
2380 REM SUBROUTINE FOR ORTHOGONAL POLYNOMIALS
2390 LET P0(1)=1
2400 LET P0(2)=X-C(2,2)
2410 FOR J=3 TO N
2420 LET J1=J-1
2430 LET J2=J-2
2440 LET P0(J)=(X-C(J,2))*P0(J1)-C(J1,3)*P0(J2)
2450 NEXT J
2460 RETURN
2470 REM SUBROUTINE FOR THE FIRST DERIVATIVES OF O. P.
2480 LET P1(1)=0
2490 LET P1(2)=1
2500 FOR J=3 TO N
```

```
2510 LET J1=J-1
2520 LET J2=J-2
2530 LET P1(J)=P0(J1)+(X-C(J,2))*P1(J1)-C(J1,3)*P1(J2)
2540 NEXT J
2550 RETURN
2560 REM SUBROUTINE FOR THE SECOND DERIVATIVES OF O. P.
2570 LET P2(1)=0
2580 LET P2(2)=0
2590 FOR J=3 TO N
2600 LET J1=J-1
2610 LET J2=J-2
2620 LET P2(J)=2*P1(J1)+(X-C(J,2))*P2(J1)-C(J1,3)*P2(J2)
2630 NEXT J
2640 RETURN
2650 END
```

Examples

1. The function $y=x^3$ is given by 6 points

  (1,1), (2,8), (3,27), (4,64), (5,125), (6,216).

The weights at all 6 points W=1. The coefficients of the series are

  73.5, 41.8, 10.5, 0.999997, $-2.22524*10^{-6}$, $-1.73569*10^{-6}$,

and the variances

  1045.2, 21.6, $3.4845*10^{-10}$, (increasing values).

The minimum of variances indicates that the series should be restricted to 4 terms. Some results of approximation in single precision:

  x=2.5,   y=15.625, y_1=18.75, y_2=15.
  x=4.5,   y=91.125, y_1=60.75, y_2=27.

2. The function $y=e^x$ is given by 6 points

  (0.95,2.58571), (0.96,2.61170), (0.97,2.63794),
  (0.98,2.66446), (0.99,2.69123), (1,2.71828).

The weights at all 6 points W=1. The coefficients of the series are

  2.65155, 2.65132, 1.32465, -3.70537, 3.88581, 115627,

and the variances

  $1.93934*10^{-7}$, $2.64701*10^{-11}$, (increasing values).

The minimum of variances and the sharp increase of coefficients indicate that the series should be restricted to 3 terms (that is, to a quadratic parabola with a constant value of its second derivative). Some results of approximation in single precision:

$x=0.955$, $y=2.59868$, $y_1=2.59834$, $y_2=2.64929$.
$x=0.975$, $y=2.65117$, $y_1=2.65132$, $y_2=2.64929$.
$x=0.990$, $y=2.69124$, $y_1=2.69106$, $y_2=2.64929$.

The second derivative is correct only in two digits, which is nevertheless a success when compared with the results of interpolation or approximation by other methods.

3. The function $y=e^x$ is given by 6 points
(0.95, 2.585709659315846), (0.96, 2.611696473423118),
(0.97, 2.637944459354153), (0.98, 2.664456241929417),
(0.99, 2.691234472349262), (1, 2.718281828459045).

The weights at all 6 points $W=1$. The coefficients of the series are
2.651390357341991, 2.651390357341991, 1.325658565672543,
0.4418789980535203, 0.1104684679226807, 0.0221204607091862,
and the variances steadily decreasing from 10^{-7} to 10^{-28}. This indicates that the series should contain all 6 terms. Some results of approximation in double precision:

$x=0.955$, $y=2.598670582919602$, $y_1=2.598670582909236$,
$y_2=2.598670577769999$.
$x=0.975$, $y=2.651167210982619$, $y_1=2.651167210986341$,
$y_2=2.651167209852904$.
$x=0.990$, $y=2.691234472349265$, $y_1=2.691234472353379$,
$y_2=2.691234473938809$.

The examples show how important it is to truncate the series after the term with the minimum variance. The results of approximations are satisfactory even in single precision and excellent in double precision.

P215 Least-Squares Approximation by Orthogonal Polynomials with Equally Spaced Abscissas and a Weight Function Equal to 1 for the Function and Its First and Second Derivatives

```
10 CLS:PRINT"P215 APPROXIMATES Y(X) AND ITS FIRST AND SECOND"
20 PRINT"DERIVATIVE Y1,Y2 BY ORTHOGONAL POLYNOMIALS WITH"
30 PRINT"EQUALLY SPACED ABSCISSAS AND WEIGHT FUNCTION=1."
40 PRINT:PRINT"THE INPUT:"
50 PRINT"IF SINGLE PRECISION O1=1, IF DOUBLE O1=2."
60 DEFINT I-O:INPUT"O1";O1:CLS
70 PRINT"IF PRINTER NOT USED O2=0, IF USED O2>0."
80 INPUT"O2";O2:CLS
90 IF O1=2 THEN DEFDBL A-H,P-Z
100 PRINT"ABSCISSAS OF POINTS DEFINING THE FUNCTION: X0(I)"
110 PRINT"ORDINATES OF POINTS DEFINING THE FUNCTION: F(I)":PRINT
120 PRINT"IF ALL DATA NEW O3>1, IF ONLY ORDINATES F(I9 ARE"
130 PRINT"NEW O3=1, IF ALL DATA ARE TAKEN FROM FILES P215A"
140 PRINT"AND P215B WRITTEN IN THE FIRST RUN O3=0."
150 INPUT"O3";O3:CLS
160 PRINT"IF ONLY FUNCTION COMPUTED O4=0, IF FUNCTION AND ITS"
170 PRINT"FIRST DERIVATIVE COMPUTED O4=1, IF FUNCTION AND ITS"
180 PRINT"FIRST AND SECOND DERIVATIVES COMPUTED O4=2."
190 INPUT"O4";O4:CLS:IF O3=0 THEN GOTO 220
200 PRINT"IF SEQUENCE OF VARIANCES COMPUTED O5>0, IF NOT O5=0."
210 INPUT"O5";O5:CLS
220 PRINT"NUMBER OF POINTS DEFINING THE FUNCTION: M>2"
230 PRINT"NUMBER OF TERMS IN THE SERIES: N<M, N>2"
240 PRINT"DEGREE OF THE HIGHEST POLYNOMIAL: N-1"
250 PRINT"IF N=M THE APPROXIMATION BECOMES INTERPOLATION."
260 INPUT"M";M:INPUT"N";N:CLS
270 PRINT"INTERVAL COVERED BY N POINTS: (U,V)"
280 INPUT"U";U:INPUT"V";V:CLS
290 REM VECTORS N(j),a(j),b(j) ARE STORED IN MATRIX C(J,L),
300 REM L=1,2,3. POLYNOMIALS p(j,(xi)) ARE STORED
310 REM IN MATRIX P(J,I)
320 IF O2=0 THEN GOTO 360
330 LPRINT "P215: O1=";O1;"O2=";O2;"O3=";O3;"O4=";O4
340 IF O3>0 THEN LPRINT "O5=";O5
350 LPRINT "M=";M;"N=";N
360 DIM A(N),C(N,3),S(N)
370 DIM P(N,M),P0(N),P1(N),P2(N),X0(M)
380 IF O3<2 THEN GOTO 970
390 IF O2>0 THEN LPRINT "INTERVAL U=";U;"V=";V
400 LET H=(V-U)/(M-1)
410 LET X0(1)=U
420 FOR I=2 TO M
430 LET I1=I-1
440 LET X0(I)=X0(I1)+H
450 NEXT I
460 REM COMPUTATION OF MATRICES C(J,L) AND P(J,I)
470 LET C(1,1)=M
480 FOR I=1 TO M
490 LET P(1,I)=1
500 NEXT I
510 LET C(2,2)=0
520 FOR I=1 TO M
530 LET C(2,2)=C(2,2)+X0(I)
540 NEXT I
550 LET C(2,2)=C(2,2)/C(1,1)
560 FOR I=1 TO M
570 LET P(2,I)=X0(I)-C(2,2)
580 NEXT I
```

```
590 LET N1=N+1
600 FOR J=3 TO N1
610 LET J1=J-1
620 LET J2=J-2
630 LET C(J1,1)=0
640 IF J=N1 THEN GOTO 660
650 LET C(J,2)=0
660 FOR I=1 TO M
670 LET B=P(J1,I)*P(J1,I)
680 LET C(J1,1)=C(J1,1)+B
690 IF J=N1 THEN GOTO 710
700 LET C(J,2)=C(J,2)+B*X0(I)
710 NEXT I
720 IF J=N1 THEN GOTO 780
730 LET C(J,2)=C(J,2)/C(J1,1)
740 LET C(J1,3)=C(J1,1)/C(J2,1)
750 FOR I=1 TO M
760 LET P(J,I)=(X0(I)-C(J,2))*P(J1,I)-C(J1,3)*P(J2,I)
770 NEXT I
780 NEXT J
790 REM WRITING VECTOR X0(I) AND MATRICES C(J,L), P(J,I)...
800 REM ...ON A FILE
810 OPEN "O",1,"P215A"
820 FOR I=1 TO M
830 PRINT #1, X0(I)
840 NEXT I
850 FOR J=1 TO N
860 FOR I=1 TO M
870 PRINT #1, P(J,I)
880 NEXT I
890 FOR L=1 TO 3
900 PRINT #1, C(J,L)
910 NEXT L
920 NEXT J
930 CLOSE 1
940 GOTO 1110
950 REM READING VECTOR X0(I) AND MATRICES C(J,L), P(J,I)...
960 REM ...FROM A FILE
970 OPEN "I",1,"P215A"
980 FOR I=1 TO M
990 INPUT #1, X0(I)
1000 NEXT I
1010 FOR J=1 TO N
1020 FOR I=1 TO M
1030 INPUT #1, P(J,I)
1040 NEXT I
1050 FOR L=1 TO 3
1060 INPUT #1, C(J,L)
1070 NEXT L
1080  NEXT J
1090 CLOSE 1
1100 IF O3=0 THEN GOTO 1700
1110 PRINT "INPUT OF ORDINATES:"
1120 IF O2>0 THEN LPRINT "THE INPUT F(I):"
1130 FOR I=1 TO M
1140 PRINT "I=";I
1150 INPUT"F(I)";F(I)
1160 NEXT I
1170 CLS:PRINT "CHECK THE ORDINATES F(I):"
1180 FOR I=1 TO M
1190 PRINT "I=";I;"F(I)=";F(I)
1200 PRINT"IF NO ERROR INPUT 0"
1210 PRINT"IF ERROR INPUT ANY NUMBER>0 AND THEN THE CORRECT F(I)"
1220 INPUT"ERROR";K
```

PROGRAM P215

```
1230 IF K>0 THEN INPUT"F(I)";F(I)
1240 CLS:NEXT I
1250 PRINT"COMPUTING"
1260 IF O2=0 THEN GOTO 1320
1270 LPRINT "I,F(I):"
1280 FOR I=1 TO M
1290 LPRINT I;F(I)
1300 NEXT I
1310 LPRINT
1320 PRINT "COEFFICIENTS OF THE SERIES:"
1330 FOR J=1 TO N
1340 LET A(J)=0
1350 FOR I=1 TO M
1360 LET A(J)=A(J)+P(J,I)*F(I)
1370 NEXT I
1380 LET A(J)=A(J)/C(J,1)
1390 PRINT J;A(J)
1400 NEXT J
1410 IF O5=0 THEN GOTO 1590
1420 PRINT "THE VARIANCES:"
1430 FOR L=2 TO N
1440 LET S(L)=0
1450 FOR I=1 TO M
1460 LET B=0
1470 FOR J=1 TO L
1480 LET B=B+A(J)*P(J,I)
1490 NEXT J
1500 LET B=B-F(I)
1510 LET B=B*B
1520 LET S(L)=S(L)+B
1530 NEXT I
1540 IF L=M THEN GOTO 1560
1550 LET S(L)=S(L)/(M-L)
1560 PRINT "L=";L;"S(L)=";S(L)
1570 NEXT L
1580 REM WRITING COEFFICIENTS A(J) AND VARIANCES S(J) ON A FILE
1590 OPEN "O",1,"P215B"
1600 PRINT #1, O5
1610 FOR J=1 TO N
1620 PRINT #1, A(J)
1630 IF J=1 THEN GOTO 1650
1640 IF O5>0 THEN PRINT #1, S(J)
1650 NEXT J
1660 CLOSE 1
1670 GOTO 1780
1680 REM READING COEFFICIENTS A(J) AND VARIANCES S(I)...
1690 REM ...FROM A FILE
1700 OPEN "I",1,"P215B"
1710 INPUT #1, O5
1720 FOR J=1 TO N
1730 INPUT #1, A(J)
1740 IF J=1 THEN GOTO 1760
1750 IF O5>0 THEN INPUT #1, S(J)
1760 NEXT J
1770 CLOSE 1
1780 IF O2=0 THEN GOTO 1890
1790 LPRINT "COEFFICIENTS A(J) AND VARIANCES S(J):"
1800 LPRINT "J=1   A(1)=";A(1)
1810 FOR J=2 TO N
1820 IF O5=0 THEN GOTO 1850
1830 LPRINT J;A(J),S(J)
1840 GOTO 1860
1850 LPRINT J;A(J)
1860 NEXT J
```

```
1870 LPRINT
1880 LPRINT "THE APPROXIMATION:"
1890 PRINT:PRINT "THE INTERPOLATION:"
1900 INPUT"X";X
1910 IF O2>0 THEN LPRINT "X =";X;"N=";N
1920 GOSUB 2220
1930 LET Y=0
1940 FOR J=1 TO N
1950 LET Y=Y+A(J)*P0(J)
1960 NEXT J
1970 PRINT "Y =";Y
1980 IF O2>0 THEN LPRINT "Y =";Y
1990 IF O4=0 THEN GOTO 2170
2000 REM COMPUTATION OF THE FIRST DERIVATIVE
2010 GOSUB 2310
2020 LET Y1=0
2030 FOR J=2 TO N
2040 LET Y1=Y1+A(J)*P1(J)
2050 NEXT J
2060 PRINT "Y1=";Y1
2070 IF O2>0 THEN LPRINT "Y1=";Y1
2080 IF O4=1 THEN GOTO 2170
2090 REM COMPUTATION OF THE SECOND DERIVATIVE
2100 GOSUB 2400
2110 LET Y2=0
2120 FOR J=3 TO N
2130 LET Y2=Y2+A(J)*P2(J)
2140 NEXT J
2150 PRINT "Y2=";Y2
2160 IF O2>0 THEN LPRINT "Y2=";Y2
2170 IF O2>0 THEN LPRINT
2180 PRINT "TO STOP PRESS BREAK":PRINT
2190 INPUT"N";N
2200 GOTO 1900
2210 REM SUBROUTINE FOR ORTHOGONAL POLYNOMIALS
2220 LET P0(1)=1
2230 LET P0(2)=X-C(2,2)
2240 FOR J=3 TO N
2250 LET J1=J-1
2260 LET J2=J-2
2270 LET P0(J)=(X-C(J,2))*P0(J1)-C(J1,3)*P0(J2)
2280 NEXT J
2290 RETURN
2300 REM SUBROUTINE FOR THE FIRST DERIVATIVES OF O. P.
2310 LET P1(1)=0
2320 LET P1(2)=1
2330 FOR J=3 TO N
2340 LET J1=J-1
2350 LET J2=J-2
2360 LET P1(J)=P0(J1)+(X-C(J,2))*P1(J1)-C(J1,3)*P1(J2)
2370 NEXT J
2380 RETURN
2390 REM SUBROUTINE FOR THE SECOND DERIVATIVES OF O. P.
2400 LET P2(1)=0
2410 LET P2(2)=0
2420 FOR J=3 TO N
2430 LET J1=J-1
2440 LET J2=J-2
2450 LET P2(J)=2*P1(J1)+(X-C(J,2))*P2(J1)-C(J1,3)*P2(J2)
2460 NEXT J
2470 RETURN
2480 END
```

Examples

Since the weight function in the examples to program P214 is chosen as 1, those examples serve this program as well.

P216 Least-Squares Approximation of a Periodic Function (Fourier Series) for the Function and Its First and Second Derivatives

```
10 CLS:PRINT"P216 APPROXIMATES Y(T) AND ITS FIRST AND SECOND"
20 PRINT"DERIVATIVE Y1,Y2 BY FOURIER SERIES."
30 PRINT:PRINT"THE INPUT:"
40 PRINT"SINGLE PRECISION ONLY."
50 PRINT"IF PRINTER NOT USED O2=0, IF USED O2>0."
60 DEFINT I-O
70 INPUT"O2";O2:CLS
80 PRINT"ABSCISSAS OF POINTS DEFINING THE FUNCTION: X0(I)"
90 PRINT"ORDINATES OF POINTS DEFINING THE FUNCTION: F(I)":PRINT
100 PRINT"IF ALL DATA NEW O3>1, IF ONLY ORDINATES F(I) ARE"
110 PRINT"NEW O3=1, IF ALL DATA ARE TAKEN FROM FILES P216A"
120 PRINT"AND P216B WRITTEN IN THE FIRST RUN O3=0."
130 INPUT"O3";O3:CLS
140 PRINT"IF ONLY FUNCTION COMPUTED O4=0, IF FUNCTION AND ITS"
150 PRINT"FIRST DERIVATIVE COMPUTED O4=1, IF FUNCTION AND ITS"
160 PRINT"FIRST AND SECOND DERIVATIVES COMPUTED O4=2."
170 INPUT"O4";O4:CLS:IF O3=0 THEN GOTO 200
180 PRINT"IF SEQUENCE OF VARIANCES COMPUTED O5>0, IF NOT O5=0."
190 INPUT"O5";O5:CLS
200 PRINT"NUMBER OF POINTS DEFINING THE FUNCTION: 2M"
210 PRINT"HIGHEST HARMONIC IN THE SERIES: N<M"
220 INPUT"M";M:INPUT"N";N:CLS
230 PRINT"POINTS DEFINING THE FUNCTION ARE EQUALLY SPACED AND"
240 PRINT"DISTRIBUTED ON THE INTERVAL (-H,H)."
250 PRINT"HALFINTERVAL: H; IF H=3.141593 (PI) THEN LET H=0."
260 INPUT"H";H:CLS
270 IF O2=0 THEN GOTO 310
280 LPRINT "P216: O2=";O2;"O3=";O3;"O4=";O4
290 IF O3>0 THEN LPRINT "O5=";O5
300 LPRINT "M=";M;"N=";N;"H=";H
310 LET M1=M-1
320 LET M2=2*M
330 IF H>0 THEN Q=3.141593/H
340 IF H>0 THEN Q2=Q*Q
350 DIM A(N),B(N),V(N),F(M2),E(M),G(M1)
360 DIM C(N,M),S(N,M),C0(N),S0(N)
370 IF O3=0 THEN GOTO 1560
380 IF O3=1 THEN GOTO 670
390 REM COMPUTATION OF MATRICES C(N,M),S(N,M)
400 LET X1=3.141593/M
410 LET C(1,1)=COS(X1)
420 LET S(1,1)=SIN(X1)
430 FOR I=2 TO M1
440 LET I1=I-1
450 LET C(1,I)=C(1,I1)*C(1,1)-S(1,I1)*S(1,1)
460 LET S(1,I)=S(1,I1)*C(1,1)+C(1,I1)*S(1,1)
470 LET C(1,M)=-1
480 LET S(1,M)=0
490 NEXT I
500 FOR I=1 TO M
510 FOR J=2 TO N
520 LET J1=J-1
530 LET C(J,I)=C(J1,I)*C(1,I)-S(J1,I)*S(1,I)
540 LET S(J,I)=S(J1,I)*C(1,I)+C(J1,I)*S(1,I)
550 NEXT J
560 NEXT I
570 REM WRITING MATRICES C(J,I),S(J,I) ON A FILE
580 OPEN "O",1,"P216A"
590 FOR I=1 TO M
```

PROGRAM P216

```
600 FOR J=1 TO N
610 PRINT #1, C(J,I),S(J,I)
620 NEXT J
630 NEXT I
640 CLOSE 1
650 GOTO 740
660 REM READING MATRICES C(J,I),S(J,I) FROM A FILE
670 OPEN "I",1,"P216A"
680 FOR I=1 TO M
690 FOR J=1 TO N
700 INPUT #1, C(J,I),S(J,I)
710 NEXT J
720 NEXT I
730 CLOSE 1
740 PRINT "INPUT OF ORDINATES:"
750 IF O2>0 THEN LPRINT "THE INPUT F(I):"
760 LET H0=H/M
770 IF H=0 THEN LET H0=X1
780 FOR I=1 TO M2
790 LET J=I-M
800 LET X1=J*H0
810 PRINT "I=";J;"X0(I)=";X1
820 INPUT"F(I)";F(I)
830 NEXT I
840 CLS:PRINT "CHECK THE ORDINATES F(I):"
850 FOR I=1TO M2
860 LET J=I-M
870 LET X1=J*H0
880 PRINT "I=";J;"X0(I)=";X1;"F(I)=";F(I)
890 PRINT"IF NO ERROR INPUT 0"
900 PRINT"IF ERROR INPUT ANY NUMBER>0 AND THEN THE CORRECT F(I)"
910 INPUT"ERROR";K
920 IF K>0 THEN INPUT"F(I)";F(I)
930 CLS:NEXT I
940 PRINT"COMPUTING"
950 REM COMPUTATION OF E(I),G(I)
960 LET E0=F(M)
970 LET E(M)=F(M2)
980 FOR I=1 TO M1
990 LET J=M+I
1000 LET K=M-I
1010 LET E(I)=F(J)+F(K)
1020 LET G(I)=F(J)-F(K)
1030 IF O2=0 THEN GOTO 1040
1040 NEXT I
1050 IF O2=0 THEN GOTO 1120
1060 FOR I=1 TO M2
1070 LET J=I-M
1080 LET X1=J*H0
1090 LPRINT "I=";J;"X0(I)=";X1;"F(I)=";F(I)
1100 NEXT I
1110 LPRINT
1120 PRINT "COEFFICIENTS OF THE SERIES"
1130 LET A0=E0+E(M)
1140 FOR I=1 TO M1
1150 LET A0=A0+E(I)
1160 NEXT I
1170 LET A0=A0/M2
1180 PRINT "A0=";A0
1190 FOR J=1 TO N
1200 LET A(J)=E0+E(M)*C(J,M)
1210 LET B(J)=0
1220 FOR I=1 TO M1
1230 LET A(J)=A(J)+E(I)*C(J,I)
```

```
1240 LET B(J)=B(J)+G(I)*S(J,I)
1250 NEXT I
1260 LET A(J)=A(J)/M
1270 LET B(J)=B(J)/M
1280 PRINT J;"A(J)=";A(J);"B(J)=";B(J)
1290 NEXT J
1300 IF O5=0 THEN GOTO 1470
1310 PRINT "THE VARIANCES:"
1320 FOR L=2 TO N
1330 LET V(L)=E0*E0+E(M)*E(M)
1340 LET D=0
1350 FOR I=1 TO M1
1360 LET D=D+E(I)*E(I)+G(I)*G(I)
1370 NEXT I
1380 LET V(L)=V(L)+D/2
1390 LET D=0
1400 FOR J=1 TO L
1410 LET D=D+A(J)*A(J)+B(J)*B(J)
1420 NEXT J
1430 LET V(L)=(V(L)-(A0*A0+D/2)*M2)/(M2-2*L-1)
1440 PRINT "L=";L;"S(L)=";V(L)
1450 NEXT L
1460 REM WRITING A(J),B(J) AND VARIANCES ON A FILE
1470 OPEN "O",1,"P216B"
1480 PRINT #1, O5,A0,A(1),B(1)
1490 FOR J=2 TO N
1500 PRINT #1, A(J),B(J)
1510 IF O5>0 THEN PRINT #1, V(J)
1520 NEXT J
1530 CLOSE 1
1540 GOTO 1630
1550 REM READING A(J),B(J) AND VARIANCES FROM A FILE
1560 OPEN "I",1,"P216B"
1570 INPUT #1, O5,A0,A(1),B(1)
1580 FOR J=2 TO N
1590 INPUT #1, A(J),B(J)
1600 IF O5>0 THEN INPUT #1, V(J)
1610 NEXT J
1620 CLOSE 1
1630 IF O2=0 THEN GOTO 1750
1640 LPRINT "COEFFICIENTS A(J),B(J) AND VARIANCES S(J)"
1650 LPRINT "J=0   A(0)=";A0
1660 LPRINT "J=1   A(1)=";A(1);"B(1)=";B(1)
1670 FOR J=2 TO N
1680 IF O5=0 THEN GOTO 1710
1690 LPRINT "J=";J;"A(J)=";A(J);"B(J)=";B(J);"S(J)=";V(J)
1700 GOTO 1720
1710 LPRINT "J=";J;"A(J)=";A(J);"B(J)=";B(J)
1720 NEXT J
1730 LPRINT
1740 LPRINT "THE APPROXIMATION:"
1750 PRINT:PRINT "THE APPROXIMATION:"
1760 INPUT"T";X
1770 IF O2>0 THEN LPRINT "T =";X;"N=";N
1780 IF H>0 THEN LET X=X*Q
1790 REM COMPUTATION OF C0(J),S0(J)
1800 LET C0(1)=COS(X)
1810 LET S0(1)=SIN(X)
1820 FOR J=2 TO N
1830 LET J1=J-1
1840 LET C0(J)=C0(J1)*C0(1)-S0(J1)*S0(1)
1850 LET S0(J)=S0(J1)*C0(1)+C0(J1)*S0(1)
1860 NEXT J
1870 LET Y=A0
```

```
1880 FOR J=1 TO N
1890 LET Y=Y+A(J)*C0(J)+B(J)*S0(J)
1900 NEXT J
1910 PRINT "Y =";Y
1920 IF O2>0 THEN LPRINT "Y =";Y
1930 IF O4=0 THEN GOTO 2110
1940 REM COMPUTATION OF THE FIRST DERIVATIVE
1950 LET Y1=0
1960 FOR J=1 TO N
1970 LET Y1=Y1+J*(B(J)*C0(J)-A(J)*S0(J))
1980 NEXT J
1990 IF H>0 THEN LET Y1=Y1*Q
2000 PRINT "Y1=";Y1
2010 IF O2>0 THEN LPRINT "Y1=";Y1
2020 IF O4=1 THEN GOTO 2110
2030 REM COMPUTATION OF THE SECOND DERIVATIVE
2040 LET Y2=0
2050 FOR J=1 TO N
2060 LET Y2=Y2-J*J*(A(J)*C0(J)+B(J)*S0(J))
2070 NEXT J
2080 IF H>0 THEN LET Y2=Y2*Q2
2090 PRINT "Y2=";Y2
2100 IF O2>0 THEN LPRINT "Y2=";Y2
2110 IF O2>0 THEN LPRINT
2120 PRINT "TO STOP PRESS BREAK":PRINT
2130 INPUT"N";N
2140 GOTO 1760
2150 END
```

Example

A periodic function with the half-interval H=4

$$y=\sin(x)+[\sin(2x)]/5+[\sin(3x)]/10+$$
$$+\cos(x)+(2/5)*\cos(2x)+[\cos(3x)]/7$$

in which

$$x=(PI)t/H$$

is approximated by a truncated Fourier series with N=3 terms (M=4). The 8 input data are

-1.18391, -1.3, -0.371726, 1.54286,

1.58391, 0.5, -0.028274, -0.742857.

The coefficients of cosine terms

$A_0=3.72529*10^{-7}$, $A_1=1$, $A_2=0.4$, $A_3=0.142858$,

the coefficients of sine terms

$B_1=1$, $B_2=0.2$, $B_3=0.100001$,
and the variances
   0.0405452, $1.90735*10^{-6}$.

Some results of approximation with N=3 terms for the function y and its first and second derivatives y_1, y_2:

   t=0.25,   y=1.7963, y_1=0.675787, y_2=-2.79389.
   t=2.1278,   y=0.409783, y_1=-0.651371, y_2=0.808570.
   t=-4,   y=-0.742858, y_1=-0.706861, y_2=0.422986.

P217 Least-Squares Approximation of an Even Periodic Function (Fourier Series) for the Function and Its First and Second Derivatives

```
10 CLS:PRINT"P217 APPROXIMATES AN EVEN FUNCTION Y(T) AND ITS FIRST"
20 PRINT"AND SECOND DERIVATIVES Y1,Y2 BY FOURIER SERIES."
30 PRINT:PRINT"THE INPUT:"
40 PRINT"SINGLE PRECISION ONLY."
50 PRINT"IF PRINTER NOT USED O2=0, IF USED O2>0."
60 DEFINT I-O
70 INPUT"O2";O2:CLS
80 PRINT"ABSCISSAS OF POINTS DEFINING THE FUNCTION: X0(I)"
90 PRINT"ORDINATES OF POINTS DEFINING THE FUNCTION: F(I)":PRINT
100 PRINT"IF ALL DATA NEW O3>1, IF ONLY ORDINATES F(I) ARE"
110 PRINT"NEW O3=1, IF ALL DATA ARE TAKEN FROM FILES P217A"
120 PRINT"AND P217B WRITTEN IN THE FIRST RUN O3=0."
130 INPUT"O3";O3:CLS
140 PRINT"IF ONLY FUNCTION COMPUTED O4=0, IF FUNCTION AND ITS"
150 PRINT"FIRST DERIVATIVE COMPUTED O4=1, IF FUNCTION AND ITS"
160 PRINT"FIRST AND SECOND DERIVATIVES COMPUTED O4=2."
170 INPUT"O4";O4:CLS:IF O3=0 THEN GOTO 200
180 PRINT"IF SEQUENCE OF VARIANCES COMPUTED O5>0, IF NOT O5=0."
190 INPUT"O5";O5:CLS
200 PRINT"NUMBER OF POINTS DEFINING THE FUNCTION: M+1"
210 PRINT"HIGHEST HARMONIC IN THE SERIES: N<M"
220 INPUT"M";M:INPUT"N";N:CLS
230 PRINT"POINTS DEFINING THE FUNCTION ARE EQUALLY SPACED AND"
240 PRINT"DISTRIBUTED ON THE HALFINTERVAL (0,H)."
250 PRINT"IF H=3.141593 (PI) THEN LET H=0."
260 INPUT"H";H:CLS
270 IF O2=0 THEN GOTO 310
280 LPRINT "P217: O2=";O2;"O3=";O3;"O4=";O4
290 IF O3>0 THEN LPRINT "O5=";O5
300 LPRINT "M=";M;"N=";N;"H=";H
310 LET M1=M-1
320 LET M2=M+1
330 IF H>0 THEN Q=3.141593/H
340 IF H>0 THEN Q2=Q*Q
350 DIM A(N),B(N),V(N),F(M2),E(M)
360 DIM C(N,M),S(N,M),C0(N),S0(N)
370 IF O3=0 THEN GOTO 1510
380 IF O3=1 THEN GOTO 670
390 REM COMPUTATION OF MATRICES C(N,M),S(N,M)
400 LET X1=3.141593/M
410 LET C(1,1)=COS(X1)
420 LET S(1,1)=SIN(X1)
430 FOR I=2 TO M1
440 LET I1=I-1
450 LET C(1,I)=C(1,I1)*C(1,1)-S(1,I1)*S(1,1)
460 LET S(1,I)=S(1,I1)*C(1,1)+C(1,I1)*S(1,1)
470 LET C(1,M)=-1
480 LET S(1,M)=0
490 NEXT I
500 FOR I=1 TO M
510 FOR J=2 TO N
520 LET J1=J-1
530 LET C(J,I)=C(J1,I)*C(1,I)-S(J1,I)*S(1,I)
540 LET S(J,I)=S(J1,I)*C(1,I)+C(J1,I)*S(1,I)
550 NEXT J
560 NEXT I
570 REM WRITING MATRICES C(J,I),S(J,I) ON A FILE
580 OPEN "O",1,"P217A"
590 FOR I=1 TO M
```

```
600 FOR J=1 TO N
610 PRINT #1, C(J,I),S(J,I)
620 NEXT J
630 NEXT I
640 CLOSE 1
650 GOTO 740
660 REM READING MATRICES C(J,I),S(J,I) FROM A FILE
670 OPEN "I",1,"P217A"
680 FOR I=1 TO M
690 FOR J=1 TO N
700 INPUT #1, C(J,I),S(J,I)
710 NEXT J
720 NEXT I
730 CLOSE 1
740 PRINT "INPUT OF ORDINATES:"
750 IF O2>0 THEN LPRINT "THE INPUT F(I):"
760 LET H0=H/M
770 IF H=0 THEN LET H0=X1
780 FOR I=1 TO M2
790 LET J=I-1
800 LET X1=J*H0
810 PRINT "I=";J;"X0(I)=";X1
820 INPUT"F(I)";F(I)
830 NEXT I
840 CLS:PRINT "CHECK THE ORDINATES F(I):"
850 FOR I=1TO M2
860 LET J=I-1
870 LET X1=J*H0
880 PRINT "I=";J;"X0(I)=";X1;"F(I)=";F(I)
890 PRINT"IF NO ERROR INPUT 0"
900 PRINT"IF ERROR INPUT ANY NUMBER>0 AND THEN THE CORRECT F(I)"
910 INPUT"ERROR";K
920 IF K>0 THEN INPUT"F(I)";F(I)
930 CLS:NEXT I
940 PRINT"COMPUTING"
950 REM COMPUTATION OF E(I)
960 LET E0=F(1)
970 LET E(M)=F(M2)
980 FOR I=1 TO M1
990 LET J=I+1
1000 LET E(I)=2*F(J)
1010 IF O2=0 THEN GOTO 1020
1020 NEXT I
1030 IF O2=0 THEN GOTO 1100
1040 FOR I=1 TO M2
1050 LET J=I-1
1060 LET X1=J*H0
1070 LPRINT "I=";J;"X0(I)=";X1;"F(I)=";F(I)
1080 NEXT I
1090 LPRINT
1100 PRINT "COEFFICIENTS OF THE SERIES"
1110 LET A0=E0+E(M)
1120 FOR I=1 TO M1
1130 LET A0=A0+E(I)
1140 NEXT I
1150 LET A0=.5*A0/M
1160 PRINT "A0=";A0
1170 FOR J=1 TO N
1180 LET A(J)=E0+E(M)*C(J,M)
1190 FOR I=1 TO M1
1200 LET A(J)=A(J)+E(I)*C(J,I)
1210 NEXT I
1220 LET A(J)=A(J)/M
1230 PRINT J;"A(J)=";A(J)
```

PROGRAM P217

```
1240 NEXT J
1250 IF O5=0 THEN GOTO 1420
1260 PRINT "THE VARIANCES:"
1270 FOR L=2 TO N
1280 LET V(L)=E0*E0+E(M)*E(M)
1290 LET D=0
1300 FOR I=1 TO M1
1310 LET D=D+E(I)*E(I)
1320 NEXT I
1330 LET V(L)=V(L)+D/2
1340 LET D=0
1350 FOR J=1 TO L
1360 LET D=D+A(J)*A(J)
1370 NEXT J
1380 LET V(L)=(V(L)-2*M*(A0*A0+D/2))/(2*M-2*L-1)
1390 PRINT "L=";L;"S(L)=";V(L)
1400 NEXT L
1410 REM WRITING A(J) AND VARIANCES ON A FILE
1420 OPEN "O",1,"P217B"
1430 PRINT #1, O5,A0,A(1)
1440 FOR J=2 TO N
1450 PRINT #1, A(J)
1460 IF O5>0 THEN PRINT #1, V(J)
1470 NEXT J
1480 CLOSE 1
1490 GOTO 1580
1500 REM READING A(J) AND VARIANCES FROM A FILE
1510 OPEN "I",1,"P217B"
1520 INPUT #1, O5,A0,A(1)
1530 FOR J=2 TO N
1540 INPUT #1, A(J)
1550 IF O5>0 THEN INPUT #1, V(J)
1560 NEXT J
1570 CLOSE 1
1580 IF O2=0 THEN GOTO 1700
1590 LPRINT "COEFFICIENTS A(J) AND VARIANCES S(J)"
1600 LPRINT "J=0   A(0)=";A0
1610 LPRINT "J=1   A(1)=";A(1)
1620 FOR J=2 TO N
1630 IF O5=0 THEN GOTO 1660
1640 LPRINT "J=";J;"A(J)=";A(J);"S(J)=";V(J)
1650 GOTO 1670
1660 LPRINT "J=";J;"A(J)=";A(J)
1670 NEXT J
1680 LPRINT
1690 LPRINT "THE APPROXIMATION:"
1700 PRINT:PRINT "THE APPROXIMATION:"
1710 INPUT"T";X
1720 IF O2>0 THEN LPRINT "T =";X;"N=";N
1730 IF H>0 THEN LET X=X*Q
1740 REM COMPUTATION OF C0(J),S0(J)
1750 LET C0(1)=COS(X)
1760 LET S0(1)=SIN(X)
1770 FOR J=2 TO N
1780 LET J1=J-1
1790 LET C0(J)=C0(J1)*C0(1)-S0(J1)*S0(1)
1800 LET S0(J)=S0(J1)*C0(1)+C0(J1)*S0(1)
1810 NEXT J
1820 LET Y=A0
1830 FOR J=1 TO N
1840 LET Y=Y+A(J)*C0(J)
1850 NEXT J
1860 PRINT "Y =";Y
1870 IF O2>0 THEN LPRINT "Y =";Y
```

```
1880 IF O4=0 THEN GOTO 2060
1890 REM COMPUTATION OF THE FIRST DERIVATIVE
1900 LET Y1=0
1910 FOR J=1 TO N
1920 LET Y1=Y1-J*A(J)*S0(J)
1930 NEXT J
1940 IF H>0 THEN LET Y1=Y1*Q
1950 PRINT "Y1=";Y1
1960 IF O2>0 THEN LPRINT "Y1=";Y1
1970 IF O4=1 THEN GOTO 2060
1980 REM COMPUTATION OF THE SECOND DERIVATIVE
1990 LET Y2=0
2000 FOR J=1 TO N
2010 LET Y2=Y2-J*J*A(J)*C0(J)
2020 NEXT J
2030 IF H>0 THEN LET Y2=Y2*Q2
2040 PRINT "Y2=";Y2
2050 IF O2>0 THEN LPRINT "Y2=";Y2
2060 IF O2>0 THEN LPRINT
2070 PRINT "TO STOP PRESS BREAK":PRINT
2080 INPUT"N";N
2090 GOTO 1710
2100 END
```

Example

An even periodic function with the half-interval H=4

$$y=1+\cos(x)+(2/5)*\cos(2x)+[\cos(3x)]/7$$

in which

$$t=(PI)t/H$$

is approximated by a truncated Fourier series with N=3 terms (M=4). The 5 input data are

2.54286, 1.60609, 0.6, 0.393908, 0.257143.

The coefficients of the series are

$A_0=1$, $A_1=1$, $A_2=0.400001$, $A_3=0.142857$,

and the variances

0.0272121, $3.8147*10^{-6}$.

Some results of approximation with N=3 terms for the function y and its first and second derivatives y_1, y_2:

t=0.5, y=2.26139, y_1=-1.05583, y_2=-1.57129.
t=1.5, y=0.967858, y_1=-1.04109, y_2=1.19455.
t=2.5, y=0.466456, y_1=-0.152514, y_2=0.201222.

P218 Least-Squares Approximation of an Odd Periodic Function (Fourier Series) for the Function and Its First and Second Derivatives

```
10 CLS:CLS:PRINT"P218 APPROXIMATES AN ODD FUNCTION Y(T) AND ITS FIRST"
20 PRINT"AND SECOND DERIVATIVES Y1,Y2 BY FOURIER SERIES."
30 PRINT:PRINT"THE INPUT:"
40 PRINT"SINGLE PRECISION ONLY."
50 PRINT"IF PRINTER NOT USED O2=0, IF USED O2>0."
60 DEFINT I-O
70 INPUT"O2";O2:CLS
80 PRINT"ABSCISSAS OF POINTS DEFINING THE FUNCTION: X0(I)"
90 PRINT"ORDINATES OF POINTS DEFINING THE FUNCTION: F(I)":PRINT
100 PRINT"IF ALL DATA NEW O3>1, IF ONLY ORDINATES F(I) ARE"
110 PRINT"NEW O3=1, IF ALL DATA ARE TAKEN FROM FILES P218A"
120 PRINT"AND P218B WRITTEN IN THE FIRST RUN O3=0."
130 INPUT"O3";O3:CLS
140 PRINT"IF ONLY FUNCTION COMPUTED O4=0, IF FUNCTION AND ITS"
150 PRINT"FIRST DERIVATIVE COMPUTED O4=1, IF FUNCTION AND ITS"
160 PRINT"FIRST AND SECOND DERIVATIVES COMPUTED O4=2."
170 INPUT"O4";O4:CLS:IF O3=0 THEN GOTO 200
180 PRINT"IF SEQUENCE OF VARIANCES COMPUTED O5>0, IF NOT O5=0."
190 INPUT"O5";O5:CLS
200 PRINT"NUMBER OF POINTS DEFINING THE FUNCTION: M-1"
210 PRINT"HIGHEST HARMONIC IN THE SERIES: N<M"
220 INPUT"M";M:INPUT"N";N:CLS
230 PRINT"POINTS DEFINING THE FUNCTION ARE EQUALLY SPACED AND"
240 PRINT"DISTRIBUTED ON THE HALFINTERVAL (0,H)."
250 PRINT"IF H=3.141593 (PI) THEN LET H=0."
260 INPUT"H";H:CLS
270 IF O2=0 THEN GOTO 310
280 LPRINT "P218: O2=";O2;"O3=";O3;"O4=";O4
290 IF O3>0 THEN LPRINT "O5=";O5
300 LPRINT "M=";M;"N=";N;"H=";H
310 LET M1=M-1
320 LET M2=2*M
330 IF H>0 THEN Q=3.141593/H
340 IF H>0 THEN Q2=Q*Q
350 DIM A(N),B(N),V(N),F(M1),G(M1)
360 DIM C(N,M),S(N,M),C0(N),S0(N)
370 IF O3=0 THEN GOTO 1390
380 IF O3=1 THEN GOTO 670
390 REM COMPUTATION OF MATRICES C(N,M),S(N,M)
400 LET X1=3.141593/M
410 LET C(1,1)=COS(X1)
420 LET S(1,1)=SIN(X1)
430 FOR I=2 TO M1
440 LET I1=I-1
450 LET C(1,I)=C(1,I1)*C(1,1)-S(1,I1)*S(1,1)
460 LET S(1,I)=S(1,I1)*C(1,1)+C(1,I1)*S(1,1)
470 LET C(1,M)=-1
480 LET S(1,M)=0
490 NEXT I
500 FOR I=1 TO M
510 FOR J=2 TO N
520 LET J1=J-1
530 LET C(J,I)=C(J1,I)*C(1,I)-S(J1,I)*S(1,I)
540 LET S(J,I)=S(J1,I)*C(1,I)+C(J1,I)*S(1,I)
550 NEXT J
560 NEXT I
570 REM WRITING MATRICES C(J,I),S(J,I) ON A FILE
580 OPEN "O",1,"P218A"
590 FOR I=1 TO M
```

```
600 FOR J=1 TO N
610 PRINT #1, C(J,I),S(J,I)
620 NEXT J
630 NEXT I
640 CLOSE 1
650 GOTO 740
660 REM READING MATRICES C(J,I),S(J,I) FROM A FILE
670 OPEN "I",1,"P218A"
680 FOR I=1 TO M
690 FOR J=1 TO N
700 INPUT #1, C(J,I),S(J,I)
710 NEXT J
720 NEXT I
730 CLOSE 1
740 PRINT "INPUT OF ORDINATES:"
750 IF O2>0 THEN LPRINT "THE INPUT F(I):"
760 LET H0=H/M
770 IF H=0 THEN LET H0=X1
780 FOR I=1 TO M1
790 LET X1=I*H0
800 PRINT "I=";I;"X0(I)=";X1
810 INPUT"F(I)";F(I)
820 NEXT I
830 CLS:PRINT "CHECK THE ORDINATES F(I):"
840 FOR I=1TO M1
850 LET X1=I*H0
860 PRINT "I=";I;"X0(I)=";X1;"F(I)=";F(I)
870 PRINT"IF NO ERROR INPUT 0"
880 PRINT"IF ERROR INPUT ANY NUMBER>0 AND THEN THE CORRECT "F(I)"
890 INPUT"ERROR";K
900 IF K>0 THEN INPUT"F(I)";F(I)
910 CLS:NEXT I
920 PRINT"COMPUTING"
930 REM COMPUTATION OF G(I)
940 FOR I=1 TO M1
950 LET G(I)=2*F(I)
960 IF O2=0 THEN GOTO 970
970 NEXT I
980 IF O2=0 THEN GOTO 1040
990 FOR I=1 TO M1
1000 LET X1=I*H0
1010 LPRINT "I=";I;"X0(I)=";X1;"F(I)=";F(I)
1020 NEXT I
1030 LPRINT
1040 PRINT "COEFFICIENTS OF THE SERIES"
1050 FOR J=1 TO N
1060 LET B(J)=0
1070 FOR I=1 TO M1
1080 LET B(J)=B(J)+G(I)*S(J,I)
1090 NEXT I
1100 LET B(J)=B(J)/M
1110 PRINT J;"B(J)=";B(J)
1120 NEXT J
1130 IF O5=0 THEN GOTO 1300
1140 PRINT "THE VARIANCES:"
1150 FOR L=2 TO N
1160 LET V(L)=0
1170 LET D=0
1180 FOR I=1 TO M1
1190 LET D=D+G(I)*G(I)
1200 NEXT I
1210 LET V(L)=V(L)+D/2
1220 LET D=0
1230 FOR J=1 TO L
```

P. 106 PROGRAM P218

```
1240 LET D=D+B(J)*B(J)
1250 NEXT J
1260 LET V(L)=(V(L)-D*M)/(M2-2*L-1)
1270 PRINT "L=";L;"S(L)=";V(L)
1280 NEXT L
1290 REM WRITING B(J) AND VARIANCES ON A FILE
1300 OPEN "O",1,"P218B"
1310 PRINT #1, O5,B(1)
1320 FOR J=2 TO N
1330 PRINT #1, B(J)
1340 IF O5>0 THEN PRINT #1, V(J)
1350 NEXT J
1360 CLOSE 1
1370 GOTO 1460
1380 REM READING B(J) AND VARIANCES FROM A FILE
1390 OPEN "I",1,"P218B"
1400 INPUT #1, O5,B(1)
1410 FOR J=2 TO N
1420 INPUT #1, B(J)
1430 IF O5>0 THEN INPUT #1, V(J)
1440 NEXT J
1450 CLOSE 1
1460 IF O2=0 THEN GOTO 1570
1470 LPRINT "COEFFICIENTS B(J) AND VARIANCES S(J)"
1480 LPRINT "J=1   B(1)=";B(1)
1490 FOR J=2 TO N
1500 IF O5=0 THEN GOTO 1530
1510 LPRINT "J=";J;"B(J)=";B(J);"S(J)=";V(J)
1520 GOTO 1540
1530 LPRINT "J=";J;"B(J)=";B(J)
1540 NEXT J
1550 LPRINT
1560 LPRINT "THE APPROXIMATION:"
1570 PRINT:PRINT "THE APPROXIMATION:"
1580 INPUT"T";X
1590 IF O2>0 THEN LPRINT "T =";X;"N=";N
1600 IF H>0 THEN LET X=X*Q
1610 REM COMPUTATION OF C0(J),S0(J)
1620 LET C0(1)=COS(X)
1630 LET S0(1)=SIN(X)
1640 FOR J=2 TO N
1650 LET J1=J-1
1660 LET C0(J)=C0(J1)*C0(1)-S0(J1)*S0(1)
1670 LET S0(J)=S0(J1)*C0(1)+C0(J1)*S0(1)
1680 NEXT J
1690 LET Y=A0
1700 FOR J=1 TO N
1710 LET Y=Y+B(J)*S0(J)
1720 NEXT J
1730 PRINT "Y =";Y
1740 IF O2>0 THEN LPRINT "Y =";Y
1750 IF O4=0 THEN GOTO 1930
1760 REM COMPUTATION OF THE FIRST DERIVATIVE
1770 LET Y1=0
1780 FOR J=1 TO N
1790 LET Y1=Y1+J*B(J)*C0(J)
1800 NEXT J
1810 IF H>0 THEN LET Y1=Y1*Q
1820 PRINT "Y1=";Y1
1830 IF O2>0 THEN LPRINT "Y1=";Y1
1840 IF O4=1 THEN GOTO 1930
1850 REM COMPUTATION OF THE SECOND DERIVATIVE
1860 LET Y2=0
1870 FOR J=1 TO N
```

```
1880 LET Y2=Y2-J*J*B(J)*S0(J)
1890 NEXT J
1900 IF H>0 THEN LET Y2=Y2*Q2
1910 PRINT "Y2=";Y2
1920 IF O2>0 THEN LPRINT "Y2=";Y2
1930 IF O2>0 THEN LPRINT
1940 PRINT "TO STOP PRESS BREAK":PRINT
1950 INPUT"N";N
1960 GOTO 1580
1970 END
```

Example

An odd periodic function with the half-interval H=4

$$y = \sin(x) + [\sin(2x)]/5 + [\sin(3x)]/10$$

in which

$$x = (PI)t/H$$

is approximated by a truncated Fourier series with N=3 terms (M=4). The 3 input data are

0.977817, 0.9, 0.577817.

The coefficients of the series are

$B_1 = 1$, $B_2 = 0.2$, $B_3 = 0.0999996$,

and the variances

0.0133336, $1.43051 * 10^{-6}$.

Some results of approximation with N=3 terms for the function y and its first and second derivatives y_1, y_2:

t=0.5,   y=0.616492, y_1=1.03792, y_2=-1.09791.
t=1.5,   y=1.02703, y_1=-0.139269, y_2=-0.706386.
t=2.5,   y=0.74419, y_1=-0.305019, y_2=-8.50129*10^{-3}.

P219 Economization of a Power Series

```
10 CLS:PRINT"P219 ECONOMIZES A POWER SERIES."
20 PRINT:PRINT"THE INPUT:"
30 PRINT"IF SINGLE PRECISION O1=1, IF DOUBLE O1=2."
40 DEFINT I-O:INPUT"O1";O1:CLS
50 PRINT"IF PRINTER NOT USED O2=0, IF USED O2>0."
60 INPUT"O2";O2:CLS
70 IF O1=2 THEN DEFDBL A,B,H
80 PRINT"COEFFICIENTS OF THE SERIES: A(J)":PRINT
90 PRINT"IF COEFFICIENTS A(J) FROM INPUT O3=0, IF FROM"
100 PRINT"FILE P219A WRITTEN IN PRECEDING RUN O3>0."
110 INPUT"O3";O3:CLS
120 PRINT"HIGHEST POWER OF THE SERIES: N, 2<N<13"
130 PRINT"INTERVAL OF CONVERGENCE OF THE SERIES: (-H,H)"
140 INPUT"N";N:INPUT"H";H:CLS
150 IF O2=0 THEN GOTO 180
160 LPRINT "P219: O1=";O1;"O2=";O2;"O3=";O3;"N=";N
170 LPRINT "H=";H
180 DIM A(13),K(10,7)
190 LET N1=N+1
200 IF O3=0 THEN GOTO 270
210 OPEN "I",1,"P219A"
220 FOR I=1 TO N1
230 INPUT #1, A(I)
240 NEXT I
250 CLOSE 1
260 GOTO 420
270 PRINT "INPUT OF COEFFICIENTS:"
280 FOR I=1 TO N1
290 LET J=I-1
300 PRINT "POWER J=";J;"I=";I
310 INPUT"A(I)";A(I)
320 NEXT I
330 CLS:PRINT "CHECK THE COEFFICIENTS:"
340 FOR I=1 TO N1
350 LET J=I-1
360 PRINT "POWER J=";J;"I=";I;"A(I)=";A(I)
370 PRINT"IF NO ERROR INPUT 0"
380 PRINT"IF ERROR INPUT ANY NUMBER>0 AND THEN THE CORRECT A(I)"
390 INPUT"ERROR";K
400 IF K>0 THEN INPUT"A(I)";A(I)
410 CLS:NEXT I
420 IF O2>0 THEN LPRINT "THE INPUT:"
430 LET H0=1
440 FOR I=1 TO N1
450 LET J=I-1
460 IF O2>0 THEN LPRINT "POWER J=";J;"I=";I;"A(I)=";A(I)
470 IF H=1 THEN GOTO 500
480 LET A(I)=A(I)*H0
490 LET H0=H0*H
500 NEXT I
510 IF O2>0 THEN LPRINT
520 LET J0=2
530 FOR I=1 TO 10
540 IF ((I/2)-INT(I/2))=0 THEN LET J0=J0+1
550 FOR J=1 TO J0
560 READ K(I,J)
570 NEXT J
580 NEXT I
590 REM ECONOMIZATION PROCESS
600 LET N2=N-2
610 LET N0=INT(N/2)
620 FOR I=1 TO N0
```

```
630 LET J=I+1
640 LET L=N1-I-I
650 LET A(L)=A(L)-(K(N2,J)/K(N2,1))*A(N1)
660 NEXT I
670 LET H0=1
680 PRINT "ECONOMIZED COEFFICIENTS:"
690 FOR I=1 TO N
700 LET B=A(I)/H0
710 LET J=I-1
720 PRINT "J=";J;"A(J)=";B
730 IF H=1 THEN GOTO 750
740 LET H0=H0*H
750 NEXT I
760 PRINT"IF THE ECONOMIZATION PROCESS IS TO BE REPEATED O4=1,"
770 PRINT"IF THE ECONOMIZED COEFFICIENTS ARE TO BE STORED IN"
780 PRINT"FILE P219A O4=2, IF STOP O4=0."
790 INPUT"O4";O4
800 IF O4><1THEN GOTO 840
810 LET N1=N
820 LET N=N-1
830 GOTO 600
840 IF O2>0 THEN GOTO 860
850 IF O4=0 THEN STOP
860 IF O4=0 THEN GOTO 880
870 OPEN "O",1,"P219A"
880 IF O2>0 THEN LPRINT "ECONOMIZED COEFFICIENTS:"
890 LET H0=1
900 FOR I=1 TO N
910 LET J=I-1
920 LET A(I)=A(I)/H0
930 IF H=1 THEN GOTO 950
940 LET H0=H0*H
950 IF O2>0 THEN LPRINT "POWER J=";J;"A(J)=";A(I)
960 IF O4=0 THEN GOTO 980
970 PRINT #1, A(I)
980 NEXT I
990 IF O4=2 THEN CLOSE 1
1000 STOP
1010 DATA 4,-3,8,-8,1,16,-20,5,32,-48,18,-1,64,-112,56,-7
1020 DATA 128,-256,160,-32,1,256,-576,432,-120,9,512,-1280
1030 DATA 1120,-400,50,-1,1024,-2816,2816,-1232,220,-11
1040 DATA 2048,-6144,6912,-3584,840,-72,1
1050 END
```

Example

An odd power series for the function y=sin(PI/2)x with coefficients

$A_1=1.5707963$, $A_3=-0.6549641$, $A_5=0.079692626$,
$A_7=-4.6817541*10^{-3}$, $A_9=1.6044118*10^{-4}$,
$A_{11}=-3.5988432*10^{-6}$,

is economized on the interval (-1,1) in double precision.

After the first economization the coefficients are

$A_1=1.570796261340552$, $A_3=-0.6459633268110313$,
$A_5=0.079688296141775$, $A_7=-4.6718572812*10^{-3}$,
$A_9=1.505443612*10^{-4}$.

After the second economization the coefficients are

$A_1=1.570790968765354$, $A_3=-0.6458927591417188$,
$A_5=0.07943425253225001$, $A_7=-4.3331324685*10^{-3}$.

P220 Padé Approximation of a Truncated Series with 8 Terms for the Function and Its First and Second Derivatives

```
10 CLS:PRINT"P220 COMPUTES Y(X) AND ITS FIRST AND SECOND"
20 PRINT"DERIVATIVES Y1,Y2 BY PADE' APPROXIMATION."
30 PRINT:PRINT"THE INPUT:"
40 PRINT"IF SINGLE PRECISION O1=1, IF DOUBLE O1=2."
50 DEFINT I-O:INPUT"O1";O1:CLS
60 PRINT"IF PRINTER NOT USED O2=0, IF USED O2>0."
70 INPUT"O2";O2:CLS
80 IF O1=2 THEN DEFDBL A-H,P-Z
90 PRINT"IF DATA FROM INPUT O3=0, IF FROM FILE P220A WRITTEN"
100 PRINT"IN THE FIRST RUN O3>0."
110 INPUT"O3";O3:CLS
120 PRINT"IF ONLY FUNCTION COMPUTED O4=0, IF FUNCTION AND ITS"
130 PRINT"FIRST DERIVATIVE COMPUTED O4=1, IF FUNCTION AND ITS"
140 PRINT"FIRST AND SECOND DERIVATIVES COMPUTED O4=2."
150 INPUT"O4";O4:CLS:IF O3>0 THEN GOTO 170
160 PRINT"COEFFICIENTS OF THE POWER SERIES: C(I) (8 TERMS)."
170 IF O2=0 THEN GOTO 190
180 LPRINT "P220: O1=";O1;"O2=";O2;"O3=";O3;"O4=";O4
190 DIM A(5),B(3),C(8)
200 IF O3>0 THEN GOTO 830
210 PRINT "INPUT OF COEFFICIENTS:"
220 FOR I=1 TO 8
230 LET J=I-1
240 PRINT "POWER J=";J;"I=";I
250 INPUT"C(I)";C(I)
260 NEXT I
270 CLS:PRINT "CHECK THE COEFFICIENTS:"
280 FOR I=1 TO 8
290 LET J=I-1
300 PRINT "POWER J=";J;"I=";I;"C(I)=";C(I)
310 PRINT"IF NO ERROR INPUT 0"
320 PRINT"IF ERROR INPUT ANY NUMBER>0 AND THEN THE CORRECT C(I)"
330 INPUT"ERROR";K
340 IF K>0 THEN INPUT"C(I)";C(I)
350 CLS:NEXT I
360 IF O2=0 THEN GOTO 440
370 LPRINT "THE INPUT:"
380 FOR I=1 TO 8
390 LET J=I-1
400 LPRINT "POWER J=";J;"I=";I;"C(I)=";C(I)
410 NEXT I
420 LPRINT
430 REM COMPUTATION OF COEFFICIENTS B(I)
440 LET D2=C(3)/C(2)
450 LET D1=C(4)/C(2)
460 LET D0=C(5)/C(2)
470 LET E2=C(4)/C(3)
480 LET E1=C(5)/C(3)
490 LET E0=C(6)/C(3)
500 LET G2=C(5)/C(4)
510 LET G1=C(6)/C(4)
520 LET G0=C(7)/C(4)
530 LET H0=(D0-E0)/(D2-E2)
540 LET H1=(D1-E1)/(D2-E2)
550 LET B(1)=(H0-(D0-G0)/(D2-G2))/((D1-G1)/(D2-G2)-H1)
560 LET B(2)=-H1*B(1)-H0
570 LET B(3)=-D2*B(2)-D1*B(1)-D0
580 REM COMPUTATION OF COEFFICIENTS A(I)
590 LET A(1)=C(1)
600 LET A(2)=C(2)+C(1)*B(1)
610 LET A(3)=C(3)+C(2)*B(1)+C(1)*B(2)
```

```
620 LET A(4)=C(4)+C(3)*B(1)+C(2)*B(2)+C(1)*B(3)
630 LET A(5)=C(5)+C(4)*B(1)+C(3)*B(2)+C(2)*B(3)
640 REM WRITING COEFFICIENTS A(I),B(I) ON A FILE
650 PRINT "COEFFICIENTS A(I),B(I):"
660 IF O2>0 THEN LPRINT "COEFFICIENTS A(I),B(I):"
670 OPEN "O",1,"P220A"
680 FOR I=1 TO 5
690 PRINT "I=";I;"A(I)=";A(I)
700 IF O2>0 THEN LPRINT "I=";I;"A(I)=";A(I)
710 PRINT #1, A(I)
720 NEXT I
730 PRINT
740 IF O2>0 THEN LPRINT
750 FOR I=1 TO 3
760 PRINT "I=";I;"B(I)=";B(I)
770 IF O2>0 THEN LPRINT "I=";I;"B(I)=";B(I)
780 PRINT #1, B(I)
790 NEXT I
800 CLOSE 1
810 GOTO 910
820 REM READING COEFFICIENTS A(I),B(I) FROM A FILE
830 OPEN "I",1,"P220A"
840 FOR I=1 TO 5
850 INPUT #1, A(I)
860 NEXT I
870 FOR I=1 TO 3
880 INPUT #1, B(I)
890 NEXT I
900 CLOSE 1
910 IF O2>0 THEN LPRINT
920 IF O2>0 THEN LPRINT "THE INTERPOLATION:"
930 PRINT:PRINT "THE INTERPOLATION:"
940 INPUT"X";X
950 IF O2>0 THEN LPRINT "X =";X
960 LET P=A(1)+X*(A(2)+X*(A(3)+X*(A(4)+X*A(5))))
970 LET Q=1+X*(B(1)+X*(B(2)+X*B(3)))
980 LET Y=P/Q
990 PRINT "Y =";Y
1000 IF O2>0 THEN LPRINT "Y =";Y
1010 IF O4=0 THEN GOTO 1150
1020 REM COMPUTATION OF THE FIRST DERIVATIVE
1030 LET P1=A(2)+X*(2*A(3)+X*(3*A(4)+4*X*A(5)))
1040 LET Q1=B(1)+X*(2*B(2)+3*X*B(3))
1050 LET Y1=(P1-P*Q1/Q)/Q
1060 PRINT "Y1=";Y1
1070 IF O2>0 THEN LPRINT "Y1=";Y1
1080 IF O4=1 THEN GOTO 1150
1090 REM COMPUTATION OF THE SECOND DERIVATIVE
1100 LET P2=2*A(3)+X*(6*A(4)+12*X*A(5))
1110 LET Q2=2*B(2)+6*X*B(3)
1120 LET Y2=(P2-P*Q2/Q)/Q-2*Y1*Q1/Q
1130 PRINT "Y2=";Y2
1140 IF O2>0 THEN LPRINT "Y2=";Y2
1150 IF O2>0 THEN LPRINT
1160 PRINT "TO STOP PRESS BREAK":PRINT
1170 GOTO 940
1180 END
```

Example

A power series for the function $y=e^x$ with coefficients

$c_0=1$, $c_1=1$, $c_2=0.5$, $c_3=0.166667$
$c_4=0.0416667$, $c_5=8.33333*10^{-3}$,
$c_6=1.38889*10^{-3}$, $c_7=1.98413*10^{-4}$,

is approximated by a rational function by Padé approximation in single precision. The coefficients of the power series in the numerator are

1, 0.499992, 0.099996, $8.33261*10^{-3}$,

and the coefficients of the power series in the denominator

1, -0.500008, 0.100004, $-8.33387*10^{-3}$.

Some results of the approximation for the function y and its first and second derivatives y_1, y_2:

x=0.5, y=1.64872, y_1=1.64872, y_2=1.64875.
x=0.75, y=2.11700, y_1=2.11703, y_2=2.11727.
x=1, y=2.71831, y_1=2.71851, y_2=2.71992.

P221 Modified Padé Approximation of a Truncated Series with 13 Terms for the Function and Its First and Second Derivatives

```
10 CLS:PRINT"P221 COMPUTES Y(X) AND ITS FIRST AND SECOND"
20 PRINT"DERIVATIVES Y1,Y2 BY MODIFIED PADE' APPROXIMATION."
30 PRINT:PRINT"THE INPUT:"
40 PRINT"IF SINGLE PRECISION O1=1, IF DOUBLE O1=2."
50 DEFINT I-O:INPUT"O1";O1:CLS
60 PRINT"IF PRINTER NOT USED O2=0, IF USED O2>0."
70 INPUT"O2";O2:CLS
80 IF O1=2 THEN DEFDBL A-H,P-Z
90 PRINT"IF DATA FROM INPUT O3=0, IF FROM FILE P221A WRITTEN"
100 PRINT"IN THE FIRST RUN O3>0."
110 INPUT"O3";O3:CLS
120 PRINT"IF ONLY FUNCTION COMPUTED O4=0, IF FUNCTION AND ITS"
130 PRINT"FIRST DERIVATIVE COMPUTED O4=1, IF FUNCTION AND ITS"
140 PRINT"FIRST AND SECOND DERIVATIVES COMPUTED O4=2."
150 INPUT"O4";O4:CLS
160 PRINT"IF FUNCTION HAS NO SYMMETRY O5=0, IF IT IS ODD O5=1"
170 PRINT"IF IT IS EVEN O5=2."
180 INPUT"O5";O5:CLS:IF O3>0 THEN GOTO 200
190 PRINT"COEFFICIENTS OF THE POWER SERIES: C(I) (13 TERMS)."
200 IF O2=0 THEN GOTO 220
210 LPRINT "P221: O1=";O1;"O2=";O2;"O3=";O3;"O4=";O4;"O5=";O5
220 DIM A(4),B(3),A0(3),B0(3),C(13)
230 IF O3>0 THEN GOTO 1090
240 PRINT "INPUT OF COEFFICIENTS:"
250 FOR I=1 TO 13
260 LET J=I-1
270 PRINT "POWER J=";J;"I=";I
280 INPUT"C(I)";C(I)
290 NEXT I
300 CLS:PRINT "CHECK THE COEFFICIENTS:"
310 FOR I=1 TO 13
320 LET J=I-1
330 PRINT "POWER J=";J;"I=";I;"C(I)=";C(I)
340 PRINT"IF NO ERROR INPUT 0"
350 PRINT"IF ERROR INPUT ANY NUMBER>0 AND THEN THE CORRECT C(I)"
360 INPUT"ERROR";K
370 IF K>0 THEN INPUT"C(I)";C(I)
380 CLS:NEXT I
390 PRINT"COMPUTING"
400 IF O2=0 THEN GOTO 490
410 LPRINT "THE INPUT:"
420 FOR I=1 TO 13
430 LET J=I-1
440 LPRINT "POWER J=";J;"I=";I;"C(I)=";C(I)
450 NEXT I
460 LPRINT
470 IF O5=1 THEN GOTO 700
480 REM COMPUTATION OF COEFFICIENTS B(I)
490 LET D2=C(5)/C(3)
500 LET D1=C(7)/C(3)
510 LET D0=C(9)/C(3)
520 LET E2=C(7)/C(5)
530 LET E1=C(9)/C(5)
540 LET E0=C(11)/C(5)
550 LET G2=C(9)/C(7)
560 LET G1=C(11)/C(7)
570 LET G0=C(13)/C(7)
580 LET H0=(D0-E0)/(D2-E2)
590 LET H1=(D1-E1)/(D2-E2)
```

P. 114

```
600 LET B(1)=(H0-(D0-G0)/(D2-G2))/((D1-G1)/(D2-G2)-H1)
610 LET B(2)=-H1*B(1)-H0
620 LET B(3)=-D2*B(2)-D1*B(1)-D0
630 REM COMPUTATION OF COEFFICIENTS A(I)
640 LET A(1)=C(1)
650 LET A(2)=C(3)+C(1)*B(1)
660 LET A(3)=C(5)+C(3)*B(1)+C(1)*B(2)
670 LET A(4)=C(7)+C(5)*B(1)+C(3)*B(2)+C(1)*B(3)
680 IF O5=2 THEN GOTO 890
690 REM COMPUTATION OF COEFFICIENTS B0(I)
700 LET D2=C(4)/C(2)
710 LET D1=C(6)/C(2)
720 LET D0=C(8)/C(2)
730 LET E2=C(6)/C(4)
740 LET E1=C(8)/C(4)
750 LET E0=C(10)/C(4)
760 LET G2=C(8)/C(6)
770 LET G1=C(10)/C(6)
780 LET G0=C(12)/C(6)
790 LET H0=(D0-E0)/(D2-E2)
800 LET H1=(D1-E1)/(D2-E2)
810 LET B0(1)=(H0-(D0-G0)/(D2-G2))/((D1-G1)/(D2-G2)-H1)
820 LET B0(2)=-H1*B0(1)-H0
830 LET B0(3)=-D2*B0(2)-D1*B0(1)-D0
840 REM COMPUTATION OF COEFFICIENTS A0(I)
850 LET A0(1)=C(2)
860 LET A0(2)=C(4)+C(2)*B0(1)
870 LET A0(3)=C(6)+C(4)*B0(1)+C(2)*B0(2)
880 REM WRITING COEFFICIENTS A(I),B(I),A0(I),B0(I) ON A FILE
890 PRINT "COEFFICIENTS A(I),B(I),A0(I),B0(I):"
900 IF O2>0 THEN LPRINT "COEFFICIENTS A(I),B(I),A0(I),B0(I):"
910 OPEN "O",1,"P221A"
920 FOR I=1 TO 4
930 PRINT "I=";I;"A(I)=";A(I)
940 IF O2>0 THEN LPRINT "I=";I;"A(I)=";A(I)
950 PRINT #1, A(I)
960 IF I=4 THEN GOTO 1050
970 PRINT "I=";I;"B(I)=";B(I)
980 PRINT "I=";I;"A0(I)=";A0(I)
990 PRINT "I=";I;"B0(I)=";B0(I)
1000 IF O2=0 THEN GOTO 1040
1010 LPRINT "I=";I;"B(I)=";B(I)
1020 LPRINT "I=";I;"A0(I)=";A0(I)
1030 LPRINT "I=";I;"B0(I)=";B0(I)
1040 PRINT #1, B(I),A0(I),B0(I)
1050 NEXT I
1060 CLOSE 1
1070 GOTO 1170
1080 REM READING COEFFICIENTS A(I),B(I),A0(I),B0(I) FROM A FILE
1090 OPEN "I",1,"P221A"
1100 FOR I=1 TO 4
1110 IF I=4 THEN GOTO 1140
1120 INPUT #1, A(I),B(I),A0(I),B0(I)
1130 GOTO 1150
1140 INPUT #1, A(I)
1150 NEXT I
1160 CLOSE 1
1170 IF O2>0 THEN LPRINT
1180 IF O2>0 THEN LPRINT "THE INTERPOLATION:"
1190 PRINT:PRINT "THE INTERPOLATION:"
1200 INPUT"X";X
1210 IF O2>0 THEN LPRINT "X =";X
1220 LET X2=X*X
1230 IF O5=1 THEN GOTO 1290
```

PROGRAM P221

```
1240 LET P4=A(1)+X2*(A(2)+X2*(A(3)+X2*A(4)))
1250 LET Q4=1+X2*(B(1)+X2*(B(2)+X2*B(3)))
1260 LET Y4=P4/Q4
1270 IF O5=2 THEN LET Y=Y4
1280 IF O5=2 THEN GOTO 1330
1290 LET P3=X*(A0(1)+X2*(A0(2)+X2*A0(3)))
1300 LET Q3=1+X2*(B0(1)+X2*(B0(2)+X2*B0(3)))
1310 LET Y=P3/Q3
1320 IF O5=0 THEN LET Y=Y+Y4
1330 PRINT "Y =";Y
1340 IF O2>0 THEN LPRINT "Y =";Y
1350 IF O4=0 THEN GOTO 1640
1360 REM COMPUTATION OF THE FIRST DERIVATIVE
1370 IF O5=1 THEN GOTO 1430
1380 LET P6=X*(2*A(2)+X2*(4*A(3)+6*X2*A(4)))
1390 LET Q6=X*(2*B(1)+X2*(4*B(2)+6*X2*B(3)))
1400 LET Y6=(P6-P4*Q6/Q4)/Q4
1410 IF O5=2 THEN LET Y1=Y6
1420 IF O5=2 THEN GOTO 1480
1430 LET P5=A0(1)+X2*(3*A0(2)+5*X2*A0(3))
1440 LET Q5=X*(2*B0(1)+X2*(4*B0(2)+6*X2*B0(3)))
1450 LET Y5=(P5-P3*Q5/Q3)/Q3
1460 LET Y1=Y5
1470 IF O5=0 THEN LET Y1=Y1+Y6
1480 PRINT "Y1=";Y1
1490 IF O2>0 THEN LPRINT "Y1=";Y1
1500 IF O4=1 THEN GOTO 1640
1510 REM COMPUTATION OF THE SECOND DERIVATIVE
1520 IF O5=1 THEN GOTO 1580
1530 LET P8=2*A(2)+X2*(12*A(3)+30*X2*A(4))
1540 LET Q8=2*B(1)+X2*(12*B(2)+30*X2*B(3))
1550 LET Y8=(P8-P4*Q8/Q4)/Q4-2*Y6*Q6/Q4
1560 IF O5=2 THEN LET Y2=Y8
1570 IF O5=2 THEN GOTO 1620
1580 LET P7=X*(6*A0(2)+20*X2*A0(3))
1590 LET Q7=2*B0(1)+X2*(12*B0(2)+30*X2*B0(3))
1600 LET Y2=(P7-P3*Q7/Q3)/Q3-2*Y5*Q5/Q3
1610 IF O5=0 THEN LET Y2=Y2+Y8
1620 PRINT "Y2=";Y2
1630 IF O2>0 THEN LPRINT "Y2=";Y2
1640 IF O2>0 THEN LPRINT
1650 PRINT "TO STOP PRESS BREAK":PRINT
1660 GOTO 1200
1670 END
```

Examples

1. A power series for the function $y=e^x$ with coefficients

    $c_0=1$, $c_1=1$, $c_2=0.5$, $c_3=0.166667$,
    $c_4=0.0416667$, $c_5=8.33333*10^{-3}$, $c_6=1.38889*10^{-3}$,
    $c_7=1.98413*10^{-4}$, $c_8=2.48016*10^{-5}$, $c_9=2.75573*10^{-6}$,
    $c_{10}=2.75573*10^{-7}$, $c_{11}=2.50521*10^{-8}$, $c_{12}=2.08768*10^{-9}$,

is approximated by a rational function by modified Padé approximation in single precision. The coefficients of the power series in the numerator for the even part of the approximated function are

   1, 0.470596, 0.0273883, $3.72338*10^{-4}$,

and the coefficients of the power series in the denominator

   1, -0.294044, $4.23734*10^{-4}$, $-3.23562*10^{-6}$.

The coefficients of the power series in the numerator for the odd part of the appoximated function are

   1, 0.129956, $2.90355*10^{-3}$,

and the coefficients of the power series in the denominator

   1, -0.0367104, $6.88615*10^{-4}$, $-7.26218*10^{-6}$.

Some results of the approximation for the function y and its first and second derivatives y_1, y_2:

   $x=1$, $y=2.71828$, $y_1=2.71828$, $y_2=2.71828$.
   $x=1.5$, $y=4.48169$, $y_1=4.48169$, $y_2=4.48170$.
   $x=3$, $y=20.0865$, $y_1=20.0900$, $y_2=20.1055$.

2. A power series for the odd function $y=\sinh(x)$ is approximated by a rational function by modified Padé approximation in single precision. The coefficients of the series are identical with the odd terms of the series of example 1.

   Some results of the approximation:

   $x=1$, $y=1.17520$, $y_1=1.54308$, $y_2=1.17520$.
   $x=2$, $y=3.62686$, $y_1=3.76221$, $y_2=3.62699$.
   $x=3$, $y=10.0187$, $y_1=10.0716$, $y_2=10.0355$.

3. A power series for the even function $y=\cosh(x)$ is approximated by a rational function by modified Padé approximation in single precision. The coefficients of the series are identical with the even terms of the series of example 1.

   Some results of the approximation:

   $x=1$, $y=1.54308$, $y_1=1.17520$, $y_2=1.54308$.
   $x=2$, $y=3.76220$, $y_1=3.62687$, $y_2=3.76222$.
   $x=3$, $y=10.0678$, $y_1=10.0184$, $y_2=10.0701$.

Compare these results with those of example 2.

P222 Least-Squares Approximation by a Rational Function with Chebyshev Polynomials for the Function and Its First and Second Derivatives

```
10 CLS:PRINT"P222 COMPUTES Y(T) AND ITS FIRST AND SECOND"
20 PRINT"DERIVATIVES Y1,Y2 BY RATIONAL APPROXIMATION."
30 PRINT:PRINT"THE INPUT:"
40 PRINT"IF SINGLE PRECISION O1=1, IF DOUBLE O1=2."
50 DEFINT I-O:INPUT"O1";O1:CLS
60 PRINT"IF PRINTER NOT USED O2=0, IF USED O2>0."
70 INPUT"O2";O2:CLS
80 IF O1=2 THEN DEFDBL A-H,P-Z
90 PRINT"ABSCISSAS OF 13 POINTS DEFINING THE FUNCTION: X1(I)"
100 PRINT"ORDINATES OF 13 POINTS DEFINING THE FUNCTION: F(I)"
110 PRINT"DERIVATIVE OF THE FUNCTION AT ITS CENTRAL POINT: F1"
120 PRINT
130 PRINT"IF ALL DATA NEW O3>1, IF ONLY ORDINATES F(I) AND"
140 PRINT"DERIVATIVE F1 NEW O3=1, IF ALL DATA FROM FILES"
150 PRINT"P222A AND P222B WRITTEN IN THE FIRST RUN O3=0."
160 INPUT"O3";O3:CLS
170 PRINT"IF ONLY FUNCTION COMPUTED O4=0, IF FUNCTION AND ITS"
180 PRINT"FIRST DERIVATIVE COMPUTED O4=1, IF FUNCTION AND ITS"
190 PRINT"FIRST AND SECOND DERIVATIVES COMPUTED O4=2."
200 INPUT"O4";O4:CLS
210 PRINT"IF FUNCTION HAS NO SYMMETRY O5=0, IF IT IS ODD O5=1,"
220 PRINT"IF IT IS EVEN O5=2."
230 INPUT"O5";O5:CLS
240 PRINT"DEGREE OF THE HIGHEST CHEBYSHEV POLYNOMIAL: M<12"
250 PRINT"IF M=12 THE APPROXIMATION BECOMES INTERPOLATION."
260 INPUT"M";M:CLS
270 PRINT"INTERVAL COVERED BY 13 POINTS: (U,V)"
280 INPUT"U";U:INPUT"V";V:CLS
290 PRINT"VECTOR a(even) STORED AS A(I), a(odd) AS A0(I),"
300 PRINT"b AS B(I), d AS B0(I)"
310 IF O2=0 THEN GOTO 340
320 LPRINT "P222: O1=";O1;"O2=";O2;"O3=";O3
330 LPRINT "O4=";O4;"O5=";O5;"M=";M;"U=";U;"V=";V
340 DIM A(7),B(3),A0(6),B0(3),F(13),E(6),G(6)
350 DIM T(13,7),T0(13),T1(13),T2(13),X0(7),X2(6),X4(6),X1(13)
360 IF O3=0 THEN GOTO 1920
370 READ X0(1),X0(2),X0(3),X0(4),X0(5),X0(6),X0(7)
380 IF O3=1 THEN GOTO 570
390 LET M0=13
400 FOR I=1 TO 7
410 LET X=X0(I)
420 GOSUB 2870
430 FOR J=1 TO 13
440 LET T(J,I)=T0(J)
450 NEXT J
460 NEXT I
470 REM WRITING TENSOR T(J,I) ON A FILE
480 OPEN "O",1,"P222A"
490 FOR J=1 TO 13
500 FOR I=1 TO 7
510 PRINT #1, T(J,I)
520 NEXT I
530 NEXT J
540 CLOSE 1
550 GOTO 640
560 REM READING TENSOR T(J,I) FROM A FILE
570 OPEN "I",1,"P222A"
580 FOR J=1 TO 13
590 FOR I=1 TO 7
```

```
600 INPUT #1, T(J,I)
610 NEXT I
620 NEXT J
630 CLOSE 1
640 LET U0=(V+U)/2
650 LET V0=(V-U)/2
660 LET X1(7)=U0
670 FOR I=1 TO 6
680 LET J=I+1
690 LET X2(I)=X0(J)*X0(J)
700 LET X4(I)=X2(I)*X2(I)
710 LET K=7+I
720 LET L=7-I
730 LET H=V0*X0(J)
740 LET X1(K)=U0+H
750 LET X1(L)=U0-H
760 NEXT I
770 IF O2>0 THEN LPRINT "THE INPUT:"
780 PRINT "INPUT OF ORDINATES:"
790 FOR I=1 TO 13
800 IF O5=0 THEN GOTO 820
810 IF I>7 THEN GOTO 850
820 PRINT "I=";I;"X1(I)=";X1(I)
830 INPUT"F(I)";F(I)
840 GOTO 880
850 LET J=14-I
860 LET F(I)=F(J)
870 IF O5=1 THEN LET F(I)=-F(I)
880 NEXT I
890 PRINT "DERIVATIVE F1:"
900 INPUT"F1";F1
910 CLS:PRINT "CHECK THE ORDINATES F(I) AND F1:"
920 FOR I=1 TO 14
930 IF I<14 THEN PRINT "I=";I;"X1(I)=";X1(I);"F(I)=";F(I)
940 IF I=14 THEN PRINT "F1=";F1
950 PRINT"IF NO ERROR INPUT 0"
960 PRINT"IF ERROR INPUT ANY NUMBER>0 AND THEN THE CORRECT VALUE"
970 INPUT"ERROR";K
980 IF K=0 THEN GOTO 1010
990 IF I<14 THEN INPUT"F(I)";F(I)
1000 IF I=14 THEN INPUT"F1";F1
1010 CLS:NEXT I
1020 PRINT"COMPUTING"
1030 IF O2=0 THEN GOTO 1110
1040 LPRINT "I,X1(I),F(I):"
1050 FOR I=1 TO 13
1060 LPRINT "I=";I;"X1(I)=";X1(I);"F(I)=";F(I)
1070 NEXT I
1080 LPRINT "F1=";F1
1090 LPRINT
1100 REM COMPUTATION OF E(I),G(I) AND G1(0)
1110 FOR I=1 TO 6
1120 LET K=7+I
1130 LET L=7-I
1140 LET E(I)=(F(K)+F(L))/2
1150 LET G(I)=(F(K)-F(L))/2
1160 NEXT I
1170 LET E0=F(7)
1180 LET G1=F1*V0
1190 IF O5=1 THEN GOTO 1520
1200 REM COMPUTATION OF COEFFICIENTS B(I)
1210 LET C=(E0/E(2)-1)/X2(2)
1220 LET D=(E0/E(4)-1)/X2(4)
1230 LET H=(E0/E(6)-1)/X2(6)
```

PROGRAM P222

```
1240 LET S=(X4(4)-X4(2))/(X2(4)-X2(2))
1250 LET W=(X4(6)-X4(2))/(X2(6)-X2(2))
1260 LET S0=(D-C)/(X2(4)-X2(2))
1270 LET W0=(H-C)/(X2(6)-X2(2))
1280 LET B(3)=(S0-W0)/(S-W)
1290 LET B(2)=S0-S*B(3)
1300 LET B(1)=C-X2(2)*B(2)-X4(2)*B(3)
1310 PRINT "B(I):";B(1);B(2);B(3)
1320 REM COMPUTATION OF COEFFICIENTS A(J)
1330 FOR J=1 TO 7
1340 LET K=J+J-1
1350 LET A(J)=T(K,1)*E0
1360 FOR I=1 TO 3
1370 LET L=I+I+1
1380 LET A(J)=A(J)+2*T(K,L)*E0
1390 LET L=L-2
1400 LET Z=1+X2(L)*(B(1)+X2(L)*B(2)+X4(L)*B(3))
1410 LET L1=L+1
1420 LET A(J)=A(J)+2*T(K,L1)*E(L)*Z
1430 NEXT I
1440 IF J=1 THEN GOTO 1460
1450 LET A(J)=2*A(J)
1460 LET A(J)=A(J)/13
1470 NEXT J
1480 PRINT "A(I):";A(1);A(2);A(3);A(4)
1490 PRINT "A(I):";A(5);A(6);A(7)
1500 IF O5=2 THEN GOTO 1800
1510 REM COMPUTATION OF COEFFICIENTS B0(I)
1520 LET C=(G1*X0(3)/G(2)-1)/X2(2)
1530 LET D=(G1*X0(5)/G(4)-1)/X2(4)
1540 LET H=(G1*X0(7)/G(6)-1)/X2(6)
1550 LET S=(X4(4)-X4(2))/(X2(4)-X2(2))
1560 LET W=(X4(6)-X4(2))/(X2(6)-X2(2))
1570 LET S0=(D-C)/(X2(4)-X2(2))
1580 LET W0=(H-C)/(X2(6)-X2(2))
1590 LET B0(3)=(S0-W0)/(S-W)
1600 LET B0(2)=S0-S*B0(3)
1610 LET B0(1)=C-X2(2)*B0(2)-X4(2)*B0(3)
1620 PRINT "B0(I):";B0(1);B0(2);B0(3)
1630 REM COMPUTATION OF COEFFICIENTS A0(J)
1640 FOR J=1 TO 6
1650 LET K=J+J
1660 LET A0(J)=0
1670 FOR I=1 TO 3
1680 LET L=I+I+1
1690 LET A0(J)=A0(J)+T(K,L)*G1*X0(L)
1700 LET L1=L-1
1710 LET L=L-2
1720 LET Z=1+X2(L)*(B0(1)+X2(L)*B0(2)+X4(L)*B0(3))
1730 LET A0(J)=A0(J)+T(K,L1)*G(L)*Z
1740 NEXT I
1750 LET A0(J)=A0(J)/3.25
1760 NEXT J
1770 PRINT "A0(I):";A0(1);A0(2);A0(3)
1780 PRINT "A0(I):";A0(4);A0(5);A0(6)
1790 REM WRITING COEFFICIENTS A(I),A0(I),B(I),B0(I) ON A FILE
1800 OPEN "O",1,"P222B"
1810 FOR I=1 TO 7
1820 PRINT #1, A(I)
1830 IF I>6 THEN GOTO 1870
1840 PRINT #1, A0(I)
1850 IF I>3 THEN GOTO 1870
1860 PRINT #1, B(I),B0(I)
1870 NEXT I
```

```
1880 PRINT #1, U0,V0
1890 CLOSE 1
1900 GOTO 2020
1910 REM READING COEFFICIENTS A(I),A0(I),B(I),B0(I) FROM A FILE
1920 OPEN "I",1,"P222B"
1930 FOR I=1 TO 7
1940 INPUT #1, A(I)
1950 IF I>6 THEN GOTO 1990
1960 INPUT #1, A0(I)
1970 IF I>3 THEN GOTO 1990
1980 INPUT #1, B(I),B0(I)
1990 NEXT I
2000 INPUT #1, U0,V0
2010 CLOSE 1
2020 IF O2>0 THEN LPRINT
2030 IF O2>0 THEN LPRINT "THE APPROXIMATION:"
2040 PRINT:PRINT "THE APPROXIMATION:"
2050 INPUT "X";X
2060 IF O2>0 THEN LPRINT "T =";X;"M=";M
2070 LET X=(X-U0)/V0
2080 LET X9=X*X
2090 LET V2=V0*V0
2100 LET M0=M
2110 LET M1=INT((M+1)/2)
2120 LET M2=INT(M/2)+1
2130 GOSUB 2870
2140 IF O5=1 THEN GOTO 2240
2150 LET P4=0
2160 FOR J=1 TO M2
2170 LET K=J+J-1
2180 LET P4=P4+A(J)*T0(K)
2190 NEXT J
2200 LET Q4=1+X9*(B(1)+X9*(B(2)+X9*B(3)))
2210 LET Y4=P4/Q4
2220 IF O5=2 THEN LET Y=Y4
2230 IF O5=2 THEN GOTO 2320
2240 LET P3=0
2250 FOR J=1 TO M1
2260 LET K=J+J
2270 LET P3=P3+A0(J)*T0(K)
2280 NEXT J
2290 LET Q3=1+X9*(B0(1)+X9*(B0(2)+X9*B0(3)))
2300 LET Y=P3/Q3
2310 IF O5=0 THEN LET Y=Y+Y4
2320 PRINT "Y =";Y
2330 IF O2>0 THEN LPRINT "Y =";Y
2340 IF O4=0 THEN GOTO 2820
2350 REM COMPUTATION OF THE FIRST DERIVATIVE
2360 GOSUB 2960
2370 IF O5=1 THEN GOTO 2470
2380 LET P6=0
2390 FOR J=2 TO M2
2400 LET K=J+J-1
2410 LET P6=P6+A(J)*T1(K)
2420 NEXT J
2430 LET Q6=X*(2*B(1)+X9*(4*B(2)+6*X9*B(3)))
2440 LET Y6=(P6-P4*Q6/Q4)/Q4
2450 IF O5=2 THEN LET Y1=Y6/V0
2460 IF O5=2 THEN GOTO 2560
2470 LET P5=0
2480 FOR J=1 TO M1
2490 LET K=J+J
2500 LET P5=P5+A0(J)*T1(K)
2510 NEXT J
```

PROGRAM P222

```
2520 LET Q5=X*(2*B0(1)+X9*(4*B0(2)+6*X9*B0(3)))
2530 LET Y5=(P5-P3*Q5/Q3)/Q3
2540 LET Y1=Y5/V0
2550 IF O5=0 THEN LET Y1=Y1+Y6
2560 PRINT "Y1=";Y1
2570 IF O2>0 THEN LPRINT "Y1=";Y1
2580 IF O4=1 THEN GOTO 2820
2590 REM COMPUTATION OF THE SECOND DERIVATIVE
2600 GOSUB 3050
2610 IF O5=1 THEN GOTO 2710
2620 LET P8=0
2630 FOR J=1 TO M2
2640 LET K=J+J-1
2650 LET P8=P8+A(J)*T2(K)
2660 NEXT J
2670 LET Q8=2*B(1)+X9*(12*B(2)+30*X9*B(3))
2680 LET Y8=(P8-P4*Q8/Q4)/Q4-2*Y6*Q6/Q4
2690 IF O5=2 THEN LET Y2=Y8/V2
2700 IF O5=2 THEN GOTO 2800
2710 LET P7=0
2720 FOR J=2 TO M1
2730 LET K=J+J
2740 LET P7=P7+A0(J)*T2(K)
2750 NEXT J
2760 LET Q7=2*B0(1)+X9*(12*B0(2)+30*X9*B0(3))
2770 LET Y2=(P7-P3*Q7/Q3)/Q3-2*Y5*Q5/Q3
2780 LET Y2=Y2/V2
2790 IF O5=0 THEN LET Y2=Y2+Y8
2800 PRINT "Y2=";Y2
2810 IF O2>0 THEN LPRINT "Y2=";Y2
2820 IF O2>0 THEN LPRINT
2830 PRINT "TO STOP PRESS BREAK":PRINT
2840 INPUT"M";M
2850 GOTO 2050
2860 REM SUBROUTINE FOR CHEBYSHEV POLYNOMIALS
2870 LET T0(1)=1
2880 LET T0(2)=X
2890 FOR L=3 TO M0
2900 LET L1=L-1
2910 LET L2=L-2
2920 LET T0(L)=2*X*T0(L1)-T0(L2)
2930 NEXT L
2940 RETURN
2950 REM SUBROUTINE FOR THE FIRST DERIVATIVES OF CH. P.
2960 LET T1(1)=0
2970 LET T1(2)=1
2980 FOR L=3 TO M0
2990 LET L1=L-1
3000 LET L2=L-2
3010 LET T1(L)=2*T0(L1)+2*X*T1(L1)-T1(L2)
3020 NEXT L
3030 RETURN
3040 REM SUBROUTINE FOR THE SECOND DERIVATIVES OF CH. P.
3050 LET T2(1)=0
3060 LET T2(2)=0
3070 FOR L=3 TO M0
3080 LET L1=L-1
3090 LET L2=L-2
3100 LET T2(L)=4*T1(L1)+2*X*T2(L1)-T2(L2)
3110 NEXT L
3120 RETURN
3130 DATA 0,.2393156642875578,.4647231720437685
3140 DATA .6631226582407952,.8229838658936564
3150 DATA .9350162426854149,.9927088740980540
3160 END
```

Examples

1. An odd function y=arctan(x) is approximated in single precision by a rational function on the interval (-1,1). The degree of the highest Chebyshev polynomial in the numerator is denoted M.

   Some results of the approximation for the function y and its first and second derivatives:

   x=0.5,   y=0.463647, y_1=0.799991, y_2=-0.639905,   M=11.
   x=0.75, y=0.643502, y_1=0.639999, y_2=-0.614579,   M=11.
   x=1,   y=0.785397, y_1=0.499901, y_2=-0.504290,   M=11.
   x=1,   y=0.785400, y_1=0.500284, y_2=-0.489233,   M=9.

2. An even function y=cos(x) is approximated in single precision by a rational function on the interval (-PI/2,PI/2).

   Some results of the approximation:

   x=0.25, y=0.968212, y_1=-0.255921, y_1=-0.876930, M=12.
   x=0.5,   y=0.876794, y_1=-0.474293, y_2=-0.847700, M=12.

3. A function y=arctan(x) is approximated in single precision by a rational function on the interval (0,1).

   Some results of the approximation:

   x=0.1, y=0.0996689, y_1=0.880709, y_2=0.083652,   M=12.
   x=0.5, y=0.463648, y_1=0.799997, y_2=-0.218029, M=12.
   x=1,   y=0.785398, y_1=0.625073, y_2=-0.306684, M=12.

Numerical results show that this rational approximation is not quite satisfactory, especially when derivatives are computed.

P223 Numerical Differentiation: First and Second Derivatives of a Function Defined by 3, 5, or 7 Points with Equally Spaced Abscissas

```
10 CLS:PRINT"P223 COMPUTES FIRST AND SECOND DERIVATIVE Y1,Y2"
20 PRINT"OF A FUNCTION DEFINED BY N EQUALLY SPACED POINTS."
30 PRINT:PRINT"THE INPUT:"
40 PRINT"IF SINGLE PRECISION O1=1, IF DOUBLE O1=2."
50 PRINT"USE DOUBLE PRECISION IF SECOND DERIVATIVE IS COMPUTED."
60 DEFINT I-O:INPUT"O1";O1:CLS
70 PRINT"IF PRINTER NOT USED O2=0, IF USED O2>0."
80 INPUT"O2";O2:CLS
90 IF O1=2 THEN DEFDBL F,Y
100 PRINT"IF ONLY FIRST DERIVATIVE COMPUTED O4=1, IF ALSO"
110 PRINT"SECOND DERIVATIVE COMPUTED O4=2."
120 INPUT"O4";O4:CLS
130 PRINT"NUMBER OF POINTS DEFINING THE FUNCTION: N=3,5,7"
140 PRINT"ABSCISSA WHERE THE DERIVATIVE IS COMPUTED: M=1,2,...N"
150 PRINT"INTERVAL BETWEEN TWO ABSCISSAS: H"
160 INPUT"N";N:INPUT"M";M:INPUT"H";H:CLS
170 IF O2=0 THEN GOTO 200
180 LPRINT "P223: O1=";O1;"O2=";O2;"O4=";O4
190 LPRINT "N=";N;"M=";M;"H=";H
200 DIM F(7),K(7,7),L(7,7)
210 LET N0=3
220 FOR I=1 TO N0
230 FOR J=1 TO N0
240 READ K(I,J),L(I,J)
250 NEXT J
260 NEXT I
270 READ K0,L0
280 IF N=N0 THEN GOTO 310
290 LET N0=N0+2
300 GOTO 220
310 LET H2=H*H
320 PRINT "INPUT OF ORDINATES:"
330 FOR I=1 TO N
340 PRINT "I=";I
350 INPUT"F(I)";F(I)
360 NEXT I
370 CLS:PRINT "CHECK THE ORDINATES:"
380 FOR I=1 TO N
390 PRINT"I=";I;"F(I)=";F(I)
400 PRINT"IF NO ERROR INPUT 0"
410 PRINT"IF ERROR INPUT ANY NUMBER>0 AND THEN THE CORRECT F(I)"
420 INPUT"ERROR";K
430 IF K>0 THEN INPUT"F(I)";F(I)
440 CLS:NEXT I
450 IF O2=0 THEN GOTO 700
460 LPRINT "THE INPUT:"
470 FOR I=1 TO N
480 LPRINT "I=";I;"F(I)=";F(I)
490 NEXT I
500 GOTO 680
510 REM INPUT OF ONE POINT ON THE RIGHT
520 FOR I=2 TO N
530 LET J=I-1
540 LET F(J)=F(I)
550 NEXT I
560 INPUT"F(N)";F(N)
570 IF O2>0 THEN LPRINT "F(N)=";F(N)
580 GOTO 710
590 REM INPUT OF ONE POINT ON THE LEFT
```

```
600 LET K=N-1
610 FOR I=N TO 1 STEP -1
620 LET J=I+1
630 LET F(J)=F(I)
640 NEXT I
650 INPUT"F(1)";F(1)
660 IF O2>0 THEN LPRINT "F(1)=";F(1)
670 GOTO 710
680 LPRINT
690 LPRINT "THE DERIVATIVES:"
700 PRINT:PRINT "THE DERIVATIVES:"
710 LET Y1=0
720 LET Y2=0
730 FOR I=1 TO N
740 LET Y1=Y1+K(M,I)*F(I)
750 IF O4=1 THEN GOTO 770
760 LET Y2=Y2+L(M,I)*F(I)
770 NEXT I
780 LET Y1=Y1/(K0*H)
790 PRINT "N=";N;"M=";M;"Y1=";Y1
800 IF O2>0 THEN LPRINT "N=";N;"M=";M;"Y1=";Y1
810 IF O4=1 THEN GOTO 850
820 LET Y2=Y2/(L0*H2)
830 PRINT "            Y2=";Y2
840 IF O2>0 THEN LPRINT "            Y2=";Y2
850 IF O2>0 THEN LPRINT
860 PRINT "IF NO NEW POINT O3=0"
870 PRINT "IF ONE NEW POINT ON THE RIGHT O3=1"
880 PRINT "OF ONE NEW POINT ON THE LEFT O3=-1"
890 PRINT "TO STOP PRESS BREAK":PRINT
900 INPUT"O3";O3:INPUT"M";M
910 IF O3=0 THEN GOTO 710
920 IF O3=1 THEN GOTO 520
930 GOTO 600
940 DATA -3,1,4,-2,-1,1,-1,1,0,-2,1,1,1,1,-4,-2,3,1,2,1
950 DATA -25,35,48,-104,-36,114,16,-56,-3,11,-3,11,-10,-20
960 DATA 18,6,-6,4,1,-1,1,-1,-8,16,0,-30,8,16,-1,-1,-1,-1
970 DATA 6,4,-18,6,10,-20,3,11,3,11,-16,-56,36,114,-48,-104
980 DATA 25,35,12,12,-147,1624,360,-6264,-450,10530,400,-10160
990 DATA -225,5940,72,-1944,-10,274,-10,274,-77,-294,150,-510
1000 DATA -100,940,50,-570,-15,186,2,-26,2,-26,-24,456,-35,-840
1010 DATA 80,400,-30,30,8,-24,-1,4,-1,4,9,-54,-45,540,0,-980,45
1020 DATA 540,-9,-54,1,4,1,4,-8,-24,30,30,-80,400,35,-840,24,456
1030 DATA -2,-26,-2,-26,15,186,-50,-570,100,940,-150,-510,77
1040 DATA -294,10,274,10,274,-72,-1944,225,5940,-400,-10160,450
1050 DATA 10530,-360,-6264,147,1624,60,360
1060 END
```

Examples

1. The function $y=e^x$ is given in single precision by 3 points with abscissas 0.96, 0.97, 0.98. The function and its numerically computed first and second derivatives y_1, y_2 at these points are:

    x=0.96,   y=2.61170, y_1=2.61002, y_2=2.79665.
    x=0.97,   y=2.63794, y_1=2.63799, y_2=2.79665.
    x=0.98,   y=2.66446, y_1=2.66595, y_2=2.79665.

2. The function $y=e^x$ is given in single precision by 5 points with abscissas 0.95, 0.96, 0.97, 0.98, 0.99. The function and its numerically computed derivatives are:

    x=0.95,   y=2.58571, y_1=2.58900, y_2=1.65145.
    x=0.96,   y=2.61170, y_1=2.61049, y_2=2.54949.
    x=0.97,   y=2.63794, y_1=2.63800, y_2=2.84592.
    x=0.98,   y=2.66446, y_1=2.66548, y_2=2.55267.
    x=0.99,   y=2.69123, y_1=2.68707, y_2=1.67211.

3. The function $y=e^x$ is given in single precision by 7 points with abscissas 0.94, 0.95, 0.96, 0.97, 0.98, 0.99, 1. The function and its numerically computed derivatives are:

    x=0.94,   y=2.55998, y_1=2.55330, y_2=5.53725.
    x=0.95,   y=2.58571, y_1=2.58769, y_2=2.33057.
    x=0.96,   y=2.61170, y_1=2.61066, y_2=2.49198.
    x=0.97,   y=2.63794, y_1=2.63803, y_2=2.86730.
    x=0.98,   y=2.66446, y_1=2.66523, y_2=2.47892.
    x=0.99,   y=2.69123, y_1=2.68878, y_2=2.49905.
    x=1.00,   y=2.71828, y_1=2.72725, y_2=6.30697.

4. The function $y=e^x$ is given in double precision by 3 points with abscissas 0.96, 0.97, 0.98. The function and its numerically computed derivatives are:

  x=0.96,   y=2.611696473423118, y_1=2.611608819266048,
    y_2=2.637966508929392.
  x=0.97,   y=2.637944459354153, y_1=2.637988484278581,
    y_2=2.637966508929392.
  x=0.98,   y=2.664456241929417, y_1=2.664368149291103,
    y_2=2.637966508929392.

5. The function $y=e^x$ is given in double precision by 5 points with abscissas 0.95, 0.96, 0.97, 0.98, 0.99. The function and its numerically computed derivatives are:

x=0.95,   y=2.585709659315846, y_1=2.585709711852566,
       y_2=2.585711976619965.
x=0.96,   y=2.611696473423118, y_1=2.611696533115934,
       y_2=2.611696383709276.
x=0.97,   y=2.637944459354153, y_1=2.637944517437447,
       y_2=2.63794458966876.
x=0.98,   y=2.664456241929417, y_1=2.664456594503227,
       y_2=2.664456594503227.
x=0.99,   y=2.691234472349262, y_1=2.691234527209534,
       y_2=2.691232398206572.

6. The function $y=e^x$ is given in double precision by 7 points with abscissas 0.94,0.95,0.96,0.97,0.98,0.99,1. The function and its numerically computed derivatives are:
x=0.94,   y=2.559981418329271, y_1=2.559981570915078,
       y_2=2.559981528968025.
x=0.95,   y=2.585709659315846, y_1=2.585709813436434,
       y_2=2.585709770589675.
x=0.96,   y=2.611696473423118, y_1=2.611696629092258,
       y_2=2.611696585837764.
x=0.97,   y=2.637944459354152, y_1=2.637944616587885,
       y_2=2.637944572928024.
x=0.98,   y=2.664456241929417, y_1=2.664456400743439,
       y_2=2.664456356617276.
x=0.99,   y=2.691234472349262, y_1=2.691234632759266,
       y_2=2.691234588199879.
x=1.00,   y=2.718281828459045, y_1=2.718281990481541,
       y_2=2.718281945515134.

Thus the accuracy of numerical differentiation depends more upon the accuracy of input data than upon the number of points.

CHAPTER 3: EVALUATION OF DEFINITE INTEGRALS

P301 Integration of a Numerically Defined Integrand by Composite Simpson's Rule

```
10 CLS:PRINT"P301 EVALUATES A DEFINITE INTEGRAL WITH NUMERICALLY"
20 PRINT"DEFINED INTEGRAND BY COMPOSITE SIMPSON RULE."
30 PRINT:PRINT"THE INPUT:"
40 PRINT"SINGLE PRECISION ONLY."
50 PRINT"IF PRINTER NOT USED O2=0, IF USED O2>0."
60 DEFINT I-O:INPUT"O2";O2:CLS
70 PRINT"NUMBER OF EQUALLY SPACED POINTS DEFINING"
80 PRINT"THE INTEGRAND: N (AN ODD NUMBER)"
90 PRINT"INTERVAL OF INTEGRATION: (A,B)"
100 INPUT"N";N:INPUT"A";A:INPUT"B";B:CLS
110 PRINT"ORDINATES OF POINTS DEFINING THE INTEGRAND: F(I)"
120 IF O2=0 THEN GOTO 140
130 LPRINT "P301: O2=";O2;"N=";N;"A=";A;"B=";B
140 DIM F(N)
150 PRINT "INPUT OF ORDINATES:"
160 FOR I=1 TO N
170 PRINT "I=";I
180 INPUT"F(I)";F(I)
190 NEXT I
200 CLS:PRINT "CHECK THE ORDINATES:"
210 FOR I=1 TO N
220 PRINT "I=";I;"F(I)=";F(I)
230 PRINT"IF NO ERROR INPUT 0"
240 PRINT"IF ERROR INPUT ANY NUMBER>0 AND THEN THE CORRECT F(I)"
250 INPUT"ERROR";K
260 IF K>0 THEN INPUT"F(I)";F(I)
270 CLS:NEXT I
280 IF O2=0 THEN GOTO 350
290 LPRINT "THE INPUT:"
300 FOR I=1 TO N
310 LPRINT "I=";I;"F(I)=";F(I)
320 NEXT I
330 LPRINT
340 REM THE INTEGRATION
350 LET N0=N-1
360 LET Q=0
370 FOR I=2 TO N0 STEP 2
380 LET Q=Q+F(I)
390 NEXT I
400 LET Q=4*Q+F(1)+F(N)
410 LET Q0=0
420 LET N0=N0-1
430 FOR I=3 TO N0 STEP 2
440 LET Q0=Q0+F(I)
450 NEXT I
460 LET Q=(Q+2*Q0)*(B-A)/(3*N-3)
470 PRINT "VALUE OF THE INTEGRAL Q=";Q
480 IF O2>0 THEN LPRINT "VALUE OF THE INTEGRAL Q=";Q
490 END
```

Examples

1. An integral with the integrand given numerically by a set of 11 equally spaced points
   (0,0), (1,1), (2,8), (3,27), (4,64), (5,125), (6,216),
   (7,343), (8,512), (9,729), (10,1000)
is evaluated. Its value is Q=2500 (an exact value, the integrand is $f(x)=x^3$, the limits are 0,10).

2. An integral with the integrand given numerically by a set of 7 equally spaced points
   (0,0), (0.261799,0.258819), (0.523599,0.5),
   (0.785398,0.707107), (1.047198,0.866025),
   (1.308997,0.966926), (1.570796,1)
is evaluated. Its value is Q=1.00038 (the exact value is 1, the integrand $f(x)=\sin(x)$, the limits are 0, PI/2=1.570796).

P302 Integration of a Numerically Defined Integrand by Modified Composite Simpson's Rule

```
10 CLS:PRINT"P302 EVALUATES A DEFINITE INTEGRAL WITH NUMERICALLY"
20 PRINT"DEFINED INTEGRAND BY MODIFIED COMPOSITE SIMPSON RULE."
30 PRINT:PRINT"THE INPUT:"
40 PRINT"SINGLE PRECISION ONLY."
50 PRINT"IF PRINTER NOT USED O2=0, IF USED O2>0."
60 DEFINT I-O:INPUT"O2";O2:CLS
70 PRINT"NUMBER OF ARBITRARILY SPACED POINTS DEFINING"
80 PRINT"THE INTEGRAND: N (ANY NUMBER>4)"
90 INPUT"N";N:CLS
100 PRINT"ABSCISSAS OF POINTS DEFINING THE INTEGRAND: X(I)"
110 PRINT"ORDINATES OF POINTS DEFINING THE INTEGRAND: F(I)"
120 IF O2=0 THEN GOTO 140
130 LPRINT "P302: O2=";O2;"N=";N
140 DIM F(N),X(N)
150 PRINT "INPUT OF ABSCISSAS AND ORDINATES:"
160 FOR I=1 TO N
170 PRINT "I=";I
180 INPUT"X(I)";X(I):INPUT"F(I)";F(I)
190 NEXT I
200 CLS:PRINT "CHECK THE INPUT:"
210 FOR I=1 TO N
220 PRINT "I=";I;"X(I)=";X(I);"F(I)=";F(I)
230 PRINT"IF NO ERROR IN X(I) INPUT 0"
240 PRINT"IF ERROR INPUT ANY NUMBER>0 AND THEN THE CORRECT X(I)"
250 INPUT"ERROR";K
260 IF K>0 THEN INPUT"X(I)";X(I)
270 PRINT:PRINT"IF NO ERROR IN F(I) INPUT 0"
280 PRINT"IF ERROR INPUT ANY NUMBER>0 AND THEN THE CORRECT F(I)"
290 INPUT"ERROR";K
300 IF K>0 THEN INPUT"F(I)";F(I)
310 CLS:NEXT I
320 IF O2=0 THEN GOTO 390
330 LPRINT "THE INPUT:"
340 FOR I=1 TO N
350 LPRINT "I=";I;"X(I)=";X(I);"F(I)=";F(I)
360 NEXT I
370 LPRINT
380 REM THE INTEGRATION
390 LET N2=N-2
400 LET Q=0
410 FOR I=2 TO N2
420 LET I0=I-1
430 LET I1=I+1
440 LET I2=I+2
450 LET B1=(F(I)-F(I0))/(X(I)-X(I0))
460 LET B2=(F(I1)-F(I))/(X(I1)-X(I))
470 LET B3=(F(I2)-F(I1))/(X(I2)-X(I1))
480 LET C1=(B2-B1)/(X(I1)-X(I0))
490 LET C2=(B3-B2)/(X(I2)-X(I))
500 LET D=(C2-C1)/(X(I2)-X(I0))
510 IF I=2 THEN GOTO 560
520 IF I=N2 THEN GOTO 590
530 LET J=I
540 LET J1=I1
550 GOTO 610
560 LET J=1
570 LET J1=3
580 GOTO 610
590 LET J=N2
600 LET J1=N
```

```
610 LET X0=(X(J1)+X(J))/2
620 LET F0=F(I0)+(X0-X(I0))*(B1+(X0-X(I))*(C1+(X0-X(I1))*D))
630 LET Q=Q+(X(J1)-X(J))*(F(J1)+F(J)+4*F0)/6
640 NEXT I
650 PRINT "VALUE OF THE INTEGRAL Q=";Q
660 IF O2>0 THEN LPRINT "VALUE OF THE INTEGRAL Q=";Q
670 END
```

Examples

1. An integral with the integrand given numerically by a set of 6 points

(0,0), (0.35,0.042875), (0.7,0.343), (1.05,1.15763),

(1.4,2.744), (2,8)

is evaluated. Its value is Q=4 (an exact value, the integrand is $f(x)=x^3$, the limits are 0,2).

2. An integral with the integrand given numerically by 6 points

(0,0), (0.35,0.342898), (0.7,0.644218),

(1.05,0.867423), (1.4,0.98545), (1.5708,1)

is evaluated. Its value is Q=0.999926 (the exact value is 1, the integrand is $f(x)=\sin(x)$, the limits 0,PI/2=1.5708).

P303 Integration of a Numerically Defined Integrand with Equally Spaced Abscissas by Composite Corrected Trapezoidal Rule

```
10 CLS:PRINT"P303 EVALUATES A DEFINITE INTEGRAL WITH NUMERICALLY"
20 PRINT"DEFINED INTEGRAND WITH EQUALLY SPACED ABSCISSAS"
30 PRINT"BY COMPOSITE CORRECTED TRAPEZOIDAL RULE."
40 PRINT:PRINT"THE INPUT:"
50 PRINT"SINGLE PRECISION ONLY."
60 PRINT"IF PRINTER NOT USED O2=0, IF USED O2>0."
70 DEFINT I-O:INPUT"O2";O2:CLS
80 PRINT"NUMBER OF EQUALLY SPACED POINTS DEFINING"
90 PRINT"THE INTEGRAND: N (ANY NUMBER)"
100 PRINT"INTERVAL OF INTEGRATION: (A,B)"
110 INPUT"N";N:INPUT"A";A:INPUT"B";B:CLS
120 PRINT"ORDINATES OF POINTS DEFINING THE INTEGRAND: F(I)"
130 IF O2=0 THEN GOTO 150
140 LPRINT "P303: O2=";O2;"N=";N;"A=";A;"B=";B
150 DIM F(N)
160 PRINT "INPUT OF ORDINATES:"
170 FOR I=1 TO N
180 PRINT "I=";I
190 INPUT"F(I)";F(I)
200 NEXT I
210 PRINT"FIRST DERIVATIVES OF THE INTEGRAND AT THE LIMITS: F1,G1"
220 INPUT"F1";F1:INPUT"G1";G1
230 CLS:PRINT "CHECK THE ORDINATES AND DERIVATIVES"
240 FOR I=1 TO N+2
250 IF I<=N THEN PRINT "I=";I;"F(I)=";F(I)
260 IF I=N+1 THEN PRINT"F1=";F1
270 IF I=N+2 THEN PRINT"G1=";G1
280 PRINT"IF NO ERROR INPUT 0"
290 PRINT"IF ERROR INPUT ANY NUMBER>0 AND THEN THE CORRECT VALUE"
300 INPUT"ERROR";K
310 IF K=0 THEN GOTO 350
320 IF I<=N THEN INPUT"F(I)";F(I)
330 IF I=N+1 THEN INPUT"F1";F1
340 IF I=N+2 THEN INPUT"G1";G1
350 CLS:NEXT I
360 IF O2=0 THEN GOTO 440
370 LPRINT "THE INPUT:"
380 FOR I=1 TO N
390 LPRINT "I=";I;"F(I)=";F(I)
400 NEXT I
410 LPRINT "F1=";F1;"G1=";G1
420 LPRINT
430 REM THE INTEGRATION
440 LET N0=N-1
450 LET H=(B-A)/N0
460 LET Q=(F(1)+F(N))/2+(F1-G1)*H/12
470 FOR I=2 TO N0
480 LET Q=Q+F(I)
490 NEXT I
500 LET Q=Q*H
510 PRINT "VALUE OF THE INTEGRAL Q=";Q
520 IF O2>0 THEN LPRINT "VALUE OF THE INTEGRAL Q=";Q
530 END
```

Examples

1. An integral over the interval (0,10) with the integrand given numerically by a set of 6 equally spaced points with the ordinates

    0, 8, 64 216, 512, 1000

and by its derivatives at both ends, 0,300, is evaluated. Its value is Q=2500 (an exact value, the integrand is $f(x)=x^3$).

2. An integral over the interval (0,1.5708) with the integrand given numerically by a set of 6 equally spaced points with the ordinates

    0, 0.309017, 0.0.587785, 0.809017, 0.951057, 1,

and by its derivatives at both ends, 1,0, is evaluated. Its value is Q=0.999989 (the exact value is 1, the integrand is $f(x)=\sin(x)$, the limits 0,PI/2).

P304 Integration of a Numerically Defined Integrand with Arbitrarily Spaced Abscissas by Composite Corrected Trapezoidal Rule

```
10 CLS:PRINT"P304 EVALUATES A DEFINITE INTEGRAL WITH NUMERICALLY"
20 PRINT"DEFINED INTEGRAND WITH ARBITRARILY SPACED ABSCISSAS"
30 PRINT"BY COMPOSITE CORRECTED TRAPEZOIDAL RULE."
40 PRINT:PRINT"THE INPUT:"
50 PRINT"SINGLE PRECISION ONLY."
60 PRINT"IF PRINTER NOT USED O2=0, IF USED O2>0."
70 DEFINT I-O:INPUT"O2";O2:CLS
80 PRINT"NUMBER OF ARBITRARILY SPACED POINTS DEFINING"
90 PRINT"THE INTEGRAND: N (ANY NUMBER)"
100 INPUT"N";N:CLS
110 PRINT"ABSCISSAS OF POINTS DEFINING THE INTEGRAND: X(I)"
120 PRINT"ORDINATES OF POINTS DEFINING THE INTEGRAND: F(I)"
130 PRINT"FIRST DERIVATIVES OF THE INTEGRAND AT X(I): F1(I)"
140 IF O2=0 THEN GOTO 160
150 LPRINT "P304: O2=";O2;"N=";N
160 DIM F(N),F1(N),X(N)
170 PRINT "INPUT OF ABSCISSAS, ORDINATES AND DERIVATIVES:"
180 FOR I=1 TO N
190 PRINT "I=";I
200 INPUT"X(I)";X(I):INPUT"F(I)";F(I):INPUT"F1(I)";F1(I)
210 NEXT I
220 CLS:PRINT "CHECK THE INPUT:"
230 FOR I=1 TO N
240 PRINT "I=";I;"X(I)=";X(I);"F(I)=";F(I);"F1(I)=";F1(I)
250 PRINT"IF NO ERROR IN X(I) INPUT 0"
260 PRINT"IF ERROR INPUT ANY NUMBER>0 AND THEN THE CORRECT X(I)"
270 INPUT"ERROR";K
280 IF K>0 THEN INPUT"X(I)";X(I)
290 PRINT:PRINT"IF NO ERROR IN F(I) INPUT 0"
300 PRINT"IF ERROR INPUT ANY NUMBER>0 AND THEN THE CORRECT F(I)"
310 INPUT"ERROR";K
320 IF K>0 THEN INPUT"F(I)";F(I)
330 PRINT:PRINT"IF NO ERROR IN F1(I) INPUT 0"
340 PRINT"IF ERROR INPUT ANY NUMBER>0 AND THEN THE CORRECT F1(I)"
350 INPUT"ERROR";K
360 IF K>0 THEN INPUT"F1(I)";F1(I)
370 CLS:NEXT I
380 IF O2=0 THEN GOTO 450
390 LPRINT "THE INPUT:"
400 FOR I=1 TO N
410 LPRINT "I=";I;"X(I)=";X(I);"F(I)=";F(I);"F1(I)=";F1(I)
420 NEXT I
430 LPRINT
440 REM THE INTEGRATION
450 LET N0=N-1
460 LET Q=0
470 FOR I=1 TO N0
480 LET J=I+1
490 LET H=X(J)-X(I)
500 LET Q=Q+(H/2)*(F(I)+F(J)+H*(F1(I)-F1(J))/6)
510 NEXT I
520 PRINT "VALUE OF THE INTEGRAL Q=";Q
530 IF O2>0 THEN LPRINT "VALUE OF THE INTEGRAL Q=";Q
540 END
```

Examples

1. An integral with the integrand given numerically by a set of 6 points

   $x=0$, $f(x)=0$, $f_1=0$,
   $x=1$, $f(x)=1$, $f_1=3$,
   $x=5$, $f(x)=125$, $f_1=75$,
   $x=6$, $f(x)=216$, $f_1=108$,
   $x=8$, $f(x)=512$, $f_1=192$,
   $x=10$, $f(x)=1000$, $f_1=300$

(f_1 denotes the first derivative) is evaluated. Its value is $Q=2500$ (an exact value, the integrand is $f(x)=x^3$, the limits $0,10$).

2. An integral with the integrand given numerically by a set of 8 points

   $x=0$, $f(x)=f_1=1$,
   $x=2$, $f(x)=f_1=7.38906$,
   $x=4$, $f(x)=f_1=54.5982$,
   $x=6$, $f(x)=f_1=403.429$,
   $x=7$, $f(x)=f_1=1096.63$,
   $x=8$, $f(x)=f_1=2980.96$,
   $x=9$, $f(x)=f_1=8103.08$,
   $x=10$, $f(x)=f_1=22026.5$.

is evaluated. Its value is $Q=21980$ (the exact value is $Q=e^{10}-e^0=22025.5$, the integrand is $f(x)=e^x$, the limits $0,10$, the relative error $(22025.5-21980)/22025.5=0.00207$).

P305 Composite Gaussian Integration of an Integral over a Finite Interval

```
10 CLS:PRINT"P305 EVALUATES A DEFINITE INTEGRAL OVER A FINITE"
20 PRINT"INTERVAL BY COMPOSITE GAUSSIAN QUADRATURE."
30 PRINT:PRINT"PRESS BREAK, DEFINE THE INTEGRAND ON LINES"
40 PRINT"330-350 AND RUN AGAIN."
50 PRINT:PRINT"THE INPUT:"
60 PRINT"IF SINGLE PRECISION O1=1, IF DOUBLE O1=2."
70 DEFINT I-O:INPUT"O1";O1:CLS
80 PRINT"IF PRINTER NOT USED O2=0, IF USED O2>0."
90 INPUT"O2";O2:CLS
100 IF O1=2 THEN DEFDBL A-H,Q-Z
110 PRINT"NUMBER OF SUBINTERVALS: M>=1"
120 PRINT"ACCURACY OF QUADRATURE INCREASES WITH"
130 PRINT"INCREASING M, BUT THE INTEGRATION LASTS LONGER."
140 INPUT"M";M:CLS
150 PRINT"INTERVAL OF INTEGRATION: (A,B)"
160 INPUT"A";A:INPUT"B";B:CLS
170 PRINT"MAXIMUM RELATIVE ERROR OF QUADRATURE: P"
180 PRINT"IF P=0 ACCURACY OF INTEGRATION IS NOT TESTED."
190 PRINT"IF P>0 MAXIMUM NUMBER OF SUBINTERVALS: 16*M"
200 INPUT"P";P:CLS
210 IF O2=0 THEN GOTO 240
220 LPRINT "P305: O1=";O1;"O2=";O2;"M=";M
230 LPRINT "A=";A;"B=";B;"ACCURACY P=";P
240 LET H=(B-A)/M
250 LET H2=H/2
260 DIM U(16),W(16)
270 REM READ ABSCISSAS AND WEIGHTS FOR GAUSSIAN QUADRATURE
280 PRINT "READING DATA AND INTEGRATING"
290 FOR I=1 TO 16
300 READ U(I),W(I)
310 NEXT I
320 REM WRITE THE RIGHT-HAND SIDE, 3 LINES AVAILABLE
330 DEF FNF(X)=
360 REM GAUSSIAN QUADRATURE
370 LET K=0
380 LET Q=0
390 LET A0=A-H2
400 FOR J=1 TO M
410 LET X0=H*J
420 FOR I=1 TO 16
430 LET X=H2*U(I)
440 LET X5=A0+X0+X
450 LET X6=A0+X0-X
460 LET Q=Q+W(I)*(FNF(X5)+FNF(X6))
470 NEXT I
480 NEXT J
490 LET Q=Q*H2
500 IF P=0 THEN GOTO 640
510 IF K=4 THEN GOTO 640
520 REM TESTING THE ACCURACY OF INTEGRATION
530 PRINT "Q=";Q
540 IF K=0 THEN GOTO 570
550 LET D=ABS((Q-Q0)/Q)
560 IF D<P THEN GOTO 640
570 LET Q0=Q
580 LET K=K+1
590 LET M=M+M
600 LET H=H2
610 LET H2=H2/2
620 GOTO 380
630 REM PRINTING THE RESULT
```

```
640 PRINT "VALUE OF THE INTEGRAL Q=";Q
650 IF P>0 THEN PRINT "NUMBER OF SUBINTERVALS M=";M;"K=";K
660 IF O2=0 THEN STOP
670 LPRINT "VALUE OF THE INTEGRAL Q=";Q
680 IF P>0 THEN LPRINT "NUMBER OF SUBINTERVALS M=";M;"K=";K
690 STOP
700 DATA .04830766568773832,.09654008851472780
710 DATA .14447196158279649,.09563872007927486
720 DATA .23928736225213707,.09384439908080457
730 DATA .33186860228212765,.09117387869576388
740 DATA .42135127613063535,.08765209300440381
750 DATA .50689990893222939,.08331192422694676
760 DATA .58771575724076233,.07819389578707031
770 DATA .66304426693021520,.07234579410884851
780 DATA .73218211874028968,.06582222277636185
790 DATA .79448379596794241,.05868409347853555
800 DATA .84936761373256997,.05099805926237618
810 DATA .89632115576605212,.04283589802222668
820 DATA .93490607593773969,.03427386291302143
830 DATA .96476225558750643,.02539206530926206
840 DATA .98561151154526834,.01627439473090567
850 DATA .99726386184948156,.00701861000947010
860 END
```

Example

An integral with the integrand $f(x)=1/x$ and the limits $(1,10)$ is evaluated in double precision with the accuracy $P=1*10^{-14}$. The quadrature yields the value $Q=2.302585092994046$ which is exact in all 16 digits [$Q=\log(10)$].

P306 Laguerre Integration of an Integral over a Semi-Infinite Interval

```
10 CLS:PRINT"P306 EVALUATES A DEFINITE INTEGRAL OVER"
20 PRINT"A SEMI-INFINITE INTERVAL BY LAGUERRE QUADRATURE."
30 PRINT:PRINT"PRESS BREAK, DEFINE THE INTEGRAND ON LINES"
40 PRINT"250-270 AND RUN AGAIN."
50 PRINT:PRINT"THE INPUT:"
60 PRINT"IF SINGLE PRECISION O1=1, IF DOUBLE O1=2."
70 DEFINT I-O:INPUT"O1";O1:CLS
80 PRINT"IF PRINTER NOT USED O2=0, IF USED O2>0."
90 INPUT"O2";O2:CLS
100 IF O1=2 THEN DEFDBL A-H,Q-Z
110 PRINT"IF THE UPPER LIMIT OF INTEGRAL IS INFINITY O3>0."
120 PRINT"IF THE LOWER LIMIT OF INTEGRAL IS MINUS INFINITY O3<0."
130 PRINT"THE FINITE LIMIT OF THE INTEGRAL: A"
140 INPUT"O3";O3:INPUT"A";A:CLS
150 IF O2=0 THEN GOTO 180
160 LPRINT "P306: O1=";O1;"O2=";O2;"O3=";O3
170 LPRINT "A=";A
180 DIM V(15),Z(15)
190 PRINT "READING DATA AND INTEGRATING"
200 REM READ ABSCISSAS AND WEIGHTS FOR LAGUERRE QUADRATURE
210 FOR I=1 TO 15
220 READ V(I),Z(I)
230 NEXT I
240 REM WRITE THE RIGHT-HAND SIDE, 3 LINES AVAILABLE
250 DEF FNF(X)=
280 REM LAGUERRE QUADRATURE
290 LET Q=0
300 FOR I=1 TO 15
310 IF O3>0 THEN LET X=A+V(I)
320 IF O3<0 THEN LET X=A-V(I)
330 LET Q=Q+Z(I)*FNF(X)
340 NEXT I
350 REM PRINTING THE RESULT
360 PRINT "VALUE OF THE INTEGRAL Q=";Q
370 IF O2=0 THEN STOP
380 LPRINT "VALUE OF THE INTEGRAL Q=";Q
390 STOP
400 DATA .093307812017,.239578170311
410 DATA .492691740302,.560100842793
420 DATA 1.215595412071,.887008262919
430 DATA 2.269949526204,1.22366440215
440 DATA 3.667622721751,1.57444872163
450 DATA 5.425336627414,1.94475197653
460 DATA 7.565916226613,2.34150205664
470 DATA 10.120228568019,2.77404192683
480 DATA 13.130282482176,3.25564334640
490 DATA 16.654407708330,3.80631171423
500 DATA 20.776478899449,4.45847775384
510 DATA 25.623894226729,5.27001778443
520 DATA 31.407519169754,6.35956346973
530 DATA 38.530683306486,8.03178763212
540 DATA 48.026085572686,11.5277721009
550 END
```

Examples

1. An integral with the integrand $f(x)=xe^{-x^2}$ over a semi-infinite interval from 0 to infinity is evaluated in single precision. The quadrature yields the exact value $Q=\frac{1}{2}$. This exactness is a consequence of the simple form of the integrand, which is most suitable for Laguerre integration.

2. An integral with the integrand $f(x)=e^{-x^3}$ over a semi-infinite interval from 0 to infinity is evaluated in single precision. The quadrature yields the value $Q=0.88353$. The exact value is

$$Q = \Gamma(3)/3 = 0.89298.$$

The relative error is thus $(0.89298-0.88353)/0.89298=0.0106$ (cf. example 2 to P307).

P307 Composite Gauss-Laguerre Integration of an Integral over a Semi-Infinite Interval

```
10 CLS:PRINT"P307 EVALUATES DEFINITE INTEGRAL OVER A SEMI-INFINITE"
20 PRINT"INTERVAL BY COMPOSITE GAUSS-LAGUERRE QUADRATURE."
30 PRINT:PRINT"PRESS BREAK, DEFINE THE INTEGRAND ON LINES"
40 PRINT"420-440 AND RUN AGAIN."
50 PRINT:PRINT"THE INPUT:"
60 PRINT"IF SINGLE PRECISION O1=1, IF DOUBLE O1=2."
70 DEFINT I-O:INPUT"O1";O1:CLS
80 PRINT"IF PRINTER NOT USED O2=0, IF USED O2>0."
90 INPUT"O2";O2:CLS
100 IF O1=2 THEN DEFDBL A-H,Q-Z
110 PRINT"IF THE UPPER LIMIT OF INTEGRAL IS INFINITY O3>0."
120 PRINT"IF THE LOWER LIMIT OF INTEGRAL IS MINUS INFINITY O3<0."
130 PRINT:PRINT"THE FINITE LIMIT OF THE INTEGRAL: A"
140 INPUT"O3";O3:INPUT"A";A:CLS
150 PRINT"CHOSEN LIMIT FOR GAUSSIAN QUADRATURE: B"
160 PRINT"CHOOSE B<0 IF O3<0.":PRINT
170 PRINT"NUMBER OF SUBINTERVALS: M>=1"
180 PRINT"ACCURACY OF GAUSSIAN QUADRATURE INCREASES WITH"
190 PRINT"INCREASING M, BUT THE INTEGRATION LASTS LONGER.":PRINT
200 PRINT"MAXIMUM RELATIVE ERROR OF GAUSSIAN QUADRATURE: P"
210 PRINT"IF P=0 ACCURACY OF INTEGRATION IS NOT TESTED."
220 PRINT"IF P>0 MAXIMUM NUMBER OF SUBINTERVALS: 16*M":PRINT
230 PRINT"TO ACHIVE MAXIMUM ACCURACY OF INTEGRATION CHOOSE P>0"
240 PRINT"AND ABS(B) SO HIGH THAT Q(LAGUERRE) INFLUENCES NO"
250 PRINT"MORE THAN LAST TWO SIGNIFICANT DIGITS OF Q(GAUSS)."
260 INPUT"B";B:INPUT"M";M:INPUT"P";P:CLS
270 IF O2=0 THEN GOTO 300
280 LPRINT "P307: O1=";O1;"O2=";O2;"O3=";O3;"M=";M
290 LPRINT "A=";A;"B=";B;"ACCURACY P=";P
300 IF O3>0 THEN LET H=(B-A)/M
310 IF O3<0 THEN LET H=(A-B)/M
320 LET H2=H/2
330 DIM U(16),W(16),V(15),Z(15)
340 PRINT "READING DATA AND INTEGRATING"
350 REM READ ABSCISSAS AND WEIGHTS FOR GAUSSIAN AND
360 REM LAGUERRE QUADRATURE
370 FOR I=1 TO 15
380 READ U(I),W(I),V(I),Z(I)
390 NEXT I
400 READ U(16),W(16)
410 REM WRITE THE RIGHT-HAND SIDE, 3 LINES AVAILABLE
420 DEF FNF(X)=
450 REM GAUSSIAN QUADRATURE
460 LET K=0
470 LET Q=0
480 IF O3>0 THEN LET A0=A-H2
490 IF O3<0 THEN LET A0=B-H2
500 FOR J=1 TO M
510 LET X0=H*J
520 FOR I=1 TO 16
530 LET X=H2*U(I)
540 LET X5=A0+X0+X
550 LET X6=A0+X0-X
560 LET Q=Q+W(I)*(FNF(X5)+FNF(X6))
570 NEXT I
580 NEXT J
590 LET Q=Q*H2
600 IF P=0 THEN GOTO 740
610 IF K=4 THEN GOTO 740
620 REM TESTING THE ACCURACY OF INTEGRATION
630 PRINT "Q(GAUSS)=";Q
```

```
640 IF K=0 THEN GOTO 670
650 LET D=ABS((Q-Q0)/Q)
660 IF D<P THEN GOTO 740
670 LET Q0=Q
680 LET K=K+1
690 LET M=M+M
700 LET H=H2
710 LET H2=H2/2
720 GOTO 470
730 REM LAGUERRE QUADRATURE
740 LET Q1=Q
750 LET Q=0
760 FOR I=1 TO 15
770 IF O3>0 THEN LET X=B+V(I)
780 IF O3<0 THEN LET X=B-V(I)
790 LET Q=Q+Z(I)*FNF(X)
800 NEXT I
810 IF P>0 THEN PRINT "Q(LAGUERRE)=";Q
820 LET Q=Q+Q1
830 REM PRINTING THE RESULT
840 PRINT "VALUE OF THE INTEGRAL Q=";Q
850 IF P>0 THEN PRINT "NUMBER OF SUBINTERVALS M=";M;"K=";K
860 IF O2=0 THEN STOP
870 LPRINT "VALUE OF THE INTEGRAL Q=";Q
880 IF P>0 THEN LPRINT "NUMBER OF SUBINTERVALS M=";M;"K=";K
890 STOP
900 DATA .04830766568773832,.09654008851472780
910 DATA .093307812017,.239578170311
920 DATA .14447196158279649,.09563872007927486
930 DATA .492691740302,.560100842793
940 DATA .23928736225213707,.09384439908080457
950 DATA 1.215595412071,.887008262919
960 DATA .33186860228212765,.09117387869576388
970 DATA 2.269949526204,1.22366440215
980 DATA .42135127613063535,.08765209300440381
990 DATA 3.667622721751,1.57444872163
1000 DATA .50689990893222939,.08331192422694676
1010 DATA 5.425336627414,1.94475197653
1020 DATA .58771575724076233,.07819389578707031
1030 DATA 7.565916226613,2.34150205664
1040 DATA .66304426693021520,.07234579410884851
1050 DATA 10.120228568019,2.77404192683
1060 DATA .73218211874028968,.06582222277636185
1070 DATA 13.130282482176,3.25564334640
1080 DATA .79448379596794241,.05868409347853555
1090 DATA 16.654407708330,3.80631171423
1100 DATA .84936761373256997,.05099805926237618
1110 DATA 20.776478899449,4.45847775384
1120 DATA .89632115576605212,.04283589802222668
1130 DATA 25.623894226729,5.27001778443
1140 DATA .93490607593773969,.03427386291302143
1150 DATA 31.407519169754,6.35956346973
1160 DATA .96476225558750643,.02539206530926206
1170 DATA 38.530683306486,8.03178763212
1180 DATA .98561151154526834,.01627439473090567
1190 DATA 48.026085572686,11.5277721009
1200 DATA .99726386184948156,.00701861000947010
1210 END
```

Examples

1. An integral with the integrand $f(x)=xe^{-x^2}$ over a semi-infinite interval from 0 to infinity is evaluated in single precision with the accuracy $P=4*10^{-7}$ and with the limit for Gaussian quadrature $B=6$. The integration yields the exact value $Q=½$.

2. An integral with the integrand $f(x)=e^{-x^3}$ over a semi-infinite interval from 0 to infinity is evaluated in single precision with the accuracy $P=4*10^{-7}$ and with the limit for Gaussian quadrature $B=3$. The integration yields the exact value $Q=\Gamma(3)/3=0.89298$ (cf. example 2 to P306).

P308 Hermite Integration of an Integral over an Infinite Interval

```
10 CLS:PRINT"P308 EVALUATES A DEFINITE INTEGRAL OVER AN INFINITE"
20 PRINT"INTERVAL BY HERMITE QUADRATURE."
30 PRINT:PRINT"PRESS BREAK, DEFINE THE INTEGRAND ON LINES"
40 PRINT"190-210 AND RUN AGAIN."
50 PRINT:PRINT"THE INPUT:"
60 PRINT"IF SINGLE PRECISION O1=1, IF DOUBLE O1=2."
70 DEFINT I-O:INPUT"O1";O1:CLS
80 PRINT"IF PRINTER NOT USED O2=0, IF USED O2>0."
90 INPUT"O2";O2:CLS
100 IF O1=2 THEN DEFDBL A-H,Q-Z
110 IF O2=0 THEN GOTO 130
120 LPRINT "P308: O1=";O1;"O2=";O2
130 DIM V(10),Z(10)
140 PRINT "READING DATA AND INTEGRATING"
150 REM READ ABSCISSAS AND WEIGHTS FOR HERMITE QUADRATURE
160 READ V(1),V(2),V(3),V(4),V(5),V(6),V(7),V(8),V(9),V(10)
170 READ Z(1),Z(2),Z(3),Z(4),Z(5),Z(6),Z(7),Z(8),Z(9),Z(10)
180 REM WRITE THE RIGHT-HAND SIDE, 3 LINES AVAILABLE
190 DEF FNF(X)=
220 REM HERMITE QUADRATURE
230 LET Q=0
240 FOR I=1 TO 10
250 LET Q=Q+Z(I)*(FNF(V(I))+FNF(-V(I)))
260 NEXT I
270 REM PRINTING THE RESULT
280 PRINT "VALUE OF THE INTEGRAL Q=";Q
290 IF O2=0 THEN STOP
300 LPRINT "VALUE OF THE INTEGRAL Q=";Q
310 STOP
320 DATA .2453407083009,.7374737285454,1.2340762153953
330 DATA 1.7385377121166,2.2549740020893,2.7888060584281
340 DATA 3.3478545673832,3.9447640401156,4.6036824495507
350 DATA 5.3874808900112,.4909215006667,.4938433852721
360 DATA .4999208713363,.5096790271175,.5240803509486
370 DATA .5448517423644,.5752624428525,.6222786961914
380 DATA .7043329611769,.8985919614523
390 END
```

Example

An integral with the integrand $f(x)=x^2 e^{-x^2}$ over an infinite interval is evaluated in single precision. The quadrature yields the exact value Q=0.886227. This exactness is a consequence of the simple form of the integrand which is most suitable for Hermitian integration.

P309 Composite Gauss-Laguerre Integration of an Integral over an Infinite Interval

```
10 CLS:PRINT"P309 EVALUATES A DEFINITE INTEGRAL OVER AN INFINITE"
20 PRINT"INTERVAL BY COMPOSITE GAUSS-LAGUERRE QUADRATURE."
30 PRINT:PRINT"PRESS BREAK, DEFINE THE INTEGRAND ON LINES"
40 PRINT"390-410 AND RUN AGAIN."
50 PRINT:PRINT"THE INPUT:"
60 PRINT"IF SINGLE PRECISION O1=1, IF DOUBLE O1=2."
70 DEFINT I-O:INPUT"O1";O1:CLS
80 PRINT"IF PRINTER NOT USED O2=0, IF USED O2>0."
90 INPUT"O2";O2:CLS
100 IF O1=2 THEN DEFDBL A-H,Q-Z
110 PRINT"NUMBER OF SUBINTERVALS: M>=1"
120 PRINT"ACCURACY OF GAUSSIAN QUADRATURE INCREASES WITH"
130 PRINT"INCREASING M, BUT THE INTEGRATION LASTS LONGER."
140 PRINT
150 PRINT"CHOSEN LIMITS FOR GAUSSIAN QUADRATURE: (A,B)"
160 PRINT"CHOOSE INTEGER A,B; A<-1, B>1"
170 PRINT
180 PRINT"MAXIMUM RELATIVE ERROR OF GAUSSIAN QUADRATURE: P"
190 PRINT"IF P=0 ACCURACY OF INTEGRATION IS NOT TESTED."
200 PRINT"IF P>0 MAXIMUM NUMBER OF SUBINTERVALS: 16*M":PRINT
210 PRINT"TO ACHIVE MAXIMUM ACCURACY OF INTEGRATION CHOOSE P>0"
220 PRINT"AND ABS(A),B SO HIGH THAT Q(LAGUERRE) INFLUENCES NO"
230 PRINT"MORE THAN LAST TWO SIGNIFICANT DIGITS OF Q(GAUSS)."
240 INPUT"M";M:INPUT"A";A:INPUT"B";B:INPUT"P";P:CLS
250 IF O2=0 THEN GOTO 280
260 LPRINT "P309:  O1=";O1;"O2=";O2;"M=";M
270 LPRINT "A=";A;"B=";B;"ACCURACY P=";P
280 LET H=(B-A)/M
290 LET H2=H/2
300 DIM U(16),W(16),V(15),Z(15)
310 PRINT "READING DATA AND INTEGRATING"
320 REM READ ABSCISSAS AND WEIGHTS FOR GAUSSIAN AND...
330 REM ...LAGUERRE QUADRATURE
340 FOR I=1 TO 15
350 READ U(I),W(I),V(I),Z(I)
360 NEXT I
370 READ U(16),W(16)
380 REM WRITE THE RIGHT-HAND SIDE, 3 LINES AVAILABLE
390 DEF FNF(X)=
420 REM GAUSSIAN QUADRATURE
430 LET K=0
440 LET Q=0
450 LET A0=A-H2
460 FOR J=1 TO M
470 LET X0=H*J
480 FOR I=1 TO 16
490 LET X=H2*U(I)
500 LET X5=A0+X0+X
510 LET X6=A0+X0-X
520 LET Q=Q+W(I)*(FNF(X5)+FNF(X6))
530 NEXT I
540 NEXT J
550 LET Q=Q*H2
560 IF P=0 THEN GOTO 700
570 IF K=4 THEN GOTO 700
580 REM TESTING THE ACCURACY OF INTEGRATION
590 PRINT "Q(GAUSS)=";Q
600 IF K=0 THEN GOTO 630
610 LET D=ABS((Q-Q0)/Q)
620 IF D<P THEN GOTO 700
630 LET Q0=Q
```

```
640 LET K=K+1
650 LET M=M+M
660 LET H=H2
670 LET H2=H2/2
680 GOTO 440
690 REM LAGUERRE QUADRATURES
700 LET Q1=Q
710 LET Q=0
720 FOR I=1 TO 15
730 LET X5=B+V(I)
740 LET X6=A-V(I)
750 LET Q=Q+Z(I)*(FNF(X5)+FNF(X6))
760 NEXT I
770 IF P>0 THEN PRINT "Q(LAGUERRE)=";Q
780 LET Q=Q+Q1
790 REM PRINTING THE RESULT
800 PRINT "VALUE OF THE INTEGRAL Q=";Q
810 IF P>0 THEN PRINT "NUMBER OF SUBINTERVALS M=";M;"K=";K
820 IF O2=0 THEN STOP
830 LPRINT "VALUE OF THE INTEGRAL Q=";Q
840 IF P>0 THEN LPRINT "NUMBER OF SUBINTERVALS M=";M;"K=";K
850 STOP
860 DATA .04830766568773832,.09654008851472780
870 DATA .093307812017,.239578170311
880 DATA .14447196158279649,.09563872007927486
890 DATA .492691740302,.560100842793
900 DATA .23928736225213707,.09384439908080457
910 DATA 1.215595412071,.887008262919
920 DATA .33186860228212765,.09117387869576388
930 DATA 2.269949526204,1.22366440215
940 DATA .42135127613063535,.08765209300440381
950 DATA 3.667622721751,1.57444872163
960 DATA .50689990893222939,.08331192422694676
970 DATA 5.425336627414,1.94475197653
980 DATA .58771575724076233,.07819389578707031
990 DATA 7.565916226613,2.34150205664
1000 DATA .66304426693021520,.07234579410884851
1010 DATA 10.120228568019,2.77404192683
1020 DATA .73218211874028968,.06582222277636185
1030 DATA 13.130282482176,3.25564334640
1040 DATA .79448379596794241,.05868409347853555
1050 DATA 16.654407708330,3.80631171423
1060 DATA .84936761373256997,.05099805926237618
1070 DATA 20.776478899449,4.45847775384
1080 DATA .89632115576605212,.04283589802222668
1090 DATA 25.623894226729,5.27001778443
1100 DATA .93490607593773969,.03427386291302143
1110 DATA 31.407519169754,6.35956346973
1120 DATA .96476225558750643,.02539206530926206
1130 DATA 38.530683306486,8.03178763212
1140 DATA .98561151154526834,.01627439473090567
1150 DATA 48.026085572686,11.52777721009
1160 DATA .99726386184948156,.00701861000947010
1170 END
```

Example

An integral with the integrand $f(x)=x^2 e^{-x^2}$ over an infinite interval is evaluated with the accuracy $P=4*10^{-7}$ and with the limits of Gaussian quadrature $A=-5$, $B=5$. The integration yields the exact value $Q=0.886227$.

P310 Romberg Integration

```
10 CLS:PRINT"P310 EVALUATES A DEFINITE INTEGRAL OVER A FINITE"
20 PRINT"INTERVAL BY ROMBERG QUADRATURE."
30 PRINT:PRINT"PRESS BREAK, DEFINE THE INTEGRAND ON LINES"
40 PRINT"200-220 AND RUN AGAIN."
50 PRINT:PRINT"THE INPUT:"
60 PRINT"IF SINGLE PRECISION O1=1, IF DOUBLE O1=2."
70 DEFINT I-O:INPUT"O1";O1:CLS
80 PRINT"IF PRINTER NOT USED O2=0, IF USED O2>0."
90 INPUT"O2";O2:CLS
100 IF O1=2 THEN DEFDBL A-H,P-Z
110 PRINT"INTERVAL OF INTEGRATION: (A,B)"
120 INPUT"A";A:INPUT"B";B:CLS
130 PRINT"MAXIMUM NUMBER OF STEPS: K0"
140 PRINT"PRESS BREAK WHEN THE REQUIRED ACCURACY IS REACHED."
150 INPUT"K0";K0:CLS
160 IF O2=0 THEN GOTO 190
170 LPRINT "P310: O1=";O1;"O2=";O2;"K0=";K0
180 LPRINT "A=";A;"B=";B
190 DIM T(K0)
200 DEF FNF(X)=
230 REM TRAPEZOIDAL RULE
240 LET H=B-A
250 LET S=(FNF(A)+FNF(B))*H/2
260 PRINT
270 PRINT "VALUE OF THE INTEGRAL Q:"
280 IF O2>0 THEN LPRINT
290 IF O2>0 THEN LPRINT "VALUE OF THE INTEGRAL Q:"
300 PRINT S
310 PRINT
320 IF O2>0 THEN LPRINT S
330 IF O2>0 THEN LPRINT
340 LET A=A-H
350 LET N=1
360 LET K0=K0-1
370 FOR K=1 TO K0
380 LET Q=0
390 LET A=A+H/2
400 FOR M=1 TO N
410 LET X=A+H*M
420 LET Q=Q+FNF(X)
430 NEXT M
440 LET N=N+N
450 LET H=H/2
460 LET T(1)=S/2+H*Q
470 PRINT T(1)
480 IF O2>0 THEN LPRINT T(1)
490 REM EXTRAPOLATION TO THE LIMIT
500 LET P=1
510 FOR M=1 TO K
520 LET P=4*P
530 LET M1=M+1
540 LET Q=T(M1)
550 LET T(M1)=(P*T(M)-S)/(P-1)
560 PRINT T(M1)
570 IF O2>0 THEN LPRINT T(M1)
580 LET S=Q
590 NEXT M
600 PRINT
610 IF O2>0 THEN LPRINT
620 LET S=T(1)
630 NEXT K
640 END
```

Examples

1. An integral with the integrand $f(x)=x^3$ and the limits (0,10) is evaluated in single precision. The quadrature has already yielded the exact value Q=2500 by the second step.

2. An integral with the integrand $f(x)=1/(1+x^2)$ and the limits (0,1) is evaluated in double precision. The quadrature yields the value Q=0.7853981633974484 in 8 steps. It differs from the exact value Q=arctan(1)=(PI/4) by 1 in the sixteenth decimal place.

P311 Chebyshev-Gauss Integration of a Singular Integral

```
10 CLS:PRINT"P311 EVALUATES A SINGULAR INTEGRAL"
20 PRINT"BY CHEBYSHEV-GAUSS QUADRATURE."
30 PRINT"INTERVAL OF INTEGRATION: (-1,1)"
40 PRINT:PRINT"PRESS BREAK, DEFINE THE INTEGRAND ON LINES"
50 PRINT"150-170 AND RUN AGAIN."
60 PRINT:PRINT"THE INPUT:"
70 PRINT"SINGLE PRECISION ONLY."
80 PRINT"IF PRINTER NOT USED O2=0, IF USED O2>0."
90 DEFINT I-O:INPUT"O2";O2:CLS
100 PRINT"NUMBER OF ABSCISSAS: N"
110 INPUT"N";N:CLS
120 IF O2=0 THEN GOTO 150
130 LPRINT "P311: O2=";O2;"N=";N
140 REM WRITE THE RIGHT-HAND SIDE, 3 LINES AVAILABLE
150 DEF FNF(X)=
180 REM THE QUADRATURE
190 LET Q=0
200 LET N0=INT(N/2)
210 LET C=N/2-N0
220 IF C>0 THEN LET Q=FNF(0)
230 LET C=1.570796/N
240 FOR I=1 TO N0
250 LET X=COS(C*(I+I-1))
260 LET Q=Q+FNF(X)+FNF(-X)
270 NEXT I
280 LET Q=2*C*Q
290 REM PRINTING THE RESULT
300 PRINT "VALUE OF THE INTEGRAL Q=";Q
310 IF O2=0 THEN STOP
320 LPRINT "VALUE OF THE INTEGRAL Q=";Q
330 END
```

Examples

1. A singular integral with the function $g(x)=x^2$ in the integrand is evaluated over the interval $(-1,1)$. The integration yields the exact value $Q=(PI)/2=1.57080$.

2. A singular integral with the function x^8 in the integrand is evaluated over the interval $(-1,1)$. The integration yields the exact value $Q=(35/108)*(PI)=0.85903$.

P312 Integration of an Integral with Logarithmic Singularity

```
10 CLS:PRINT"P312 EVALUATES AN INTEGRAL WITH LOGARITHMIC"
20 PRINT"SINGULARITY."
30 PRINT"INTERVAL OF INTEGRATION: (0,1)"
40 PRINT:PRINT"PRESS BREAK, DEFINE THE INTEGRAND ON LINES"
50 PRINT"170-190 AND RUN AGAIN."
60 PRINT:PRINT"THE INPUT:"
70 PRINT"SINGLE PRECISION ONLY."
80 PRINT"IF PRINTER NOT USED O2=0, IF USED O2>0."
90 DEFINT I-O:INPUT"O2";O2:CLS
100 IF O2=0 THEN GOTO 120
110 LPRINT "P312: O2=";O2
120 DIM V(4),Z(4)
130 REM READ ABSCISSAS AND WEIGHTS FOR QUADRATURE
140 READ V(1),V(2),V(3),V(4)
150 READ Z(1),Z(2),Z(3),Z(4)
160 REM WRITE THE RIGHT-HAND SIDE, 3 LINES AVAILABLE
170 DEF FNF(X)=
200 REM THE QUADRATURE
210 LET Q=0
220 FOR I=1 TO 4
230 LET Q=Q+Z(I)*FNF(V(I))
240 NEXT I
250 REM PRINTING THE RESULT
260 PRINT "VALUE OF THE INTEGRAL Q=";Q
270 IF O2=0 THEN STOP
280 LPRINT "VALUE OF THE INTEGRAL Q=";Q
290 STOP
300 DATA .041448,.245275,.556165,.848982
310 DATA -.383464,-.386875,-.190435,-.039225
320 END
```

Examples

1. A singular integral with the integrand $f(x)=\log(x)/(1+x)$ is evaluated over the interval $(0,1)$. The integration yields the exact value $Q=-(PI)^2/12=-0.822466$.

2. A singular integral with the integrand
$$f(x)=\log(x)*\log(1+x)$$
is evaluated over the interval $(0,1)$. The integration yields the exact value $Q=2-\log(4)-(PI)^2/12=-0.208761$.

CHAPTER 4: ORDINARY DIFFERENTIAL EQUATIONS

P401 Fourth-Order Taylor-Series Method for a Set of Differential Equations of First Order

```
10 CLS:PRINT"P401 INTEGRATES A SET OF DIFFERENTIAL EQUATIONS"
20 PRINT"OF FIRST ORDER BY TAYLOR-SERIES METHOD.":PRINT
30 PRINT"PRESS BREAK, DEFINE Y1 (DIFFERENTIAL EQUATIONS) ON"
40 PRINT"LINES 1810-1850 AND HIGHER DERIVATIVES Y2-Y4 ON LINES"
50 PRINT"1890-2030. IF EQUATIONS HAVE A SINGULAR POINT AT THE"
60 PRINT"INITIAL ABSCISSA X0 DEFINE Y1-Y4 ALSO BY SPECIAL"
70 PRINT"EQUATIONS ON LINES 2080-2270 (O3=2) OR NUMERICALLY IN"
80 PRINT"THE INPUT (O3>2). THEN RUN AGAIN."
90 PRINT:PRINT"THE INPUT:"
100 PRINT"IF SINGLE PRECISION O1=1, IF DOUBLE O1=2."
110 DEFINT I-O:INPUT"O1";O1:CLS
120 PRINT"IF PRINTER NOT USED O2=0, IF USED O2>0."
130 INPUT"O2";O2:CLS
140 IF O1=2 THEN DEFDBL A-H,Q-Z
150 PRINT"IF DERIVATIVES Y1-Y4 AT THE INITIAL ABSCISSA X0 FROM"
160 PRINT"DIFFERENTIAL EQUATIONS O3=1, IF THESE EQUATIONS HAVE"
170 PRINT"A SINGULAR POINT AT X0 THE DERIVATIVES Y1-Y4 ARE"
180 PRINT"GIVEN EITHER BY SPECIAL EQUATIONS AND O3=2 OR"
190 PRINT"NUMERICALLY IN THE INPUT AND O3>2,"
200 PRINT"IF INITIAL DATA FROM FILE P401A O3=0.":PRINT
210 PRINT"FILE P401A IS WRITTEN IN THE PRECEDING RUN AND SIMPLI-"
220 PRINT"FIES THE CONTINUATION OF INTEGRATION IN SUCCESSIVE RUNS."
230 INPUT"O3";O3:CLS
240 PRINT"IF RESULTS PRINTED AT INTERVALS H0 O4=0 (X7 IS THE"
250 PRINT"END ABSCISSA), IF PRINTED AFTER M STEPS O4>0 (M7 IS"
260 PRINT"THE TOTAL NUMBER OF STEPS)."
270 INPUT"O4";O4:IF O4>0 THEN GOTO 290
280 INPUT"H0";H0:INPUT"X7";X7:GOTO 300
290 INPUT"M";M:INPUT"M7";M7
300 CLS:PRINT"ACCURACY OF INTEGRATION: P1"
310 PRINT"ACCURACY IN THE SECOND CONTROL OF STEP SIZE H: P2"
320 PRINT"IF P2=0 THE SECOND CONTROL IS NOT APPLIED."
330 INPUT"P1";P1:INPUT"P2";P2:CLS
340 PRINT"NUMBER OF DIFFERENTIAL EQUATIONS: N"
350 INPUT"N";N:CLS
360 PRINT "INPUT OF INITIAL DATA:"
370 IF O2=0 THEN GOTO 430
380 LPRINT "P401: O1=";O1;"O2=";O2;"O3=";O3;"O4=";O4
390 LPRINT "P1=";P1;"P2=";P2
400 IF O4=0 THEN LPRINT "N=";N;"H0=";H0;"X7=";X7
410 IF O4>0 THEN LPRINT "N=";N;"M=";M;"M7=";M7
420 LPRINT "INPUT OF INITIAL DATA:"
430 LET P7=P1
440 REM P7 IS THE INPUT OF P1; USED FOR CORRECTING P1
450 DIM Y(N),Y1(N),Y2(N),Y3(N),Y4(N),Y9(N),C(N)
460 IF O3=0 THEN GOTO 560
470 INPUT"X0";X
480 IF O2>0 THEN LPRINT "X0=";X
490 FOR I=1 TO N
500 PRINT "I=";I
510 INPUT"Y(I)";Y(I)
520 IF O2>0 THEN LPRINT "I=";I;"Y(I)=";Y(I)
530 NEXT I
540 GOTO 720
550 REM READ INITIAL DATA FROM A FILE
560 OPEN "I",1,"P401A"
570 INPUT #1, X
580 PRINT "X0=";X
```

P. 151

```
590 IF O2>0 THEN LPRINT "X0=";X
600 FOR I=1 TO N
610 INPUT #1, Y(I)
620 IF O2>0 THEN LPRINT "I=";I;"Y(I)=";Y(I)
630 PRINT "I=";I;"Y(I)=";Y(I)
640 NEXT I
650 CLOSE 1
660 LET P3=P2/8
670 REM P3 IS A LOWER LIMIT FOR INCREASING P1
680 LET P1=24*P1
690 REM THIS SUBSTITUTION AVOIDS MULTIPLICATIONS ON L.970
700 LET P4=P1
710 REM P4 IS THE INPUT OF P1; USED FOR CORRECTING P1
720 IF O2=0 THEN GOTO 750
730 LPRINT
740 IF O3<3 THEN LPRINT "RESULTS OF INTEGRATION:"
750 LET X6=X
760 LET K=0
770 LET O6=0
780 REM STEP SIZE IN THE FIRST STEP WHEN MAX ABS(Y4(I))=0
790 LET H=SQR(SQR(P1))
800 IF O4=0 THEN LET H1=0
810 IF O4>0 THEN LET K1=0
820 IF O3<3 THEN PRINT "RESULTS OF INTEGRATION:"
830 IF O4>0 THEN GOTO 870
840 IF X>=X7 THEN GOTO 1510
850 IF H1=H0 THEN GOSUB 1510
860 GOTO 890
870 IF K1=M THEN GOSUB 1510
880 REM STEP SIZE DETERMINED BY MAX ABS(Y4(I))
890 LET K=K+1
900 LET K1=K1+1
910 GOSUB 1780
920 LET P0=0
930 FOR I=1 TO N
940 LET P8=ABS(Y4(I))
950 IF P8>P0 THEN LET P0=P8
960 NEXT I
970 IF P0>0 THEN LET H=SQR(SQR(P1/P0))
980 IF O4>0 THEN GOTO 1100
990 IF X7>(X+H) THEN GOTO 1030
1000 LET H=X7-X
1010 LET X=X7
1020 GOTO 1120
1030 LET H1=H1+H
1040 IF H1<H0 THEN GOTO 1100
1050 LET X6=X6+H0
1060 LET H=X6-X
1070 LET X=X6
1080 LET H1=H0
1090 GOTO 1120
1100 LET X=X+H
1110 REM THE INTEGRATION
1120 FOR I=1 TO N
1130 LET Y(I)=Y(I)+H*(Y1(I)+H*(Y2(I)+H*(Y3(I)+Y4(I)*H/4)/3)/2)
1140 NEXT I
1150 IF P2=0 THEN GOTO 830
1160 REM SECOND CONTROL OF H; Y9(I) IS TEMP-STORAGE OF Y1(I)
1170 FOR I=1 TO N
1180 LET Y9(I)=Y1(I)
1190 NEXT I
1200 LET O5=1
1210 GOSUB 1780
1220 REM COMPUTATION OF CORRECTIONS C(I)
```

```
1230 REM P8,P9 USED FOR DETERMINING MAX ABS(C(I))
1240 LET O5=0
1250 LET P9=0
1260 FOR I=1 TO N
1270 LET C(I)=H*(Y1(I)-Y9(I)-H*(Y2(I)+H*(Y3(I)+Y4(I)*H/3)/2))/5
1280 LET P8=ABS(C(I))
1290 IF P8>P9 THEN LET P7=P8
1300 NEXT I
1310 IF P9>P2 THEN GOTO 1430
1320 REM CORRECTION OF Y(I)
1330 FOR I=1 TO N
1340 LET Y(I)=Y(I)+C(I)
1350 NEXT I
1360 LET O6=1
1370 REM CORRECTION (INCREASE) OF P1
1380 IF P9>P3 THEN GOTO 830
1390 LET P1=P1/.4096
1400 IF P1>P4 THEN LET P1=P4
1410 GOTO 830
1420 REM CORRECTION (DECREASE) OF P1
1430 LET H1=H1-H
1440 LET X=X-H
1450 LET P1=.4096*P1
1460 FOR I=1 TO N
1470 LET Y1(I)=Y9(I)
1480 NEXT I
1490 GOTO 970
1500 REM SUBROUTINE FOR PRINTING THE RESULTS
1510 LET H1=0
1520 LET K1=0
1530 PRINT "AT X=";X;"AND K=";K;"Y(I):"
1540 FOR I=1 TO N
1550 PRINT "I=";I;"Y(1)=";Y(I)
1560 NEXT I
1570 PRINT
1580 IF O2=0 THEN GOTO 1640
1590 LPRINT "AT X=";X;"AND K=";K;"Y(I):"
1600 FOR I=1 TO N
1610 LPRINT "I=";I;"Y(1)=";Y(I)
1620 NEXT I
1630 LPRINT
1640 IF O4>0 THEN GOTO 1670
1650 IF X>=X7 THEN GOTO 1700
1660 RETURN
1670 IF K>=M7 THEN GOTO 1700
1680 RETURN
1690 REM WRITING RESULTS ON A FILE
1700 OPEN "O",1,"P401A"
1710 PRINT #1, X
1720 FOR I=1 TO N
1730 PRINT #1, Y(I)
1740 NEXT I
1750 CLOSE 1
1760 STOP
1770 REM SUBROUTINE FOR THE DERIVATIVES
1780 IF O3>1 THEN GOTO 2060
1790 IF O6=1 THEN GOTO 1880
1800 REM DEFINE THE FIRST DERIVATIVES
1810 LET Y1(1)=
1860 IF O5=1 THEN RETURN
1870 REM DEFINE THE SECOND, THIRD AND FOURTH DERIVATIVES
1880 LET O6=0
1890 LET Y2(1)=
1900 LET Y3(1)=
```

```
1910 LET Y4(1)=
2040 RETURN
2050 DEFINE FOUR DERIVATIVES AT X0
2060 IF O3>2 THEN GOTO 2290
2070 LET O3=1
2080 LET Y1(1)=
2090 LET Y2(1)=
2100 LET Y3(1)=
2110 LET Y4(1)=
2280 RETURN
2290 LET O3=1
2300 PRINT "INPUT OF NUMERICAL VALUES OF Y1-Y4 AT X0:"
2310 IF O2>0 THEN LPRINT "NUMERICAL VALUES OF Y1-Y4 AT X0:"
2320 FOR I=1 TO N
2330 PRINT "I=";I;"Y1(I),Y2(I):"
2340 INPUT"Y1(I)";Y1(I):INPUT"Y2(I)";Y2(I)
2350 PRINT "I=";I;"Y3(I),Y4(I):"
2360 INPUT"Y3(I)";Y3(I):INPUT"Y4(I)";Y4(I)
2370 IF O2=0 THEN GOTO 2400
2380 LPRINT "I=";I;"Y1(I)=";Y1(I);"Y2(I)=";Y2(I)
2390 LPRINT "      Y3(I)=";Y3(I);"Y4(I)=";Y4(I)
2400 NEXT I
2410 PRINT "RESULTS OF INTEGRATION:"
2420 IF O2=0 THEN GOTO 2450
2430 LPRINT
2440 LPRINT "RESULTS OF INTEGRATION:"
2450 RETURN
2460 END
```

Examples

1. One differential equation $y'=y$ with the initial condition $y=1$ at $x=0$ is integrated in single precision from $x=0$ to $x=2$; the results are printed at intervals of 0.2. Accuracy of integration $P_1=1*10^{-6}$, $P_2=0$. Some results of integration (K is the number of steps):

$x=0.2$, $y=1.22140$, K=7,
$x=1.0$, $y=2.71828$, K=38,
$x=2.0$, $y=7.38905$, K=87.

2. The same differential equation is integrated in double precision from $x=0$ to $x=1$; the results printed at intervals of 0.2. Accuracy of integration $P_1=1*10^{-10}$, $P_2=4*10^{-11}$. Some results (rounded to 10 digits):

```
x=0.2,   y=1.221402758, K=65,
x=0.6,   y=1.822118800, K=206,
x=1.0,   y=2.718281828, K=362.
```

The integral of the differential equation in both examples is
$y=e^x$.

P402 Fourth-Order Taylor-Series Method for a Set of Differential Equations of Second Order

```
10 CLS:PRINT"P402 INTEGRATES A SET OF DIFFERENTIAL EQUATIONS"
20 PRINT"OF SECOND ORDER BY TAYLOR-SERIES METHOD.":PRINT
30 PRINT"PRESS BREAK, DEFINE Y2 (DIFFERENTIAL EQUATIONS) ON"
40 PRINT"LINES 1820-1860 AND HIGHER DERIVATIVES Y3-Y4 ON LINES"
50 PRINT"1900-1990. IF EQUATIONS HAVE A SINGULAR POINT AT THE"
60 PRINT"INITIAL ABSCISSA X0 DEFINE Y2-Y4 ALSO BY SPECIAL EQUATIONS"
70 PRINT"ON LINES 2040-2180 (O3=2) OR NUMERICALLY IN THE INPUT"
80 PRINT"(O3>2). THEN RUN AGAIN."
90 PRINT:PRINT"THE INPUT:"
100 PRINT"IF SINGLE PRECISION O1=1, IF DOUBLE O1=2."
110 DEFINT I-O:INPUT"O1";O1:CLS
120 PRINT"IF PRINTER NOT USED O2=0, IF USED O2>0."
130 INPUT"O2";O2:CLS
140 IF O1=2 THEN DEFDBL A-H,Q-Z
150 PRINT"IF DERIVATIVES Y2-Y4 AT THE INITIAL ABSCISSA X0 FROM"
160 PRINT"DIFFERENTIAL EQUATIONS O3=1, IF THESE EQUATIONS HAVE"
170 PRINT"A REGULAR SINGULAR POINT AT X0 THE DERIVATIVES Y2-Y4"
180 PRINT"ARE GIVEN EITHER BY SPECIAL EQUATIONS AND O3=2 OR"
190 PRINT"NUMERICALLY IN THE INPUT AND O3>2,"
200 PRINT"IF INITIAL DATA FROM FILE P402A O3=0.":PRINT
210 PRINT"FILE P402A IS WRITTEN IN THE PRECEDING RUN AND SIMPLI-"
220 PRINT"FIES THE CONTINUATION OF INTEGRATION IN SUCCESSIVE RUNS."
230 INPUT"O3";O3:CLS
240 PRINT"IF RESULTS PRINTED AT INTERVALS H0 O4=0 (X7 IS THE"
250 PRINT"END ABSCISSA), IF PRINTED AFTER M STEPS O4>0 (M7 IS"
260 PRINT"THE TOTAL NUMBER OF STEPS)."
270 INPUT"O4";O4:IF O4>0 THEN GOTO 290
280 INPUT"H0";H0:INPUT"X7";X7:GOTO 300
290 INPUT"M";M:INPUT"M7";M7
300 CLS:PRINT"ACCURACY OF INTEGRATION: P1"
310 PRINT"ACCURACY IN THE SECOND CONTROL OF STEP SIZE H: P2"
320 PRINT"IF P2=0 THE SECOND CONTROL IS NOT APPLIED."
330 INPUT"P1";P1:INPUT"P2";P2:CLS
340 PRINT"NUMBER OF DIFFERENTIAL EQUATIONS: N"
350 INPUT"N";N:CLS
360 PRINT "INPUT OF INITIAL DATA:"
370 IF O2=0 THEN GOTO 430
380 LPRINT "P402: O1=";O1;"O2=";O2;"O3=";O3;"O4=";O4
390 LPRINT "P1=";P1;"P2=";P2
400 IF O4=0 THEN LPRINT "N=";N;"H0=";H0;"X7=";X7
410 IF O4>0 THEN LPRINT "N=";N;"M=";M;"M7=";M7
420 LPRINT "INPUT OF INITIAL DATA:"
430 LET P7=P1
440 REM P7 IS THE INPUT OF P1; USED FOR CORRECTING P1
450 DIM Y(N),Y1(N),Y2(N),Y3(N),Y4(N),Y9(N),C(N)
460 IF O3=0 THEN GOTO 560
470 INPUT"X0";X
480 IF O2>0 THEN LPRINT "X0=";X
490 FOR I=1 TO N
500 PRINT "I=";I
510 INPUT"Y(I)";Y(I):INPUT"Y1(I)";Y1(I)
520 IF O2>0 THEN LPRINT "I=";I;"Y(I)=";Y(I);"Y1(I)=";Y1(I)
530 NEXT I
540 GOTO 720
550 REM READ INITIAL DATA FROM A FILE
560 OPEN "I",1,"P402A"
570 INPUT #1, X
580 PRINT "X0=";X
590 IF O2>0 THEN LPRINT "X0=";X
600 FOR I=1 TO N
```

```
610 INPUT #1, Y(I),Y1(I)
620 PRINT "I=";I;"Y(I)=";Y(I);"Y1(I)=";Y1(I)
630 IF O2>0 THEN LPRINT "I=";I;"Y(I)=";Y(I);"Y1(I)=";Y1(I)
640 NEXT I
650 CLOSE 1
660 LET P3=P2/8
670 REM P3 IS A LOWER LIMIT FOR INCREASING P1
680 LET P1=24*P1
690 REM THIS SUBSTITUTION AVOIDS MULTIPLICATIONS ON L.970
700 LET P4=P1
710 REM P4 IS INPUT OF P1; USED FOR CORRECTING P1
720 IF O2=0 THEN GOTO 750
730 LPRINT
740 IF O3<3 THEN LPRINT "RESULTS OF INTEGRATION:"
750 LET X6=X
760 LET K=0
770 LET O6=0
780 REM STEP SIZE IN THE FIRST STEP WHEN MAX ABS(Y4(I))=0
790 LET H=SQR(SQR(P1))
800 IF O4=0 THEN LET H1=0
810 IF O4>0 THEN LET K1=0
820 IF O3<3 THEN PRINT "RESULTS OF INTEGRATION:"
830 IF O4>0 THEN GOTO 870
840 IF X>=X7 THEN GOTO 1520
850 IF H1=H0 THEN GOSUB 1520
860 GOTO 890
870 IF K1=M THEN GOSUB 1520
880 REM STEP SIZE DETERMINED BY MAX ABS(Y4(I))
890 LET K=K+1
900 LET K1=K1+1
910 GOSUB 1790
920 LET P0=0
930 FOR I=1 TO N
940 LET P8=ABS(Y4(I))
950 IF P8>P0 THEN LET P0=P8
960 NEXT I
970 IF P0>0 THEN LET H=SQR(SQR(P1/P0))
980 IF O4>0 THEN GOTO 1100
990 IF X7>=(X+H) THEN GOTO 1030
1000 LET H=X7-X
1010 LET X=X7
1020 GOTO 1120
1030 LET H1=H1+H
1040 IF H1<H0 THEN GOTO 1100
1050 LET X6=X6+H0
1060 LET H=X6-X
1070 LET X=X6
1080 LET H1=H0
1090 GOTO 1120
1100 LET X=X+H
1110 REM THE INTEGRATION
1120 FOR I=1 TO N
1130 LET Y(I)=Y(I)+H*(Y1(I)+H*(Y2(I)+H*(Y3(I)+Y4(I)*H/4)/3)/2)
1140 LET Y1(I)=Y1(I)+H*(Y2(I)+H*(Y3(I)+Y4(I)*H/3)/2)
1150 NEXT I
1160 IF P2=0 THEN GOTO 830
1170 REM SECOND CONTROL OF H; Y9(I) IS TEMP-STORAGE OF Y2(I)
1180 FOR I=1 TO N
1190 LET Y9(I)=Y2(I)
1200 NEXT I
1210 LET O5=1
1220 GOSUB 1790
1230 REM COMPUTATION OF CORRECTIONS C(I)
1240 REM P8,P9 USED FOR DETERMINING MAX ABS(C(I))
```

PROGRAM P402

```
1250 LET O5=0
1260 LET P9=0
1270 FOR I=1 TO N
1280 LET C(I)=H*H*(Y2(I)-Y9(I)-H*(Y3(I)+H*Y4(I)/2))/20
1290 LET P8=ABS(C(I))
1300 IF P8>P9 THEN LET P9=P8
1310 NEXT I
1320 IF P9>P2 THEN GOTO 1440
1330 REM CORRECTION OF Y(I)
1340 FOR I=1 TO N
1350 LET Y(I)=Y(I)+C(I)
1360 NEXT I
1370 LET O6=1
1380 REM CORRECTION (INCREASE) OF P1
1390 IF P9>P3 THEN GOTO 830
1400 LET P1=P1/.4096
1410 IF P1>P4 THEN LET P1=P4
1420 GOTO 830
1430 REM CORRECTION (DECREASE) OF P1
1440 LET H1=H1-H
1450 LET X=X-H
1460 LET P1=.4096*P1
1470 FOR I=1 TO N
1480 LET Y2(I)=Y9(I)
1490 NEXT I
1500 GOTO 970
1510 REM SUBROUTINE FOR PRINTING THE RESULTS
1520 LET H1=0
1530 LET K1=0
1540 PRINT "AT X=";X;"AND K=";K;"Y(I),Y1(I):"
1550 FOR I=1 TO N
1560 PRINT "I=";I;"Y(I)=";Y(I);"Y1(I)=";Y1(I)
1570 NEXT I
1580 PRINT
1590 IF O2=0 THEN GOTO 1650
1600 LPRINT "AT X=";X;"AND K=";K;"Y(I),Y1(I):"
1610 FOR I=1 TO N
1620 LPRINT "I=";I;"Y(I)=";Y(I);"Y1(I)=";Y1(I)
1630 NEXT I
1640 LPRINT
1650 IF O4>0 THEN GOTO 1680
1660 IF X>=X7 THEN GOTO 1710
1670 RETURN
1680 IF K>=M7 THEN GOTO 1710
1690 RETURN
1700 REM WRITING RESULTS ON A FILE
1710 OPEN "O",1,"P402A"
1720 PRINT #1, X
1730 FOR I=1 TO N
1740 PRINT #1, Y(I),Y1(I)
1750 NEXT I
1760 CLOSE 1
1770 STOP
1780 REM SUBROUTINE FOR THE DERIVATIVES
1790 IF O3>1 THEN GOTO 2020
1800 IF O6=1 THEN GOTO 1890
1810 REM DEFINE THE SECOND DERIVATIVES
1820 LET Y2(1)=
1870 IF O5=1 THEN RETURN
1880 REM DEFINE THE THIRD AND FOURTH DERIVATIVES
1890 LET O6=0
1900 LET Y3(1)=
1910 LET Y4(1)=
2000 RETURN
```

```
2010 DEFINE THREE DERIVATIVES AT X0
2020 IF O3>2 THEN GOTO 2200
2030 LET O3=1
2040 LET Y2(1)=
2050 LET Y3(1)=
2060 LET Y4(1)=
2190 RETURN
2200 LET O3=1
2210 PRINT "INPUT OF NUMERICAL VALUES OF Y2-Y4 AT X0:"
2220 IF O2>0 THEN LPRINT "NUMERICAL VALUES OF Y2-Y4 AT X0:"
2230 FOR I=1 TO N
2240 PRINT "I=";I;"Y2(I),Y3(I),Y4(I):"
2250 INPUT"Y2(I)";Y2(I):INPUT"Y3(I)";Y3(I):INPUT"Y4(I)";Y4(I)
2260 IF O2=0 THEN GOTO 2280
2270 LPRINT "I=";I;"Y2(I)=";Y2(I);"Y3(I)=";Y3(I);"Y4(I)=";Y4(I)
2280 NEXT I
2290 PRINT "RESULTS OF INTEGRATION:"
2300 IF O2=0 THEN GOTO 2330
2310 LPRINT
2320 LPRINT "RESULTS OF INTEGRATION:"
2330 RETURN
2340 END
```

Examples

1. One differential equation $y''=-y$ with the initial conditions $y=1$, $y'=0$ at $x=0$ is integrated in single precision from $x=0$ to $x=1.5$, the results are printed at intervals of 0.25. Accuracy of integration $P_1=1*10^{-6}$, $P_2=0$. Some results of integration (K is the number of steps):

        x=0.25,   y=0.968912,   y'=-0.247404,   K=8,

        x=0.75,   y=0.731689,   y'=-0.681638,   K=24,

        x=1.50,   y=0.0707378,   y'=-0.997493,   K=45.

2. The same equation is integrated in double precision from $x=0$ to $x=1.75$ and the results printed at intervals of 0.25. Accuracy of integration $P_1=1*10^{-10}$, $P_2=4*10^{-11}$. Some results (rounded to 9 decimal places):

        x=0.25,   y=0.968912422,   y'=-0.247403959,   K=79,

        x=0.75,   y=0.731688869,   y'=-0.681638760,   K=232,

        x=1.50,   y=0.070737203,   y'=-0.007494984,   K=419,

        x=1.75,   y=-0.178246054,   y'=-0.983985941,   K=458.

The integral of the differential equation in both examples is
    $y=\cos(x)$, $y'=-\sin(x)$.

P403 Fourth-Order Standard Runge-Kutta Method for a Set of Differential Equations of First Order

```
10 CLS:PRINT"P403 INTEGRATES A SET OF DIFFERENTIAL EQUATIONS"
20 PRINT"OF FIRST ORDER BY RUNGE-KUTTA METHOD.":PRINT
30 PRINT"PRESS BREAK, DEFINE DIFFERENTIAL EQUATIONS ON LINES"
40 PRINT"1660-1750. IF EQUATIONS HAVE A SINGULAR POINT AT THE"
50 PRINT"INITIAL ABSCISSA X0 DEFINE Y1 ALSO BY SPECIAL EQUATIONS"
60 PRINT"ON LINES 1800-1890 (O3=2) OR NUMERICALLY IN THE INPUT"
70 PRINT"(O3>2). IN ALL THESE DEFINITIONS THE DEPENDENT VARIABLES"
80 PRINT"ARE DENOTED Y(I,1), I=1,2,...N. THEN RUN AGAIN."
90 PRINT:PRINT"THE INPUT:"
100 PRINT"IF SINGLE PRECISION O1=1, IF DOUBLE O1=2."
110 DEFINT I-O:INPUT"O1";O1:CLS
120 PRINT"IF PRINTER NOT USED O2=0, IF USED O2>0."
130 INPUT"O2";O2:CLS
140 IF O1=2 THEN DEFDBL A-H,Q-Z
150 PRINT"IF DERIVATIVES Y1 AT THE INITIAL ABSCISSA X0 FROM"
160 PRINT"DIFFERENTIAL EQUATIONS O3=1, IF THESE EQUATIONS HAVE"
170 PRINT"A SINGULAR POINT AT X0 THE DERIVATIVES ARE GIVEN"
180 PRINT"EITHER BY SPECIAL EQUATIONS AND O3=2 OR NUMERICALLY"
190 PRINT"IN THE INPUT AND O3>2, IF INITIAL DATA FROM FILE"
200 PRINT"P403A O3=0.":PRINT
210 PRINT"FILE P403A IS WRITTEN IN THE PRECEDING RUN AND SIMPLI-"
220 PRINT"FIES THE CONTINUATION OF INTEGRATION IN SUCCESSIVE RUNS."
230 INPUT"O3";O3:CLS
240 PRINT"ACCURACY OF INTEGRATION: P"
250 PRINT"IF P=0 NO CONTROL OF ACCURACY IS APPLIED.":INPUT"P";P:CLS
260 PRINT"STEP SIZE OF INTEGRATION: H"
270 PRINT"INTERVAL OF PRINTING THE RESULTS: H0"
280 PRINT"H0 IS AN INTEGER MULTIPLE OF H."
290 INPUT"H";H:INPUT"H0";H0:CLS
300 PRINT"NUMBER OF DIFFERENTIAL EQUATIONS: N"
310 PRINT"MAXIMUM NUMBER OF INTERVAL HALVING: L0"
320 PRINT"END ABSCISSA OF INTEGRATION: X7"
330 INPUT"N";N:INPUT"L0";L0:INPUT"X7";X7:CLS
340 PRINT "INPUT OF INITIAL DATA:"
350 DIM A(4),B(3),Y(N,5),Y1(N)
360 IF O2=0 THEN GOTO 410
370 LPRINT "P403: O1=";O1;"O2=";O2;"O3=";O3
380 LPRINT "P=";P;"H=";H;"H0=";H0
390 LPRINT "N=";N;"L0=";L0;"X7=";X7
400 LPRINT "INPUT OF INITIAL DATA:"
410 IF O3=0 THEN GOTO 510
420 INPUT"X0";X
430 IF O2>0 THEN LPRINT "X0=";X
440 FOR I=1 TO N
450 PRINT "I=";I
460 INPUT"Y(I)";Y(I,1)
470 IF O2>0 THEN LPRINT "I=";I;"Y(I)=";Y(I,1)
480 NEXT I
490 GOTO 610
500 REM READ INITIAL DATA FROM A FILE
510 OPEN "I",1,"P403A"
520 INPUT #1, X
530 PRINT "X0=";X
540 IF O2>0 THEN LPRINT "X0=";X
550 FOR I=1 TO N
560 INPUT #1, Y(I,1)
570 PRINT "I=";I;"Y(I)=";Y(I,1)
580 IF O2>0 THEN LPRINT "I=";I;"Y(I)=";Y(I,1)
590 NEXT I
600 CLOSE 1
610 IF O2=0 THEN GOTO 640
```

```
620 LPRINT
630 IF O3<3 THEN LPRINT "RESULTS OF INTEGRATION:"
640 LET X6=X+H0
650 LET L0=2[L0
660 LET K=0
670 FOR I=1 TO N
680 LET Y(I,2)=Y(I,1)
690 LET Y(I,3)=Y(I,1)
700 NEXT I
710 REM STANDARD RUNGE-KUTTA METHOD
720 IF O3<3 THEN PRINT "RESULTS OF INTEGRATION:"
730 IF P=0 THEN GOTO 780
740 LET L9=1
750 LET L9=L9+L9
760 LET L=0
770 LET X1=X
780 LET A(1)=H/6
790 LET A(2)=H/3
800 LET A(3)=A(2)
810 LET A(4)=A(1)
820 LET B(1)=H/2
830 LET B(2)=B(1)
840 LET B(3)=H
850 LET K=K+1
860 FOR J=1 TO 4
870 GOSUB 1650
880 IF J=2 THEN GOTO 910
890 IF J=4 THEN GOTO 910
900 LET X=X+B(1)
910 FOR I=1 TO N
920 LET Y(I,3)=Y(I,3)+A(J)*Y1(I)
930 IF J=4 THEN GOTO 960
940 LET Y(I,1)=Y(I,2)+B(J)*Y1(I)
950 GOTO 1020
960 LET Y(I,1)=Y(I,3)
970 IF P=0 THEN GOTO 1010
980 IF L>0 THEN GOTO 1010
990 LET Y(I,4)=Y(I,3)
1000 LET Y(I,5)=Y(I,2)
1010 LET Y(I,2)=Y(I,3)
1020 NEXT I
1030 NEXT J
1040 IF P=0 THEN GOTO 1340
1050 REM CONTROL OF ACCURACY BY REPEATEDLY HALVING H
1060 IF L=0 THEN GOTO 1200
1070 IF L<L9 THEN GOTO 1300
1080 LET P9=0
1090 FOR I=1 TO N
1100 LET P8=ABS(Y(I,3)-Y(I,4))
1110 IF P8>P9 THEN LET P9=P8
1120 NEXT I
1130 IF P9<P THEN GOTO 1340
1140 LET L=0
1150 LET L9=L9+L9
1160 IF L9=L0 THEN GOTO 1320
1170 FOR I=1 TO N
1180 LET Y(I,4)=Y(I,3)
1190 NEXT I
1200 LET H=H/2
1210 LET X=X1
1220 FOR I=1 TO N
1230 LET Y(I,1)=Y(I,5)
1240 LET Y(I,2)=Y(I,5)
1250 LET Y(I,3)=Y(I,5)
```

PROGRAM P403

```
1260 NEXT I
1270 IF L=L9 THEN GOTO 750
1280 LET L=L+1
1290 GOTO 780
1300 LET L=L+1
1310 GOTO 850
1320 PRINT "MAXIMUM NUMBER OF HALVING TO";L0;"STEPS REACHED"
1330 PRINT "ERRORS STILL HIGHER THAN P=";P
1340 LET X1=1.0001*X
1350 IF X1>X6 THEN GOSUB 1410
1360 IF P=0 THEN GOTO 850
1370 LET H=H+H
1380 REM STEP SIZE H TENTATIVELY DOUBLED
1390 GOTO 740
1400 REM SUBROUTINE FOR PRINTING THE RESULTS
1410 LET X=X6
1420 LET X6=X6+H0
1430 PRINT "AT X=";X;"AND K=";K;"Y(I):"
1440 FOR I=1 TO N
1450 PRINT "I=";I;"Y(I)=";Y(I,1)
1460 NEXT I
1470 PRINT
1480 IF O2=0 THEN GOTO 1540
1490 LPRINT "AT X=";X;"AND K=";K;"Y(I):"
1500 FOR I=1 TO N
1510 LPRINT "I=";I;"Y(I)=";Y(I,1)
1520 NEXT I
1530 LPRINT
1540 IF X>=X7 THEN GOTO 1570
1550 RETURN
1560 REM WRITING RESULTS ON A FILE
1570 OPEN "O",1,"P403A"
1580 PRINT #1, X
1590 FOR I=1 TO N
1600 PRINT #1, Y(I,1)
1610 NEXT I
1620 CLOSE 1
1630 STOP
1640 REM SUBROUTINE FOR D-EQS
1650 IF O3>1 THEN GOTO 1780
1660 LET Y1(1)=
1760 RETURN
1770 DEFINE SPECIAL D-EQS AT X0
1780 IF O3>2 THEN GOTO 1910
1790 LET O3=1
1800 LET Y1(1)=Y(1,1)
1900 RETURN
1910 PRINT "INPUT OF NUMERICAL VALUES OF Y1(I) AT X0:"
1920 IF O2>0 THEN LPRINT "NUMERICAL VALUES OF Y1(I) AT X0:"
1930 LET O3=1
1940 FOR I=1 TO N
1950 PRINT "I=";I;"Y1(I):"
1960 INPUT"Y1(I)";Y1(I)
1970 IF O2=0 THEN GOTO 1990
1980 LPRINT "I=";I;"Y1(I)=";Y1(I)
1990 NEXT I
2000 IF O2=0 THEN GOTO 2040
2010 PRINT "RESULTS OF INTEGRATION:"
2020 LPRINT
2030 LPRINT "RESULTS OF INTEGRATION:"
2040 RETURN
2050 END
```

Examples

1. One differential equation $y'=y$ with the initial condition $y=1$ at $x=0$ is integrated in single precision from $x=0$ to $x=2$, interval of integration $H=0.2$, interval of printing $H_0=0.2$. Accuracy of integration $P=1*10^{-6}$, maximum number of interval halving $L_0=10$. Some results of integration (K is the number of steps):

      $x=0.2$, $y=1.22140$, $K=7$,
      $x=1.0$, $y=2.71828$, $K=31$,
      $x=2.0$, $y=7.38906$, $K=61$.

2. The same differential equation is integrated in double precision from $x=0$ to $x=2$, $H=0.01$, $H_0=0.2$, $P=0$. Some results (rounded to 10 digits):

      $x=0.2$, $y=1.221402758$, $K=20$,
      $x=1.0$, $y=2.718281828$, $K=100$,
      $x=2.0$, $y=7.389056098$, $K=200$.

The integral of the differential equation in both examples is $y=e^x$.

P404   Gill's Version of Fourth-Order Runge-Kutta Method for a Set of Differential Equations of First Order

```
10 CLS:PRINT"P404 INTEGRATES A SET OF DIFFERENTIAL EQUATIONS"
20 PRINT"OF FIRST ORDER BY GILL'S VERSION OF RUNGE-KUTTA METHOD."
30 PRINT:PRINT"PRESS BREAK, DEFINE DIFFERENTIAL EQUATIONS ON"
40 PRINT"LINES 1650-1740. IF EQUATIONS HAVE A SINGULAR POINT AT"
50 PRINT"INITIAL ABSCISSA X0 DEFINE Y1 ALSO BY SPECIAL EQUATIONS"
60 PRINT"ON LINES 1790-1880 (O3=2) OR NUMERICALLY IN THE INPUT"
70 PRINT"(O3>2). IN ALL THESE DEFINITIONS THE DEPENDENT VARIABLES"
80 PRINT"ARE DENOTED Y(I,1), I=1,2,...N. THEN RUN AGAIN."
90 PRINT:PRINT"THE INPUT:"
100 PRINT"IF SINGLE PRECISION O1=1, IF DOUBLE O1=2."
110 DEFINT I-O:INPUT"O1";O1:CLS
120 PRINT"IF PRINTER NOT USED O2=0, IF USED O2>0."
130 INPUT"O2";O2:CLS
140 IF O1=2 THEN DEFDBL A-H,Q-Z
150 PRINT"IF DERIVATIVES Y1 AT THE INITIAL ABSCISSA X0 FROM"
160 PRINT"DIFFERENTIAL EQUATIONS O3=1, IF THESE EQUATIONS HAVE"
170 PRINT"A SINGULAR POINT AT X0 THE DERIVATIVES ARE GIVEN"
180 PRINT"EITHER BY SPECIAL EQUATIONS AND O3=2 OR NUMERICALLY"
190 PRINT"IN THE INPUT AND O3>2, IF INITIAL DATA FROM FILE"
200 PRINT"P404A O3=0.":PRINT
210 PRINT"FILE P404A IS WRITTEN IN THE PRECEDING RUN AND SIMPLI-"
220 PRINT"FIES THE CONTINUATION OF INTEGRATION IN SUCCESSIVE RUNS."
230 INPUT"O3";O3:CLS
240 PRINT"ACCURACY OF INTEGRATION: P"
250 PRINT"IF P=0 NO CONTROL OF ACCURACY IS APPLIED.":INPUT"P";P:CLS
260 PRINT"STEP SIZE OF INTEGRATION: H"
270 PRINT"INTERVAL OF PRINTING THE RESULTS: H0"
280 PRINT"H0 IS AN INTEGER MULTIPLE OF H."
290 INPUT"H";H:INPUT"H0";H0:CLS
300 PRINT"NUMBER OF DIFFERENTIAL EQUATIONS: N"
310 PRINT"MAXIMUM NUMBER OF INTERVAL HALVING: L0"
320 PRINT"END ABSCISSA OF INTEGRATION: X7"
330 INPUT"N";N:INPUT"L0";L0:INPUT"X7";X7:CLS
340 PRINT "INPUT OF INITIAL DATA:"
350 DIM A(4),B(4),C(4),Y(N,4),Y1(N)
360 IF O2=0 THEN GOTO 410
370 LPRINT "P404: O1=";O1;"O2=";O2;"O3=";O3
380 LPRINT "P=";P;"H=";H;"H0=";H0
390 LPRINT "N=";N;"L0=";L0;"X7=";X7
400 LPRINT "INPUT OF INITIAL DATA:"
410 IF O3=0 THEN GOTO 510
420 INPUT"X0";X
430 IF O2>0 THEN LPRINT "X0=";X
440 FOR I=1 TO N
450 PRINT "I=";I
460 INPUT"Y(I)";Y(I,1)
470 IF O2>0 THEN LPRINT "I=";I;"Y(I)=";Y(I,1)
480 NEXT I
490 GOTO 610
500 REM READ INITIAL DATA FROM A FILE
510 OPEN "I",1,"P404A"
520 INPUT #1, X
530 PRINT "X0=";X
540 IF O2>0 THEN LPRINT "X0=";X
550 FOR I=1 TO N
560 INPUT #1, Y(I,1)
570 PRINT "I=";I;"Y(I)=";Y(I,1)
580 IF O2>0 THEN LPRINT "I=";I;"Y(I)=";Y(I,1)
590 NEXT I
600 CLOSE 1
610 IF O2=0 THEN GOTO 640
```

```
620 LPRINT
630 IF O3<3 THEN LPRINT "RESULTS OF INTEGRATION:"
640 LET X6=X+H0
650 LET L0=2[L0
660 LET K=0
670 LET H1=H/2
680 FOR I=1 TO N
690 LET Y(I,3)=Y(I,1)
700 LET Y(I,2)=0
710 NEXT I
720 REM GILL'S VERSION OF RUNGE-KUTTA METHOD
730 IF O3<3 THEN PRINT "RESULTS OF INTEGRATION:"
740 LET A(1)=.5
750 LET A(2)=.29289321881345248
760 LET A(3)=1.7071067811866548
770 LET A(4)=.16666666666666667
780 LET B(1)=2
790 LET B(2)=1
800 LET B(3)=1
810 LET B(4)=2
820 LET C(1)=.5
830 LET C(2)=A(2)
840 LET C(3)=A(3)
850 LET C(4)=.5
860 IF P=0 THEN GOTO 910
870 LET L9=1
880 LET L9=L9+L9
890 LET L=0
900 LET X1=X
910 LET K=K+1
920 FOR J=1 TO 4
930 GOSUB 1640
940 IF J=1 THEN GOTO 970
950 IF J=3 THEN GOTO 970
960 LET X=X+H1
970 FOR I=1 TO N
980 LET D=H*Y1(I)
990 LET E=A(J)*(D-B(J)*Y(I,2))
1000 LET Y(I,1)=Y(I,1)+E
1010 LET Y(I,2)=Y(I,2)+3*E-C(J)*D
1020 NEXT I
1030 NEXT J
1040 IF P=0 THEN GOTO 1320
1050 REM CONTROL OF ACCURACY BY REPEATEDLY HALVING H
1060 IF L=0 THEN GOTO 1170
1070 IF L<L9 THEN GOTO 1250
1080 LET P9=0
1090 FOR I=1 TO N
1100 LET P8=ABS(Y(I,1)-Y(I,4))
1110 IF P8>P9 THEN LET P9=P8
1120 NEXT I
1130 IF P9<P THEN GOTO 1290
1140 LET L=0
1150 LET L9=L9+L9
1160 IF L9=L0 THEN GOTO 1270
1170 LET H=H1
1180 LET H1=H1/2
1190 LET X=X1
1200 FOR I=1 TO N
1210 LET Y(I,4)=Y(I,1)
1220 LET Y(I,1)=Y(I,3)
1230 NEXT I
1240 IF L=L9 THEN GOTO 870
1250 LET L=L+1
```

PROGRAM P404

```
1260 GOTO 910
1270 PRINT "MAXIMUM NUMBER OF HALVING TO";L0;"STEPS REACHED"
1280 PRINT "ERRORS STILL HIGHER THAN P=";P
1290 FOR I=1 TO N
1300 LET Y(I,3)=Y(I,1)
1310 NEXT I
1320 LET X1=1.0001*X
1330 IF X1>X6 THEN GOSUB 1400
1340 IF P=0 THEN GOTO 910
1350 LET H1=H
1360 LET H=H+H
1370 REM STEP SIZE H TENTATIVELY DOUBLED
1380 GOTO 870
1390 REM SUBROUTINE FOR PRINTING THE RESULTS
1400 LET X=X6
1410 LET X6=X6+H0
1420 PRINT "AT X=";X;"AND K=";K;"Y(I):"
1430 FOR I=1 TO N
1440 PRINT "I=";I;"Y(I)=";Y(I,1)
1450 NEXT I
1460 PRINT
1470 IF O2=0 THEN GOTO 1530
1480 LPRINT "AT X=";X;"AND K=";K;"Y(I):"
1490 FOR I=1 TO N
1500 LPRINT "I=";I;"Y(I)=";Y(I,1)
1510 NEXT I
1520 LPRINT
1530 IF X>=X7 THEN GOTO 1560
1540 RETURN
1550 REM WRITING RESULTS ON A FILE
1560 OPEN "O",1,"P404A"
1570 PRINT #1, X
1580 FOR I=1 TO N
1590 PRINT #1, Y(I,1)
1600 NEXT I
1610 CLOSE 1
1620 STOP
1630 REM SUBROUTINE FOR D-EQS
1640 IF O3>1 THEN GOTO 1770
1650 LET Y1(1)=
1750 RETURN
1760 DEFINE SPECIAL D-EQS AT X0
1770 IF O3>2 THEN GOTO 1900
1780 LET O3=1
1790 LET Y1(1)=Y(1,1)
1890 RETURN
1900 PRINT "INPUT OF NUMERICAL VALUES OF Y1(I) AT X0:"
1910 IF O2>0 THEN LPRINT "NUMERICAL VALUES OF Y1(I) AT X0:"
1920 LET O3=1
1930 FOR I=1 TO N
1940 PRINT "I=";I;"Y1(I):"
1950 INPUT"Y1(I)";Y1(I)
1960 IF O2=0 THEN GOTO 1980
1970 LPRINT "I=";I;"Y1(I)=";Y1(I)
1980 NEXT I
1990 PRINT "RESULTS OF INTEGRATION:"
2000 IF O2=0 THEN GOTO 2030
2010 LPRINT
2020 LPRINT "RESULTS OF INTEGRATION:"
2030 RETURN
2040 END
```

Examples

1. One differential equation $y'=y$ with the initial condition $y=1$ at $x=0$ is integrated in single precision from $x=0$ to $x=2$, interval of integration $H=0.1$, interval of printing $H_0=0.2$. Accuracy of integration $P=1*10^{-6}$, maximum number of interval halving $L_0=10$. Some results of integration (K is the number of steps):

    $x=0.2$, $y=1.22140$, $K=6$,
    $x=1.0$, $y=2.71828$, $K=30$,
    $x=2.0$, $y=7.38906$, $K=60$.

2. The same differential equation is integrated in double precision from $x=0$ to $x=2$, $H=0.1$, $H_0=0.2$, $P=1*10^{-10}$, $L_0=10$. Some results (rounded to 10 digits):

    $x=0.2$, $y=1.221402758$, $K=55$,
    $x=1.0$, $y=2.718281828$, $K=247$,
    $x=2.0$, $y=7.389056099$, $K=487$.

The integral of the differential equation in both examples is $y=e^x$.

P405 Third-Order Predictor-Corrector Method for a Set of Differential Equations of First Order

```
10 CLS:PRINT"P405 INTEGRATES A SET OF DIFFERENTIAL EQUATIONS OF"
20 PRINT"FIRST ORDER BY THIRD ORDER PREDICTOR-CORRECTOR METHOD."
30 PRINT:PRINT"PRESS BREAK, DEFINE DIFFERENTIAL EQUATIONS ON"
40 PRINT"LINES 2240-2330. IF EQUATIONS HAVE A SINGULAR POINT AT"
50 PRINT"INITIAL ABSCISSA X0 DEFINE Y1 ALSO BY SPECIAL EQUATIONS"
60 PRINT"ON LINES 2380-2470 (O3=2) OR NUMERICALLY IN THE INPUT"
70 PRINT"(O3>2). IN ALL THESE DEFINITIONS THE DEPENDENT VARIABLES"
80 PRINT"ARE DENOTED Y(I,1), THEIR DERIVATIVES Y1(I,1),"
90 PRINT"I=1,2,...N. THEN RUN AGAIN."
100 PRINT:PRINT "THE INPUT:"
110 PRINT"IF SINGLE PRECISION O1=1, IF DOUBLE O1=2."
120 DEFINT I-O:INPUT"O1";O1:CLS
130 PRINT"IF PRINTER NOT USED O2=0, IF USED O2>0."
140 INPUT"O2";O2:CLS
150 IF O1=2 THEN DEFDBL A-H,Q-Z
160 PRINT"IF DERIVATIVES Y1 AT THE INITIAL ABSCISSA X0 FROM"
170 PRINT"DIFFERENTIAL EQUATIONS O3=1, IF THESE EQUATIONS HAVE"
180 PRINT"A SINGULAR POINT AT X0 THE DERIVATIVES ARE GIVEN"
190 PRINT"EITHER BY SPECIAL EQUATIONS AND O3=2 OR NUMERICALLY"
200 PRINT"IN THE INPUT AND O3>2, IF INITIAL DATA FROM FILE"
210 PRINT"P405A O3=0.":PRINT
220 PRINT"FILE P405A IS WRITTEN IN THE PRECEDING RUN AND SIMPLI-"
230 PRINT"FIES THE CONTINUATION OF INTEGRATION IN SUCCESSIVE RUNS"
240 INPUT"O3";O3:CLS
250 PRINT"IF RESULTS PRINTED AT INTERVALS H0 O4=0 (X7 IS THE"
260 PRINT"END ABSCISSA), IF PRINTED AFTER M STEPS O4>0 (M7 IS"
270 PRINT"THE TOTAL NUMBER OF STEPS)"
280 INPUT"O4";O4:IF O4>0 THEN GOTO 300
290 INPUT"H0";H0:INPUT"X7";X7:GOTO 310
300 INPUT"M";M:INPUT"M7";M7
310 CLS:PRINT"ACCURACY OF INTEGRATION IN STARTER: P0"
320 PRINT"ACCURACY OF INTEGRATION: P1"
330 PRINT"ACCURACY IN SECOND CONTROL OF STEP SIZE H: P2"
340 PRINT"IF P2=0 SECOND CONTROL IS NOT APPLIED."
350 INPUT"P0";P0:INPUT"P1";P1:INPUT"P2";P2:CLS
360 PRINT"NUMBER OF DIFFERENTIAL EQUATIONS: N"
370 INPUT"N";N:CLS
380 PRINT "INPUT OF INITIAL DATA:"
390 IF O2=0 THEN GOTO 440
400 LPRINT "P405: O1=";O1;"O2=";O2;"O3=";O3;"O4=";O4
410 LPRINT "P0=";P0;"P1=";P1;"P2=";P2
420 IF O4=0 THEN LPRINT "N=";N;"H0=";H0;"X7=";X7
430 IF O4>0 THEN LPRINT "N=";N;"M=";M;"M7=";M7
440 DIM Y(N,3),Y1(N,5)
450 LET P3=P2/8
460 REM P3 IS A LOWER LIMIT FOR INCREASING P1
470 LET P1=3*P1
480 REM THIS SUBSTUTUTION AVOIDS MULTIPLICATIONS ON L.1180
490 LET P4=P1
500 REM P4 IS THE INPUT OF P1; USED FOR CORRECTING P1
510 IF O2>0 THEN LPRINT "INPUT OF INITIAL DATA:"
520 IF O3=0 THEN GOTO 620
530 INPUT"X0";X
540 IF O2>0 THEN LPRINT "X0=";X
550 FOR I=1 TO N
560 PRINT "I=";I
570 INPUT"Y(I)";Y(I,1)
580 IF O2>0 THEN LPRINT "I=";I;"Y(I)=";Y(I,1)
590 NEXT I
600 GOTO 720
610 REM READ INITIAL DATA FROM A FILE
```

```
620 OPEN "I",1,"P405A"
630 INPUT #1, X
640 PRINT "X0=";X
650 IF O2>0 THEN LPRINT "X0=";X
660 FOR I=1 TO N
670 INPUT #1, Y(I,1)
680 PRINT "I=";I;"Y(I)=";Y(I,1)
690 IF O2>0 THEN LPRINT "I=";I;"Y(I)=";Y(I,1)
700 NEXT I
710 CLOSE 1
720 IF O2=0 THEN GOTO 750
730 LPRINT
740 IF O3<3 THEN LPRINT "RESULTS OF INTEGRATION:"
750 LET X6=X+H0
760 LET K=1
770 LET K1=1
780 LET P5=1/3
790 REM STEP SIZE IN STARTER WHEN MAX ABS(Y1(I,1))=0:
800 LET H=P0[P5
810 IF O3<3 THEN PRINT "RESULTS OF INTEGRATION:"
820 REM STARTER - STEP 1
830 GOSUB 2230
840 LET P9=0
850 FOR I=1 TO N
860 LET P8=ABS(Y1(I,1))
870 IF P8>P9 THEN LET P9=P8
880 LET Y1(I,4)=0
890 NEXT I
900 IF P9>0 THEN LET H=P0/P9
910 LET X=X+H
920 LET L=1
930 GOTO 1340
940 REM STARTER - STEP 2
950 GOSUB 2230
960 FOR I=1 TO N
970 LET Y1(I,4)=Y1(I,1)-Y1(I,2)
980 NEXT I
990 LET H=1.5*H1
1000 LET Q=1.5
1010 LET X=X+H
1020 LET L=2
1030 GOTO 1340
1040 IF O4=0 THEN GOTO 1070
1050 IF K1=M THEN GOSUB 1960
1060 REM COMPUTATION OF THE STEP SIZE H
1070 GOSUB 2230
1080 LET K=K+1
1090 LET K1=K1+1
1100 LET H3=H1+H2
1110 LET P9=0
1120 FOR I=1 TO N
1130 LET Y1(I,4)=Y1(I,1)-Y1(I,2)
1140 LET Y1(I,5)=(Y1(I,4)/H1-(Y1(I,1)-Y1(I,3))/H3)/H2
1150 LET P8=ABS(Y1(I,5))
1160 IF P8>P9 THEN LET P9=P8
1170 NEXT I
1180 IF P9>0 THEN LET H=(P1/P9)[P5
1190 LET Q=H/H1
1200 IF Q<=1.5 THEN GOTO 1230
1210 LET H=1.5*H1
1220 LET Q=1.5
1230 LET X=X+H
1240 IF O4>0 THEN GOTO 1340
1250 IF X7<X6 THEN LET X6=X7
```

```
1260 IF X<X6 THEN GOTO 1340
1270 LET X5=X
1280 LET H9=H
1290 LET H=X6+H-X
1300 LET X=X6
1310 LET Q=H/H1
1320 LET L=5
1330 REM PREDICTOR
1340 LET H5=H/2
1350 LET H6=H*H/6
1360 LET H7=H5*Q
1370 FOR I=1 TO N
1380 LET C=H7*Y1(I,4)
1390 IF L=4 THEN GOTO 1430
1400 LET Y(I,2)=Y(I,1)
1410 LET Y1(I,3)=Y1(I,2)
1420 LET Y1(I,2)=Y1(I,1)
1430 LET Y(I,1)=Y(I,2)+H*Y1(I,2)+C
1440 NEXT I
1450 IF L<3 THEN GOTO 1510
1460 LET C=H6*(3*H1+2*H)
1470 FOR I=1 TO N
1480 LET Y(I,1)=Y(I,1)+C*Y1(I,5)
1490 NEXT I
1500 REM CORRECTOR
1510 GOSUB 2230
1520 FOR I=1 TO N
1530 LET Y(I,3)=Y(I,1)
1540 LET Y(I,1)=Y(I,2)+H5*(Y1(I,1)+Y1(I,2))
1550 NEXT I
1560 IF L=1 THEN GOTO 1850
1570 LET H8=H6/(H+H1)
1580 FOR I=1 TO N
1590 LET C=H8*(Y1(I,1)-Y1(I,2)-Q*(Y1(I,2)-Y1(I,3)))
1600 LET Y(I,1)=Y(I,1)-C
1610 NEXT I
1620 IF L=2 THEN GOTO 1840
1630 IF L=4 THEN GOTO 1840
1640 IF L=5 THEN GOTO 1890
1650 IF P2=0 THEN GOTO 1850
1660 REM SECOND CONTROL OF H
1670 LET P9=0
1680 FOR I=1 TO N
1690 LET P8=ABS(Y(I,1)-Y(I,3))
1700 IF P8>P9 THEN LET P9=P8
1710 NEXT I
1720 IF P9<P2 THEN GOTO 1800
1730 LET P1=.421875*P1
1740 LET L=4
1750 LET X=X-H
1760 LET Q=.75*Q
1770 LET H=.75*H
1780 LET X=X+H
1790 GOTO 1340
1800 IF P9>P3 THEN GOTO 1850
1810 LET P1=P1/.421875
1820 IF P1>P4 THEN LET P1=P4
1830 GOTO 1850
1840 LET L=3
1850 LET H2=H1
1860 LET H1=H
1870 IF L=1 THEN GOTO 950
1880 GOTO 1040
1890 LET L=4
```

```
1900 GOSUB 1960
1910 LET H=H9
1920 LET Q=H/H1
1930 LET X=X5
1940 GOTO 1340
1950 REM SUBROUTINE FOR PRINTING THE RESULTS
1960 LET X6=X6+H0
1970 LET K1=0
1980 PRINT "AT X=";X;"AND K=";K;"Y(I):"
1990 FOR I=1 TO N
2000 PRINT "I=";I;"Y(I)=";Y(I,1)
2010 NEXT I
2020 PRINT
2030 IF O2=0 THEN GOTO 2090
2040 LPRINT "AT X=";X;"AND K=";K;"Y(I):"
2050 FOR I=1 TO N
2060 LPRINT "I=";I;"Y(I)=";Y(I,1)
2070 NEXT I
2080 LPRINT
2090 IF O4>0 THEN GOTO 2120
2100 IF X>=X7 THEN GOTO 2150
2110 RETURN
2120 IF K>=M7 THEN GOTO 2150
2130 RETURN
2140 REM WRITING RESULTS ON A FILE
2150 OPEN "O",1,"P405A"
2160 PRINT #1, X
2170 FOR I=1 TO N
2180 PRINT #1, Y(I,1)
2190 NEXT I
2200 CLOSE 1
2210 STOP
2220 REM SUBROUTINE FOR D-EQS
2230 IF O3>1 THEN GOTO 2360
2240 LET Y1(1,1)=
2340 RETURN
2350 DEFINE SPECIAL D-EQS AT X0
2360 IF O3>2 THEN GOTO 2490
2370 LET O3=1
2380 LET Y1(1,1)=
2480 RETURN
2490 PRINT "INPUT OF NUMERICAL VALUES OF Y1 AT X0:"
2500 IF O2>0 THEN LPRINT "NUMERICAL VALUES OF Y1 AT X0:"
2510 LET O3=1
2520 FOR I=1 TO N
2530 PRINT "I=";I;"Y1(I):"
2540 INPUT"Y2(I)";Y1(I,1)
2550 IF O2=0 THEN GOTO 2570
2560 LPRINT "I=";I;"Y1(I)=";Y1(I,1)
2570 NEXT I
2580 PRINT "RESULTS OF INTEGRATION:"
2590 IF O2=0 THEN GOTO 2620
2600 LPRINT
2610 LPRINT "RESULTS OF INTEGRATION:"
2620 RETURN
2630 END
```

Examples

1. One differential equation $y'=y$ with the initial condition $y=1$ at $x=0$ is integrated in single precision from $x=0$ to $x=2$, the results are printed at intervals $H_0=0.2$. Accuracy of integration $P_0=1*10^{-4}$, $P_1=1*10^{-6}$, $P_2=4*10^{-7}$. Some results of integration (K is the number of steps):

        $x=0.2$, $y=1.22140$, $K=22$,
        $x=1.0$, $y=2.71829$, $K=75$,
        $x=2.0$, $y=7.38907$, $K=202$.

2. The same differential equation is integrated in double precision from $x=0$ to $x=2$, $H_0=0.2$, $P_0=1*10^{-6}$, $P_1=1*10^{-8}$, $P_2=1*10^{-10}$. Some results (rounded to 10 digits):

        $x=0.2$, $y=1.221402758$, $K=71$,
        $x=1.0$, $y=2.718281833$, $K=321$,
        $x=2.0$, $y=7.389056114$, $K=743$.

The integral of the differential equation in both examples is $y=e^x$.

P406 Fourth-Order Predictor-Corrector Method for a Set of Differential Equations of First Order

```
10 CLS:PRINT"P406 INTEGRATES A SET OF DIFFERENTIAL EQUATIONS OF"
20 PRINT"FIRST ORDER BY FOURTH ORDER PREDICTOR-CORRECTOR METHOD."
30 PRINT:PRINT"PRESS BREAK, DEFINE Y1,Y2 (DIFFERENTIAL EQUATIONS"
40 PRINT"AND THEIR DERIVATIVES) ON LINES 2290-2480. IF EQUATIONS"
50 PRINT"HAVE A SINGULAR POINT AT THE INITIAL ABSCISSA X0 DEFINE"
60 PRINT"Y1,Y2 ALSO BY SPECIAL EQUATIONS ON LINES 2530-2720 (O3=2)"
70 PRINT"OR NUMERICALLY IN THE INPUT (O3>2). IN ALL THESE DEFINI-"
80 PRINT"TIONS THE DEPENDENT VARIABLES ARE DENOTED Y(I,1), THEIR"
90 PRINT"DERIVATIVES Y1(I,1),Y2(I,1),I=1,2,...N. THEN RUN AGAIN."
100 PRINT:PRINT"THE INPUT:"
110 PRINT"IF SINGLE PRECISION O1=1, IF DOUBLE O1=2."
120 DEFINT I-O:INPUT"O1";O1:CLS
130 PRINT"IF PRINTER NOT USED O2=0, IF USED O2>0."
140 INPUT"O2";O2:CLS
150 IF O1=2 THEN DEFDBL A-H,Q-Z
160 PRINT"IF DERIVATIVES Y1,Y2 AT THE INITIAL ABSCISSA X0 FROM"
170 PRINT"DIFFERENTIAL EQUATIONS O3=1, IF THESE EQUATIONS HAVE"
180 PRINT"A SINGULAR POINT AT X0 THE DERIVATIVES ARE GIVEN"
190 PRINT"EITHER BY SPECIAL EQUATIONS AND O3=2 OR NUMERICALLY"
200 PRINT"IN THE INPUT AND O3>2, IF INITIAL DATA FROM FILE"
210 PRINT"P406A O3=0.":PRINT
220 PRINT"FILE P406A IS WRITTEN IN THE PRECEDING RUN AND SIMPLI-"
230 PRINT"FIES THE CONTINUATION OF INTEGRATION IN SUCCESSIVE RUNS."
240 INPUT"O3";O3:CLS
250 PRINT"IF RESULTS PRINTED AT INTERVALS H0 O4=0 (X7 IS THE"
260 PRINT"END ABSCISSA), IF PRINTED AFTER M STEPS O4>0 (M7 IS"
270 PRINT"THE TOTAL NUMBER OF STEPS)."
280 INPUT"O4";O4:IF O4>0 THEN GOTO 300
290 INPUT"H0";H0:INPUT"X7";X7:GOTO 310
300 INPUT"M";M:INPUT"M7";M7
310 CLS:PRINT"ACCURACY OF INTEGRATION IN STARTER: P0"
320 PRINT"ACCURACY OF INTEGRATION: P1"
330 PRINT"ACCURACY IN SECOND CONTROL OF STEP SIZE H: P2"
340 PRINT"IF P2=0 THE SECOND CONTROL IS NOT APPLIED."
350 INPUT"P0";P0:INPUT"P1";P1:INPUT"P2";P2:CLS
360 PRINT"NUMBER OF DIFFERENTIAL EQUATIONS: N"
370 INPUT"N";N:CLS
380 PRINT "INPUT OF INITIAL DATA:"
390 IF O2=0 THEN GOTO 440
400 LPRINT "P406: O1=";O1;"O2=";O2;"O3=";O3;"O4=";O4
410 LPRINT "P0=";P0;"P1=";P1;"P2=";P2
420 IF O4=0 THEN LPRINT "N=";N;"H0=";H0;"X7=";X7
430 IF O4>0 THEN LPRINT "N=";N;"M=";M;"M7=";M7
440 DIM Y(N,3),Y1(N,3),Y2(N,5)
450 LET P3=P2/8
460 REM P3 IS A LOWER LIMIT FOR INCREASING P1
470 LET P1=12*P1
480 REM THIS SUBSTUTUTION AVOIDS MULTIPLICATIONS ON L.1190
490 LET P4=P1
500 REM P4 IS THE INPUT OF P1; USED FOR CORRECTING P1
510 LET P5=.75
520 LET P5=P5*P5*P5*P5
530 IF O2>0 THEN LPRINT "INPUT OF INITIAL DATA:"
540 IF O3=0 THEN GOTO 640
550 INPUT"X0";X
560 IF O2>0 THEN LPRINT "X0=";X
570 FOR I=1 TO N
580 PRINT "I=";I
590 INPUT"Y(I)";Y(I,1)
600 IF O2>0 THEN LPRINT "I=";I;"Y(I)=";Y(I,1)
610 NEXT I
```

PROGRAM P406

```
620 GOTO 740
630 REM READ INITIAL DATA FROM A FILE
640 OPEN "I",1,"P406A"
650 INPUT #1, X
660 PRINT "X0=";X
670 IF O2>0 THEN LPRINT "X0=";X
680 FOR I=1 TO N
690 INPUT #1, Y(I,1)
700 PRINT "I=";I;"Y(I)=";Y(I,1)
710 IF O2>0 THEN LPRINT "I=";I;"Y(I)=";Y(I,1)
720 NEXT I
730 CLOSE 1
740 IF O2=0 THEN GOTO 770
750 LPRINT
760 IF O3<3 THEN LPRINT "RESULTS OF INTEGRATION:"
770 LET X6=X+H0
780 LET K=1
790 LET K1=1
800 REM STEP SIZE IN STARTER WHEN MAX ABS(Y2(I,1))=0:
810 LET H=SQR(2*P0)
820 IF O3<3 THEN PRINT "RESULTS OF INTEGRATION:"
830 REM STARTER - STEP 1
840 GOSUB 2280
850 LET P9=0
860 FOR I=1 TO N
870 LET P8=ABS(Y2(I,1))
880 IF P8>P9 THEN LET P9=P8
890 LET Y2(I,4)=0
900 NEXT I
910 IF P9>0 THEN LET H=SQR(2*P0/P9)
920 LET X=X+H
930 LET L=1
940 GOTO 1350
950 REM STARTER - STEP 2
960 GOSUB 2280
970 FOR I=1 TO N
980 LET Y2(I,4)=Y2(I,1)-Y2(I,2)
990 NEXT I
1000 LET H=1.5*H1
1010 LET Q=1.5
1020 LET X=X+H
1030 LET L=2
1040 GOTO 1350
1050 IF O4=0 THEN GOTO 1080
1060 IF K1=M THEN GOSUB 2010
1070 REM COMPUTATION OF THE STEP SIZE H
1080 GOSUB 2280
1090 LET K=K+1
1100 LET K1=K1+1
1110 LET H3=H1+H2
1120 LET P9=0
1130 FOR I=1 TO N
1140 LET Y2(I,4)=Y2(I,1)-Y2(I,2)
1150 LET Y2(I,5)=(Y2(I,4)/H1-(Y2(I,1)-Y2(I,3))/H3)/H2
1160 LET P8=ABS(Y2(I,5))
1170 IF P8>P9 THEN LET P9=P8
1180 NEXT I
1190 IF P9>0 THEN LET H=SQR(SQR(P1/P9))
1200 LET Q=H/H1
1210 IF Q<=1.5 THEN GOTO 1240
1220 LET H=1.5*H1
1230 LET Q=1.5
1240 LET X=X+H
1250 IF O4>0 THEN GOTO 1350
```

```
1260 IF X7<X6 THEN LET X6=X7
1270 IF X<X6 THEN GOTO 1350
1280 LET X5=X
1290 LET H9=H
1300 LET H=X6+H-X
1310 LET X=X6
1320 LET Q=H/H1
1330 LET L=5
1340 REM PREDICTOR
1350 LET H5=H/2
1360 LET H6=H/6
1370 LET H7=H6*Q
1380 LET H6=H6*H
1390 FOR I=1 TO N
1400 LET C=H7*Y2(I,4)
1410 IF L=4 THEN GOTO 1460
1420 LET Y(I,2)=Y(I,1)
1430 LET Y1(I,2)=Y1(I,1)
1440 LET Y2(I,3)=Y2(I,2)
1450 LET Y2(I,2)=Y2(I,1)
1460 LET Y(I,1)=Y(I,2)+H*(Y1(I,2)+H5*Y2(I,2)+C)
1470 NEXT I
1480 IF L<3 THEN GOTO 1550
1490 LET C=H5*H6*(2*H1+H)
1500 LET C1=H6*(3*H1+2*H)
1510 FOR I=1 TO N
1520 LET Y(I,1)=Y(I,1)+C*Y2(I,5)
1530 NEXT I
1540 REM CORRECTOR
1550 GOSUB 2280
1560 LET H4=H/3
1570 FOR I=1 TO N
1580 LET Y(I,3)=Y(I,1)
1590 LET Y(I,1)=Y(I,2)+H*(Y1(I,2)+H4*(Y2(I,2)+Y2(I,1)/2))
1600 NEXT I
1610 IF L=1 THEN GOTO 1900
1620 LET H8=H6/(H+H1)
1630 FOR I=1 TO N
1640 LET C=H8*(Y2(I,1)-Y2(I,2)-Q*(Y2(I,2)-Y2(I,3)))
1650 LET Y(I,1)=Y(I,1)-C*H5
1660 NEXT I
1670 IF L=2 THEN GOTO 1890
1680 IF L=4 THEN GOTO 1890
1690 IF L=5 THEN GOTO 1940
1700 IF P2=0 THEN GOTO 1900
1710 REM SECOND CONTROL OF H
1720 LET P9=0
1730 FOR I=1 TO N
1740 LET P8=ABS(Y(I,1)-Y(I,3))
1750 IF P8>P9 THEN LET P9=P8
1760 NEXT I
1770 IF P9<P2 THEN GOTO 1850
1780 LET P1=P1*P5
1790 LET L=4
1800 LET X=X-H
1810 LET Q=.75*Q
1820 LET H=.75*H
1830 LET X=X+H
1840 GOTO 1350
1850 IF P9>P3 THEN GOTO 1900
1860 LET P1=P1/P5
1870 IF P1>P4 THEN LET P1=P4
1880 GOTO 1900
1890 LET L=3
```

PROGRAM P406

```
1900 LET H2=H1
1910 LET H1=H
1920 IF L=1 THEN GOTO 960
1930 GOTO 1050
1940 LET L=4
1950 GOSUB 2010
1960 LET H=H9
1970 LET Q=H/H1
1980 LET X=X5
1990 GOTO 1350
2000 REM SUBROUTINE FOR PRINTING THE RESULTS
2010 LET X6=X6+H0
2020 LET K1=0
2030 PRINT "AT X=";X;"AND K=";K;"Y(I):"
2040 FOR I=1 TO N
2050 PRINT "I=";I;"Y(I)=";Y(I,1)
2060 NEXT I
2070 PRINT
2080 IF O2=0 THEN GOTO 2140
2090 LPRINT "AT X=";X;"AND K=";K;"Y(I):"
2100 FOR I=1 TO N
2110 LPRINT "I=";I;"Y(I)=";Y(I,1)
2120 NEXT I
2130 LPRINT
2140 IF O4>0 THEN GOTO 2170
2150 IF X>=X7 THEN GOTO 2200
2160 RETURN
2170 IF K>=M7 THEN GOTO 2200
2180 RETURN
2190 REM WRITING RESULTS ON A FILE
2200 OPEN "O",1,"P406A"
2210 PRINT #1, X
2220 FOR I=1 TO N
2230 PRINT #1, Y(I,1)
2240 NEXT I
2250 CLOSE 1
2260 STOP
2270 REM SUBROUTINE FOR D-EQS
2280 IF O3>1 THEN GOTO 2510
2290 LET Y1(1,1)=
2300 LET Y2(1,1)=
2490 RETURN
2500 DEFINE SPECIAL D-EQS AT X0
2510 IF O3>2 THEN GOTO 2740
2520 LET O3=1
2530 LET Y1(1,1)=
2540 LET Y2(1,1)=
2730 RETURN
2740 PRINT "INPUT OF NUMERICAL VALUES OF Y1,Y2 AT X0:"
2750 IF O2>0 THEN LPRINT "NUMERICAL VALUES OF Y1,Y2 AT X0:"
2760 LET O3=1
2770 FOR I=1 TO N
2780 PRINT "I=";I;"Y1(I),Y2(I):"
2790 INPUT"Y1(I)";Y1(I,1):INPUT"Y2(I)";Y2(I,1)
2800 IF O2=0 THEN GOTO 2820
2810 LPRINT "I=";I;"Y1(I)=";Y1(I,1);"Y2(I)=";Y2(I,1)
2820 NEXT I
2830 PRINT "RESULTS OF INTEGRATION:"
2840 IF O2=0 THEN GOTO 2870
2850 LPRINT
2860 LPRINT "RESULTS OF INTEGRATION:"
2870 RETURN
2880 END
```

Examples

1. One differential equation $y'=y$ with the initial condition $y=1$ at $x=0$ is integrated in single precision from $x=0$ to $x=2$, the results are printed at intervals $H_0=0.2$. Accuracy of integration $P_0=1*10^{-6}$, $P_1=1*10^{-6}$, $P_2=0$. Some results of integration (K is the number of steps):

   $x=0.2$, $y=1.22140$, $K=10$,
   $x=1.0$, $y=2.71828$, $K=23$,
   $x=2.0$, $y=7.38907$, $K=44$.

2. The same differential equation is integrated in double precision from $x=0$ to $x=2$, $H_0=0.2$, $P_0=1*10^{-6}$, $P_1=1*10^{-8}$, $P_2=1*10^{-10}$. Some results (rounded to 10 digits):

   $x=0.2$, $y=1.221402758$, $K=19$,
   $x=1.0$, $y=2.718281829$, $K=93$,
   $x=2.0$, $y=7.389056102$, $K=210$.

The integral of the differential equation in both preceding examples is $y=e^x$.

3. One differential equation $y'=1/x$ with the initial condition $y=0$ at $x=1$ is integrated in double precision from $x=1$ to $x=10$, $H_0=1$, $P_0=1*10^{-13}$, $P_1=2*10^{-13}$, $P_2=1*10^{-13}$. Some results of integration (rounded to 12 decimal places):

   $x=2$, $y=0.693147180560$, $K=788$,
   $x=4$, $y=1.386294361120$, $K=1522$,
   $x=7$, $y=1.945910149056$, $K=2115$,
   $x=10$, $y=2.302585092995$, $K=2492$.

The integral of this differential equation is $y=\log(x)$. Maximum error $1*10^{-12}$ occurs at $x=7$ and $x=10$. Computer time 67:59 min.

P407 Fourth-Order Predictor-Corrector Method for a Set of Differential Equations of Second Order

```
10 CLS:PRINT"P407 INTEGRATES A SET OF DIFFERENTIAL EQUATIONS OF"
20 PRINT"SECOND ORDER BY FOURTH ORDER PREDICTOR-CORRECTOR METHOD."
30 PRINT:PRINT"PRESS BREAK, DEFINE DIFFERENTIAL EQUATIONS ON"
40 PRINT"LINES 2330-2420, IF EQUATIONS HAVE A SINGULAR POINT AT"
50 PRINT"INITIAL ABSCISSA X0 DEFINE Y2 ALSO BY SPECIAL EQUATIONS"
60 PRINT"ON LINES 2470-2560 (O3=2) OR NUMERICALLY IN THE INPUT"
70 PRINT"(O3>2). IN ALL THESE DEFINITIONS THE DEPENDENT VARIABLES"
80 PRINT"ARE DENOTED Y(I,1), THEIR DERIVATIVES Y1(I,1),Y2(I,1),"
90 PRINT"I=1,2,...N. THEN RUN AGAIN."
100 PRINT:PRINT"THE INPUT:"
110 PRINT"IF SINGLE PRECISION O1=1, IF DOUBLE O1=2."
120 DEFINT I-O:INPUT"O1";O1:CLS
130 PRINT"IF PRINTER NOT USED O2=0, IF USED O2>0."
140 INPUT"O2";O2:CLS
150 IF O1=2 THEN DEFDBL A-H,O-Z
160 PRINT"IF DERIVATIVES Y2 AT THE INITIAL ABSCISSA X0 FROM"
170 PRINT"DIFFERENTIAL EQUATIONS O3=1, IF THESE EQUATIONS HAVE"
180 PRINT"A SINGULAR POINT AT X0 THE DERIVATIVES ARE GIVEN"
190 PRINT"EITHER BY SPECIAL EQUATIONS AND O3=2 OR NUMERICALLY"
200 PRINT"IN THE INPUT AND O3>2, IF INITIAL DATA FROM FILE"
210 PRINT"P407A O3=0,":PRINT
220 PRINT"FILE P407A IS WRITTEN IN THE PRECEDING RUN AND SIMPLI-"
230 PRINT"FIES THE CONTINUATION OF INTEGRATION IN SUCCESSIVE RUNS."
240 INPUT"O3";O3:CLS
250 PRINT"IF RESULTS PRINTED AT INTERVALS H0 O4=0 (X7 IS THE"
260 PRINT"END ABSCISSA), IF PRINTED AFTER M STEPS O4>0 (M7 IS"
270 PRINT"THE TOTAL NUMBER OF STEPS)."
280 INPUT"O4";O4:IF O4>0 THEN GOTO 300
290 INPUT"H0";H0:INPUT"X7";X7:GOTO 310
300 INPUT"M";M:INPUT"M7";M7
310 CLS:PRINT"ACCURACY OF INTEGRATION IN STARTER: P0"
320 PRINT"ACCURACY OF INTEGRATION: P1"
330 PRINT"ACCURACY IN SECOND CONTROL OF STEP SIZE H: P2"
340 PRINT"IF P2=0 THE SECOND CONTROL IS NOT APPLIED."
350 INPUT"P0";P0:INPUT"P1";P1:INPUT"P2";P2:CLS
360 PRINT"NUMBER OF DIFFERENTIAL EQUATIONS: N"
370 INPUT"N";N:CLS
380 PRINT "INPUT OF INITIAL DATA:"
390 IF O2=0 THEN GOTO 440
400 LPRINT "P407: O1=";O1;"O2=";O2;"O3=";O3;"O4=";O4
410 LPRINT "P0=";P0;"P1=";P1;"P2=";P2
420 IF O4=0 THEN LPRINT "N=";N;"H0=";H0;"X7=";X7
430 IF O4>0 THEN LPRINT "N=";N;"M=";M;"M7=";M7
440 DIM Y(N,3),Y1(N,3),Y2(N,5)
450 LET P3=P2/8
460 REM P3 IS A LOWER LIMIT FOR INCREASING P1
470 LET P1=12*P1
480 REM THIS SUBSTUTUTION AVOIDS MULTIPLICATIONS ON L.1190
490 LET P4=P1
500 REM P4 IS THE INPUT OF P1; USED FOR CORRECTING P1
510 LET P5=.75
520 LET P5=P5*P5*P5*P5
530 IF O2>0 THEN LPRINT "INPUT OF INITIAL DATA:"
540 IF O3=0 THEN GOTO 640
550 INPUT"X0";X
560 IF O2>0 THEN LPRINT "X0=";X
570 FOR I=1 TO N
580 PRINT "I=";I
590 INPUT"Y(I)";Y(I,1):INPUT"Y1(I)";Y1(I,1)
600 IF O2>0 THEN LPRINT "I=";I;"Y(I)=";Y(I,1);"Y1(I)=";Y1(I,1)
610 NEXT I
```

```
620 GOTO 740
630 REM READ INITIAL DATA FROM A FILE
640 OPEN "I",1,"P407A"
650 INPUT #1, X
660 PRINT "X0=";X
670 IF O2>0 THEN LPRINT "X0=";X
680 FOR I=1 TO N
690 INPUT #1, Y(I,1),Y1(I,1)
700 PRINT "I=";I;"Y(I)=";Y(I,1);"Y1(I)=";Y1(I,1)
710 IF O2>0 THEN LPRINT "I=";I;"Y(I)=";Y(I,1);"Y1(I)=";Y1(I,1)
720 NEXT I
730 CLOSE 1
740 IF O2=0 THEN GOTO 770
750 LPRINT
760 IF O3<3 THEN LPRINT "RESULTS OF INTEGRATION:"
770 LET X6=X+H0
780 LET K=1
790 LET K1=1
800 REM STEP SIZE IN STARTER WHEN MAX ABS(Y2(I,1))=0:
810 LET H=SQR(2*P0)
820 IF O3<3 THEN PRINT "RESULTS OF INTEGRATION:"
830 REM STARTER - STEP 1
840 GOSUB 2320
850 LET P9=0
860 FOR I=1 TO N
870 LET P8=ABS(Y2(I,1))
880 IF P8>P9 THEN LET P9=P8
890 LET Y2(I,4)=0
900 NEXT I
910 IF P9>0 THEN LET H=SQR(2*P0/P9)
920 LET X=X+H
930 LET L=1
940 GOTO 1350
950 REM STARTER - STEP 2
960 GOSUB 2320
970 FOR I=1 TO N
980 LET Y2(I,4)=Y2(I,1)-Y2(I,2)
990 NEXT I
1000 LET H=1.5*H1
1010 LET Q=1.5
1020 LET X=X+H
1030 LET L=2
1040 GOTO 1350
1050 IF O4=0 THEN GOTO 1080
1060 IF K1=M THEN GOSUB 2050
1070 REM COMPUTATION OF THE STEP SIZE H
1080 GOSUB 2320
1090 LET K=K+1
1100 LET K1=K1+1
1110 LET H3=H1+H2
1120 LET P9=0
1130 FOR I=1 TO N
1140 LET Y2(I,4)=Y2(I,1)-Y2(I,2)
1150 LET Y2(I,5)=(Y2(I,4)/H1-(Y2(I,1)-Y2(I,3))/H3)/H2
1160 LET P8=ABS(Y2(I,5))
1170 IF P8>P9 THEN LET P9=P8
1180 NEXT I
1190 IF P9>0 THEN LET H=SQR(SQR(P1/P9))
1200 LET Q=H/H1
1210 IF Q<=1.5 THEN GOTO 1240
1220 LET H=1.5*H1
1230 LET Q=1.5
1240 LET X=X+H
1250 IF O4>0 THEN GOTO 1350
```

```
1260 IF X7<X6 THEN LET X6=X7
1270 IF X<X6 THEN GOTO 1350
1280 LET X5=X
1290 LET H9=H
1300 LET H=X6+H-X
1310 LET X=X6
1320 LET Q=H/H1
1330 LET L=5
1340 REM PREDICTOR
1350 LET H5=H/2
1360 LET H6=H/6
1370 LET H7=H6*Q
1380 LET H6=H6*H
1390 FOR I=1 TO N
1400 LET C=H7*Y2(I,4)
1410 IF L=4 THEN GOTO 1460
1420 LET Y(I,2)=Y(I,1)
1430 LET Y1(I,2)=Y1(I,1)
1440 LET Y2(I,3)=Y2(I,2)
1450 LET Y2(I,2)=Y2(I,1)
1460 LET Y(I,1)=Y(I,2)+H*(Y1(I,2)+H5*Y2(I,2)+C)
1470 LET Y1(I,1)=Y1(I,2)+H*Y2(I,2)+3*C
1480 NEXT I
1490 IF L<3 THEN GOTO 1570
1500 LET C=H5*H6*(2*H1+H)
1510 LET C1=H6*(3*H1+2*H)
1520 FOR I=1 TO N
1530 LET Y(I,1)=Y(I,1)+C*Y2(I,5)
1540 LET Y1(I,1)=Y1(I,1)+C1*Y2(I,5)
1550 NEXT I
1560 REM CORRECTOR
1570 GOSUB 2320
1580 LET H4=H/3
1590 FOR I=1 TO N
1600 LET Y(I,3)=Y(I,1)
1610 LET Y(I,1)=Y(I,2)+H*(Y1(I,2)+H4*(Y2(I,2)+Y2(I,1)/2))
1620 LET Y1(I,1)=Y1(I,2)+H5*(Y2(I,2)+Y2(I,1))
1630 NEXT I
1640 IF L=1 THEN GOTO 1940
1650 LET H8=H6/(H+H1)
1660 FOR I=1 TO N
1670 LET C=H8*(Y2(I,1)-Y2(I,2)-Q*(Y2(I,2)-Y2(I,3)))
1680 LET Y(I,1)=Y(I,1)-C*H5
1690 LET Y1(I,1)=Y1(I,1)-C
1700 NEXT I
1710 IF L=2 THEN GOTO 1930
1720 IF L=4 THEN GOTO 1930
1730 IF L=5 THEN GOTO 1980
1740 IF P2=0 THEN GOTO 1940
1750 REM SECOND CONTROL OF H
1760 LET P9=0
1770 FOR I=1 TO N
1780 LET P8=ABS(Y(I,1)-Y(I,3))
1790 IF P8>P9 THEN LET P9=P8
1800 NEXT I
1810 IF P9<P2 THEN GOTO 1890
1820 LET P1=P1*P5
1830 LET L=4
1840 LET X=X-H
1850 LET Q=.75*Q
1860 LET H=.75*H
1870 LET X=X+H
1880 GOTO 1350
1890 IF P9>P3 THEN GOTO 1940
```

```
1900 LET P1=P1/P5
1910 IF P1>P4 THEN LET P1=P4
1920 GOTO 1940
1930 LET L=3
1940 LET H2=H1
1950 LET H1=H
1960 IF L=1 THEN GOTO 960
1970 GOTO 1050
1980 LET L=4
1990 GOSUB 2050
2000 LET H=H9
2010 LET Q=H/H1
2020 LET X=X5
2030 GOTO 1350
2040 REM SUBROUTINE FOR PRINTING THE RESULTS
2050 LET X6=X6+H0
2060 LET K1=0
2070 PRINT "AT X=";X;"AND K=";K;"Y(I),Y1(I):"
2080 FOR I=1 TO N
2090 PRINT "I=";I;"Y(I)=";Y(I,1);"Y1(I)=";Y1(I,1)
2100 NEXT I
2110 PRINT
2120 IF O2=0 THEN GOTO 2180
2130 LPRINT "AT X=";X;"AND K=";K;"Y(I),Y1(I):"
2140 FOR I=1 TO N
2150 LPRINT "I=";I;"Y(I)=";Y(I,1);"Y1(I)=";Y1(I,1)
2160 NEXT I
2170 LPRINT
2180 IF O4>0 THEN GOTO 2210
2190 IF X>=X7 THEN GOTO 2240
2200 RETURN
2210 IF K>=M7 THEN GOTO 2240
2220 RETURN
2230 REM WRITING RESULTS ON A FILE
2240 OPEN "O",1,"P407A"
2250 PRINT #1, X
2260 FOR I=1 TO N
2270 PRINT #1, Y(I,1),Y1(I,1)
2280 NEXT I
2290 CLOSE 1
2300 STOP
2310 REM SUBROUTINE FOR D-EQS
2320 IF O3>1 THEN GOTO 2450
2330 LET Y2(1,1)=
2430 RETURN
2440 DEFINE SPECIAL D-EQS AT X0
2450 IF O3>2 THEN GOTO 2580
2460 LET O3=1
2470 LET Y2(1,1)=
2570 RETURN
2580 PRINT "INPUT OF NUMERICAL VALUES OF Y2 AT X0:"
2590 IF O2>0 THEN LPRINT "NUMERICAL VALUES OF Y2 AT X0:"
2600 LET O3=1
2610 FOR I=1 TO N
2620 PRINT "I=";I;"Y2(I):"
2630 INPUT"Y2(I)";Y2(I,1)
2640 IF O2=0 THEN GOTO 2660
2650 LPRINT "I=";I;"Y2(I)=";Y2(I,1)
2660 NEXT I
2670 PRINT "RESULTS OF INTEGRATION:"
2680 IF O2=0 THEN GOTO 2710
2690 LPRINT
2700 LPRINT "RESULTS OF INTEGRATION:"
2710 RETURN
2720 END
```

Examples

1. One differential equation $y''=-y$ with the initial conditions $y=1$, $y'=0$ at $x=0$ is integrated in single precision from $x=0$ to $x=1.5$, the results are printed at intervals $H_0=0.25$. Accuracy of integration $P_0=1*10^{-6}$, $P_1=1*10^{-6}$, $P_2=0$. Some results of integration (K is the number of steps):

         $x=0.25$, $y=0.968913$, $y'=-0.247404$, $K=11$,
         $x=1.00$, $y=0.540300$, $y'=-0.841477$, $K=21$,
         $x=1.50$, $y=0.0707297$, $y'=-0.997512$, $K=27$.

2. The same differential equation is integrated in double precision from $x=0$ to $x=1.5$, $H_0=0.25$, $P_0=1*10^{-8}$, $P_1=1*10^{-10}$, $P_2=4*10^{-12}$. Some results (rounded to 8 decimal places):

         $x=0.25$, $y=0.96891242$, $y'=-0.24740396$, $K=43$,
         $x=1.00$, $y=0.54030230$, $y'=-0.84147099$, $K=144$,
         $x=1.50$, $y=0.07073719$, $y'=-0.99749450$, $K=217$.

The integral of the differential equation in both examples is
      $y=\cos(x)$, $y'=-\sin(x)$.

CHAPTER 5: BOUNDARY-VALUE PROBLEMS OF ORDINARY DIFFERENTIAL EQUATIONS

P501 Lagrange Interpolation for Boundary-Value Problems of One Ordinary Differential Equation of Second Order

```
10 CLS:PRINT"P501 COMPUTES THE FIRST DERIVATIVE AT THE BOUNDARY A"
20 PRINT"FROM THE VALUE OF THE FUNCTION AT THE BOUNDARY B"
30 PRINT"BY LAGRANGE INTERPOLATION."
40 PRINT:PRINT"THE INPUT:"
50 PRINT"SINGLE PRECISION ONLY."
60 PRINT"IF PRINTER NOT USED O2=0, IF USED O2>0."
70 DEFINT I-O
80 INPUT"O2";O2:CLS
90 PRINT"NUMBER N TAKES THE VALUES 2,3,4,... IN THE 1-ST,"
100 PRINT"2-ND, 3-RD,... RUN"
110 INPUT"N";N:CLS
120 IF N>2 THEN GOTO 150
130 PRINT"THE GIVEN BOUNDARY VALUE AT THE BOUNDARY B: Y0"
140 INPUT"Y0";Y0:CLS
150 PRINT"THE INPUT PAIRS D(I),Y(I), I=1,2,... ARE THE CHOSEN"
160 PRINT"DERIVATIVES D(I) AT THE BOUNDARY A AND THE CORRES-"
170 PRINT"PONDING FUNCTIONS Y(I) AT THE BOUNDARY B."
180 IF O2=0 THEN GOTO 200
190 LPRINT "P501: O2=";O2;"N=";N
200 DIM C(4),D(4),Y(4)
210 IF N>2 THEN GOTO 310
220 INPUT"D(1)";D(1):INPUT"Y(1)";Y(1)
230 INPUT"D(2)";D(2):INPUT"Y(2)";Y(2):CLS
240 IF O2=0 THEN GOTO 280
250 LPRINT "Y0=";Y0
260 LPRINT "D(1)=";D(1);"Y(1)=";Y(1)
270 LPRINT "D(2)=";D(2);"Y(2)=";Y(2)
280 GOSUB 920
290 LET D0=D(1)+(Y0-Y(1))*(D(1)-D(2))/(Y(1)-Y(2))
300 GOTO 850
310 IF N>3 THEN GOTO 380
320 LET M=2
330 GOSUB 990
340 INPUT"D(3)";D(3):INPUT"Y(3)";Y(3):CLS
350 IF O2>0 THEN LPRINT "D(3)=";D(3);"Y(3)=";Y(3)
360 GOSUB 920
370 GOTO 580
380 IF N>4 THEN GOTO 450
390 LET M=3
400 GOSUB 990
410 INPUT"D(4)";D(4):INPUT"Y(4)";Y(4):CLS
420 IF O2>0 THEN LPRINT "D(4)=";D(4);"Y(4)=";Y(4)
430 GOSUB 920
440 GOTO 580
450 LET M=4
460 GOSUB 990
470 FOR I=1 TO 3
480 LET J=I+1
490 LET Y(I)=Y(J)
500 LET D(I)=D(J)
510 NEXT I
520 INPUT"D(N)";D(4):INPUT"Y(N)";Y(4):CLS
530 IF O2>0 THEN LPRINT "D(N)=";D(4);"Y(N)=";Y(4)
540 LET N=4
550 GOSUB 920
560 REM LAGRANGE INTERPOLATION
```

PROGRAM P501

```
570 REM COMPUTATION OF C(I)
580 FOR J=1 TO N
590 LET C0=1
600 FOR I=1 TO N
610 IF I=J THEN GOTO 630
620 LET C0=C0*(Y(J)-Y(I))
630 NEXT I
640 LET C(J)=D(J)/C0
650 NEXT J
660 REM INTERPOLATION
670 LET Z=1
680 FOR I=1 TO N
690 LET X1=Y0-Y(I)
700 IF X1=0 THEN GOTO 720
710 LET Z=Z*X1
720 NEXT I
730 LET K=0
740 LET D0=0
750 FOR I=1 TO N
760 LET X1=Y0-Y(I)
770 IF X1=0 THEN GOTO 800
780 LET D0=D0+C(I)/X1
790 GOTO 820
800 LET Y1=C(I)
810 LET K=1
820 NEXT I
830 IF K=1 THEN D0=Y1
840 LET D0=D0*Z
850 PRINT "INITIAL DERIVATIVE D0=";D0
860 PRINT"THIS COMPUTED INITIAL DERIVATIVE D0 IS THE INPUT Y1"
870 PRINT"IN THE NEXT INTEGRATION OF THE DIFFERENTIAL EQUATION."
880 IF O2=0 THEN GOTO 910
890 LPRINT "INITIAL DERIVATIVE D0=";D0
900 LPRINT
910 STOP
920 OPEN "O",1,"P501A"
930 PRINT #1,Y0
940 FOR I=1 TO N
950 PRINT #1, D(I),Y(I)
960 NEXT I
970 CLOSE 1
980 RETURN
990 OPEN "I",1,"P501A"
1000 INPUT #1, Y0
1010 FOR I=1 TO M
1020 INPUT #1, D(I),Y(I)
1030 NEXT I
1040 CLOSE 1
1050 RETURN
1060 END
```

Example

The differential equation of second order
$$y''=-(1+y'^2)/y$$
is to be integrated with the boundary conditions $y(0)=1$, $y(1)=2$. The integration is carried out with the help of P407 (accuracy of integration $P_0=P_1=1*10^{-6}$, $P_2=4*10^{-7}$, results of integration printed at intervals $H_0=0.2$).

In the first two integrations the initial data are chosen $y(0)=1$, $y'(0)=0.5$ and $y(0)=1$, $y'(0)=1$, corresponding boundary values are $y(1)=1.00001$ and $y(1)=1.41423$. Program P501 yields the initial condition $y'(0)=1.70706$.

In the third integration with the initial data $y(0)=0$, $y'(0)=1.70706$ the boundary value is $y(1)=1.84776$. Program P501 yields now $y'(0)=1.99994$.

In the fourth integration with the initial data $y(0)=0$, $y'(0)=1.99994$ the boundary value is $y(1)=2$. The results of this final integration are (K is the number of steps):

```
        x=0.2,   y=1.32664,   y'=1.35679,   K=41,
        x=0.4,   y=1.56204,   y'=1.02430,   K=50,
        x=0.6,   y=1.74356,   y'=0.80297,   K=56,
        x=0.8,   y=1.88680,   y'=0.63616,   K=61,
        x=1.0,   y=2.00000,   y'=0.50002,   K=66.
```

CHAPTER 6: NONLINEAR EQUATIONS

P601 Roots of a Quadratic Equation

```
10 CLS:PRINT"P601 COMPUTES THE ROOTS OF A QUADRATIC EQUATION."
20 PRINT:PRINT"THE INPUT:"
30 PRINT"IF SINGLE PRECISION O1=1, IF DOUBLE O1=2."
40 DEFINT I-O:INPUT"O1";O1:CLS
50 PRINT"IF PRINTER NOT USED O2=0, IF USED O2>0."
60 INPUT"O2";O2:CLS
70 IF O1=2 THEN DEFDBL A-H,P-Z
80 PRINT"THE EQUATION: A*X[2+B*X+C=0"
90 PRINT "INPUT OF COEFFICIENTS:"
100 INPUT"A";A:INPUT"B";B:INPUT"C";C:CLS
110 IF O2=0 THEN GOTO 140
120 LPRINT "P601: O1=";O1;"O2=";O2
130 LPRINT "A=";A;"B=";B;"C=";C
140 LET B=(B/A)/2
150 LET C=C/A
160 LET A=B*B-C
170 IF A<0 THEN GOTO 240
180 IF A>0 THEN GOTO 320
190 PRINT "DOUBLE ROOT: X1,2=";-B
200 IF O2=0 THEN STOP
210 LPRINT "DOUBLE ROOT: X1,2=";-B
220 STOP
230 REM COMPUTATION OF COMPLEX ROOTS
240 LET A=-A
250 LET D=SQR(A)
260 IF O1=2 THEN GOSUB 420
270 PRINT "COMPLEX ROOTS: RE(X1,2)=";-B;"   IM(X1,2)=+-";D
280 IF O2=0 THEN STOP
290 LPRINT "COMPLEX ROOTS: RE(X1,2)=";-B;"   IM(X1,2)=+-";D
300 STOP
310 REM COMPUTATION OF REAL ROOTS
320 LET D=SQR(A)
330 IF O1=2 THEN GOSUB 420
340 IF B=0 THEN LET A=D
350 IF B<>0 THEN LET A=-B-SGN(B)*D
360 LET B=C/A
370 PRINT "REAL ROOTS: X1=";A;"    X2=";B
380 IF O2=0 THEN STOP
390 LPRINT "REAL ROOTS: X1=";A;"    X2=";B
400 STOP
410 REM DOUBLE PRECISION FOR SQR(A)
420 LET X9=D/1.D16
430 LET X8=D
440 LET D=(D+A/D)/2
450 IF ABS(D-X8)>X9 THEN GOTO 430
460 RETURN
470 END
```

Examples

1. The quadratic equation $x^2-6x+9=0$ is solved in single precision. It has one double root $x_{1,2}=3$.

2. The quadratic equation $x^2-3x+2=0$ is solved in single precision. It has two real roots $x_1=2$, $x_2=1$.

3. The quadratic equation $3x^2-6x+15=0$ is solved in single precision. It has two complex conjugate roots $x_{1,2}=1+-2i$.

4. The quadratic equation $x^2+x+1=0$ is solved in double precision. It has two complex conjugate roots
$x_{1,2}=-0.5+-0.8660254037844387i$.

P602 Roots of a Cubic Equation

```
10 CLS:PRINT"P602 COMPUTES THE ROOTS OF A CUBIC EQUATION."
20 PRINT:PRINT"THE INPUT:"
30 PRINT"SINGLE PRECISION ONLY. IF DOUBLE PRECISION REQUIRED"
40 PRINT"USE PROGRAM P606; THE ROOTS ARE THE INITIAL ESTIMATES."
50 PRINT:PRINT"IF PRINTER NOT USED O2=0, IF USED O2>0."
60 DEFINT I-O
70 INPUT"O2";O2:CLS
80 PRINT"THE EQUATION: A*X[3+B*X[2+C*X+D=0"
90 PRINT "INPUT OF COEFFICIENTS:"
100 INPUT"A";A:INPUT"B";B:INPUT"C";C:INPUT"D";D:CLS
110 PRINT"WARNING: THE ROUNDOFF ERRORS MAY DISTORT RESULTS,"
120 PRINT"FOR INSTANCE GENERATE A SMALL IMAGINARY PART OF A ROOT"
130 PRINT"THAT IS IN FACT REAL, EVEN WHEN THE COEFFICIENTS"
140 PRINT"OF THE EQUATION AND ITS ROOTS ARE ALL INTEGERS."
150 PRINT
160 IF O2=0 THEN GOTO 200
170 LPRINT "P602: O2=";O2
180 LPRINT "A=";A;"B=";B;"C=";C;"D=";D
190 REM REDUCTION TO THE REDUCED FORM
200 LET F=(B/A)/3
210 LET Q=F*F
220 LET P=(C/A)/3-Q
230 LET Q=F*(Q-(C/A)/2)+(D/A)/2
240 LET G=Q*Q+P*P*P
250 IF G>0 THEN GOTO 410
260 IF G=0 THEN GOTO 600
270 REM SOLUTION: ALL THREE ROOTS ARE REAL
280 LET S=2*SQR(-P)
290 LET U=-Q/SQR(-P*P*P)
300 LET U=(1.570796-ATN(U/SQR(1-U*U)))/3
310 LET X1=S*COS(U)-F
320 LET X2=S*COS(U+2.094395)-F
330 LET X3=S*COS(U+4.18879)-F
340 PRINT "THREE REAL ROOTS:"
350 PRINT "X1=";X1;"X2=";X2;"X3=";X3
360 IF O2=0 THEN STOP
370 LPRINT "THREE REAL ROOTS:"
380 LPRINT "X1=";X1;"X2=";X2;"X3=";X3
390 STOP
400 REM SOLUTION: ONE ROOT IS REAL, TWO COMPLEX CONJUGATE
410 LET G=SQR(G)
420 LET U=G-Q
430 LET H=1/3
440 LET U=(ABS(U)[H)*SGN(U)
450 LET V=-G-Q
460 LET V=(ABS(V)[H)*SGN(V)
470 LET X1=U+V-F
480 LET X2=-X1/2-(B/A)/2
490 LET X3=ABS(U-V)*SQR(3)/2
500 PRINT "ONE REAL ROOT X1=";X1
510 PRINT "TWO COMPLEX CONJUGATE ROOTS:"
520 PRINT "RE(X2,3)=";X2;"    IM(X2,3)=+-";X3
530 IF O2=0 THEN STOP
540 LPRINT "ONE REAL ROOT X1=";X1
550 LPRINT "TWO COMPLEX CONJUGATE ROOTS:"
560 LPRINT "RE(X2,3)=";X2;"    IM(X2,3)=+-";X3
570 STOP
580 REM SOLUTION: ALL THREE ROOTS ARE REAL, AT LEAST TWO...
590 REM ...ARE EQUAL
600 LET X1=2*SQR(-P)*SGN(-Q)-F
610 LET X2=-X1/2-(B/A)/2
620 PRINT "THREE REAL ROOTS:"
```

```
630 PRINT "X1=";X1;"X2=X3=";X2
640 IF O2=0 THEN STOP
650 LPRINT "THREE REAL ROOTS:"
660 LPRINT "X1=";X1;"X2=X3=";X2
670 END
```

Examples

1. The cubic equation $x^3+4x^2+3x=0$ has three real roots $x_1=2.38419*10^{-7}$, $x_2=-3$, $x_3=-1$. The first root is in fact $x_1=0$.

2. The cubic equation $x^3-5x^2+11x-15=0$ has one real root $x_1=3$ and two complex conjugate roots $x_{2,3}=1+-2i$.

3. The cubic equation $x^3-4x^2+5x-2=0$ has one real root $x_1=2$ and two complex conjugate roots $x_{2,3}=1+-3.77071*10^{-4}$. The last two are in fact a real double root $x_{2,3}=1$.

These examples show the influence of roundoff errors on the character of roots.

P603 Roots of a Biquadratic Equation

```
10 CLS:PRINT"P603 COMPUTES THE ROOTS OF A BIQUADRATIC EQUATION."
20 PRINT:PRINT"THE INPUT:"
30 PRINT"SINGLE PRECISION ONLY. IF DOUBLE PRECISION REQUIRED"
40 PRINT"USE PROGRAM P606; THE ROOTS ARE THE INITIAL ESTIMATES."
50 PRINT:PRINT"IF PRINTER NOT USED O2=0, IF USED O2>0."
60 DEFINT I-O
70 INPUT"O2";O2:CLS
80 PRINT"THE EQUATION: A*X[4+B*X[3+C*X[2+D*X+E=0"
90 PRINT "INPUT OF COEFFICIENTS:"
100 INPUT"A";A:INPUT"B";B:INPUT"C";C
110 INPUT"D";D:INPUT"E";E:CLS
120 PRINT"WARNING: THE ROUNDOFF ERRORS MAY DISTORT RESULTS,"
130 PRINT"FOR INSTANCE GENERATE A SMALL IMAGINARY PART OF A ROOT"
140 PRINT"THAT IS IN FACT REAL, EVEN WHEN THE COEFFICIENTS"
150 PRINT"OF THE EQUATION AND ITS ROOTS ARE ALL INTEGERS."
160 PRINT
170 IF O2=0 THEN GOTO 210
180 LPRINT "P603: O2=";O2
190 LPRINT "A=";A;"B=";B;"C=";C
200 LPRINT "D=";D;"E=";E
210 DIM Z(6)
220 REM REDUCTION TO CUBIC RESOLVENT EQUATION WITH...
230 REM ...COEFFICIENTS 2*P1,(P1*P1-4*R),-Q1*Q1
240 LET F1=(B/A)/4
250 LET H=F1*F1
260 LET G=C/A
270 LET P1=G-6*H
280 LET R=D/A
290 LET Q1=2*F1*(4*H-G)+R
300 LET R=E/A+F1*(F1*(G-3*H)-R)
310 REM REDUCTION TO THE REDUCED FORM OF CUBIC EQUATION...
320 REM ...WITH COEFFICIENTS 3*P,2*Q
330 LET F=2*P1/3
340 LET Q=F*F
350 LET G=P1*P1-4*R
360 LET P=G/3-Q
370 LET Q=F*(Q-G/2)-Q1*Q1/2
380 REM SOLUTION OF THE CUBIC EQUATION
390 LET G=Q*Q+P*P*P
400 IF G>0 THEN GOTO 490
410 IF G=0 THEN GOTO 590
420 LET S=2*SQR(-P)
430 LET U=-Q/SQR(-P*P*P)
440 LET U=(1.570796-ATN(U/SQR(1-U*U)))/3
450 LET Z(1)=S*COS(U)-F
460 LET Z(2)=S*COS(U+2.094395)-F
470 LET Z(3)=S*COS(U+4.18879)-F
480 GOTO 620
490 LET G=SQR(G)
500 LET U=G-Q
510 LET H=1/3
520 LET U=(ABS(U)[H)*SGN(U)
530 LET V=-G-Q
540 LET V=(ABS(V)[H)*SGN(V)
550 LET Z(1)=U+V-F
560 LET X1=-Z(1)/2-P1
570 LET X2=ABS(U-V)*SQR(3)/2
580 GOTO 750
590 LET Z(1)=2*SQR(-P)*SGN(-Q)-F
600 LET Z(2)=-Z(1)/2-P1
610 LET Z(3)=Z(2)
620 FOR I=1 TO 3
```

```
630 LET J=I+3
640 IF Z(I)<0 THEN GOTO 680
650 LET Z(I)=SQR(Z(I))
660 LET Z(J)=0
670 GOTO 700
680 LET Z(J)=SQR(-Z(I))
690 LET Z(I)=0
700 NEXT I
710 GOTO 920
720 REM COMPUTATION OF SQR(Z(I)), I=1,2,3
730 REM REAL PARTS DENOTED Z(1),Z(2),Z(3), IMAGINARY PARTS...
740 REM ...Z(4),Z(5),Z(6)
750 IF Z(1)<0 THEN GOTO 790
760 LET Z(1)=SQR(Z(1))
770 LET Z(4)=0
780 GOTO 810
790 LET Z(1)=0
800 LET Z(4)=SQR(-Z(1))
810 LET H=SQR((SQR(X1*X1+X2*X2)+ABS(X1))/2)
820 IF X1<0 THEN GOTO 860
830 LET Z(2)=H
840 LET Z(5)=X2/(2*H)
850 GOTO 880
860 LET Z(2)=X2/(2*H)
870 LET Z(5)=H
880 LET Z(3)=Z(2)
890 LET Z(6)=-Z(5)
900 REM COMPUTATION OF THE ROOTS OF BIQUADRATIC EQUATION
910 REM X1 IS THE REAL PART, X2 THE IMAGINARY PART OF A ROOT
920 IF Q1=0 THEN GOTO 980
930 LET G=Z(2)*Z(3)-Z(5)*Z(6)
940 LET H=Z(2)*Z(6)+Z(3)*Z(5)
950 LET R=Q1/(G*G+H*H)
960 LET Z(1)=-R*G
970 LET Z(4)=R*H
980 LET X1=(Z(1)+Z(2)+Z(3))/2-F1
990 LET X2=(Z(4)+Z(5)+Z(6))/2
1000 PRINT "RE(X1)=";X1;"    IM(X1)=";X2
1010 IF O2>0 THEN LPRINT "RE(X1)=";X1;"    IM(X1)=";X2
1020 LET X1=(Z(1)-Z(2)-Z(3))/2-F1
1030 LET X2=(Z(4)-Z(5)-Z(6))/2
1040 PRINT "RE(X2)=";X1;"    IM(X2)=";X2
1050 IF O2>0 THEN LPRINT "RE(X2)=";X1;"    IM(X2)=";X2
1060 LET X1=(-Z(1)+Z(2)-Z(3))/2-F1
1070 LET X2=(-Z(4)+Z(5)-Z(6))/2
1080 PRINT "RE(X3)=";X1;"    IM(X3)=";X2
1090 IF O2>0 THEN LPRINT "RE(X3)=";X1;"    IM(X3)=";X2
1100 LET X1=(-Z(1)-Z(2)+Z(3))/2-F1
1110 LET X2=(-Z(4)-Z(5)+Z(6))/2
1120 PRINT "RE(X4)=";X1;"    IM(X4)=";X2
1130 IF O2>0 THEN LPRINT "RE(X4)=";X1;"    IM(X4)=";X2
1140 END
```

Examples

1. The biquadratic equation $x^4-10x^3+35x^2-50x+24=0$ has four real roots

$$x_1=4, \quad x_2=3, \quad x_3=1, \quad x_4=2.$$

Since the roots of an algebraic equation with real coefficients are either real or complex conjugate, it is obvious that the small imaginary parts of those roots whose real parts differ from each other must in fact be generated by roundoff errors.

2. The biquadratic equation $x^4-3x^3+3x^2-3x+2=0$ has two real roots $x_1=2$, $x_2=1$ and two complex conjugate roots $x_{3,4}=+-i$.

3. The biquadratic equation $x^4-2x^3+6x^2-2x+5=0$ has two pairs of complex conjugate roots

$$x_{1,2}=1+-2i, \quad x_{3,4}=+-i.$$

4. The biquadratic equation $x^4-16x^3+90x^2-200x+125=0$ has one real root $x_1=1$ and a triple real root $x_{2,3,4}=5$.

The imaginary parts of roots that are very small relative to the real parts and the real parts that are very small relative to the imaginary parts are neglected because they are generated by roundoff errors.

P604 Preliminary Location of Real Roots of a Nonlinear Equation

```
10 CLS:PRINT"P604 EVALUATES THE FUNCTION Y=F(X) AT EQUALLY SPACED"
20 PRINT"POINTS BETWEEN THE LIMITS (U,V) TO LOCATE REAL ROOTS"
30 PRINT"OF A NONLINEAR EQUATION F(X)=0.":PRINT
40 PRINT"IF THE EQUATION IS TRANSCENDENTAL PRESS BREAK AND DEFINE"
50 PRINT"F AND ITS DERIVATIVE F1 ON LINES 1210,1220. THEN RUN AGAIN."
60 PRINT:PRINT"THE INPUT:"
70 PRINT"IF SINGLE PRECISION O1=1, IF DOUBLE O1=2."
80 PRINT"THE INPUT O1 DETERMINES THE PRECISION OF COEFFICIENTS"
90 PRINT"OF THE POLYNOMIAL IN THE INPUT AND IN FILE P604A"
100 PRINT"WRITTEN IN THIS PROGRAM AND USED AS INPUT IN P605."
110 PRINT"LOCATION OF ROOTS IS ALWAYS IN SINGLE PRECISION."
120 DEFINT I-O:INPUT"O1";O1:CLS
130 IF O1=2 THEN DEFDBL A
140 PRINT"IF PRINTER NOT USED O2=0, IF ONLY LOCATIONS OF ROOTS"
150 PRINT"PRINTED O2=1, IF ALSO X,F(X),F1(X) AT EACH POINT"
160 PRINT"PRINTED O2>1"
170 INPUT"O2";O2:CLS
180 PRINT"IF THE EQUATION IS TRANSCENDENTAL O3=0"
190 PRINT"IF THE EQUATION IS A POLYNOMIAL O3>0"
200 INPUT"O3";O3:CLS:IF O3=0 THEN GOTO 250
210 PRINT"DEGREE OF THE POLYNOMIAL: N"
220 PRINT"THERE ARE (N+1) COEFFICIENTS A(I) IN THE POLYNOMIAL,"
230 PRINT"COEFFICIENT A(I) STANDS AT (I-1)-ST POWER OF X."
240 INPUT"N";N:CLS
250 PRINT"INTERVAL OF LOCATION PROCESS: (U,V)"
260 PRINT"SPACING IN LOCATION PROCESS: H"
270 INPUT"U";U:INPUT"V";V:INPUT"H";H:CLS
280 IF O2=0 THEN GOTO 310
290 LPRINT "P604: O1=";O1;"O2=";O2;"O3=";O3
300 LPRINT "U=";U;"V=";V;"H=";H
310 IF O3=0 THEN GOTO 630
320 IF O2>0 THEN LPRINT "DEGREE OF THE POLYNOMIAL N=";N
330 LET N1=N+1
340 LET N2=N-1
350 DIM A(N1),B(N1),C(N),B1(N1)
360 PRINT "THE INPUT A(I):"
370 IF O2>0 THEN LPRINT "THE INPUT A(I):"
380 FOR I=N1 TO 1 STEP -1
390 LET J=I-1
400 PRINT "POWER=";J;"I=";I
410 INPUT"A(I)";A(I)
420 NEXT I
430 CLS:PRINT "CHECK A(I):"
440 FOR I=N1 TO 1 STEP -1
450 LET J=I-1
460 PRINT "POWER=";J;"I=";I;"A(I)=";A(I)
470 PRINT"IF NO ERROR INPUT 0"
480 PRINT"IF ERROT INPUT ANY NUMBER>0 AND THEN THE CORRECT A(I)"
490 INPUT"ERROR";K
500 IF K>0 THEN INPUT"A(I)";A(I)
510 CLS:NEXT I
520 REM WRITING THE INPUT ON A FILE
530 OPEN "O",1, "P604A"
540 FOR I=N1 TO 1 STEP -1
550 LET J=I-1
560 PRINT #1, A(I)
570 IF O2=0 THEN GOTO 590
580 LPRINT "POWER=";J;"I=";I;"A(I)=";A(I)
590 LET B1(I)=A(I)
600 NEXT I
610 CLOSE 1
```

PROGRAM P604

```
620 CLS
630 IF O2>0 THEN LPRINT
640 PRINT "LOCATION OF REAL ROOTS:"
650 IF O2>1 THEN LPRINT "LOCATION OF REAL ROOTS:"
660 LET X=U
670 LET L=0
680 GOSUB 1080
690 LET X1=X
700 LET S=F
710 LET S1=F1
720 LET X=X+H
730 GOSUB 1080
740 PRINT "X=";X;"F=";F;"F1=";F1
750 IF O2>1 THEN LPRINT "X=";X;"F=";F;"F1=";F1
760 IF SGN(F)=-SGN(S) THEN GOTO 900
770 IF SGN(S1)=SGN(F1) THEN GOTO 1020
780 IF SGN(F)=-SGN(F1) THEN GOTO 1020
790 IF L<2 THEN GOTO 1020
800 PRINT "A MULTIPLE ROOT BETWEEN X1=";X1;"AND X2=";X;"?"
810 PRINT "THE QUESTION MARK INDICATES THAT A MULTIPLE ROOT"
820 PRINT "MAY OCCUR IN THE INTERVAL. IT IS PLAUSIBLE THAT"
830 PRINT "IT DOES OCCUR ONLY IF F(X) IS CLOSE TO ZERO."
840 IF O2=0 THEN GOTO 930
850 LPRINT "A MULTIPLE ROOT BETWEEN X1=";X1;"AND X2=";X;"?"
860 LPRINT "THE QUESTION MARK INDICATES THAT A MULTIPLE ROOT"
870 LPRINT "MAY OCCUR IN THE INTERVAL. IT IS PLAUSIBLE THAT"
880 LPRINT "IT DOES OCCUR ONLY IF F(X) IS CLOSE TO ZERO."
890 GOTO 930
900 IF L<2 THEN GOTO 1020
910 PRINT "A ROOT BETWEEN X1=";X1;"AND X2=";X
920 IF O2>0 THEN LPRINT "A ROOT BETWEEN X1=";X1;"AND X2=";X
930 PRINT "X1=";X1;"F(X1)=";S;"F1(X1)=";S1
940 PRINT "X2=";X;"F(X2)=";F;"F1(X2)=";F1
950 IF O2=0 THEN GOTO 1000
960 LPRINT "X1=";X1;"F(X1)=";S;"F1(X1)=";S1
970 LPRINT "X2=";X;"F(X2)=";F;"F1(X2)=";F1
980 LPRINT
990 GOTO 1020
1000 PRINT "TO CONTINUE PRESS ENTER"
1010 INPUT A$
1020 LET X1=X
1030 LET S=F
1040 LET S1=F1
1050 IF X>V THEN PRINT "END OF LOCATION PROCESS":STOP
1060 GOTO 720
1070 REM SUBROUTINE FOR F(X) AND F1(X)
1080 IF O3=0 THEN GOTO 1210
1090 LET B(N1)=B1(N1)
1100 LET C(N)=B(N1)
1110 FOR I=N TO 1 STEP -1
1120 LET J=I+1
1130 LET K=I-1
1140 LET B(I)=B(J)*X+B1(I)
1150 IF I=1 THEN GOTO 1170
1160 LET C(K)=C(I)*X+B(I)
1170 NEXT I
1180 LET F=B(1)
1190 LET F1=C(1)
1200 GOTO 1230
1210 LET F=
1220 LET F1=
1230 LET L=L+1
1240 IF F<>0 THEN RETURN
1250 LET L=0
```

```
1260 IF F1<>0 THEN GOTO 1320
1270 PRINT "A MULTIPLE ROOT X=";X
1280 IF O2=0 THEN PRINT "TO CONTINUE PRESS ENTER"
1290 IF O2=0 THEN INPUT A$
1300 IF O2>0 THEN LPRINT "A MULTIPLE ROOT X=";X
1310 RETURN
1320 PRINT "A ROOT X=";X
1330 IF O2=0 THEN PRINT "TO CONTINUE PRESS ENTER"
1340 IF O2=0 THEN INPUT A$
1350 IF O2>0 THEN LPRINT "A ROOT X=";X
1360 RETURN
1370 END
```

Examples

1. The preliminary location of real roots of an algebraic equation $x^5-15.5x^4+77.5x^3-155x^2+124x-32=0$:

$0.4 < x_1 < 0.64$, $0.9 < x_2 < 1.15$, $1.9 < x_3 < 2.15$,
$3.9 < x_4 < 4.15$, $7.9 < x_5 < 8.15$.

2. The preliminary location of 8 real roots of a transcendental equation $\sin(x)-x/25=0$:

$-0.1 < x_1 < 0.15$, $2.9 < x_2 < 3.15$, $6.4 < x_3 < 6.65$,
$8.9 < x_4 < 9.15$, $12.9 < x_5 < 13.15$, $14.9 < x_6 < 15.15$,
$19.65 < x_7 < 19.9$, $20.9 < x_8 < 21.15$.

Because the transcendental equation is odd, for each positive root there exists a negative one of the same absolute value.

P605 Real Roots of a Nonlinear Equation (with Deflation Subroutine for an Algebraic Equation)

```
10 CLS:PRINT"P605 COMPUTES REAL ROOTS OF A NONLINEAR EQUATION F(X)=0"
20 PRINT"BY ITERATION.":PRINT
30 PRINT"IF THE EQUATION IS TRANSCENDENTAL PRESS BREAK AND DEFINE"
40 PRINT"F AND ITS FIRST AND SECOND DERIVATIVES F1,F2 ON LINES"
50 PRINT"1850-1870. THEN RUN AGAIN."
60 PRINT:PRINT"THE INPUT:"
70 PRINT"IF SINGLE PRECISION O1=1, IF DOUBLE O1=2.":PRINT
80 PRINT"IF COEFFICIENTS OF A POLYNOMIAL EQUATION ARE TAKEN"
90 PRINT"FROM FILE P604A WRITTEN IN P604 THE SAME PRECISION"
100 PRINT"MUST BE USED AS IN P604."
110 DEFINT I-O:INPUT"O1";O1:CLS
120 IF O1=2 THEN DEFDBL A-H,S-Z
130 PRINT"IF PRINTER NOT USED O2=0, IF ONLY FINAL RESULTS"
140 PRINT"PRINTED O2=1, IF RESULTS AT EACH STEP PRINTED O2>1."
150 INPUT"O2";O2:CLS
160 PRINT"IF THE EQUATION IS TRANSCENDENTAL O3=0, IF IT IS A"
170 PRINT"POLYNOMIAL AND ITS COEFFICIENTS A(I) GIVEN BY THE"
180 PRINT"INPUT O3=1, IF A(I) TAKEN FROM FILE P604A O3>1."
190 INPUT"O3";O3:CLS
200 PRINT"THE ITERATION STOPS AND THE ROOT IS TO BE RECOMPUTED"
210 PRINT"WHEN THE MAXIMUM NUMBER OF ITERATIONS M IS REACHED.":PRINT
220 PRINT"THE ITERATION STOPS WHEN ABS(F(X))<R, X BEING THE"
230 PRINT"COMPUTED ROOT AND R A CHOSEN CONSTANT."
240 INPUT"M";M:INPUT"R";R:CLS:IF O3=0 THEN GOTO 380
250 PRINT"DEGREE OF THE POLYNOMIAL: N":PRINT
260 PRINT"THERE ARE (N+1) COEFFICIENTS A(I) IN THE POLYNOMIAL,"
270 PRINT"COEFFICIENT A(I) STANDS AT (I-1)-ST POWER OF X."
280 PRINT"USE THE RESULTS FROM P604 FOR INITIAL ESTIMATE OF X."
290 INPUT"N";N:CLS
300 PRINT"IF THE EQUATION IS A POLYNOMIAL ALL ITS REAL ROOTS"
310 PRINT"MAY BE COMPUTED EITHER FROM THE ORIGINAL EQUATION"
320 PRINT"OR THE EQUATION MAY BE DEFLATED EACH TIME WHEN"
330 PRINT"A ROOT IS FOUND. AFTER ALL REAL ROOTS OF THE"
340 PRINT"POLYNOMIAL ARE FOUND FROM THE DEFLATED EQUATION"
350 PRINT"THEIR ACCURACY MAY BE INCREASED BY USING THEM AS"
360 PRINT"INITIAL ESTIMATES AND RECOMPUTING THEM FROM THE"
370 PRINT"ORIGINAL EQUATION.":PRINT
380 PRINT"IF THE METHOD WITH GLOBAL CUBIC CONVERGENCE USED O4=0,"
390 PRINT"IF NEWTON METHOD WITH QUADRATIC CONVERGENCE USED O4>0."
400 INPUT"O4";O4:CLS
410 PRINT"ADVICE: COMPUTE ALL SIMPLE ROOTS FIRST AND THEN"
420 PRINT"THE MULTIPLE ONES.":PRINT
430 IF O2=0 THEN GOTO 460
440 LPRINT "P605: O1=";O1;"O2=";O2;"O3=";O3;"O4=";O4
450 LPRINT "M=";M;"R=";R
460 IF O3=0 THEN GOTO 880
470 IF O2>0 THEN LPRINT "DEGREE OF THE POLYNOMIAL N=";N
480 LET N1=N+1
490 LET N2=N-1
500 DIM A(N1),A1(N1),B(N1),C(N),D(N2)
510 PRINT:PRINT "THE INPUT A(I):"
520 IF O2>0 THEN LPRINT "THE INPUT A(I):"
530 IF O3>1 THEN GOTO 700
540 FOR I=N1 TO 1 STEP -1
550 LET J=I-1
560 PRINT "POWER=";J;"I=";I
570 INPUT"A(I)";A(I)
580 NEXT I
590 CLS:PRINT "CHECK A(I):"
600 FOR I=N1 TO 1 STEP -1
610 LET J=I-1
```

```
620 PRINT "POWER=";J;"I=";I;"A(I)=";A(I)
630 PRINT"IF NO ERROR INPUT 0"
640 PRINT"IF ERROR INPUT ANY NUMBER>0 AND THEN THE CORRECT A(I)"
650 INPUT"ERROR";K
660 IF K>0 THEN INPUT"A(I)";A(I)
670 CLS:NEXT I
680 GOTO 790
690 REM READING THE INPUT FROM A FILE
700 OPEN "I",1,"P604A"
710 FOR I=N1 TO 1 STEP -1
720 INPUT #1, A(I)
730 LET J=I-1
740 PRINT "POWER=";J;"I=";I;"A(I)=";A(I)
750 IF O2>0 THEN LPRINT "POWER=";J;"I=";I;"A(I)=";A(I)
760 NEXT I
770 CLOSE 1
780 REM WRITING COEFFICIENTS A(I) INTO MEMORY
790 FOR I=1 TO N1
800 LET A1(I)=A(I)
810 NEXT I
820 GOTO 880
830 REM READING COEFFICIENTS A(I) FROM MEMORY
840 FOR I=1 TO N1
850 LET A(I)=A1(I)
860 NEXT I
870 REM COMPUTATION OF REAL ROOTS
880 IF O2>0 THEN LPRINT
890 IF O3>0 THEN GOTO 910
900 PRINT "WHEN ALL ROOTS COMPUTED PRESS BREAK TO STOP":PRINT
910 LET I1=1
920 LET I2=2
930 GOTO 980
940 PRINT "MAKE ANOTHER INITIAL ESTIMATE"
950 IF O2=0 THEN GOTO 980
960 LPRINT "AN IMPROPER INITIAL ESTIMATE. RERUN"
970 LPRINT
980 PRINT "INITIAL ESTIMATE OF X:"
990 INPUT"X";X
1000 IF O2=1 THEN LPRINT "INITIAL ESTIMATE X=";X
1010 LET K=0
1020 IF O2>1 THEN LPRINT "F,F1,F2 AND X AT EACH ITERATION:"
1030 GOSUB 1640
1040 IF K=M THEN GOTO 940
1050 IF ABS(F)<R THEN GOTO 1260
1060 REM ITERATION PROCESS
1070 LET K=K+1
1080 PRINT "F=";F;"F1=";F1;"F2=";F2;"X=";X
1090 IF O2<2 THEN GOTO 1110
1100 LPRINT "F=";F;"F1=";F1;"F2=";F2;"X=";X
1110 IF O4=0 THEN GOTO 1140
1120 LET X=X-F/F1
1130 GOTO 1030
1140 LET S0=F1*F1-2*F*F2
1150 IF S0>0 THEN GOTO 1180
1160 LET X=X-2*F/F1
1170 GOTO 1030
1180 LET S=SQR(S0)
1190 IF O1=1 THEN GOTO 1240
1200 LET S1=S/1.D16
1210 LET S2=S
1220 LET S=(S+S0/S)/2
1230 IF ABS(S-S2)>S1 THEN GOTO 1210
1240 LET X=X-2*F/(F1+S*SGN(F1))
1250 GOTO 1030
```

PROGRAM P605

```
1260 PRINT "THE ROOT=";X;"F(X)=";F;"NUMBER OF ITERATIONS:";K
1270 PRINT
1280 IF O2=0 THEN GOTO 1310
1290 LPRINT "THE ROOT=";X;"F(X)=";F;"NUMBER OF ITERATIONS:";K
1300 LPRINT
1310 IF I1=N THEN GOTO 1890
1320 LET K=0
1330 IF O3=0 THEN GOTO 980
1340 PRINT "WHEN ALL REAL ROOTS ARE COMPUTED LET O5=0."
1350 PRINT "IF O5=1 A FURTHER ROOT IS COMPUTED FROM ORIGINAL EQ."
1360 PRINT "BUT O5=1 MUST NOT BE CHOSEN ONCE O5>1 HAS BEEN USED."
1370 PRINT "IF O5>1 A FURTHER ROOT COMPUTED FROM DEFLATED EQ."
1380 INPUT"O5";O5
1390 IF O5=0 THEN GOTO 1600
1400 IF O5=1 THEN GOTO 980
1410 REM THE DEFLATION
1420 PRINT "COEFFICIENTS OF THE DEFLATED POLYNOMIAL:"
1430 PRINT "I=";N1;"A(I)=";A(N1)
1440 IF O2<2 THEN GOTO 1480
1450 REM THE DEFLATION
1460 LPRINT "COEFFICIENTS OF THE DEFLATED POLYNOMIAL:"
1470 LPRINT "I=";N1;"A(I)=";A(N1)
1480 FOR I=N TO I1 STEP -1
1490 LET J=I+1
1500 LET A(I)=A(J)*X+A(I)
1510 PRINT "I=";I;"A(I)=";A(I)
1520 IF O2>1 THEN LPRINT "I=";I;"A(I)=";A(I)
1530 NEXT I
1540 LET I1=I1+1
1550 LET I2=I2+1
1560 IF O5=0 THEN GOTO 1600
1570 PRINT
1580 IF O2>1 THEN LPRINT
1590 GOTO 980
1600 PRINT "ALL REAL ROOTS ARE COMPUTED. THE END"
1610 IF O2>0 THEN LPRINT "ALL REAL ROOTS COMPUTED. THE END"
1620 STOP
1630 REM SUBROUTINE FOR F(X),F1(X) AND F2(X)
1640 IF O3=0 THEN GOTO 1850
1650 LET B(N1)=A(N1)
1660 LET C(N)=B(N1)
1670 LET D(N2)=C(N)
1680 FOR I=N TO I1 STEP -1
1690 LET J=I+1
1700 LET L1=I-1
1710 LET L2=I-2
1720 LET B(I)=B(J)*X+A(I)
1730 IF I=I1 THEN GOTO 1770
1740 LET C(L1)=C(I)*X+B(I)
1750 IF I=I2 THEN GOTO 1770
1760 LET D(L2)=D(L1)*X+C(L1)
1770 NEXT I
1780 LET F=B(I1)
1790 LET F1=C(I1)
1800 IF I1<N THEN GOTO 1830
1810 LET F2=0
1820 RETURN
1830 LET F2=2*D(I1)
1840 RETURN
1850 LET F=
1860 LET F1=
1870 LET F2=
1880 RETURN
1890 PRINT "ALL ROOTS OF THE POLYNOMIAL COMPUTED."
```

```
1900 PRINT "IF THE ROOTS ARE TO BE RECOMPUTED FROM THE ORIGINAL"
1910 PRINT"POLYNOMIAL LET O5=1."
1920 INPUT"O5";O5
1930 IF O5<>1 THEN GOTO 2000
1940 PRINT
1950 PRINT "ALL ROOTS ARE RECOMPUTED FROM ORIGINAL POLYNOMIAL:"
1960 IF O2=0 THEN GOTO 840
1970 LPRINT
1980 LPRINT "ALL ROOTS ARE RECOMPUTED FROM ORIGIVAL POLYNOMIAL:"
1990 GOTO 840
2000 IF O2>0 THEN LPRINT "ALL ROOTS OF THE POLYNOMIAL COMPUTED."
2010 END
```

Examples

1. The real roots of an algebraic equation
$$x^5-15.5x^4+77.5x^3-155x^2+124x-32=0$$
are computed in single precision by the method with global cubic convergence. Once a root is found, the equation is deflated (maximum number of iterations M=10, maximum error $R=1*10^{-5}$). The initial estimate X0 is taken from the results of example 1 to P604; the roots x_i (i=1,2,...5) and the number of iterations N are:

   X0=0.55, x_1=0.5,   N=2,
   X0=1.1, x_2=1,   N=2,
   X0=2.1, x_3=2,   N=2,
   X0=4.1, x_4=4,   N=1,
   X0=7.9, x_5=8,   N=1.

All five roots are then recomputed from the original equation. The results are identical.

2. The roots of the same algebraic equation are computed in single precision by Newton's method with quadratic convergence from successively deflated polynomials. The results are identical, but the number of iterations is higher: N=3, 3, 3, 2, 1.

3. The real roots of the transcendental equation
$$\sin(x)-x/25=0$$
are computed in single precision by the method with global cubic convergence. Maximum number of iterations M=25, maximum error

P. 200 PROGRAM P605

$R=1*10^{-6}$ The initial estimates X0 are taken from example 2 to P604; the roots x_i (i=1,2,...8) are:

  X0=0, x_1=0,   N=0,
  X0=3, x_2=3.02048, N=2,
  X0=6.5, x_3=6.54821, N=2,
  X0=9, x_4=9.05419, N=2,
  X0=13, x_5=13.1188, N=2,
  X0=15, x_6=15.0614, N=2,
  X0=19.75, x_7=19.7611, N=1,
  X0=21, x_8=20.9943, N=1.

Because the transcendental equation is odd, for each positive root there exists a negative one of the same absolute value.

P606 Real and Complex Roots of an Algebraic Equation (with Deflation Subroutine)

```
10 CLS:PRINT"P606 COMPUTES REAL AND COMPLEX ROOTS OF A POLYNOMIAL"
20 PRINT"BY ITERATION METHOD WITH GLOBAL CUBIC CONVERGENCE."
30 PRINT:PRINT"THE INPUT:"
40 PRINT"IF SINGLE PRECISION O1=1, IF DOUBLE O1=2."
50 DEFINT I-O:INPUT"O1";O1:CLS
60 IF O1=2 THEN DEFDBL A-D,F,G,S-Y
70 PRINT"IF PRINTER NOT USED O2=0, IF ONLY FINAL RESULTS"
80 PRINT"PRINTED O2=1, IF RESULTS AT EACH STEP PRINTED O2>1."
90 INPUT"O2";O2:CLS
100 PRINT"NUMBER OF REAL ROOTS: M0. IF M0=0 ALL ROOTS ARE"
110 PRINT"COMPUTED IN COMPLEX ARITHMETIC. IF M0>0 REAL ROOTS"
120 PRINT"ARE LOCATED AND THEN COMPUTED IN REAL ARITHMETIC."
130 INPUT"M0";M0:CLS:IF M0=0 THEN GOTO 170
140 PRINT"INTERVAL OF LOCATION PROCESS: (E1,E2)"
150 PRINT"SPACING IN LOCATION PROCESS: H"
160 INPUT"E1";E8:INPUT"E2";E9:INPUT"H";H:CLS
170 PRINT"THE ITERATION STOPS AND THE ROOT IS TO BE RECOMPUTED"
180 PRINT"FROM ANOTHER INITIAL ESTIMATE WHEN THE MAXIMUM NUMBER"
190 PRINT"OF ITERATIONS M IS REACHED.":PRINT
200 PRINT"THE ITERATION STOPS WHEN ABS(F(Z))<R, Z BEING"
210 PRINT"A COMPLEX ROOT WHOSE REAL PART IS X AND IMAGINARY"
220 PRINT"PART Y, AND R A CHOSEN CONSTANT."
230 INPUT"M";M:INPUT"R";R:CLS
240 PRINT"DEGREE OF THE POLYNOMIAL: N"
250 PRINT"THERE ARE (N+1) COEFFICIENTS A(I) IN THE POLYNOMIAL,"
260 PRINT"COEFFICIENT A(I) STANDS AT (I-1)-ST POWER OF X."
270 INPUT"N";N:CLS:IF M0=0 THEN GOTO 380
280 PRINT"AFTER ALL REAL ROOTS HAVE BEEN LOCATED THE PROGRAM"
290 PRINT"ASKS FOR THEIR NUMBER M0. NEVER INCLUDE THE ROOTS"
300 PRINT"THAT ARE NOT CERTAIN; THE OMITTED ONES WILL BE"
310 PRINT"DETERMINED IN COMPLEX ARITHMETIC.":PRINT
320 PRINT"USE THE RESULTS FROM LOCATION PROCESS AS INITIAL"
330 PRINT"ESTIMATES AND COMPUTE THE ROOTS FROM DEFLATED EQUATION."
340 PRINT"WHEN ALL ROOTS ARE REAL INCREASE THE ACCURACY BY"
350 PRINT"USING THEM AS INITIAL ESTIMATES AND RECOMPUTE THEM"
360 PRINT"FROM THE ORIGINAL EQUATION IN REAL ARITHMETIC.":PRINT
370 PRINT"TO CONTINUE PRESS ENTER.":INPUT A$:CLS
380 PRINT"WHEN SOME ROOTS ARE COMPLEX OR WHEN IN THE INPUT M0"
390 PRINT"CHOSEN ZERO AND THE LOCATION PROCESS IS SKIPPED, ALL"
400 PRINT"ROOTS ARE COMPUTED IN COMPLEX ARITHMETIC. REAL PART OF"
410 PRINT"THE POLYNOMIAL AND ITS DERIVATIVES F,F1,F2. IMAGINARY"
420 PRINT"PART G,G1,G2. AS INITIAL ESTIMATE PUT X CLOSE TO ZERO"
430 PRINT"AND Y=0. ITERATION ALWAYS CONVERGES TO SOME ROOT."
440 PRINT:PRINT"TO CONTINUE PRESS ENTER.":INPUT A$:CLS
450 PRINT"ONCE A ROOT IS FOUND THE EQUATION IS DEFLATED. WHEN"
460 PRINT"ALL ROOTS ARE DETERMINED THEIR EXACT VALUES ARE USED"
470 PRINT"AS INITIAL ESTIMATES AND RECOMPUTED FROM THE ORIGINAL"
480 PRINT"EQUATION IN COMPLEX ARITHMETIC.":PRINT
490 PRINT"IF THIS ITERATION CONVERGES TO A ROOT THAT HAS BEEN"
500 PRINT"RECOMPUTED CHOOSE O6>0 AND RECOMPUTE IT AGAIN WITH"
510 PRINT"THE SAME INITIAL ESTIMATE.":PRINT
520 PRINT"ADVICE: COMPUTE ALL SIMPLE ROOTS FIRST AND THEN"
530 PRINT"THE MULTIPLE ONES.":PRINT
540 IF O2=0 THEN GOTO 570
550 LPRINT "P606: O1=";O1;"O2=";O2;"M0=";M0;"M=";M;"R=";R
560 LPRINT "DEGREE OF THE POLYNOMIAL N=";N
570 LET N1=N+1
580 LET N2=N-1
590 LET K0=1
600 DIM A(N1),A0(N1),B1(N1),B2(N1),C1(N),C2(N),D1(N2),D2(N2)
610 DIM E2(N1),E3(N1),E4(N)
```

PROGRAM P606

```
620 PRINT "THE INPUT A(I):"
630 IF O2>0 THEN LPRINT "THE INPUT A(I):"
640 FOR I=N1 TO 1 STEP -1
650 LET J=I-1
660 PRINT "POWER=";J;"I=";I
670 INPUT"A(I)";A(I)
680 NEXT I
690 CLS:PRINT "CHECK A(I):"
700 FOR I=N1 TO 1 STEP -1
710 LET J=I-1
720 PRINT "POWER=";J;"I=";I;"A(I)=";A(I)
730 PRINT"IF NO ERROR INPUT 0"
740 PRINT"IF ERROR INPUT ANY NUMBER>0 AND THEN THE CORRECT A(I)"
750 INPUT"ERROR";K
760 IF K>0 THEN INPUT"A(I)";A(I)
770 CLS:NEXT I
780 FOR I=N1 TO 1 STEP -1
790 LET J=I-1
800 IF O2>0 THEN LPRINT "POWER=";J;"I=";I;"A(I)=";A(I)
810 LET A0(I)=A(I)
820 LET E2(I)=A(I)
830 NEXT I
840 IF O2>0 THEN LPRINT
850 IF M0=0 THEN GOTO 1610
860 PRINT:PRINT "LOCATION OF REAL ROOTS:"
870 IF O2>0 THEN LPRINT "LOCATION OF REAL ROOTS:"
880 IF O2>0 THEN LPRINT "E1=";E8;"E2=";E9;"H=";H
890 LET Z=E8
900 LET I0=0
910 GOSUB 1310
920 LET Z1=Z
930 LET E5=E
940 LET E6=E1
950 LET Z=Z+H
960 GOSUB 1310
970 PRINT "X=";Z;"F(X)=";E;"F1(X)=";E1
980 IF O2>1 THEN LPRINT "X=";Z;"F(X)=";E;"F1(X)=";E1
990 IF SGN(E)=-SGN(E5) THEN GOTO 1130
1000 IF SGN(E6)=SGN(E1) THEN GOTO 1250
1010 IF SGN(E)=-SGN(E1) THEN GOTO 1250
1020 IF I0<2 THEN GOTO 1250
1030 PRINT "A MULTIPLE ROOT BETWEEN X1=";Z1;"AND X2=";Z;"?"
1040 PRINT "THE QUESTION MARK INDICATES THAT A MULTIPLE ROOT"
1050 PRINT "MAY OCCUR IN THE INTERVAL. IT IS PLAUSIBLE THAT"
1060 PRINT "IT DOES OCCUR ONLY IF F(X) IS CLOSE TO ZERO."
1070 IF O2=0 THEN GOTO 1160
1080 LPRINT "A MULTIPLE ROOT BETWEEN X1=";Z1;"AND X2=";Z;"?"
1090 LPRINT "THE QUESTION MARK INDICATES THAT A MULTIPLE ROOT"
1100 LPRINT "MAY OCCUR IN THE INERVAL. IT IS PLAUSIBLE THAT"
1110 LPRINT "IT DOES OCCUR ONLY IF F(X) IS CLOSE TO ZERO."
1120 GOTO 1160
1130 IF I0<2 THEN GOTO 1250
1140 PRINT "ROOT BETWEEN X1=";Z1;"AND X2=";Z
1150 IF O2>0 THEN LPRINT "ROOT BETWEEN X1=";Z1;"AND X2=";Z
1160 PRINT "X1=";Z1;"F(X1)=";E5;"F1(X1)=";E6
1170 PRINT "X2=";Z;"F(X2)=";E;"F1(X2)=";E1
1180 IF O2=0 THEN GOTO 1230
1190 LPRINT "X1=";Z1;"F(X1)=";E5;"F1(X1)=";E6
1200 LPRINT "X2=";Z;"F(X2)=";E;"F1(X2)=";E1
1210 LPRINT
1220 GOTO 1250
1230 PRINT "TO CONTINUE PRESS ENTER"
1240 INPUT A$
1250 LET Z1=Z
```

```
1260 LET E5=E
1270 LET E6=E1
1280 IF Z>E9 THEN GOTO 1570
1290 GOTO 950
1300 REM SUBROUTINE FOR F(X) AND F1(X)
1310 LET E3(N1)=E2(N1)
1320 LET E4(N)=E3(N1)
1330 FOR I=N TO 1 STEP -1
1340 LET J=I+1
1350 LET K=I-1
1360 LET E3(I)=E3(J)*Z+E2(I)
1370 IF I=1 THEN GOTO 1390
1380 LET E4(K)=E4(I)*Z+E3(I)
1390 NEXT I
1400 LET E=E3(1)
1410 LET E1=E4(1)
1420 LET I0=I0+1
1430 IF E<>0 THEN RETURN
1440 LET I0=0
1450 IF E1<>0 THEN GOTO 1510
1460 PRINT "A MULTIPLE ROOT X=";Z
1470 IF O2=0 THEN PRINT "TO CONTINUE PRESS ENTER"
1480 IF O2=0 THEN INPUT A$
1490 IF O2>0 THEN LPRINT "A MULTIPLE ROOT X=";Z
1500 RETURN
1510 PRINT "ROOT X=";Z
1520 IF O2=0 THEN PRINT "TO CONTINUE PRESS ENTER"
1530 IF O2=0 THEN INPUT A$
1540 IF O2>0 THEN LPRINT "ROOT X=";Z
1550 RETURN
1560 REM COMPUTATION OF REAL ROOTS
1570 IF O2>0 THEN LPRINT
1580 PRINT "HOW MANY REAL ROOTS?"
1590 INPUT M0
1600 LET Y=0
1610 LET I0=0
1620 LET I3=0
1630 LET I1=1
1640 LET I2=2
1650 IF M0>0 THEN GOTO 1720
1660 LET O5=4
1670 GOTO 2170
1680 PRINT "MAKE ANOTHER INITIAL ESTIMATE"
1690 IF O2=0 THEN GOTO 1720
1700 LPRINT "AN IMPROPER INITIAL ESTIMATE. RERUN"
1710 LPRINT
1720 PRINT "INITIAL ESTIMATE OF X:"
1730 INPUT"X";X
1740 IF O2=1 THEN LPRINT "INITIAL ESTIMATE X=";X
1750 LET K=0
1760 IF O2>1 THEN LPRINT "F,F1,F2 AND X AT EACH ITERATION:"
1770 GOSUB 3420
1780 IF K=M THEN GOTO 1680
1790 IF ABS(F)<R THEN GOTO 1930
1800 REM ITERATION PROCESS IN REAL ARITHMETIC
1810 LET K=K+1
1820 PRINT "F=";F;"F1=";F1;"F2=";F2;"X=";X
1830 IF O2<2 THEN GOTO 1850
1840 LPRINT "F=";F;"F1=";F1;"F2=";F2;"X=";X
1850 LET W0=F1*F1-2*F*F2
1860 IF W0>0 THEN GOTO 1890
1870 LET X=X-2*F/F1
1880 GOTO 1770
1890 LET W=SQR(W0)
```

PROGRAM P606

```
1900 IF O1=2 THEN GOSUB 3740
1910 LET X=X-2*F/(F1+W*SGN(F1))
1920 GOTO 1770
1930 PRINT "ROOT=";X;"F(X)=";F;"NUMBER OF ITERATIONS:";K
1940 LET I3=I3+1
1950 PRINT
1960 IF O2=0 THEN GOTO 1990
1970 LPRINT "ROOT=";X;"F(X)=";F;"NUMBER OF ITERATIONS:";K
1980 LPRINT
1990 IF I3<N THEN GOTO 2090
2000 IF I0>0 THEN GOTO 3790
2010 LET I0=1
2020 PRINT "ALL ROOTS ARE REAL"
2030 PRINT "RECOMPUTE THEM FROM ORIGINAL EQUATION"
2040 IF O2>0 THEN LPRINT "ROOTS FROM ORIGINAL EQUATION:"
2050 FOR I=1 TO N1
2060 LET A(I)=A0(I)
2070 NEXT I
2080 GOTO 1620
2090 LET K=0
2100 PRINT "IF FURTHER ROOT FROM ORIGINAL EQUATION O5=1"
2110 PRINT "IF FURTHER ROOT FROM DEFLATED EQUATION O5=2"
2120 INPUT"O5";O5
2130 IF I3=M0 THEN LET O5=3
2140 IF O5=1 THEN GOTO 1630
2150 GOSUB 3050
2160 IF O5=2 THEN GOTO 1720
2170 LET R=R*R
2180 LET I0=0
2190 REM COMPUTATION OF COMPLEX ROOTS
2200 IF O2>0 THEN LPRINT
2210 GOTO 2260
2220 PRINT "MAKE ANOTHER INITIAL ESTIMATE"
2230 IF O2=0 THEN GOTO 2260
2240 LPRINT "AN IMPROPER INITIAL ESTIMATE. RERUN"
2250 LPRINT
2260 PRINT "INITIAL ESTIMATE OF REAL AND IMAGINARY PART X,Y:"
2270 INPUT"X";X:INPUT"Y";Y
2280 IF O2=1 THEN LPRINT "INITIAL ESTIMATE X=";X;"Y=";Y
2290 LET K=0
2300 IF O2>1 THEN LPRINT "F,F1,F2,X;G,G1,G2,Y AT EACH ITERATION:"
2310 GOSUB 3420
2320 IF K=M THEN GOTO 2220
2330 LET F0=F*F+G*G
2340 IF F0<R THEN GOTO 2710
2350 REM ITERATION PROCESS IN COMPLEX ARITHMETIC
2360 LET K=K+1
2370 PRINT "F=";F;"F1=";F1;"F2=";F2;"X=";X
2380 PRINT "    G=";G;"G1=";G1;"G2=";G2;"Y=";Y
2390 IF O2<2 THEN GOTO 2420
2400 LPRINT "F=";F;"F1=";F1;"F2=";F2;"X=";X
2410 LPRINT "    G=";G;"G1=";G1;"G2=";G2;"Y=";Y
2420 LET S0=F1*F1-G1*G1-2*(F*F2-G*G2)
2430 LET T0=2*(F1*G1-F2*G-F*G2)
2440 IF T0=0 THEN GOTO 2570
2450 LET W0=S0*S0+T0*T0
2460 LET W=SQR(W0)
2470 IF O1=2 THEN GOSUB 3740
2480 LET W0=(W+ABS(S0))/2
2490 LET W=SQR(W0)
2500 IF O1=2 THEN GOSUB 3740
2510 IF S0<0 THEN GOTO 2550
2520 LET F1=F1-W*K0
2530 LET G1=G1-K0*T0/(2*W)
```

```
2540 GOTO 2650
2550 LET F1=F1-K0*T0/(2*W)
2560 GOTO 2640
2570 LET W0=ABS(G0)
2580 LET W=SQR(W0)
2590 IF O1=2 THEN GOSUB 3740
2600 IF S0<0 THEN GOTO 2640
2610 LET F1=F1-W*K0
2620 IF F1=0 THEN GOTO 2690
2630 GOTO 2650
2640 LET G1=G1-W*K0
2650 LET G0=(F1*F1+G1*G1)/2
2660 LET X=X-(F*F1+G*G1)/G0
2670 LET Y=Y-(F1*G-F*G1)/G0
2680 GOTO 2310
2690 LET X=X-K0*F/W
2700 GOTO 2310
2710 PRINT "ROOT: X=";X;"Y=";Y
2720 PRINT "F(Z)=";F;"G(Z)=";G;"NUMBER OF ITERATIONS:";K
2730 PRINT
2740 IF I0<2 THEN GOTO 2810
2750 PRINT "IF THE ROOT IS RECOMPUTED FOR THE FIRST TIME O6=0"
2760 PRINT "IF THE SAME ROOT IS RECOMPUTED ONCE AGAIN O6><0"
2770 INPUT"O6";O6
2780 IF O6=0 THEN GOTO 2810
2790 LET K0=-K0
2800 GOTO 2260
2810 IF O2=0 THEN GOTO 2850
2820 LPRINT "ROOT: X=";X;"Y=";Y
2830 LPRINT "F(Z)=";F;"G(Z)=";G;"NUMBER OF ITERATIONS:";K
2840 LPRINT
2850 LET K=0
2860 PRINT "IF JUST COMPUTED ROOT IS REAL O5=3"
2870 PRINT "IF IT IS COMPLEX O5=4"
2880 INPUT"O5";O5
2890 LET I3=I3+1
2900 IF O5=4 THEN LET I3=I3+1
2910 IF I3<N THEN GOTO 3040
2920 IF I0>0 THEN GOTO 3790
2930 PRINT "ALL ROOTS ARE COMPUTED"
2940 PRINT "RECOMPUTE THEM FROM ORIGINAL EQUATION"
2950 LET I0=2
2960 LET I1=1
2970 LET I2=2
2980 LET I3=0
2990 FOR I=1 TO N1
3000 LET A(I)=A0(I)
3010 NEXT I
3020 IF O2>0 THEN LPRINT "ROOTS FROM ORIGINAL EQUATION"
3030 GOTO 2260
3040 IF I0=2 THEN GOTO 2260
3050 PRINT "COEFFICIENTS OF THE DEFLATED POLYNOMIAL:"
3060 PRINT "I=";N1;"A(I)=";A(N1)
3070 IF O2<2 THEN GOTO 3100
3080 LPRINT "COEFFICIENTS OF THE DEFLATED POLYNOMIAL:"
3090 LPRINT "I=";N1;"A(I)=";A(N1)
3100 IF O5=4 THEN GOTO 3260
3110 REM THE DEFLATION BY A REAL ROOT
3120 FOR I=N TO I1 STEP -1
3130 LET J=I+1
3140 LET A(I)=A(J)*X+A(I)
3150 PRINT "I=";I;"A(I)=";A(I)
3160 IF O2>1 THEN LPRINT "I=";I;"A(I)=";A(I)
3170 NEXT I
```

PROGRAM P606

```
3180 LET I1=I1+1
3190 LET I2=I2+1
3200 PRINT
3210 IF I3=M0 THEN RETURN
3220 IF O2>1 THEN LPRINT
3230 IF O5<3 THEN RETURN
3240 GOTO 2260
3250 REM THE DEFLATION BY A COMPLEX ROOT
3260 LET U=X+X
3270 LET V=-X*X-Y*Y
3280 LET A(N)=U*A(N1)+A(N)
3290 PRINT "I=";N;"A(I)=";A(N)
3300 IF O2>1 THEN LPRINT "I=";N;"A(I)=";A(N)
3310 FOR I=N2 TO I1 STEP -1
3320 LET L1=I+1
3330 LET L2=I+2
3340 LET A(I)=V*A(L2)+U*A(L1)+A(I)
3350 PRINT "I=";I;"A(I)=";A(I)
3360 IF O2>1 THEN LPRINT "I=";I;"A(I)=";A(I)
3370 NEXT I
3380 LET I1=I1+2
3390 LET I2=I2+2
3400 GOTO 3200
3410 REM SUBROUTINE FOR F(Z),F1(Z),F2(Z) AND G(Z),G1(Z),G2(Z)
3420 LET B1(N1)=A(N1)
3430 IF I0=2 THEN LET O5=4
3440 IF O5>2 THEN LET B2(N1)=0
3450 LET C1(N)=B1(N1)
3460 IF O5>2 THEN LET C2(N)=0
3470 LET D1(N2)=C1(N)
3480 IF O5>2 THEN LET D2(N2)=0
3490 FOR I=N TO I1 STEP -1
3500 LET J=I+1
3510 LET L1=I-1
3520 LET L2=I-2
3530 LET B1(I)=B1(J)*X-B2(J)*Y+A(I)
3540 IF O5>2 THEN LET B2(I)=B2(J)*X+B1(J)*Y
3550 IF I=I1 THEN GOTO 3610
3560 LET C1(L1)=C1(I)*X-C2(I)*Y+B1(I)
3570 IF O5>2 THEN LET C2(L1)=C2(I)*X+C1(I)*Y+B2(I)
3580 IF I=I2 THEN GOTO 3610
3590 LET D1(L2)=D1(L1)*X-D2(L1)*Y+C1(L1)
3600 IF O5>2 THEN LET D2(L2)=D2(L1)*X+D1(L1)*Y+C2(L1)
3610 NEXT I
3620 LET F=B1(I1)
3630 IF O5>2 THEN LET G=B2(I1)
3640 LET F1=C1(I1)
3650 IF O5>2 THEN LET G1=C2(I1)
3660 IF I1<N THEN GOTO 3700
3670 LET F2=0
3680 LET G2=0
3690 RETURN
3700 LET F2=2*D1(I1)
3710 IF O5>2 THEN LET G2=2*D2(I1)
3720 RETURN
3730 REM SUBROUTINE FOR DOUBLE PRECISION SQR(W0)
3740 LET W9=W/1.D16
3750 LET W8=W
3760 LET W=(W+W0/W)/2
3770 IF ABS(W-W8)>W9 THEN GOTO 3750
3780 RETURN
3790 PRINT "ALL ROOTS COMPUTED"
3800 IF O2>0 THEN LPRINT "ALL ROOTS COMPUTED"
3810 END
```

Examples

1. The roots of the biquadratic equation (cf. example 1 to P603)

$$x^4 - 10x^3 + 35x^2 - 50x + 24 = 0$$

are computed in complex arithmetic (M0=0) in single precision from successively deflated polynomials. Maximum number of iterations M=15, maximum error $R = 1*10^{-5}$. The initial estimate for all roots is X=0, Y=0. The roots x_i (i=1,...4) and the number of iterations N are:

$x_1 = 1$, $\quad y_1 = -5.82077*10^{-11}$, $\quad$ N=4,
$x_2 = 3$, $\quad y_2 = -2.27374*10^{-13}$, $\quad$ N=6,
$x_3 = 2$, $\quad y_3 = 0$, $\quad$ N=1,
$x_4 = 4$, $\quad y_4 = 0$, $\quad$ N=1.

Since the imaginary parts y_i are far below the maximum error R, they are likely to have been generated by roundoff errors. All four roots are recomputed from the original equation, which confirms that they are real: 1, 2, 3, 4.

2. The roots of the biquadratic equation (cf. example 3 to P603)

$$x^4 - 2x^3 + 6x^2 - 2x + 5 = 0$$

are computed in complex arithmetic (M0=0) in single precision from succesively deflated polynomials. Maximum number of iterations M=15, maximum error $R = 4*10^{-6}$. The initial estimate for all roots is X=0, Y=0. The roots:

$x_1 = -4.14439*10^{-8}$, $\quad y_1 = -1$, $\quad$ N=3,
$x_3 = 1$, $\quad y_3 = -2$, $\quad$ N=1.

The roots are complex conjugate. Recomputation from the original equation confirms that

$Re(x_{1,2}) = 0$, $\quad Im(x_{1,2}) = +-1$,
$Re(x_{3,4}) = 1$, $\quad Im(x_{3,4}) = +-2$.

3. The roots of a polynomial equation of eighth degree $x^8 - 1 = 0$ are first located and then computed in real and complex arithmetic (M0>0) in double precision from successively deflated polynomials. Maximum number of iterations M=15, maximum error $R = 4*10^{-11}$. The preliminary location of real roots yields:

$-1.1 < x_1 < -0.85$,
$0.9 < x_2 < 1.15$.

The computation in real arithmetic with initial estimates -0.9, 0.9 yields (N is the number of iterations)

$x_1 = -1$, $\quad$ N=3,
$x_2 = 1$, $\quad$ N=3.

P. 208   PROGRAM P606

The computation in complex arithmetic with initial estimates for all roots X=0.1, Y=0 yields

$x_{3,4}=3.252049830462919*10^{-12}$, $y_{3,4}=-0.9999999999985947$, N=3,
$x_{5,6}=0.7071067811839278$, $y_{5,6}=-0.7071067811849215$, N=8,
$x_{7,8}=-0.7071067811871799$, $y_{7,8}=-0.7071067811881736$, N=1.

All 8 roots are recomputed from the original equation. The final values rounded to 10 decimal places are:

$Re(x_1)=-1$, $Im(x_1)=0$,
$Re(x_2)=1$, $Im(x_2)=0$,
$Re(x_{3,4})=0$, $Im(x_{3,4})=+-1$,
$Re(x_{5,6})=0.7071067812$, $Im(x_{5,6})=+-0.7071067812$,
$Re(x_{7,8})=-0.7071607812$, $Im(x_{7,8})=+-0.7071067812$.

4. The roots of a biquadratic equation
$$x^4-16x^3+90x^2-200x+125=0$$
are first located and then computed in single precision, M=15, R=4*10^{-5}. If the location process starts from E1=0 with the step H=0.25, it indicates one root at x=1 and a multiple root at x=5. Since the complex roots must be conjugate, the equation has 4 real roots. The computation with the initial estimates 1, 5, 5, 5 in the deflated equation gives the correct solution, $x_1=1$, $x_{2,3,4}=5$, but if the initial estimate of the multiple root is not close enough, the computation results in an inaccurate value, either real or complex conjugate. Therefore use a very small step in the location process, and when it turns out that a root may be multiple, compute it from the original equation.

P607 Real Roots of Two Nonlinear Equations by Fixed-Point Iteration

```
10 CLS:PRINT"P607 COMPUTES REAL ROOTS OF TWO NONLINEAR EQUATIONS"
20 PRINT"BY FIXED-POINT ITERATION.":PRINT
30 PRINT"PRESS BREAK AND DEFINE TWO IERATION FUNCTIONS F,G"
40 PRINT"ON LINES 400,410."
50 PRINT:PRINT"THE INPUT:"
60 PRINT"SINGLE PRECISION ONLY."
70 PRINT"IF PRINTER NOT USED O2=0, IF ONLY FINAL RESULTS"
80 PRINT"PRINTED O2=1, IF RESULTS AT EACH STEP PRINTED O2>1."
90 DEFINT I-O:INPUT"O2";O2:CLS
100 PRINT"THE ITERATION STOPS AND THE ROOTS ARE TO BE RECOMPUTED"
110 PRINT"FROM ANOTHER INITIAL ESTIMATE WHEN THE MAXIMUM NUMBER"
120 PRINT"OF ITERATIONS M IS REACHED.":PRINT
130 PRINT"THE ITERATION STOPS WHEN ABS(F-X)<=R AND ABS(G-Y)<R,"
140 PRINT"X,Y BEING THE ROOTS AND R A CHOSEN CONSTANT."
150 INPUT"M";M:INPUT"R";R:CLS
160 IF O2=0 THEN GOTO 240
170 LPRINT "P607: O2=";O2;"M=";M;"R=";R
180 GOTO 240
190 REM COMPUTATION OF ROOTS
200 PRINT "MAKE ANOTHER INITIAL ESTIMATE"
210 IF O2=0 THEN GOTO 240
220 LPRINT "AN IMPROPER INITIAL ESTIMATE. RERUN"
230 LPRINT
240 PRINT "INITIAL ESTIMATE OF X AND Y:"
250 INPUT"X";X:INPUT"Y";Y
260 IF O2>0 THEN LPRINT "INITIAL ESTIMATE X=";X;"Y=";Y
270 LET K=0
280 IF O2>1 THEN LPRINT "F,G AT EACH ITERATION K:"
290 GOSUB 400
300 IF K=M THEN GOTO 200
310 IF ABS(F-X)>R THEN GOTO 330
320 IF ABS(G-Y)<R THEN GOTO 430
330 PRINT "F=";F;"    G=";G;"    K=";K
340 IF O2>1 THEN LPRINT "F=";F;"    G=";G;"    K=";K
350 LET X=F
360 LET Y=G
370 LET K=K+1
380 GOTO 290
390 REM SUBROUTINE: EQUATIONS FOR FIXED-POINT ITERATIONS
400 LET F=
410 LET G=
420 RETURN
430 PRINT "THE SOLUTION: X=";F;"Y=";G;"NUMBER OF ITERATIONS:";K
440 IF O2=0 THEN STOP
450 LPRINT "THE SOLUTION: X=";F;"Y=";G
460 LPRINT "NUMBER OF ITERATIONS:";K
470 END
```

Example

A system of two nonlinear equations
$$x^2+y^2-10x+8=0,$$
$$x-10y+xy^2+8=0$$
is solved in fixed-point iteration. The iteration functions are
$$x=F(x,y), \quad y=G(x,y)$$
where
$$F(x,y)=(x^2+y^2)/10+0.8$$
$$G(x,y)=x(1+y^2)/10+0.8.$$
The maximum number of iterations M=25, maximum error R=4*10^{-7}, initial estimates x=y=0.5. After 15 iterations the solution is
$$x=1, \quad y=1.$$

P608 Real Roots of Two Nonlinear Equations by the Newton or Modified Newton Method

```
10  CLS:PRINT"P608 COMPUTES REAL ROOTS OF TWO NONLINEAR EQUATIONS"
20  PRINT"BY NEWTON METHOD OR MODIFIED NEWTON METHOD.":PRINT
30  PRINT"PRESS BREAK AND DEFINE THE EQUATIONS:"
40  PRINT"WHEN NEWTON METHOD APPLIED THE EQUATIONS F(X,Y),G(X,Y)"
50  PRINT"AND THEIR DERIVATIVES WITH RESPECT TO X, F1,G1, AND WITH"
60  PRINT"RESPECT TO Y, F2,G2, ARE TO BE DEFINED ON LINES 610-660."
70  PRINT"WHEN MODIFIED NEWTON METHOD APPLIED THE EQUATION F(X,Y)"
80  PRINT"AND ITS DERIVATIVE F1(X,Y) ARE TO BE DEFINED ON LINES"
90  PRINT"690,700, AND THE EQUATION G(X1,Y) AND ITS DERIVATIVE"
100 PRINT"G2(X1,Y) ON LINES 740,750. IN THE LATTER TWO EQUATIONS"
110 PRINT"THE INDEPEDENT VARIABLES ARE X1,Y.":PRINT
120 PRINT"THEN RUN AGAIN. TO CONTINUE PRESS ENTER.":INPUT A$
130 CLS:PRINT"THE INPUT:"
140 PRINT"SINGLE PRECISION ONLY."
150 PRINT"IF PRINTER NOT USED O2=0, IF ONLY FINAL RESULTS"
160 PRINT"PRINTED O2=1, IF RESULTS AT EACH STEP PRINTED O2>1."
170 INPUT"O2";O2:CLS
180 PRINT"IF NEWTON METHOD APPLIED O3=1."
190 PRINT"IF MODIFIED NEWTON METHOD APPLIED O3=2."
200 INPUT"O3";O3:CLS
210 PRINT"THE ITERATION STOPS AND THE ROOTS ARE TO BE RECOMPUTED"
220 PRINT"FROM ANOTHER INITIAL ESTIMATE WHEN THE MAXIMUM NUMBER"
230 PRINT"OF ITERATIONS M IS REACHED.":PRINT
240 PRINT"THE ITERATION STOPS WHEN ABS(X1-X)<=R AND ABS(Y1-Y)<R,"
250 PRINT"X1,X; Y1,Y BEING TWO SUCCESSIVE APPROXIMATIONS TO THE"
260 PRINT"ROOTS X,Y, AND R A CHOSEN CONSTANT."
270 INPUT"M";M:INPUT"R";R:CLS
280 IF O2=0 THEN GOTO 360
290 LPRINT "P608: O2=";O2;"O3=";O3;"M=";M;"R=";R
300 GOTO 360
310 REM COMPUTATION OF ROOTS
320 PRINT "MAKE ANOTHER INITIAL ESTIMATE"
330 IF O2=0 THEN GOTO 360
340 LPRINT "AN IMPROPER INITIAL ESTIMATE. RERUN"
350 LPRINT
360 PRINT "INITIAL ESTIMATE OF X AND Y:"
370 INPUT"X";X:INPUT"Y";Y
380 IF O2>0 THEN LPRINT "INITIAL ESTIMATE X=";X;"Y=";Y
390 LET K=0
400 IF O2>1 THEN LPRINT "X,Y AT EACH ITERATION K:"
410 IF O3=2 THEN GOTO 470
420 GOSUB 610
430 LET S=F1*G2-F2*G1
440 LET X1=X-(F*G2-G*F2)/S
450 LET Y1=Y-(G*F1-F*G1)/S
460 GOTO 510
470 GOSUB 690
480 LET X1=X-F/F1
490 GOSUB 740
500 LET Y1=Y-G/G2
510 IF K=M THEN GOTO 320
520 IF ABS(X1-X)>R THEN GOTO 540
530 IF ABS(Y1-Y)<R THEN GOTO 770
540 PRINT "X1=";X1;"   Y1=";Y1;"   K=";K
550 IF O2>1 THEN LPRINT "X1=";X1;"   Y1=";Y1;"   K=";K
560 LET X=X1
570 LET Y=Y1
580 LET K=K+1
590 GOTO 410
600 REM SUBROUTINE: EQUATIONS FOR NEWTON METHOD
610 LET F=
```

```
620 LET F1=
630 LET F2=
640 LET G=
650 LET G1=
660 LET G2=
670 RETURN
680 REM SUBROUTINE: EQUATIONS (1) FOR MODIFIED NEWTON METHOD
690 LET F=
700 LET F1=
710 RETURN
720 REM SUBROUTINE: EQUATIONS (2) FOR MODIFIED NEWTON METHOD
730 REM THE INDEPENDENT VARIABLES ARE HERE X1,Y
740 LET G=
750 LET G2=
760 RETURN
770 PRINT "THE SOLUTION: X=";X1;"Y=";Y1;
780 PRINT "NUMBER OF ITERATIONS:"K
790 IF O2=0 THEN STOP
800 LPRINT "THE SOLUTION: X=";X1;"Y=";Y1
810 LPRINT "NUMBER OF ITERATIONS:";K
820 END
```

Examples

1. A system of two nonlinear equations

$f(x,y) = e^x + xy - 1 = 0$,

$g(x,y) = \sin(xy) + x + y - 1 = 0$,

is solved by Newton's method. The partial derivatives f_1, g_1 with respect to x and f_2, g_2 with respect to y are:

$f_1 = e^x + y$, $g_1 = y*\cos(xy) + 1$,

$f_2 = x$, $g_2 = x*\cos(xy) + 1$.

The maximum number of iterations M=25, maximum error $R = 4*10^{-7}$, initial estimates $x = 0.1$, $y = 0.5$. After 3 iterations the solution is

$x = 0$, $y = 1$.

2. The same system of equations is solved by a modified Newton's method using the same parameters. After 3 iterations the solution again is

$x = 0$, $y = 1$.

CHAPTER 7: LINEAR ALGEBRA II

P701 Preliminary Location of Eigenvalues by the Gershgorin Method

```
10 CLS:PRINT"P701 COMPUTES THE CENTERS, THE RADII OF CIRCLES WITHIN"
20 PRINT"WHICH AN EIGENVALUE LIES AND THE LOWER AND UPPER LIMITS"
30 PRINT"BETWEEN WHICH ALL EIGENVALUES LIE BY GERSHGORIN METHOD."
40 PRINT:PRINT"THE INPUT:"
50 PRINT"IF SINGLE PRECISION O1=1, IF DOUBLE O1=2."
60 PRINT"DOUBLE PRECISION USED ONLY IN THE INPUT OF MATRIX"
70 PRINT"ELEMENTS, BUT NOT IN THE LOCATION PROCESS."
80 DEFINT I-O:INPUT"O1";O1:CLS
90 PRINT"IF PRINTER NOT USED O2=0, IF USED O2>0."
100 INPUT"O2";O2:IF O1=2 THEN DEFDBL A
110 CLS:PRINT"ORDER OF THE MATRIX: N"
120 INPUT"N";N:PRINT
130 PRINT"ELEMENTS OF THE MATRIX: A(I,J)"
140 IF O2=0 THEN GOTO 160
150 LPRINT "P701: O1=";O1;"O2=";O2;"N=";N
160 DIM A(N,N),R(N)
170 PRINT "THE INPUT A(I,J):"
180 IF O2>0 THEN LPRINT "THE INPUT A(I,J):"
190 REM INPUT A(I,J)
200 FOR I=1 TO N
210 FOR J=1 TO N
220 PRINT "I=";I;"J=";J
230 INPUT"A(I,J)";A(I,J)
240 NEXT J
250 NEXT I
260 CLS:PRINT "CHECK A(I,J):"
270 FOR I=1 TO N
280 FOR J=1 TO N
290 PRINT "I=";I;"J=";J;"A(I,J)=";A(I,J)
300 PRINT"IF NO ERROR INPUT 0"
310 PRINT"IF ERROR INPUT ANY NUMBER>0 AND THEN THE CORRECT A(I,J)"
320 INPUT"ERROR";K
330 IF K>0 THEN INPUT"A(I,J)";A(I,J)
340 CLS:NEXT J
350 NEXT I
360 REM WRITING INPUT ON A FILE
370 OPEN "O",1,"P701A"
380 FOR I=1 TO N
390 FOR J=1 TO N
400 PRINT #1, A(I,J)
410 IF O2>0 THEN LPRINT "I=";I;"J=";J;"A(I,J)=";A(I,J)
420 NEXT J
430 IF O2>0 THEN LPRINT
440 NEXT I
450 CLOSE 1
460 IF O2>0 THEN LPRINT
470 IF O2>0 THEN LPRINT
480 PRINT "CENTERS AND RADII OF CIRCLES:"
490 IF O2>0 THEN LPRINT "CENTERS AND RADII OF CIRCLES:"
500 FOR I=1 TO N
510 LET R(I)=0
520 FOR J=1 TO N
530 IF J=I THEN GOTO 550
540 LET R(I)=R(I)+ABS(A(I,J))
550 NEXT J
560 LET B1=A(I,I)
570 PRINT "THE CENTER=";B1;"RADIUS=";R(I)
580 IF O2>0 THEN LPRINT "THE CENTER=";B1;"RADIUS=";R(I)
```

P. 214   PROGRAM P701

```
590 NEXT I
600 IF O2>0 THEN LPRINT
610 LET E1=1.E38
620 LET E2=-E1
630 FOR I=1 TO N
640 LET B1=A(I,I)-R(I)
650 LET B2=A(I,I)+R(I)
660 IF B1<E1 THEN LET E1=B1
670 IF B2>E2 THEN LET E2=B2
680 NEXT I
690 PRINT "THE LOWER LIMIT OF EIGENVALUES=";E1
700 PRINT "THE UPPER LIMIT OF EIGENVALUES=";E2
710 IF O2=0 THEN STOP
720 LPRINT "THE LOWER LIMIT OF EIGENVALUES=";E1
730 LPRINT "THE UPPER LIMIT OF EIGENVALUES=";E2
740 END
```

Examples

1. All the eigenvalues of a matrix of third order
$$\begin{bmatrix} 0 & 1 & 0 \\ 1 & 0 & 0 \\ 0 & 0 & 0 \end{bmatrix}$$
are located within the lower limit -1 and upper limit 1.

2. All the eigenvalues of a matrix of third order
$$\begin{bmatrix} 2 & -2 & 3 \\ 1 & 1 & 1 \\ 1 & 3 & -1 \end{bmatrix}$$
are located within the lower limit -5 and upper limit 7.

3. All the eigenvalues of a matrix of third order
$$\begin{bmatrix} 6.125 & -16.375 & -10.750 \\ 1.375 & -2.125 & -2.250 \\ -0.50 & 3.500 & 3.000 \end{bmatrix}$$
are located within the lower limit -21 and upper limit 33.25.

4. The elements of a square matrix of order n are as follows
$$a_{i,i+1}=0.5, \quad a_{i+1,1}=0.5, \quad i=1,2,\ldots n-1$$
$$a_{i,j}=0, \quad \text{for all other } i,j$$
For any n all the eigenvalues are located within the lower limit -1 and upper limit 1.

5. All the eigenvalues of a matrix of fourth order
$$\begin{bmatrix} 4 & -1 & -1 & -1 \\ -1 & 4 & -1 & -1 \\ -1 & -1 & 4 & -1 \\ -1 & -1 & -1 & 4 \end{bmatrix}$$
are located within the lower limit 1 and upper limit 7.

6. All the eigenvalues of a matrix of third order
$$\begin{bmatrix} 1 & 0 & 0 \\ 0 & 0 & 1 \\ 0 & 1 & 0 \end{bmatrix}$$
are located within the lower limit -1 and upper limit 1.

P702 Coefficients of a Secular Equation by the Krylov Method

```
10 CLS:PRINT"P702 COMPUTES THE COEFFICIENTS OF SECULAR EQUATION"
20 PRINT"BY KRYLOV METHOD.":PRINT
30 PRINT"WARNING: FOR CERTAIN MATRICES KRYLOV METHOD MAY GIVE"
40 PRINT"INCORRECT RESULTS OR IT MAY BREAK DOWN. SEE EXAMPLES."
50 PRINT:PRINT"THE INPUT:"
60 PRINT"IF SINGLE PRECISION O1=1, IF DOUBLE O1=2."
70 DEFINT I-P:INPUT"O1";O1:CLS
80 PRINT"IF PRINTER NOT USED O2=0, IF USED O2>0."
90 INPUT"O2";O2:CLS
100 IF O1=2 THEN DEFDBL A-H,Q-Z
110 PRINT"IF NEW INPUT DATA FOR MATRIX ELEMENTS O3>0,"
120 PRINT"IF THE DATA FROM FILE P701A WRITTEN IN P701"
130 PRINT"OR IN THE PRECEDING RUN OF P702 O3=0."
140 INPUT"O3";O3:CLS
150 PRINT"ORDER OF THE MATRIX: N"
160 INPUT"N";N:CLS
170 PRINT"ELEMENTS OF THE MATRIX: A(I,J)"
180 IF O2=0 THEN GOTO 200
190 LPRINT "P702: O1=";O1;"O2=";O2;"O3=";O3;"N=";N
200 LET N1=N+1
210 DIM A(N,N),B(N,N1),C0(N),P(N),S(N1)
220 LET O4=1
230 LET O5=1
240 LET O6=1
250 REM INPUT A(I,J)
260 PRINT "THE INPUT A(I,J):"
270 IF O2>0 THEN LPRINT "THE INPUT A(I,J):"
280 IF O3=0 THEN GOTO 570
290 FOR I=1 TO N
300 FOR J=1 TO N
310 PRINT "I=";I;"J=";J
320 INPUT"A(I,J)";A(I,J)
330 NEXT J
340 NEXT I
350 CLS:PRINT "CHECK A(I,J):"
360 FOR I=1 TO N
370 FOR J=1 TO N
380 PRINT "I=";I;"J=";J;"A(I,J)=";A(I,J)
390 PRINT"IF NO ERROR INPUT 0"
400 PRINT"IF ERROR INPUT ANY NUMBER>0 AND THEN THE CORRECT A(I,J)"
410 INPUT"ERROR";K
420 IF K>0 THEN INPUT"A(I,J)";A(I,J)
430 CLS:NEXT J
440 NEXT I
450 REM WRITING INPUT ON A FILE
460 OPEN "O",1,"P701A"
470 FOR I=1 TO N
480 FOR J=1 TO N
490 PRINT #1, A(I,J)
500 IF O2>0 THEN LPRINT "I=";I;"J=";J;"A(I,J)=";A(I,J)
510 NEXT J
520 IF O2>0 THEN LPRINT
530 NEXT I
540 CLOSE 1
550 GOTO 680
560 REM READING A(I,J) FROM A FILE
570 OPEN "I",1,"P701A"
580 FOR I=1 TO N
590 FOR J=1 TO N
600 INPUT #1, A(I,J)
610 PRINT "I=";I"J=";J;"A(I,J)=";A(I,J)
620 IF O2>0 LPRINT "I=";I;"J=";J;"A(I,J)=";A(I,J)
```

```
630 NEXT J
640 PRINT
650 IF O2>0 THEN LPRINT
660 NEXT I
670 CLOSE 1
680 IF O2>0 THEN LPRINT
690 PRINT"COMPUTING"
700 REM COMPUTATION OF MATRIX B(N,N1)
710 LET H1=5/7
720 LET B(1,1)=1
730 LET H2=H1
740 FOR I=2 TO N
750 LET B(I,1)=0
760 IF I>O4 THEN GOTO 790
770 LET H2=H2*H1
780 LET B(I,1)=.15+H2
790 NEXT I
800 FOR J=2 TO N1
810 FOR I=1 TO N
820 LET J1=J-1
830 LET B(I,J)=0
840 FOR L=1 TO N
850 LET B(I,J)=B(I,J)+A(I,L)*B(L,J1)
860 NEXT L
870 IF J<N1 THEN GOTO 890
880 LET B(I,J)=-B(I,J)
890 NEXT I
900 NEXT J
910 REM ** LU-DECOMPOSITION **
920 REM LU-MATRICES ARE STORED AS MATRIX B(N,N)
930 IF O6=0 THEN GOTO 1000
940 LET P(1)=1
950 LET C0(1)=B(1,1)
960 FOR J=2 TO N
970 LET P(J)=0
980 LET C0(J)=B(J,1)
990 NEXT J
1000 LET I=1
1010 IF O6>0 THEN GOSUB 1450
1020 REM COMPUTATION OF L(I,1), I=2,...N
1030 FOR J=2 TO N
1040 LET B(J,1)=B(J,1)/B(1,1)
1050 NEXT J
1060 LET I=I+1
1070 IF I=N THEN GOTO 1180
1080 IF O6=0 THEN GOTO 1180
1090 REM PREPARATION FOR PIVOTING
1100 FOR J=I TO N
1110 LET C0(J)=B(J,I)
1120 FOR L=1 TO I-1
1130 LET C0(J)=C0(J)-B(J,L)*B(L,I)
1140 NEXT L
1150 NEXT J
1160 GOSUB 1450
1170 REM COMPUTATION OF U(I,J), J=I,...N
1180 LET K=I-1
1190 FOR J=I TO N
1200 FOR L=1 TO K
1210 LET B(I,J)=B(I,J)-B(I,L)*B(L,J)
1220 NEXT L
1230 IF B(I,I)=0 THEN GOTO 1350
1240 NEXT J
1250 IF I=N THEN GOTO 1640
1260 REM COMPUTATION OF L(J,I), J=I+1,...N
```

```
1270 FOR J=I+1 TO N
1280 FOR L=1 TO K
1290 LET B(J,I)=B(J,I)-B(J,L)*B(L,I)
1300 NEXT L
1310 LET B(J,I)=B(J,I)/B(I,I)
1320 NEXT J
1330 GOTO 1060
1340 REM ERROR HANDLING SUBROUTINE
1350 LET O4=O4+1
1360 CLS:PRINT"AN ERROR INHERENT TO KRYLOV METHOD HAS OCCURRED;"
1370 IF O4<=N THEN GOTO 1420
1380 LET O5=O5+1
1390 IF O5>2 THEN PRINT "NO FURTHER CORRECTION POSSIBLE":STOP
1400 LET O4=1
1410 LET O6=0
1420 PRINT "SELF-CORRECTION TAKES PLACE":PRINT
1430 GOTO 720
1440 REM SUBROUTINE FOR PARTIAL PIVOTING
1450 LET K=I+1
1460 LET M=0
1470 LET C=ABS(C0(I))
1480 FOR L=K TO N
1490 IF ABS(C0(L))<=C THEN GOTO 1540
1500 LET C=ABS(C0(L))
1510 LET P(1)=-P(1)
1520 LET P(K)=L
1530 LET M=1
1540 NEXT L
1550 IF M=0 THEN RETURN
1560 LET L=P(K)
1570 FOR J=1 TO N
1580 LET C=B(I,J)
1590 LET B(I,J)=B(L,J)
1600 LET B(L,J)=C
1610 NEXT J
1620 RETURN
1630 REM ADJUSTMENT OF B(I,N1) TO PIVOTING
1640 IF O6=0 THEN GOTO 1750
1650 FOR I=2 TO N
1660 IF P(I)=0 THEN GOTO 1720
1670 LET J=P(I)
1680 LET K=I-1
1690 LET C=B(K,N1)
1700 LET B(K,N1)=B(J,N1)
1710 LET B(J,N1)=C
1720 NEXT I
1730 REM SOLUTION OF THE FIRST TRIANGULAR SYSTEM
1740 REM VECTOR Y IS STORED AS COLUMN J=N1 OF B(I,J)
1750 FOR I=2 TO N
1760 LET K=I-1
1770 FOR J=1 TO K
1780 LET B(I,N1)=B(I,N1)-B(I,J)*B(J,N1)
1790 NEXT J
1800 NEXT I
1810 REM SOLUTION OF THE SECOND TRIANGULAR SYSTEM
1820 LET S(N)=B(N,N1)/B(N,N)
1830 LET M=N-1
1840 FOR I=M TO 1 STEP -1
1850 LET K=I+1
1860 LET S(I)=B(I,N1)
1870 FOR J=K TO N
1880 LET S(I)=S(I)-B(I,J)*S(J)
1890 NEXT J
1900 LET S(I)=S(I)/B(I,I)
```

```
1910 NEXT I
1920 LET S(N1)=1
1930 IF O4=1 THEN CLS
1940 PRINT "COEFFICIENTS OF SECULAR EQUATION:"
1950 IF O2>0 THEN LPRINT "COEFFICIENTS OF SECULAR EQUATION:"
1960 OPEN "O",1,"P702A"
1970 PRINT #1, N
1980 FOR I=N1 TO 1 STEP -1
1990 PRINT "I=";I;"S(I)=";S(I)
2000 IF O2>0 THEN LPRINT "I=";I;"S(I)=";S(I)
2010 PRINT #1, S(I)
2020 NEXT I
2030 CLOSE 1
2040 PRINT"THERE ARE (N+1) COEFFICIENTS OF THE SECULAR"
2050 PRINT"EQUATION BELONGING TO A MATRIX OF N-TH ORDER."
2060 PRINT"COEFFICIENT S(I) STANDS AT (I-1)-ST POWER OF"
2070 PRINT"THE EIGENVALUE.":PRINT
2080 PRINT"THE COEFFICIENTS ARE WRITTEN ON FILE P702A."
2090 PRINT"THE MATRIX ELEMENTS TAKEN FROM INPUT ARE WRITTEN"
2100 PRINT"ON FILE P701A."
2110 END
```

Examples

In the first three examples the secular equation has the form
$$S_4 E^3 + S_3 E^2 + S_2 E + S_1 = 0.$$

1. The coefficients of the secular equation for the matrix of example 1 to P701 are computed in double precision:
$$S_4=1, \quad S_3=0, \quad S_2=-1, \quad S_1=0.$$

2. The coefficients of the secular equation for the matrix of example 2 to P701 are computed in single precision:
$$S_4=1, \quad S_3=-2, \quad S_2=-5, \quad S_1=6.$$
A self-correction of the program is applied when these coefficients are computed.

3. The coefficients of the secular equation for the matrix of example 3 to P701 are computed in single precision:
$$S_4=1, \quad S_3=-7, \quad S_2=24, \quad S_1=-18.$$

4. The coefficients of the secular equation for the matrix of example 4 to P701 are computed in double precision. For n=5 the coefficients are as follows

$S_6=1$, $S_5=0$, $S_4=-1$, $S_3=0$, $S_2=0.1875$, $S_1=0$.
For n=21 the coefficients are

$S_{22}=1$, $S_{21}=0$, $S_{20}=-5$, $S_{19}=0$, $S_{18}=10.6875$, $S_{17}=0$,
$S_{16}=-12.75$, $S_{15}=0$, $S_{14}=9.296875$, $S_{13}=0$,
$S_{12}=-4.265625$, $S_{11}=0$, $S_{10}=1.221923828125$, $S_9=0$,
$S_8=-0.20947265625$, $S_7=0$, $S_6=0.0196380615234375$, $S_5=0$,
$S_4=-8.392333984375*10^{-4}$, $S_3=0$,
$S_2=1.049041748046875*10^{-5}$, $S_1=0$.

5. The coefficients of the secular equation for the matrix of example 5 to P701 are computed in double precision:

$S_5=1$, $S_4=-0.25$, $S_3=1$, $S_2=-154.25$, $S_1=152.4$.

Analytically computed coefficients are:

$S_5=1$, $S_4=-16$, $S_3=90$, $S_2=-200$, $S_1=125$.

This example demonstrates the limits of reliability of the Krylov method (cf. example 5 to P703).

6. The Krylov method is unable to compute the coefficients of the secular equation for the matrix of example 6 to P701 (the matrix generated by the method is singular for any choice of the vector Y). Analytically computed coefficients are:

$S_4=1$, $S_3=-1$, $S_2=-1$, $S_1=1$.

P703 Real and Complex Eigenvalues of a Real Matrix from a Secular Equation

```
10 CLS:PRINT"P703 COMPUTES REAL AND COMPLEX EIGENVALUES"
20 PRINT"OF A MATRIX FROM SECULAR EQUATION BY ITERATION."
30 PRINT:PRINT"THE INPUT:"
40 PRINT"IF SINGLE PRECISION O1=1, IF DOUBLE O1=2."
50 DEFINT I-O:INPUT"O1";O1:CLS
60 IF O1=2 THEN DEFDBL A-D,F,G,S-Y
70 PRINT"IF PRINTER NOT USED O2=0, IF ONLY FINAL RESULTS"
80 PRINT"PRINTED O2=1, IF RESULTS AT EACH STEP PRINTED O2>1."
90 INPUT"O2";O2:CLS
100 PRINT"INTERVAL OF LOCATION PROCESS: (E1,E2)"
110 PRINT"USE BOTH LIMITS E1,E2 FROM P701 AS THE INPUT."
120 PRINT"SPACING IN LOCATION PROCESS: H"
130 INPUT"E1";E8:INPUT"E2";E9:INPUT"H";H:CLS
140 PRINT"THE ITERATION STOPS AND THE EIGENVALUE IS TO BE"
150 PRINT"RECOMPUTED FROM ANOTHER INITIAL ESTIMATE WHEN THE"
160 PRINT"MAXIMUM NUMBER OF ITERATIONS M IS REACHED.":PRINT
170 PRINT"THE ITERATION STOPS WHEN ABS(F(E))<R, E BEING"
180 PRINT"A COMPLEX EIGENVALUE WHOSE REAL PART IS X AND"
190 PRINT"IMAGINARY PART Y, AND R A CHOSEN CONSTANT."
200 INPUT"M";M:INPUT"R";R:CLS
210 PRINT"ORDER OF THE MATRIX: N"
220 PRINT"THERE ARE (N+1) COEFFICIENTS S(I) IN THE SECULAR"
230 PRINT"EQUATION, COEFFICIENT S(I) STANDS AT (I-1)-ST POWER"
240 PRINT"OF THE EIGENVALUE. THE VALUES N, S(I) READ FROM FILE P702A."
250 PRINT"AFTER ALL REAL EIGENVALUES HAVE BEEN LOCATED THE"
260 PRINT"PROGRAM ASKS FOR THEIR NUMBER M0.":PRINT
270 PRINT"USE THE RESULTS FROM LOCATION PROCESS AS INITIAL"
280 PRINT"ESTIMATES AND COMPUTE THE EIGENVALUES FROM DEFLATED"
290 PRINT"SECULAR EQUATION. WHEN ALL ROOTS ARE REAL INCREASE"
300 PRINT"THE ACCURACY BY USING THEM AS INITIAL ESTIMATES AND"
310 PRINT"RECOMPUTE THEM FROM THE ORIGINAL EQUATION IN REAL"
320 PRINT"ARITHMETIC.":PRINT
330 PRINT"TO CONTINUE PRESS ENTER.":INPUT A$:CLS
340 PRINT"WHEN SOME EIGENVALUES ARE COMPLEX THEY ARE COMPUTED"
350 PRINT"IN COMPLEX ARITHMETIC. REAL PART OF THE SECULAR"
360 PRINT"EQUATION AND ITS DERIVATIVES F,F1,F2. IMAGINARY"
370 PRINT"PART G,G1,G2. AS INITIAL ESTIMATE PUT X CLOSE TO ZERO"
380 PRINT"AND Y=0. ITERATION ALWAYS CONVERGES TO SOME EIGENVALUE."
390 PRINT:PRINT"ONCE AN EIGENVALUE IS FOUND THE EQUATION IS DEFLATED."
400 PRINT"WHEN ALL EIGENVALUES ARE DETERMINED THEY ARE USED AS"
410 PRINT"INITIAL ESTIMATES AND RECOMPUTED FROM THE ORIGINAL"
420 PRINT"EQUATION IN COMPLEX ARITHMETIC.":PRINT
430 PRINT"TO CONTINUE PRESS ENTER.":INPUT A$:CLS
440 PRINT"IF THIS ITERATION CONVERGES TO AN EIGENVALUE THAT HAS"
450 PRINT"BEEN RECOMPUTED CHOOSE O6>0 AND RECOMPUTE IT AGAIN"
460 PRINT"WITH THE SAME INITIAL ESTIMATE.":PRINT
470 PRINT"ADVICE: COMPUTE ALL SIMPLE ROOTS FIRST AND THEN"
480 PRINT"THE MULTIPLE ONES.":PRINT
490 IF O2=0 THEN GOTO 510
500 LPRINT "P703: O1=";O1;"O2=";O2;"M=";M;"R=";R
510 PRINT "INPUT OF COEFFICIENTS S(I) OF SECULAR EQUATION:"
520 IF O2=0 THEN GOTO 550
530 LPRINT "INPUT OF COEFFICIENTS S(I) OF SECULAR EQUATION:"
540 REM READING THE INPUT FROM A FILE
550 OPEN "I",1,"P702A"
560 INPUT #1, N
570 LET N1=N+1
580 LET N2=N-1
590 LET K0=1
600 DIM A(N1),A0(N1),B1(N1),B2(N1),C1(N),C2(N),D1(N2),D2(N2)
610 DIM E2(N1),E3(N1),E4(N),S(N)
```

PROGRAM P703

```
620 FOR I=N1 TO 1 STEP -1
630 INPUT #1, A(I)
640 LET A0(I)=A(I)
650 PRINT "I=";I;"S(I)=";A(I)
660 IF O2>0 THEN LPRINT "I=";I;"S(I)=";A(I)
670 LET E2(I)=A(I)
680 NEXT I
690 CLOSE 1
700 IF O2>0 THEN LPRINT
710 PRINT:PRINT "LOCATION OF REAL EIGENVALUES:"
720 IF O2>0 THEN LPRINT "LOCATION OF REAL EIGENVALUES:"
730 IF O2>0 THEN LPRINT "E1=";E8;"E2=";E9;"H=";H
740 LET Z=E8
750 LET I0=0
760 GOSUB 1160
770 LET Z1=Z
780 LET E5=E
790 LET E6=E1
800 LET Z=Z+H
810 GOSUB 1160
820 PRINT "X=";Z;"F(X)=";E;"F1(X)=";E1
830 IF O2>1 THEN LPRINT "X=";Z;"F(X)=";E;"F1(X)=";E1
840 IF SGN(E)=-SGN(E5) THEN GOTO 980
850 IF SGN(E6)=SGN(E1) THEN GOTO 1100
860 IF SGN(E)=-SGN(E1) THEN GOTO 1100
870 IF I0<2 THEN GOTO 1100
880 PRINT "A MULTIPLE EV. BETWEEN X1=";Z1;"AND X2=";Z;"?"
890 PRINT "THE QUESTION MARK INDICATES THAT A MULTIPLE EV."
900 PRINT "MAY OCCUR IN THE INTERVAL. IT IS PLAUSIBLE THAT"
910 PRINT "IT DOES OCCUR ONLY IF F(X) IS CLOSE TO ZERO."
920 IF O2=0 THEN GOTO 1010
930 LPRINT "A MULTIPLE EV. BETWEEN X1=";Z1;"AND X2=";Z;"?"
940 LPRINT "THE QUESTION MARK INDICATES THAT A MULTIPLE EV."
950 LPRINT "MAY OCCUR IN THE INTERVAL. IT IS PLAUSIBLE THAT"
960 LPRINT "IF DOES OCCUT ONLY IF F(X) IS CLOSE TO ZERO."
970 GOTO 1010
980 IF I0<2 THEN GOTO 1100
990 PRINT "EIGENVALUE BETWEEN X1=";Z1;"AND X2=";Z
1000 IF O2>0 THEN LPRINT "EIGENVALUE BETWEEN X1=";Z1;"AND X2=";Z
1010 PRINT "X1=";Z1;"F(X1)=";E5;"F1(X1)=";E6
1020 PRINT "X2=";Z;"F(X2)=";E;"F1(X2)=";E1
1030 IF O2=0 THEN GOTO 1080
1040 LPRINT "X1=";Z1;"F(X1)=";E5;"F1(X1)=";E6
1050 LPRINT "X2=";Z;"F(X2)=";E;"F1(X2)=";E1
1060 LPRINT
1070 GOTO 1100
1080 PRINT "TO CONTINUE PRESS ENTER"
1090 INPUT A$
1100 LET Z1=Z
1110 LET E5=E
1120 LET E6=E1
1130 IF Z>E9 THEN GOTO 1420
1140 GOTO 800
1150 REM SUBROUTINE FOR F(X) AND F1(X)
1160 LET E3(N1)=E2(N1)
1170 LET E4(N)=E3(N1)
1180 FOR I=N TO 1 STEP -1
1190 LET J=I+1
1200 LET K=I-1
1210 LET E3(I)=E3(J)*Z+E2(I)
1220 IF I=1 THEN GOTO 1240
1230 LET E4(K)=E4(I)*Z+E3(I)
1240 NEXT I
1250 LET E=E3(1)
```

```
1260 LET E1=E4(1)
1270 LET I0=I0+1
1280 IF E<>0 THEN RETURN
1290 LET I0=0
1300 IF E1<>0 THEN GOTO 1360
1310 PRINT "A MULTIPLE EV. X=";Z
1320 IF O2=0 THEN PRINT "TO CONTINUE PRESS ENTER"
1330 IF O2=0 THEN INPUT A$
1340 IF O2>0 THEN LPRINT "A MULTIPLE EV. X=";Z
1350 RETURN
1360 PRINT "EIGENVALUE X=";Z
1370 IF O2=0 THEN PRINT "TO CONTINUE PRESS ENTER"
1380 IF O2=0 THEN INPUT A$
1390 IF O2>0 THEN LPRINT "EIGENVALUE X=";Z
1400 RETURN
1410 REM COMPUTATION OF REAL EIGENVALUES
1420 IF O2>0 THEN LPRINT
1430 PRINT "HOW MANY REAL EIGENVALUES?"
1440 INPUT M0
1450 LET Y=0
1460 LET I0=0
1470 LET I3=0
1480 LET I1=1
1490 LET I2=2
1500 IF M0>0 THEN GOTO 1570
1510 LET O5=4
1520 GOTO 2030
1530 PRINT "MAKE ANOTHER INITIAL ESTIMATE"
1540 IF O2=0 THEN GOTO 1570
1550 LPRINT "AN IMPROPER INITIAL ESTIMATE. RERUN"
1560 LPRINT
1570 PRINT "INITIAL ESTIMATE OF X:"
1580 INPUT"X";X
1590 IF O2=1 THEN LPRINT "INITIAL ESTIMATE X=";X
1600 LET K=0
1610 IF O2>1 THEN LPRINT "F,F1,F2 AND X AT EACH ITERATION:"
1620 GOSUB 3280
1630 IF K=M THEN GOTO 1530
1640 IF ABS(F)<R THEN GOTO 1780
1650 REM ITERATION PROCESS IN REAL ARITHMETIC
1660 LET K=K+1
1670 PRINT "F=";F;"F1=";F1;"F2=";F2;"X=";X
1680 IF O2<2 THEN GOTO 1700
1690 LPRINT "F=";F;"F1=";F1;"F2=";F2;"X=";X
1700 LET W0=F1*F1-2*F*F2
1710 IF W0>0 THEN GOTO 1740
1720 LET X=X-2*F/F1
1730 GOTO 1620
1740 LET W=SQR(W0)
1750 IF O1=2 THEN GOSUB 3600
1760 LET X=X-2*F/(F1+W*SGN(F1))
1770 GOTO 1620
1780 PRINT "EIGENVALUE=";X;"F(X)=";F;"NUMBER OF ITERATIONS:";K
1790 LET I3=I3+1
1800 LET S(I3)=X
1810 PRINT
1820 IF O2=0 THEN GOTO 1850
1830 LPRINT "EIGENVALUE=";X;"F(X)=";F;"NUMBER OF ITERATIONS:";K
1840 LPRINT
1850 IF I3<N THEN GOTO 1950
1860 IF I0>0 THEN GOTO 3650
1870 LET I0=1
1880 PRINT "ALL EIGENVALUES ARE REAL"
1890 PRINT "RECOMPUTE THEM FROM ORIGINAL EQUATION"
```

PROGRAM P703

```
1900 IF O2>0 THEN LPRINT "EIGENVALUES FROM ORIGINAL EQUATION:"
1910 FOR I=1 TO N1
1920 LET A(I)=A0(I)
1930 NEXT I
1940 GOTO 1470
1950 LET K=0
1960 PRINT "IF FURTHER EIGENVALUE FROM ORIGINAL EQUATION O5=1"
1970 PRINT "IF FURTHER EIGENVALUE FROM DEFLATED EQUATION O5=2"
1980 INPUT"O5";O5
1990 IF I3=M0 THEN LET O5=3
2000 IF O5=1 THEN GOTO 1480
2010 GOSUB 2910
2020 IF O5=2 THEN GOTO 1570
2030 LET R=R*R
2040 LET I0=0
2050 REM COMPUTATION OF COMPLEX EIGENVALUES
2060 IF O2>0 THEN LPRINT
2070 GOTO 2120
2080 PRINT "MAKE ANOTHER INITIAL ESTIMATE"
2090 IF O2=0 THEN GOTO 2120
2100 LPRINT "AN IMPROPER INITIAL ESTIMATE. RERUN"
2110 LPRINT
2120 PRINT "INITIAL ESTIMATE OF REAL AND IMAGINARY PART X,Y:"
2130 INPUT"X";X:INPUT"Y";Y
2140 IF O2=1 THEN LPRINT "INITIAL ESTIMATE X=";X;"Y=";Y
2150 LET K=0
2160 IF O2>1 THEN LPRINT "F,F1,F2,X;G,G1,G2,Y AT EACH ITERATION:"
2170 GOSUB 3280
2180 IF K=M THEN GOTO 2080
2190 LET F0=F*F+G*G
2200 IF F0<R THEN GOTO 2570
2210 REM ITERATION PROCESS IN COMPLEX ARITHMETIC
2220 LET K=K+1
2230 PRINT "F=";F;"F1=";F1;"F2=";F2;"X=";X
2240 PRINT "     G=";G;"G1=";G1;"G2=";G2;"Y=";Y
2250 IF O2<2 THEN GOTO 2280
2260 LPRINT "F=";F;"F1=";F1;"F2=";F2;"X=";X
2270 LPRINT "     G=";G;"G1=";G1;"G2=";G2;"Y=";Y
2280 LET S0=F1*F1-G1*G1-2*(F*F2-G*G2)
2290 LET T0=2*(F1*G1-F2*G-F*G2)
2300 IF T0=0 THEN GOTO 2430
2310 LET W0=S0*S0+T0*T0
2320 LET W=SQR(W0)
2330 IF O1=2 THEN GOSUB 3600
2340 LET W0=(W+ABS(S0))/2
2350 LET W=SQR(W0)
2360 IF O1=2 THEN GOSUB 3600
2370 IF S0<0 THEN GOTO 2410
2380 LET F1=F1-W*K0
2390 LET G1=G1-K0*T0/(2*W)
2400 GOTO 2510
2410 LET F1=F1-K0*T0/(2*W)
2420 GOTO 2500
2430 LET W0=ABS(S0)
2440 LET W=SQR(W0)
2450 IF O1=2 THEN GOSUB 3600
2460 IF S0<0 THEN GOTO 2500
2470 LET F1=F1-W*K0
2480 IF F1=0 THEN GOTO 2550
2490 GOTO 2510
2500 LET G1=G1-W*K0
2510 LET G0=(F1*F1+G1*G1)/2
2520 LET X=X-(F*F1+G*G1)/G0
2530 LET Y=Y-(F1*G-F*G1)/G0
```

```
2540 GOTO 2170
2550 LET X=X-K0*F/W
2560 GOTO 2170
2570 PRINT "EIGENVALUE: X=";X;"Y=";Y
2580 PRINT "F(Z)=";F;"G(Z)=";G;"NUMBER OF ITERATIONS:";K
2590 PRINT
2600 IF I0<2 THEN GOTO 2670
2610 PRINT "IF THE EIGENVALUE IS RECOMPUTED FIRST TIME O6=0"
2620 PRINT "IF THE SAME EIGENVALUE RECOMPUTED ONCE AGAIN O6>0"
2630 INPUT"O6";O6
2640 IF O6=0 THEN GOTO 2670
2650 LET K0=-K0
2660 GOTO 2120
2670 IF O2=0 THEN GOTO 2710
2680 LPRINT "EIGENVALUE: X=";X;"Y=";Y
2690 LPRINT "F(Z)=";F;"G(Z)=";G;"NUMBER OF ITERATIONS:";K
2700 LPRINT
2710 LET K=0
2720 PRINT "IF JUST COMPUTED EIGENVALUE IS REAL O5=3"
2730 PRINT "IF IT IS COMPLEX O5=4"
2740 INPUT"O5";O5
2750 LET I3=I3+1
2760 IF O5=4 THEN LET I3=I3+1
2770 IF I3<N THEN GOTO 2900
2780 IF I0>0 THEN GOTO 3650
2790 PRINT "ALL EIGENVALUES ARE COMPUTED"
2800 PRINT "RECOMPUTE THEM FROM ORIGINAL EQUATION"
2810 LET I0=2
2820 LET I1=1
2830 LET I2=2
2840 LET I3=0
2850 FOR I=1 TO N1
2860 LET A(I)=A0(I)
2870 NEXT I
2880 IF O2>0 THEN LPRINT "EIGENVALUES FROM ORIGINAL EQUATION"
2890 GOTO 2120
2900 IF I0=2 THEN GOTO 2120
2910 PRINT "COEFFICIENTS OF THE DEFLATED SECULAR EQUATION:"
2920 PRINT "I=";N1;"A(I)=";A(N1)
2930 IF O2<2 THEN GOTO 2960
2940 LPRINT "COEFFICIENTS OF THE DEFLATED SECULAR EQUATION:"
2950 LPRINT "I=";N1;"A(I)=";A(N1)
2960 IF O5=4 THEN GOTO 3120
2970 REM THE DEFLATION BY A REAL EIGENVALUE
2980 FOR I=N TO I1 STEP -1
2990 LET J=I+1
3000 LET A(I)=A(J)*X+A(I)
3010 PRINT "I=";I;"S(I)=";A(I)
3020 IF O2>1 THEN LPRINT "I=";I;"S(I)=";A(I)
3030 NEXT I
3040 LET I1=I1+1
3050 LET I2=I2+1
3060 PRINT
3070 IF I3=M0 THEN RETURN
3080 IF O2>1 THEN LPRINT
3090 IF O5<3 THEN RETURN
3100 GOTO 2120
3110 REM THE DEFLATION BY A COMPLEX EIGENVALUE
3120 LET U=X+X
3130 LET V=-X*X-Y*Y
3140 LET A(N)=U*A(N1)+A(N)
3150 PRINT "I=";N;"S(I)=";A(N)
3160 IF O2>1 THEN LPRINT "I=";N;"S(I)=";A(N)
3170 FOR I=N2 TO I1 STEP -1
```

PROGRAM P703

```
3180 LET L1=I+1
3190 LET L2=I+2
3200 LET A(I)=V*A(L2)+U*A(L1)+A(I)
3210 PRINT "I=";I;"A(I)=";A(I)
3220 IF O2>1 THEN LPRINT "I=";I;"A(I)=";A(I)
3230 NEXT I
3240 LET I1=I1+2
3250 LET I2=I2+2
3260 GOTO 3060
3270 REM SUBROUTINE FOR F(Z),F1(Z),F2(Z) AND G(Z),G1(Z),G2(Z)
3280 LET B1(N1)=A(N1)
3290 IF I0=2 THEN LET O5=4
3300 IF O5>2 THEN LET B2(N1)=0
3310 LET C1(N)=B1(N1)
3320 IF O5>2 THEN LET C2(N)=0
3330 LET D1(N2)=C1(N)
3340 IF O5>2 THEN LET D2(N2)=0
3350 FOR I=N TO I1 STEP -1
3360 LET J=I+1
3370 LET L1=I-1
3380 LET L2=I-2
3390 LET B1(I)=B1(J)*X-B2(J)*Y+A(I)
3400 IF O5>2 THEN LET B2(I)=B2(J)*X+B1(J)*Y
3410 IF I=I1 THEN GOTO 3470
3420 LET C1(L1)=C1(I)*X-C2(I)*Y+B1(I)
3430 IF O5>2 THEN LET C2(L1)=C2(I)*X+C1(I)*Y+B2(I)
3440 IF I=I2 THEN GOTO 3470
3450 LET D1(L2)=D1(L1)*X-D2(L1)*Y+C1(L1)
3460 IF O5>2 THEN LET D2(L2)=D2(L1)*X+D1(L1)*Y+C2(L1)
3470 NEXT I
3480 LET F=B1(I1)
3490 IF O5>2 THEN LET G=B2(I1)
3500 LET F1=C1(I1)
3510 IF O5>2 THEN LET G1=C2(I1)
3520 IF I1<N THEN GOTO 3560
3530 LET F2=0
3540 LET G2=0
3550 RETURN
3560 LET F2=2*D1(I1)
3570 IF O5>2 THEN LET G2=2*D2(I1)
3580 RETURN
3590 REM SUBROUTINE FOR DOUBLE PRECISION SQR(W0)
3600 LET W9=W/1.D16
3610 LET W8=W
3620 LET W=(W+W0/W)/2
3630 IF ABS(W-W8)>W9 THEN GOTO 3610
3640 RETURN
3650 PRINT "ALL EIGENVALUES COMPUTED"
3660 IF O2>0 THEN LPRINT "ALL EIGENVALUES COMPUTED"
3670 IF I0=2 THEN STOP
3680 REM REAL EIGENVALUES WRITTEN ON A FILE
3690 OPEN "O",1,"P703A"
3700 PRINT #1, N
3710 FOR I=1 TO N
3720 PRINT #1, S(I)
3730 NEXT I
3740 CLOSE 1
3750 PRINT "ALL EIGENVALUES ARE REAL AND WRITTEN ON FILE P703A"
3760 IF O2=0 THEN GOTO 3780
3770 LPRINT "ALL EIGENVALUES ARE REAL AND WRITTEN ON FILE P703A."
3780 END
```

Examples

1. Preliminary location of eigenvalues E_i of the matrix of example 1 to P701:
$$E_1=-1, \quad E_2=0, \quad E_3=1.$$
The eigenvalues are computed in single precision in real arithmetic (maximum error $R=1*10^{-5}$):
$$E_1=-1, \quad E_2=0, \quad E_3=1.$$

2. Preliminary location of eigenvalues E_i of the matrix of example 2 to P701:
$$-2.20<E_1<-1.85, \quad 0.95<E_2<1.30, \quad 2.70<E_3<3.05.$$
The eigenvalues are computed in single precision in real arithmetic:
$$E_1=-2, \quad E_2=1, \quad E_3=3.$$

3. Preliminary location of eigenvalues E_i of the matrix of example 3 to P701:
$$0.9<E_1<1.2.$$
There is only one real eigenvalue computed in single precision in real arithmetic
$$E_1=1,$$
and a pair of complex conjugate eigenvalues computed in single precision in complex arithmetic
$$Re(E_{2,3})=3, \quad Im(E_{2,3})=+-3.$$

4. The eigenvalues of the matrix of example 4 to P701 are computed in double-precision arithmetic with maximum error $R=2*10^{-15}$ For n=5:
$$E_{1,2}=+-0.86602540378444, \quad E_{3,4}=+-0.5, \quad E_5=0.$$
For n=21 the eigenvalues rounded to 12 decimal places are:
$$E_{1,2}=+-0.989821441881, \quad E_{3,4}=+-0.959492973615,$$
$$E_{5,6}=+-0.909631995355, \quad E_{7,8}=+-0.841253532831,$$
$$E_{9,10}=+-0.755749574355, \quad E_{11,12}=+-0.654860733945,$$
$$E_{13,14}=+-0.540640817456, \quad E_{15,16}=+-0.415415013002,$$
$$E_{17,18}=+-0.281732556841, \quad E_{19,20}=+-0.142314838273,$$
$$E_{21}=0.$$
These eigenvalues can be computed by the formula:
$$E_i=\cos[PI*i/(n+1)], \quad i=1,2,...21.$$
For n=5 the eigenvalues computed numerically are exact in all 15 decimal places. For n=21 the maximum error is $2*10^{-12}$ (for i=3,4).

PROGRAM P703

5. Preliminary location of eigenvalues E_i of the matrix of example 5 to P701 indicates only two real eigenvalues

$E_1 = 1$, $E_2 = 5$

although there must be four real eigenvalues because the matrix is real and symmetric. Numerical computation in double precision (maximum error $R = 1 * 10^{-13}$) yields two real eigenvalues

$E_1 = 1$, $E_2 = 5$

and a pair of two complex conjugate eigenvalues

$Re(E_{3,4}) = -2.875$, $Im(E_{3,4}) = +-4.715334028465004$.

The eigenvalues computed from the true secular equation (cf. example 5 to P702) are:

$E_1 = 1$, $E_{2,3,4} = 5$

6. The eigenvalues of the matrix of example 6 to P701 computed from the analytically computed secular equation (cf. example 6 to P702) are:

$E_{1,2} = 1$, $E_3 = -1$

It seems that the Krylov method may not be able to compute the coefficients of the secular equation for multiple eigenvalues.

P704 Orthonormal Eigenvectors of a Real Matrix with Simple Real Eigenvalues

```
10 CLS:PRINT"P704 TESTS THE ACCURACY OF EIGENVALUES E(I) COMPUTED"
20 PRINT"IN P703 BY SUBSTITUTING EACH OF THEM INTO DET(A-I*E):"
30 PRINT"DEVIATIONS FROM ZERO INDICATE ERRORS IN THE COMPUTED"
40 PRINT"EIGENVALUES. THEN IT COMPUTES THE SET OF ORTHONORMAL"
50 PRINT"EIGENVECTORS V0(I,J) TO SIMPLE REAL EIGENVALUES E(I)."
60 PRINT:PRINT"THE INPUT:"
70 PRINT"IF SINGLE PRECISION O1=1, IF DOUBLE O1=2."
80 DEFINT I-P:INPUT"O1";O1:CLS
90 IF O1=2 THEN DEFDBL A-H,Q-Z
100 PRINT"IF PRINTER NOT USED O2=0, IF ONLY ORTHONORMAL"
110 PRINT"EIGENVECTORS PRINTED O2=1, IF ALSO NOT NORMALIZED"
120 PRINT"EIGENVECTORS PRINTED O2>1."
130 INPUT"O2";O2:CLS
140 PRINT"ORDER OF THE MATRIX N AND ITS ELEMENTS A(I,J) ARE"
150 PRINT"READ FROM FILE P702A, THE EIGENVALUES E(I) FROM"
160 PRINT"FILE P703A.":PRINT
170 PRINT"IF THE RANK OF MATRIX A-I*E IS LOWER THAN N-1 FOR AN"
180 PRINT"EIGENVALUE, THE PROGRAM CANNOT COMPUTE THE CORRESPONDING"
190 PRINT"EIGENVECTOR. IT CONTINUES TO COMPUTE THE NEXT ONE.":PRINT
200 IF O2=0 THEN GOTO 230
210 LPRINT "P704: O1=";O1;"O2=";O2
220 REM READING THE INPUT FROM TWO FILES
230 OPEN "I",1,"P703A"
240 INPUT #1, N0
250 LET N1=N0-1
260 DIM A0(N0,N0),A1(N0,N0),A(N0,N0),B(N1),C0(N0),E(N0)
270 DIM K7(N0),P(N0),S(N0,N0),X(N1)
280 PRINT "ORDER OF THE MATRIX=";N0
290 PRINT "THE INPUT: EIGENVALUES E(I):"
300 IF O2=0 THEN GOTO 340
310 LPRINT "ORDER OF THE MATRIX=";N0
320 LPRINT
330 LPRINT "THE INPUT: EIGENVALUES E(I):"
340 FOR I=1 TO N0
350 INPUT #1, E(I)
360 PRINT "I=";I;"E(I)=";E(I)
370 IF O2>0 THEN LPRINT "I=";I;"E(I)=";E(I)
380 LET K7(I)=0
390 NEXT I
400 CLOSE 1
410 PRINT "THE INPUT: MATRIX A(I,J):"
420 IF O2=0 THEN GOTO 450
430 LPRINT
440 LPRINT "THE INPUT: MATRIX A(I,J):"
450 OPEN "I",1,"P701A"
460 FOR I=1 TO N0
470 FOR J=1 TO N0
480 INPUT #1, A0(I,J)
490 LET A1(I,J)=A0(I,J)
500 LET A(I,J)=A0(I,J)
510 PRINT "I=";I"J=";J;"A(I,J)=";A0(I,J)
520 IF O2>0 LPRINT "I=";I;"J=";J;"A(I,J)=";A0(I,J)
530 NEXT J
540 PRINT
550 IF O2>0 THEN LPRINT
560 NEXT I
570 CLOSE 1
580 LET O6=1
590 PRINT "COMPUTATION OF DET(A-I*E). DEVIATIONS FROM ZERO"
600 PRINT "INDICATE ERRORS IN THE COMPUTATION OF EIGENVALUES"
610 IF O2=0 THEN GOTO 640
```

P. 230   PROGRAM P704

```
620 LPRINT "COMPUTATION OF DET(A-I*E), DEVIATIONS FROM ZERO"
630 LPRINT "INDICATE ERRORS IN THE COMPUTATION OF EIGENVALUES"
640 LET N=N0
650 LET O5=0
660 LET I0=0
670 LET I0=I0+1
680 FOR I=1 TO N0
690 FOR J=1 TO N0
700 LET A(I,J)=A0(I,J)
710 NEXT J
720 LET A(I,I)=A(I,I)-E(I0)
730 NEXT I
740 GOSUB 1680
750 PRINT "I=";I0;"DET(A-I*E(I))=";D
760 IF O2>0 THEN LPRINT "I=";I0;"DET(A-I*E(I))=";D
770 IF I0<N0 THEN GOTO 670
780 IF O2>0 THEN LPRINT
790 PRINT "EIGENVECTORS V(I,J):"
800 IF O2>1 THEN LPRINT "EIGENVECTORS V(I,J):"
810 LET N=N1
820 LET O5=1
830 LET I0=0
840 LET J1=0
850 LET I0=I0+1
860 LET J1=J1+1
870 LET O6=1
880 FOR I=1 TO N0
890 LET A1(I,I)=A0(I,I)-E(I0)
900 NEXT I
910 LET J0=N0
920 FOR I=1 TO N0
930 LET K=I
940 IF I=J0 THEN GOTO 1040
950 IF I>J0 THEN LET K=I-1
960 FOR J=1 TO N0
970 LET L=J
980 IF J=J0 THEN GOTO 1020
990 IF J>J0 THEN LET L=J-1
1000 LET A(K,L)=A1(I,J)
1010 GOTO 1030
1020 LET B(K)=-A1(I,J)
1030 NEXT J
1040 NEXT I
1050 GOSUB 1680
1060 IF D<>0 THEN GOTO 1180
1070 LET O6=-O6
1080 IF O6<0 THEN GOTO 880
1090 LET J0=J0-1
1100 IF J0>0 THEN GOTO 920
1110 PRINT"EIGENVECTOR TO E(I), I=";I0;"CANNOT BE COMPUTED"
1120 PRINT
1130 LET K7(I0)=7
1140 IF O2<2 THEN GOTO 1300
1150 LPRINT"EIGENVECTOR TO E(I), I=";I0;"CANNOT BE COMPUTED"
1160 LPRINT
1170 GOTO 1300
1180 FOR I=1 TO N0
1190 LET K=I
1200 IF I=J0 THEN GOTO 1240
1210 IF I>J0 THEN LET K=I-1
1220 LET S(I,I0)=X(K)
1230 GOTO 1250
1240 LET S(I,I0)=1
1250 PRINT "I=";I;"J=";J1;"V(I,J)=";S(I,I0)
```

```
1260 IF O2>1 THEN LPRINT "I=";I;"J=";J1;"V(I,J)=";S(I,I0)
1270 NEXT I
1280 IF O2>1 THEN LPRINT
1290 PRINT
1300 IF I0<N0 THEN GOTO 850
1310 PRINT "ORTHONORMAL EIGENVECTORS V0(I,J):"
1320 IF O2=0 THEN GOTO 1350
1330 LPRINT
1340 LPRINT "ORTHONORMAL EIGENVECTORS V0(I,J):"
1350 FOR J=1 TO N0
1360 IF K7(J)=0 THEN GOTO 1410
1370 PRINT"EIGENVECTOR TO E(I), I=";I0;"CANNOT BE COMPUTED"
1380 IF O2=0 THEN GOTO 1610
1390 LPRINT"EIGENVECTOR TO E(I), I=";I0;"CANNOT BE COMPUTED"
1400 GOTO 1600
1410 LET S0=0
1420 FOR I=1 TO N0
1430 LET S0=S0+S(I,J)*S(I,J)
1440 NEXT I
1450 LET S1=SQR(S0)
1460 IF O1=1 THEN GOTO 1520
1470 REM DOUBLE PRECISION FOR SQR(S0)
1480 LET X9=S1/1.D16
1490 LET X8=S1
1500 LET S1=(S1+S0/S1)/2
1510 IF ABS(S1-X8)>X9 THEN GOTO 1490
1520 FOR I=1 TO N0
1530 LET S(I,J)=S(I,J)/S1
1540 PRINT "I=";I;"J=";J;"V0(I,J)=";S(I,J)
1550 IF O2>0 THEN LPRINT "I=";I;"J=";J;"V0(I,J)=";S(I,J)
1560 NEXT I
1570 IF O2>0 THEN GOTO 1600
1580 PRINT "TO CONTINUE PRESS ENTER"
1590 INPUT A$
1600 IF O2>0 THEN LPRINT
1610 PRINT
1620 NEXT J
1630 PRINT "THE END"
1640 STOP
1650 REM SUBROUTINE FOR A SYSTEM OF LINEAR ALGEBRAIC EQUATIONS
1660 REM ** LU-DECOMPOSITION **
1670 REM LU-MATRICES ARE STORED AS MATRIX A
1680 LET P(1)=1
1690 IF O6<0 THEN GOTO 1750
1700 LET C0(1)=A(1,1)
1710 FOR J=2 TO N
1720 LET P(J)=0
1730 LET C0(J)=A(J,1)
1740 NEXT J
1750 LET I=1
1760 IF O6>0 THEN GOSUB 2120
1770 REM COMPUTATION OF L(I,1), I=2,...N
1780 FOR J=2 TO N
1790 LET A(J,1)=A(J,1)/A(1,1)
1800 NEXT J
1810 LET I=I+1
1820 IF I=N THEN GOTO 1930
1830 IF O6<0 THEN GOTO 1930
1840 REM PREPARATION FOR PIVOTING
1850 FOR J=I TO N
1860 LET C0(J)=A(J,I)
1870 FOR L=1 TO I-1
1880 LET C0(J)=C0(J)-A(J,L)*A(L,I)
1890 NEXT L
```

PROGRAM P704

```
1900 NEXT J
1910 GOSUB 2120
1920 REM COMPUTATION OF U(I,J), J=I,...N
1930 LET K=I-1
1940 FOR J=I TO N
1950 FOR L=1 TO K
1960 LET A(I,J)=A(I,J)-A(I,L)*A(L,J)
1970 NEXT L
1980 IF A(I,I)<>0 THEN GOTO 2010
1990 LET D=0
2000 RETURN
2010 NEXT J
2020 IF I=N THEN GOTO 2310
2030 REM COMPUTATION OF L(J,I), J=I+1,...N
2040 FOR J=I+1 TO N
2050 FOR L=1 TO K
2060 LET A(J,I)=A(J,I)-A(J,L)*A(L,I)
2070 NEXT L
2080 LET A(J,I)=A(J,I)/A(I,I)
2090 NEXT J
2100 GOTO 1810
2110 REM SUBROUTINE FOR PARTIAL PIVOTING
2120 LET K=I+1
2130 LET M=0
2140 LET C=ABS(C0(I))
2150 FOR L=K TO N
2160 IF ABS(C0(L))<=C THEN GOTO 2210
2170 LET C=ABS(C0(L))
2180 LET P(1)=-P(1)
2190 LET P(K)=L
2200 LET M=1
2210 NEXT L
2220 IF M=0 THEN RETURN
2230 LET L=P(K)
2240 FOR J=1 TO N
2250 LET C=A(I,J)
2260 LET A(I,J)=A(L,J)
2270 LET A(L,J)=C
2280 NEXT J
2290 RETURN
2300 REM COMPUTATION OF THE DETERMINANT
2310 LET D=A(1,1)
2320 FOR I=2 TO N
2330 LET D=D*A(I,I)
2340 NEXT I
2350 LET D=D*P(1)
2360 IF O5=0 THEN RETURN
2370 IF D=0 THEN RETURN
2380 REM ADJUSTMENT OF B(I) TO PIVOTING
2390 IF O6<0 THEN GOTO 2500
2400 FOR I=2 TO N
2410 IF P(I)=0 THEN GOTO 2470
2420 LET J=P(I)
2430 LET K=I-1
2440 LET C=B(K)
2450 LET B(K)=B(J)
2460 LET B(J)=C
2470 NEXT I
2480 REM SOLUTION OF THE FIRST TRIANGULAR SYSTEM
2490 REM VECTOR Y IS STORED AS VECTOR B
2500 FOR I=2 TO N
2510 LET K=I-1
2520 FOR J=1 TO K
2530 LET B(I)=B(I)-A(I,J)*B(J)
```

```
2540 NEXT J
2550 NEXT I
2560 REM SOLUTION OF THE SECOND TRIANGULAR SYSTEM
2570 LET X(N)=B(N)/A(N,N)
2580 LET M=N-1
2590 FOR I=M TO 1 STEP -1
2600 LET K=I+1
2610 LET X(I)=B(I)
2620 FOR J=K TO N
2630 LET X(I)=X(I)-A(I,J)*X(J)
2640 NEXT J
2650 LET X(I)=X(I)/A(I,I)
2660 NEXT I
2670 RETURN
2680 END
```

Examples

1. The value of the determinant det(A-I*E) for the matrix of example 1 to P701 is exactly zero for all three eigenvalues computed in P703.

The eigenvectors are:
V_1=(0, 0, 1), V_2=(1, 1, 0), V_3=(-1, 1, 0).
Orthonormal eigenvectors are:
$V0_1$=(0, 0, 1),
$V0_2$=(0.707107, 0.707107, 0),
$V0_3$=(-0.707107, 0.707107, 0).

2. The value of the determinant det(A-I*E) for the matrix of example 2 to P701 is exactly zero for all three eigenvalues computed in P703.

The eigenvectors are:
V_1=(-0.785714,-0.0714286,1),
V_2=(-1,1,1), V_3=(1,1,1).
Orthonormal eigenvectors are:
$V0_1$=(-0.616849,-0.0560772,0.785081),
$V0_2$=(-0.57735,0.57735,0.57735),
$V0_3$=(0.57735,0.57735,0.57735).

3. Since a pair of eigenvalues of the matrix of example 3 to P701 is complex conjugate, this program cannot compute the eigenvectors.

P. 234   PROGRAM P704

4. The value of the determinant det(A-I*E) for the matrix of example 4 to P701 is exactly zero for all n eigenvalues. It is computed in double precision for n=5, 14, 20, 21.

The eigenvectors and orthonormal eigenvectors are computed for n=5 and n=21. To save space they are given here explicitly only for n=5.

Eigenvectors rounded to 13 decimal places are:

V_1=(1, 1.7320508075688, 2, 1.7320508075688, 1),
V_2=(1, -1.7320508075688, 2, -1.7320508075688, 1),
V_3=(-1, -1, 0, 1, 1),
V_4=(-1, 1, 0, -1, 1),
V_5=(1, 0, -1, 0, 1).

Orthonormal eigenvectors rounded to 13 decimal places for n=5 are:

V_1=(0.2886751345948, 0.5, 0.5773502691896, 0.5,
       0.2886751345948),
V_2=(0.2886751345948, -0.5, 0.5773502691896, -0.5,
       0.2886751345948),
V_3=(-0.5, -0.5, 0, 0.5, 0.5),
V_4=(-0.5, 0.5, 0, -0.5, 0.5),
V_5=(0.5773502691896, 0, -0.5773502691896, 0,
       0.5773502691896).

CHAPTER 8: SPECIAL FUNCTIONS

P801 The Polynomial Function and Its First and Second Derivatives

```
10 CLS:PRINT"P801 EVALUATES A POLYNOMIAL FUNCTION AND ITS FIRST AND"
20 PRINT"SECOND DERIVATIVES AT A GIVEN X."
30 PRINT:PRINT"INPUT:"
40 PRINT"IF SINGLE PRECISION O1=1, IF DOUBLE O1=2."
50 DEFINT I-O:INPUT"O1";O1:CLS
60 IF O1=2 THEN DEFDBL A-D,F,X
70 PRINT"IF PRINTER NOT USED O2=0, IF USED O2>0."
80 INPUT"O2";O2:CLS
90 PRINT"IF COEFFICIENTS OF THE POLYNOMIAL FROM INPUT O3=0,"
100 PRINT"IF FROM INPUT AND SAVED ON FILE P801A O3=1, IF FROM"
110 PRINT"FILE P801A SAVED IN THE PRECEDING RUN O3>2."
120 INPUT"O3";O3:CLS
130 PRINT"IF ONLY FUNCTION COMPUTED O4=0, IF FUNCTION AND ITS"
140 PRINT"FIRST DERIVATIVE COMPUTED O4=1, IF FUNCTION AND ITS"
150 PRINT"FIRST AND SECOND DERIVATIVES COMPUTED O4=2"
160 INPUT"O4";O4:CLS
170 PRINT"DEGREE OF THE POLYNOMIAL: N"
180 INPUT"N";N:PRINT
190 PRINT"THERE ARE (N+1) COEFFICIENTS IN A POLYNOMIAL OF N-TH"
200 PRINT"DEGREE. COEFFICIENT A(I) STANDS AT (N-1)-ST POWER OF X."
210 PRINT"POLYNOMIAL FUNCTION AND ITS DERIVATIVES: F,F1,F2"
220 IF O2=0 THEN GOTO 250
230 LPRINT "P801: O1=";O1;"O2=";O2;"O3=";O3;"O4=";O4
240 LPRINT "POLYNOMIAL FUNCTION OF DEGREE N=";N
250 LET N1=N+1
260 LET N2=N-1
270 DIM A(N1),B(N1),C(N),D(N2)
280 PRINT "THE INPUT A(I):"
290 IF O2>0 THEN LPRINT "THE INPUT A(I):"
300 IF O3>1 THEN GOTO 530
310 FOR I=N1 TO 1 STEP -1
320 LET J=I-1
330 PRINT "POWER=";J;"I=";I
340 INPUT"A(I)";A(I)
350 NEXT I
360 CLS:PRINT "CHECK A(I):"
370 FOR I=N1 TO 1 STEP -1
380 LET J=I-1
390 PRINT "POWER=";J;"I=";I;"A(I)=";A(I)
400 PRINT"IF NO ERROR INPUT 0"
410 PRINT"IF ERROR INPUT ANY NUMBER>0 AND THEN THE CORRECT A(I)"
420 INPUT"ERROR";K
430 IF K>0 THEN INPUT"A(I)";A(I)
440 CLS:NEXT I
450 IF O3=1 THEN OPEN "O",1,"P801A"
460 FOR I=N1 TO 1 STEP -1
470 IF O3=1 THEN PRINT #1, A(I)
480 LET J=I-1
490 IF O2>0 THEN LPRINT "POWER=";J;"I=";I;"A(I)=";A(I)
500 NEXT I
510 IF O3=1 THEN CLOSE 1
520 IF O3<2 THEN GOTO 610
530 OPEN "I",1,"P801A"
540 FOR I=N1 TO 1 STEP -1
550 INPUT #1, A(I)
560 LET J=I-1
570 PRINT "POWER=";J;"I=";I;"A(I)=";A(I)
580 IF O2>0 THEN LPRINT "POWER=";J;"I=";I;"A(I)=";A(I)
```

```
590 NEXT I
600 CLOSE 1
610 IF O2=0 THEN GOTO 640
620 LPRINT
630 LPRINT "THE EVALUATION:"
640 CLS:PRINT "THE EVALUATION:"
650 INPUT"X";X
660 REM COMPUTATION OF F(X),F1(X),F2(X)
670 LET B(N1)=A(N1)
680 LET C(N)=B(N1)
690 LET D(N2)=C(N)
700 FOR I=N TO 1 STEP -1
710 LET J=I+1
720 LET L1=I-1
730 LET L2=I-2
740 LET B(I)=B(J)*X+A(I)
750 IF O4=0 THEN GOTO 810
760 IF I=1 THEN GOTO 810
770 LET C(L1)=C(I)*X+B(I)
780 IF O4=1 THEN GOTO 810
790 IF I=2 THEN GOTO 810
800 LET D(L2)=D(L1)*X+C(L1)
810 NEXT I
820 LET F=B(1)
830 IF O4>0 THEN GOTO 870
840 PRINT "X=";X;"F(X)=";F
850 IF O2>0 THEN LPRINT "X=";X;"F(X)=";F
860 GOTO 650
870 LET F1=C(1)
880 IF O4>1 THEN GOTO 920
890 PRINT "X=";X;"F(X)=";F;"F1(X)=";F1
900 IF O2>0 THEN LPRINT "X=";X;"F(X)=";F;"F1(X)=";F1
910 GOTO 650
920 IF N>1 THEN GOTO 950
930 LET F2=0
940 GOTO 960
950 LET F2=2*D(1)
960 PRINT "X=";X;"F(X)=";F;"F1(X)=";F1;"F2(X)=";F2
970 IF O2>0 THEN LPRINT"X=";X;"F(X)=";F;"F1(X)=";F1;"F2(X)=";F2
980 PRINT "TO STOP PRESS BREAK":PRINT
990 GOTO 650
1000 END
```

Examples

1. The polynomial function
   $$f(x) = x^4 + x^3 + x^2 + x + 1$$
   is evaluated in single precision with its first and second derivatives f_1, f_2:

   $x=7$, $\quad f=2801$, $\quad f_1=1534$, $\quad f_2=632$,

   $x=2.12345$, $\quad f=37.5387$, $\quad f_1=57.0729$, $\quad f_2=68.8492$.

2. The same function is evaluated in double precision:

   $x=12.12345678901234$, $\quad f=23544.57817513661$,

   $f_1=7593.717172582461$, $\quad f_2=1838.479194914668$.

P802 Orthogonal Polynomials of Legendre, Laguerre, Hermite, and Chebyshev of the First and Second Kind, Their First and Second Derivatives, and Zeros of Chebyshev Polynomials of the First Kind

```
10 CLS:PRINT"P802 EVALUATES ORTHOGONAL POLYNOMIALS OF LEGENDRE,"
20 PRINT"LAGUERRE, HERMITE, CHEBYSHEV OF FIRST AND SECOND KIND,"
30 PRINT"THEIR FIRST AND SECOND DERIVATIVES AT A GIVEN X AND"
40 PRINT"ZEROS OF CHEBYSHEV POLYNOMIALS OF FIRST KIND."
50 PRINT:PRINT"THE INPUT:"
60 PRINT"IF SINGLE PRECISION O1=1, IF DOUBLE O1=2."
70 DEFINT I-O:INPUT"O1";O1:CLS
80 IF O1=2 THEN DEFDBL A-H,P-Z
90 PRINT"IF PRINTER NOT USED O2=0, IF USED O2>0."
100 INPUT"O2";O2:CLS
110 PRINT"IF ONLY FUNCTION COMPUTED O4=0, IF FUNCTION AND ITS"
120 PRINT"FIRST DERIVATIVE COMPUTED O4=1, IF FUNCTION AND ITS"
130 PRINT"FIRST AND SECOND DERIVATIVES COMPUTED O4=2."
140 INPUT"O4";O4:CLS
150 PRINT"HIGHEST DEGREE OF POLYNOMIALS OCCURRING IN THE"
160 PRINT"PRESENT COMPUTATION: N1"
170 INPUT"N1";N1:CLS
180 LET N1=N1+1
190 IF N1<2 THEN LET N1=2
200 DIM F(N1),F1(N1),F2(N1)
210 IF O2=0 THEN GOTO 230
220 LPRINT "P802: O1=";O1;"O2=";O2;"O4=";O4
230 LET N0=-1
240 PRINT"O3 IS AN INTEGER VARIABLE DETERMINING THE FUNCTION."
250 PRINT"DEGREE OF THE POLYNOMIAL: N<=N1"
260 PRINT"INDEPENDENT VARIABLE: X"
270 PRINT"ORTHOGONAL FUNCTION AND ITS DERIVATIVES: F,F1,F2"
280 PRINT:PRINT "LEGENDRE POLYNOMIALS: O3=1"
290 PRINT "LAGUERRE POLYNOMIALS: O3=2"
300 PRINT "HERMITE POLYNOMIALS: O3=3"
310 PRINT "CHEBYSHEV POLYNOMIALS OF FIRST KIND: O3=4"
320 PRINT "CHEBYSHEV POLYNOMIALS OF SECOND KIND: O3=5"
330 PRINT "ZEROS OF CHEBYSHEV POLYNOMIALS OF FIRST KIND: O3=>6"
340 INPUT"O3";O3
350 IF O3>0 GOTO 370
360 PRINT"ERROR IN O3":GOTO 340
370 IF O3<6 THEN GOTO 560
380 CLS:PRINT "ZEROS OF CHEBYSHEV POLYNOMIALS OF FIRST KIND:"
390 INPUT"N";N
400 IF O2=0 THEN GOTO 440
410 LPRINT
420 LPRINT "ZEROS OF CHEBYSHEV POLYNOMIALS OF FIRST KIND:"
430 LPRINT "DEGREE=";N
440 LET L1=INT(N/2)
450 IF O1=2 THEN GOTO 1560
460 LET A=1.570796/N
470 FOR L=1 TO L1
480 LET X0=COS(A*(L+L-1))
490 PRINT "ZERO=+-";X0
500 IF O2>0 THEN LPRINT "ZERO=+-";X0
510 NEXT L
520 IF (N/2-INT(N/2))=0 THEN GOTO 1390
530 PRINT "ZERO=0"
540 IF O2>0 THEN LPRINT "ZERO=0"
550 GOTO 1390
560 INPUT"N";N:INPUT"X";X:CLS
570 LET N1=N+1
580 IF O3=1 THEN PRINT "LEGEDRE:"
```

```
590 IF O3=2 THEN PRINT "LAGUERRE:"
600 IF O3=3 THEN PRINT "HERMITE:"
610 IF O3=4 THEN PRINT "CHEBYSHEV 1-ST KIND:"
620 IF O3=5 THEN PRINT "CHEBYSHEV 2-ND KIND:"
630 PRINT "DEGREE=";N;"X=";X
640 IF O2=0 THEN GOTO 720
650 LPRINT
660 IF O3=1 THEN LPRINT "LEGEDRE:"
670 IF O3=2 THEN LPRINT "LAGUERRE:"
680 IF O3=3 THEN LPRINT "HERMITE:"
690 IF O3=4 THEN LPRINT "CHEBYSHEV 1-ST KIND:"
700 IF O3=5 THEN LPRINT "CHEBYSHEV 2-ND KIND:"
710 LPRINT "DEGREE=";N;"X=";X
720 IF O3=5 THEN GOTO 940
730 LET F(1)=1
740 LET F1(1)=0
750 IF N>0 THEN GOTO 790
760 PRINT "F=1, F1=F2=0"
770 IF O2>0 THEN LPRINT "F=1, F1=F2=0"
780 GOTO 1390
790 ON O3 GOTO 800,830,860,800
800 LET F(2)=X
810 LET F1(2)=1
820 GOTO 880
830 LET F(2)=1-X
840 LET F1(2)=-1
850 GOTO 880
860 LET F(2)=X+X
870 LET F1(2)=2
880 IF N>1 THEN GOTO 920
890 PRINT "F=";F(2);"F1=";F1(2);"F2=0"
900 IF O2>0 THEN LPRINT "F=";F(2);"F1=";F1(2);"F2=0"
910 GOTO 1390
920 IF O3=4 THEN GOTO 1050
930 GOTO 1090
940 LET F(1)=0
950 LET F(2)=1
960 LET F1(1)=0
970 LET F1(2)=0
980 IF N>1 THEN GOTO 1050
990 IF N=0 THEN PRINT "F=F1=F2=0"
1000 IF N=1 THEN PRINT "F=1, F1=F2=0"
1010 PRINT:IF O2=0 THEN GOTO 1390
1020 IF N=0 THEN LPRINT "F=F1=F2=0"
1030 IF N=1 THEN LPRINT "F=1, F1=F2=0"
1040 GOTO 1390
1050 LET A=2
1060 LET B=0
1070 LET C=1
1080 REM COMPUTATION OF ORTHOGONAL POLYNOMIALS
1090 FOR L=3 TO N1
1100 LET L1=L-1
1110 LET L2=L-2
1120 IF O3<4 THEN GOSUB 1420
1130 LET F(L)=(A*X+B)*F(L1)-C*F(L2)
1140 NEXT L
1150 PRINT "F=";F(N1)
1160 IF O2>0 THEN LPRINT "F=";F(N1)
1170 IF O4=0 THEN GOTO 340
1180 REM COMPUTATION OF FIRST DERIVATIVES OF O.P.
1190 FOR L=3 TO N1
1200 LET L1=L-1
1210 LET L2=L-2
1220 IF O3<4 THEN GOSUB 1420
```

```
1230 LET F1(L)=A*F(L1)+(A*X+B)*F1(L1)-C*F1(L2)
1240 NEXT L
1250 PRINT "F1=";F1(N1)
1260 IF O2>0 THEN LPRINT "F1=";F1(N1)
1270 IF O4=1 THEN GOTO 340
1280 REM COMPUTATION OF SECOND DERIVATIVES OF O.P.
1290 LET F2(1)=0
1300 LET F2(2)=0
1310 FOR L=3 TO N1
1320 LET L1=L-1
1330 LET L2=L-2
1340 IF O3<4 THEN GOSUB 1420
1350 LET F2(L)=2*A*F1(L1)+(A*X+B)*F2(L1)-C*F2(L2)
1360 NEXT L
1370 PRINT "F2=";F2(N1)
1380 IF O2>0 THEN LPRINT "F2=";F2(N1)
1390 PRINT:PRINT"TO CONTINUE PRESS ENTER, TO STOP PRESS BREAK"
1400 INPUT A$:CLS:GOTO 240
1410 REM SUBROUTINE FOR THE COEFFICIENTS A,B,C
1420 ON O3 GOTO 1470,1510
1430 LET A=2
1440 LET B=0
1450 LET C=L1+L1-2
1460 RETURN
1470 LET C=1-1/L1
1480 LET A=C+1
1490 LET B=0
1500 RETURN
1510 LET A=-1/L1
1520 LET B=A+2
1530 LET C=A+1
1540 RETURN
1550 REM ZEROS OF CHEBYSHEV POLYNOMIALS IN DOUBLE PRECISION
1560 LET A=1.5707963267948966/N
1570 FOR L=1 TO L1
1580 LET X=A*(L+L-1)
1590 LET X2=X*X
1600 LET X0=1-X2/650
1610 FOR I=12 TO 1 STEP -1
1620 LET K=(I+I)*(I+I-1)
1630 LET X0=1-X0*X2/K
1640 NEXT I
1650 PRINT "ZERO=+-";X0
1660 IF O2>0 THEN LPRINT "ZERO=+-";X0
1670 NEXT L
1680 IF (N/2-INT(N/2))=0 THEN GOTO 1390
1690 PRINT "ZERO=0"
1700 IF O2>0 THEN LPRINT "ZERO=0"
1710 GOTO 1390
1720 END
```

Examples

1. An orthogonal polynomial is denoted f, its first and second derivatives f_1, f_2, its order n, and argument x.

Legendre: x=0.77, n=5, f=-0.419321, f_1=0.152885, f_2=31.479.
Laguerre: x=0.77, n=3, f=-0.496739, f_1=-0.98645, f_2=2.23.
Hermite: x=0.77, n=5, f=28.0164, f_1=-108.347, f_2=-447.019.
Chebyshev of first kind:
       x=0.77, n=5, f=-0.949805, f_1=-2.45157, f_2=53.6906.
Chebyshev of second kind:
       x=0.77, n=5, f=-0.490314, f_1=10.7381, f_2=89.8368.
Legendre in double precision:
   x=0.77, n=5, f=-0.4193212390229323, f_1=0.1528848547053824,
       f_2=31.47894794528347.

2. Zeros of Chebyshev polynomials of first kind:
   n=5,   +-0.951057, +-0.587785, 0.
   n=6,   +-0.965926, +-0.707107, +-0.25882.

P803 The Hypergeometric Series, Confluent Hypergeometric Function, Their First and Second Derivatives, the Exponential Integral and Logarithmic Integral

```
10 CLS:PRINT"P803 EVALUATES HYPERGEOMETRIC SERIES, CONFLUENT"
20 PRINT"HYPERGEOMETRIC FUNCTION AND THEIR FIRST AND SECOND"
30 PRINT"DERIVATIVES AND EXPONENTIAL INTEGRAL, LOGARITHMIC"
40 PRINT"INTEGRAL AT A GIVEN X."
50 PRINT:PRINT"THE INPUT:"
60 PRINT"IF SINGLE PRECISION O1=1, IF DOUBLE O1=2."
70 DEFINT I-O:INPUT"O1";O1:CLS
80 IF O1=2 THEN DEFDBL A-H,P,S-Z
90 PRINT"IF PRINTER NOT USED O2=0, IF USED O2>0."
100 INPUT"O2";O2:CLS
110 PRINT"IF ONLY FUNCTION COMPUTED O4=0, IF FUNCTION AND ITS"
120 PRINT"FIRST DERIVATIVE COMPUTED O4=1, IF FUNCTION AND ITS"
130 PRINT"FIRST AND SECOND DERIVATIVES COMPUTED O4=2."
140 INPUT"O4";O4:CLS
150 PRINT"THE FUNCTIONS ARE EVALUATED BY SUMMATION OF THE SERIES."
160 PRINT"IT STOPS WHEN THE MAXIMUM NUMBER OF TERMS M IS REACHED."
170 PRINT"IT ALSO STOPS WHEN THE ABSOLUTE VALUE OF THE RATIO OF"
180 PRINT"THE COMPUTED TERM TO THE SUM IS SMALLER THAN A CHOSEN"
190 PRINT"CONSTANT R."
200 INPUT"M";M:INPUT"R";R:CLS
210 IF O2=0 THEN GOTO 230
220 LPRINT "P803: O1=";O1;"O2=";O2;"O4=";O4;"M=";M;"R=";R
230 IF O1=1 THEN GOTO 300
240 PRINT "READING DATA"
250 LET P3=2.302585092994046
260 DIM U(16),W(16)
270 FOR I9=1 TO 16
280 READ U(I9),W(I9)
290 NEXT I9
300 PRINT"O3 IS AN INTEGER VARIABLE DETERMINING THE FUNCTION."
310 PRINT"PARAMETERS A,B,C DETERMINE THE HYPERGEOMETRIC SERIES."
320 PRINT"PARAMETERS A,C DETERMINE THE CONFLUENT H. F."
330 PRINT"EXPONENTIAL INTEGRAL IS DEFINED FOR X>0."
340 PRINT"LOGARITHMIC INTEGRAL IS DEFINED FOR X>1."
350 PRINT"FUNCTION AND ITS DERIVATIVES: F,F1,F2"
360 PRINT"INDEPENDENT VARIABLE: X":PRINT
370 PRINT "HYPERGEOMETRIC SERIES: O3=1"
380 PRINT "CONFLUENT HYPERGEOMETRIC FUNCTION: O3=2"
390 PRINT "EXPONENTIAL INTEGRAL: O3=3"
400 PRINT "LOGARITHMIC INTEGRAL: O3=4"
410 IF O2>0 THEN LPRINT
420 INPUT"O3";O3
430 IF O3<1 THEN GOTO 450
440 CLS:ON O3 GOTO 460,500,540,560
450 PRINT"ERROR IN O3":GOTO 410
460 INPUT"PARAMETERS: A";A:INPUT"B";B:INPUT"C";C
470 IF O2=0 THEN GOTO 570
480 LPRINT "HYPERGEOMETRIC SERIES: A=";A;"B=";B;"C=";C
490 GOTO 570
500 INPUT"PARAMETERS: A";A:INPUT"C";C
510 IF O2=0 THEN GOTO 570
520 LPRINT "CONFLUENT HYPERGEOMETRIC FUNCTION: A=";A;"C=";C
530 GOTO 570
540 IF O2>0 THEN LPRINT "EXPONENTIAL INTEGRAL:"
550 GOTO 570
560 IF O2>0 THEN LPRINT "LOGARITHMIC INTEGRAL:"
570 PRINT:INPUT "X";X
580 IF O2>0 THEN LPRINT "X";X
590 IF O3<4 THEN GOTO 660
```

```
600 LET O3=3
610 IF O1=1 THEN GOTO 640
620 GOSUB 1190
630 GOTO 660
640 LET X=LOG(X)
650 REM SUMMATION OF THE SERIES
660 LET L=0
670 LET K=1
680 ON O3 GOTO 690,730,770
690 LET S0=X*A*B/C
700 LET F=1+S0
710 LET S=S0*X*(A+K)*(B+K)/((C+K)*(K+1))
720 GOTO 800
730 LET S0=X*A/C
740 LET F=1+S0
750 LET S=S0*X*(A+K)/((C+K)*(K+1))
760 GOTO 800
770 LET S0=X
780 LET F=S0
790 LET S=S0*X*K/((K+1)*(K+1))
800 LET F=F+S
810 IF S=0 THEN GOTO 870
820 IF K=M GOTO 870
830 LET K=K+1
840 IF ABS(S/F)<R THEN GOTO 870
850 LET S0=S
860 ON O3 GOTO 710,750,790
870 IF L>0 THEN GOTO 950
880 IF O3<3 THEN GOTO 920
890 IF O1=1 THEN LET X=LOG(X)
900 IF O1=2 THEN GOSUB 1190
910 LET F=F+X+.5772156649015329
920 PRINT "F=";F;"NUMBER OF TERMS=";K
930 IF O2>0 THEN LPRINT "F=";F;"NUMBER OF TERMS=";K
940 GOTO 1060
950 LET D=1
960 IF L>1 THEN GOTO 1020
970 IF O3=1 THEN LET D=B-1
980 LET F1=(A-1)*D*F/(C-1)
990 PRINT "F1=";F1;"NUMBER OF TERMS=";K
1000 IF O2>0 THEN LPRINT "F1=";F1;"NUMBER OF TERMS=";K
1010 GOTO 1070
1020 IF O3=1 THEN LET D=(B-1)*(B-2)
1030 LET F2=(A-2)*(A-1)*D*F/((C-2)*(C-1))
1040 PRINT "F2=";F2;"NUMBER OF TERMS=";K
1050 IF O2>0 THEN LPRINT "F2=";F2;"NUMBER OF TERMS=";K
1060 IF O3=3 THEN GOTO 1160
1070 IF O4=0 THEN GOTO 1160
1080 IF L=2 THEN GOTO 1160
1090 LET L=L+1
1100 LET A=A+1
1110 LET B=B+1
1120 LET C=C+1
1130 GOTO 670
1140 IF O4=1 THEN GOTO 1160
1150 GOTO 670
1160 PRINT:PRINT"TO CONTINUE PRESS ENTER, TO STOP PRESS BREAK"
1170 INPUT A$:CLS:GOTO 300
1180 REM SUBROUTINE FOR G=LOG(X) IN DOUBLE PRECISION
1190 LET X0=X
1200 IF X0<1 THEN LET X0=1/X0
1210 LET N9=0
1220 IF X0<10 THEN GOTO 1260
1230 LET N9=N9+1
```

```
1240 LET X0=X0/10
1250 IF X0>10 THEN GOTO 1230
1260 LET H8=(X0-1)/2
1270 LET H9=(X0+1)/2
1280 LET G=0
1290 FOR I9=1 TO 16
1300 LET X9=H8*U(I9)
1310 LET G=G+W(I9)/(H9-X9*X9/H9)
1320 NEXT I9
1330 LET G=2*G*H8+N9*P3
1340 IF X<1 THEN LET G=-G
1350 LET X=G
1360 RETURN
1370 DATA .04830766568773832,.09654008851472780
1380 DATA .14447196158279649,.09563872007927486
1390 DATA .23928736225213707,.09384439908080457
1400 DATA .33186860228212765,.09117387869576388
1410 DATA .42135127613063535,.08765209300440381
1420 DATA .50689990893222939,.08331192422694676
1430 DATA .58771575724076233,.07819389578707031
1440 DATA .66304426693021520,.07234579410884851
1450 DATA .73218211874028968,.06582222277636185
1460 DATA .79448379596794241,.05868409347853555
1470 DATA .84936761373256997,.05099805926237618
1480 DATA .89632115576605212,.04283589802222668
1490 DATA .93490607593773969,.03427386291302143
1500 DATA .96476225558750643,.02539206530926206
1510 DATA .98561151154526834,.01627439473090567
1520 DATA .99726386184948156,.00701861000947010
1530 END
```

Examples

1. Hypergeometric series
   $F(5,6;1;0.115)=14.0981$,
   its first derivative $F_1=279.467$,
   its second derivative $F_2=5199.12$.

2. Confluent hypergeometric function
   $M(1,1;1)=2.71828$,
   its first derivative $M_1=2.71828$,
   its second derivative $M_2=2.71828$.

3. Exponential integral Ei(0.77)=1.26361.
   Ei(1.99)=4.91738.

4. Logarithmic integral li(2.75)=1.92665.
   li(400)=85.4179.

All these functions are computed for maximum number of terms M=250 and maximum error R=4*10^{-6}.

P804 Incomplete Elliptic Integrals of the First, Second, and Third Kind, Bessel Functions $J_n(x)$ and Modified Bessel Functions $I_n(x)$ of Integer Order, the Incomplete Gamma Function, the Error Function, Fresnel Integrals $C(x)$, $S(x)$, and Sine and Cosine Integrals in Single Precision

```
10 CLS:PRINT"P804 EVALUATES INCOMPLETE ELLIPTIC INTEGRALS, BESSEL"
20 PRINT"FUNCTIONS Jn,In, INCOMPLETE GAMMA FUNCTION, ERROR"
30 PRINT"FUNCTION, FRESNEL INTEGRALS, SINE AND COSINE INTEGRALS"
40 PRINT"AT A GIVEN X."
50 PRINT:PRINT"THE INPUT:"
60 PRINT"SINGLE PRECISION ONLY (DOUBLE PRECISION IN P805)."
70 PRINT"IF PRINTER NOT USED O2=0, IF USED O2>0."
80 DEFINT I-O:INPUT"O2";O2:CLS
90 PRINT"MAXIMUM RELATIVE ERROR IN THE EVALUATION: P"
100 PRINT"IF P=0 THE ACCURACY OF EVALUATION NOT TESTED."
110 INPUT"P";P:CLS
120 IF O2=0 THEN GOTO 140
130 LPRINT "P804: O2=";O2;"P=";P
140 LET A=0
150 PRINT"READING DATA"
160 DIM U(16),W(16)
170 REM READ ABSCISSAS AND WEIGHTS FOR GAUSSIAN QUADRATURE
180 FOR I=1 TO 16
190 READ U(I),W(I)
200 NEXT I
210 PRINT"O3 IS AN INTEGER PARAMETER DEFINING THE FUNCTION."
220 PRINT"PARAMETERS DEFINE EACH SPECIAL FUNCTION."
230 PRINT"INDEPENDENT VARIABLE: X":PRINT
240 PRINT "INCOMPLETE ELLIPTIC INTEGRAL OF 1-ST KIND: O3=1"
250 PRINT "INCOMPLETE ELLIPTIC INTEGRAL OF 2-ND KIND: O3=2"
260 PRINT "INCOMPLETE ELLIPTIC INTEGRAL OF 3-RD KIND: O3=3"
270 PRINT "BESSEL FUNCTION Jn (n=0,1,2,...): O3=4"
280 PRINT "BESSEL FUNCTION In (n=0,1,2,...): O3=5"
290 PRINT "INCOMPLETE GAMMA FUNCTION: O3=6"
300 PRINT "ERROR FUNCTION: O3=7"
310 PRINT "FRESNEL INTEGRAL C(X): O3=8"
320 PRINT "FRESNEL INTEGRAL S(X): O3=9"
330 PRINT "SINE INTEGRAL Si(X): O3=10"
340 PRINT "COSINE INTEGRAL Ci(X): O3=11"
350 IF O2>0 THEN LPRINT
360 INPUT"O3";O3
370 IF O3<1 THEN GOTO 420
380 IF O3>11 THEN GOTO 420
390 CLS
400 ON O3 GOTO 430,430,530,600,600,690
410 GOTO 770
420 PRINT"ERROR IN O3":GOTO 360
430 INPUT"PHI";B:INPUT"M";S0
440 IF O2=0 THEN GOTO 490
450 IF O3=2 THEN GOTO 480
460 LPRINT "ELLIPTIC INTEGRAL OF 1-ST KIND, PHI=";B;"M=";S0
470 GOTO 490
480 LPRINT "ELLIPTIC INTEGRAL OF 2-ND KIND, PHI=";B;"M=";S0
490 LET S9=1
500 IF O3=1 THEN DEF FNF(X)=1/SQR(1-S0*SIN(X)[2)
510 IF O3=2 THEN DEF FNF(X)=SQR(1-S0*SIN(X)[2)
520 GOTO 930
530 INPUT"PHI";B:INPUT"N";S1:INPUT"M";S0
540 IF O2=0 THEN GOTO 570
550 LPRINT "ELLIPTIC INTEGRAL OF 3-RD KIND:"
560 LPRINT "PHI=";B;"N=";S1;"M=";S0
```

```
570 LET S9=1
580 DEF FNF(X)=1/((1-S1*SIN(X)[2)*SQR(1-S0*SIN(X)[2))
590 GOTO 930
600 INPUT"X";X9:INPUT"n";N0
610 IF O2=0 THEN GOTO 640
620 IF O3=4 THEN LPRINT "BESSEL FUNCTION Jn, X=";X9;"n=";N0
630 IF O3=5 THEN LPRINT "BESSEL FUNCTION In, X=";X9;"n=";N0
640 LET B=3.141593
650 LET S9=.3183099
660 IF O3=4 THEN DEF FNF(X)=COS(X9*SIN(X)-N0*X)
670 IF O3=5 THEN DEF FNF(X)=EXP(X9*COS(X))*COS(N0*X)
680 GOTO 930
690 INPUT"a";S0:INPUT"X";B
700 IF O2=0 THEN GOTO 720
710 LPRINT "INCOMPLETE GAMMA FUNCTION, a=";S0;"X=";B
720 LET S0=S0-1
730 LET S9=1
740 IF O1=2 THEN GOTO 930
750 DEF FNF(X)=EXP(-X)*X[S0
760 GOTO 930
770 INPUT"X";B
780 IF O2=0 THEN GOTO 840
790 IF O3=7 THEN LPRINT "ERROR FUNCTION, X=";B
800 IF O3=8 THEN LPRINT "FRESNEL INTEGRAL C(X), X=";B
810 IF O3=9 THEN LPRINT "FRESNEL INTEGRAL S(X), X=";B
820 IF O3=10 THEN LPRINT "SINE INTEGRAL Si(X), X=";B
830 IF O3=11 THEN LPRINT "COSINE INTEGRAL Ci(X), X=";B
840 LET S0=1.570796
850 LET S9=1
860 IF O3=7 THEN LET S9=1.128379
870 IF O3=7 THEN DEF FNF(X)=EXP(-X*X)
880 IF O3=8 THEN DEF FNF(X)=COS(S0*X*X)
890 IF O3=9 THEN DEF FNF(X)=SIN(S0*X*X)
900 IF O3=10 THEN DEF FNF(X)=SIN(X)/X
910 IF O3=11 THEN DEF FNF(X)=(COS(X)-1)/X
920 REM GAUSSIAN QUADRATURE
930 LET H=B-A
940 LET H2=H/2
950 LET K=0
960 LET M=1
970 LET Q=0
980 LET A0=A-H2
990 FOR J=1 TO M
1000 LET X0=H*J
1010 FOR I=1 TO 16
1020 LET X=H2*U(I)
1030 LET X1=A0+X0+X
1040 LET X2=A0+X0-X
1050 LET Q=Q+W(I)*(FNF(X1)+FNF(X2))
1060 NEXT I
1070 NEXT J
1080 LET Q=Q*H2
1090 IF P=0 THEN GOTO 1230
1100 IF K=4 THEN GOTO 1230
1110 REM TESTING THE ACCURACY OF INTEGRATION
1120 PRINT "Q=";Q*S9
1130 IF K=0 THEN GOTO 1160
1140 LET D=ABS((Q-Q0)/Q)
1150 IF D<P THEN GOTO 1230
1160 LET Q0=Q
1170 LET K=K+1
1180 LET M=M+M
1190 LET H=H2
1200 LET H2=H2/2
```

PROGRAM P804

```
1210 GOTO 970
1220 REM PRINTING THE RESULT
1230 LET Q=Q*S9
1240 IF O3<11 THEN GOTO 1260
1250 LET Q=Q+LOG(B)+.5772157
1260 PRINT "VALUE OF THE FUNCTION=";Q
1270 IF P>0 THEN PRINT "NUMBER OF SUBINTERVALS M=";M
1280 IF O2=0 THEN GOTO 1300
1290 LPRINT "VALUE OF THE FUNCTION=";Q
1300 PRINT:PRINT"TO CONTINUE PRESS ENTER, TO STOP PRESS BREAK"
1310 INPUT A$:CLS:GOTO 210
1320 DATA .0483077,.09654009
1330 DATA .1444720,.09563872
1340 DATA .2392874,.0938444
1350 DATA .3318686,.09117388
1360 DATA .4213513,.08765209
1370 DATA .5068999,.08331192
1380 DATA .5877158,.07819390
1390 DATA .6630443,.07234579
1400 DATA .7321821,.06582222
1410 DATA .7944838,.05868409
1420 DATA .8493676,.05099806
1430 DATA .8963212,.04283590
1440 DATA .9349061,.03427386
1450 DATA .9647623,.02539207
1460 DATA .9856115,.01627439
1470 DATA .9972639,.00701861
1480 END
```

Examples

1. Incomplete elliptic integral of first kind
   $F(1.0472, 0.75) = 1.21260$.
2. Incomplete elliptic integral of second kind
   $E(1.0472, 0.75) = 0.918395$.
3. Incomplete elliptic integral of third kind
   $\Pi(1.0472, 0.6, 0.75) = 1.55885$.
4. Bessel function of integer order
   $J_9(10.8) = 0.312439$.
5. Modified Bessel function of integer order
   $I_2(10) = 2281.52$.
6. Incomplete gamma function
   $\gamma(2.5, 0.9) = 0.164747$.
7. Error function
   $\mathrm{erf}(0.77) = 0.723822$.

8. Fresnel integral
    C(0.7)=0.659652.
9. Fresnel integral
    S(1.96)=0.348384.
10. Sine integral
    Si(6.7)=1.43121.
11. Cosine integral
    Ci(9.9)=-0.0367637.

The accuracy of evaluation of these functions has not been tested (P=0).

P805 Incomplete Elliptic Integrals of the First, Second, and Third Kind, Bessel Functions $J_n(x)$ and Modified Bessel Functions $I_n(x)$ of Integer Order, the Incomplete Gamma Function, the Error Function, Fresnel Integrals $C(x)$, $S(x)$, and Sine and Cosine Integrals in Double Precision

```
10 CLS:PRINT"P805 EVALUATES INCOMPLETE ELLIPTIC INTEGRALS, BESSEL"
20 PRINT"FUNCTIONS Jn,In, INCOMPLETE GAMMA FUNCTION, ERROR"
30 PRINT"FUNCTION, FRESNEL INTEGRALS, SINE AND COSINE INTEGRALS"
40 PRINT"AT A GIVEN X."
50 PRINT:PRINT"THE INPUT:"
60 PRINT"DOUBLE PRECISION ONLY (SINGLE PRECISION IN P804)."
70 PRINT"IF PRINTER NOT USED O2=0, IF USED O2>0."
80 DEFINT I-O:INPUT"O2";O2:CLS
90 DEFDBL A-H,P-Z
100 PRINT"MAXIMUM RELATIVE ERROR IN THE EVALUATION: P"
110 PRINT"IF P=0 THE ACCURACY OF EVALUATION NOT TESTED."
120 INPUT"P";P:CLS
130 IF O2=0 THEN GOTO 150
140 LPRINT "P805: O2=";O2;"P=";P
150 LET A=0
160 PRINT"READING DATA"
170 PRINT:PRINT"IF P=0 EVALUATION OF A FUNCTION LASTS 1-4 MINUTES."
180 DIM U(16),W(16)
190 READ P0,P1,P2,P3
200 REM READ ABSCISSAS AND WEIGHTS FOR GAUSSIAN QUADRATURE
210 FOR I=1 TO 16
220 READ U(I),W(I)
230 NEXT I
240 PRINT"O3 IS AN INTEGER PARAMETER DEFINING THE FUNCTION."
250 PRINT"PARAMETERS DEFINE EACH SPECIAL FUNCTION."
260 PRINT"INDEPENDENT VARIABLE: X":PRINT
270 PRINT "INCOMPLETE ELLIPTIC INTEGRAL OF 1-ST KIND: O3=1"
280 PRINT "INCOMPLETE ELLIPTIC INTEGRAL OF 2-ND KIND: O3=2"
290 PRINT "INCOMPLETE ELLIPTIC INTEGRAL OF 3-RD KIND: O3=3"
300 PRINT "BESSEL FUNCTION Jn (n=0,1,2,...): O3=4"
310 PRINT "BESSEL FUNCTION In (n=0,1,2,...): O3=5"
320 PRINT "INCOMPLETE GAMMA FUNCTION: O3=6"
330 PRINT "ERROR FUNCTION: O3=7"
340 PRINT "FRESNEL INTEGRAL C(X): O3=8"
350 PRINT "FRESNEL INTEGRAL S(X): O3=9"
360 PRINT "SINE INTEGRAL Si(X): O3=10"
370 PRINT "COSINE INTEGRAL Ci(X): O3=11"
380 IF O2>0 THEN LPRINT
390 INPUT"O3";O3
400 IF O3<1 THEN GOTO 450
410 IF O3>11 THEN GOTO 450
420 CLS
430 ON O3 GOTO 460,460,540,600,600,670
440 GOTO 740
450 PRINT"ERROR IN O3":GOTO 390
460 INPUT"PHI";B:INPUT"M";S0
470 IF O2=0 THEN GOTO 520
480 IF O3=2 THEN GOTO 510
490 LPRINT "ELLIPTIC INTEGRAL OF 1-ST KIND, PHI=";B;"M=";S0
500 GOTO 520
510 LPRINT "ELLIPTIC INTEGRAL OF 2-ND KIND, PHI=";B;"M=";S0
520 LET S9=1
530 GOTO 890
540 INPUT"PHI";B:INPUT"N";S1:INPUT"M";S0
550 IF O2=0 THEN GOTO 580
560 LPRINT "ELLIPTIC INTEGRAL OF 3-RD KIND:"
```

```
570 LPRINT "PHI=";B;"N=";S1;"M=";S0
580 LET S9=1
590 GOTO 890
600 INPUT"X";X3:INPUT"n";N0
610 IF O2=0 THEN GOTO 640
620 IF O3=4 THEN LPRINT "BESSEL FUNCTION Jn, X=";X3;"n=";N0
630 IF O3=5 THEN LPRINT "BESSEL FUNCTION In, X=";X3;"n=";N0
640 LET B=P1
650 LET S9=.3183098861837907
660 GOTO 890
670 INPUT"a";S0:INPUT"X";B
680 IF O2=0 THEN GOTO 700
690 LPRINT "INCOMPLETE GAMMA FUNCTION, a=";S0;"X=";B
700 LET S0=S0-1
710 LET S9=1
720 IF O1=2 THEN GOTO 890
730 GOTO 890
740 INPUT"X";B
750 IF O2=0 THEN GOTO 810
760 IF O3=7 THEN LPRINT "ERROR FUNCTION, X=";B
770 IF O3=8 THEN LPRINT "FRESNEL INTEGRAL C(X), X=";B
780 IF O3=9 THEN LPRINT "FRESNEL INTEGRAL S(X), X=";B
790 IF O3=10 THEN LPRINT "SINE INTEGRAL Si(X), X=";B
800 IF O3=11 THEN LPRINT "COSINE INTEGRAL Ci(X), X=";B
810 LET S0=P2
820 IF O3<11 THEN GOTO 860
830 LET X=B
840 GOSUB 2580
850 LET S8=G
860 LET S9=1
870 IF O3=7 THEN LET S9=1.128379167095512
880 REM GAUSSIAN QUADRATURE
890 LET H=B-A
900 LET H2=H/2
910 LET K=0
920 LET M=1
930 LET Q=0
940 LET A0=A-H2
950 FOR J=1 TO M
960 LET X1=A0+H*J
970 ON O3 GOTO 980,980,980,980,1280,1520,1520,1280,980,980,1280
980 FOR I=1 TO 16
990 LET X2=H2*U(I)
1000 LET X=X1+X2
1010 LET O8=0
1020 LET F=0
1030 IF O3=9 THEN LET X=X*X*P2
1040 GOSUB 1990
1050 IF O3<9 THEN GOTO 1090
1060 IF O3=10 THEN LET S=S/X
1070 LET F=F+S
1080 GOTO 1210
1090 IF O3=4 THEN GOTO 1180
1100 LET X=S*S
1110 IF O3=3 THEN LET F1=1-S1*X
1120 LET X=1-S0*X
1130 GOSUB 2970
1140 IF O3=1 THEN LET F=F+1/R
1150 IF O3=2 THEN LET F=F+R
1160 IF O3=3 THEN LET F=F+1/(F1*R)
1170 GOTO 1210
1180 LET X=X3*S-N0*X
1190 GOSUB 2010
1200 LET F=F+C
```

P. 252   PROGRAM P805

```
1210 IF O8=1 THEN GOTO 1250
1220 LET O8=1
1230 LET X=X1-X2
1240 GOTO 1030
1250 LET Q=Q+W(I)*F
1260 NEXT I
1270 GOTO 1720
1280 FOR I=1 TO 16
1290 LET X2=H2*U(I)
1300 LET X=X1+X2
1310 LET O8=0
1320 LET F=0
1330 IF O3=8 THEN LET X=X*X*P2
1340 GOSUB 2010
1350 IF O3=5 THEN GOTO 1390
1360 IF O3=8 THEN LET F=F+C
1370 IF O3=11 THEN LET F=F+(C-1)/X
1380 GOTO 1450
1390 LET X7=X3*C
1400 LET X=X*N0
1410 GOSUB 2010
1420 LET X=X7
1430 GOSUB 2350
1440 LET F=F+E*C
1450 IF O8=1 THEN GOTO 1490
1460 LET O8=1
1470 LET X=X1-X2
1480 GOTO 1330
1490 LET Q=Q+W(I)*F
1500 NEXT I
1510 GOTO 1720
1520 FOR I=1 TO 16
1530 LET X2=H2*U(I)
1540 LET X=X1+X2
1550 LET O8=0
1560 LET F=0
1570 IF O3=7 THEN LET X=-X*X
1580 GOSUB 2350
1590 IF O3=6 THEN GOTO 1620
1600 LET F=F+E
1610 GOTO 1660
1620 LET F1=E
1630 LET X8=S0
1640 GOSUB 2920
1650 LET F=F+E/F1
1660 IF O8=1 THEN GOTO 1700
1670 LET O8=1
1680 LET X=X1-X2
1690 GOTO 1570
1700 LET Q=Q+W(I)*F
1710 NEXT I
1720 NEXT J
1730 LET Q=Q*H2
1740 IF P=0 THEN GOTO 1880
1750 IF K=4 THEN GOTO 1880
1760 REM TESTING THE ACCURACY OF INTEGRATION
1770 PRINT "Q=";Q*S9
1780 IF K=0 THEN GOTO 1810
1790 LET D=ABS((Q-Q0)/Q)
1800 IF D<P THEN GOTO 1880
1810 LET Q0=Q
1820 LET K=K+1
1830 LET M=M+M
1840 LET H=H2
```

```
1850 LET H2=H2/2
1860 GOTO 930
1870 REM PRINTING THE RESULT
1880 LET Q=Q*S9
1890 IF O3<11 THEN GOTO 1910
1900 LET Q=Q+S8+.5772156649015329
1910 PRINT "VALUE OF THE FUNCTION=";Q
1920 IF P>0 THEN PRINT "NUMBER OF SUBINTERVALS M=";M
1930 IF O2=0 THEN GOTO 1960
1940 LPRINT "VALUE OF THE FUNCTION=";Q
1950 IF P>0 THEN LPRINT "NUMBER OF SUBINTERVALS M=";M
1960 PRINT:PRINT"TO CONTINUE PRESS ENTER, TO STOP PRESS BREAK"
1970 INPUT A$:CLS:GOTO 240
1980 REM SUBROUTINE FOR S=SIN(X), C=COS(X) IN DOUBLE PRECISION
1990 LET O9=0
2000 GOTO 2040
2010 LET O9=1
2020 GOTO 2040
2030 LET O9=2
2040 LET X0=ABS(X)
2050 LET O4=0
2060 IF X0<=P0 THEN GOTO 2080
2070 LET X0=X0-P0*INT(X0/P0)
2080 IF X<=P2 THEN GOTO 2130
2090 LET O4=INT(X0/P2)
2100 IF O4=1 THEN LET X0=P1-X0
2110 IF O4=2 THEN LET X0=X0-P1
2120 IF O4=3 THEN LET X0=P0-X0
2130 LET X9=X0*X0
2140 IF O9=1 THEN GOTO 2240
2150 LET S=1-X9/600
2160 FOR I9=12 TO 1 STEP -1
2170 LET K9=(I9+I9)*(I9+I9+1)
2180 LET S=1-S*X9/K9
2190 NEXT I9
2200 LET S=S*X0
2210 IF O4>1 THEN LET S=-S
2220 IF X<0 THEN LET S=-S
2230 IF O9=0 THEN RETURN
2240 LET C=1-X9/650
2250 FOR I9=12 TO 1 STEP -1
2260 LET K9=(I9+I9)*(I9+I9-1)
2270 LET C=1-C*X9/K9
2280 NEXT I9
2290 IF O4=1 THEN LET C=-C
2300 IF O4=2 THEN LET C=-C
2310 RETURN
2320 DATA 6.283185307179586,3.141592653589793
2330 DATA 1.570796326794897
2340 REM SUBROUTINE FOR E=EXP(X) IN DOUBLE PRECISION
2350 IF X>-88 THEN GOTO 2380
2360 LET E=0
2370 RETURN
2380 LET X0=ABS(X)
2390 LET N9=0
2400 IF X0<P3 THEN GOTO 2440
2410 LET X0=X0/P3
2420 LET N9=INT(X0)
2430 LET X0=(X0-N9)*P3
2440 LET E=1+X0/23
2450 FOR I9=22 TO 1 STEP -1
2460 LET E=1+E*X0/I9
2470 NEXT I9
2480 LET E0=1
```

P. 254 **PROGRAM P805**

```
2490 IF N9=0 THEN GOTO 2530
2500 FOR I9=1 TO N9
2510 LET E0=10*E0
2520 NEXT I9
2530 LET E=E*E0
2540 IF X<0 THEN LET E=1/E
2550 RETURN
2560 DATA 2.302585092994046
2570 REM SUBROUTINE FOR G=LOG(X) IN DOUBLE PRECISION
2580 LET X0=X
2590 IF X0<1 THEN LET X0=1/X0
2600 LET N9=0
2610 IF X0<10 THEN GOTO 2650
2620 LET N9=N9+1
2630 LET X0=X0/10
2640 IF X0>10 THEN GOTO 2620
2650 LET H8=(X0-1)/2
2660 LET H9=(X0+1)/2
2670 LET G=0
2680 FOR I9=1 TO 16
2690 LET X9=H8*U(I9)
2700 LET G=G+W(I9)/(H9-X9*X9/H9)
2710 NEXT I9
2720 LET G=2*G*H8+N9*P3
2730 IF X<1 THEN LET G=-G
2740 RETURN
2750 DATA .04830766568773832,.09654008851472780
2760 DATA .14447196158279649,.09563872007927486
2770 DATA .23928736225213707,.09384439908080457
2780 DATA .33186860228212765,.09117387869576388
2790 DATA .42135127613063535,.08765209300440381
2800 DATA .50689990893222939,.08331192422694676
2810 DATA .58771575724076233,.07819389578707031
2820 DATA .66304426693021520,.07234579410884851
2830 DATA .73218211874028968,.06582222277636185
2840 DATA .79448379596794241,.05868409347853555
2850 DATA .84936761373256997,.05099805926237618
2860 DATA .89632115576605212,.04283589802222668
2870 DATA .93490607593773769,.03427386291302143
2880 DATA .96476225558750643,.02539206530926206
2890 DATA .98561151154526834,.01627439473090567
2900 DATA .99726386184948156,.00701861000947010
2910 REM SUBROUTINE FOR E=X[X8 IN DOUBLE PRECISION
2920 GOSUB 2580
2930 LET X=X8*G
2940 GOSUB 2350
2950 RETURN
2960 REM SUBROUTINE FOR R=SQR(X) IN DOUBLE PRECISION
2970 LET R=SQR(X)
2980 LET X9=R/1.D16
2990 LET X8=R
3000 LET R=(R+X/R)/2
3010 IF ABS(R-X8)>X9 THEN GOTO 2990
3020 RETURN
3030 END
```

Examples

1. Incomplete elliptic integral of first kind
   $F(1.047197551196598, 0.75) = 1.21259661525498$.
2. Incomplete elliptic integral of second kind
   $E(1.047197551196598, 0.75) = 0.9183932943163255$.
3. Incomplete elliptic integral of third kind
   $\Pi(1.047197551196598, 0.6, 0.75) = 1.558838789270273$.
4. Bessel function of integer order
   $J_0(3.4) = -0.3642955967620006$.
5. Modified Bessel function of integer order
   $I_1(1) = 0.5651591039924852$.
6. Incomplete gamma function
   $\gamma(2.5, 0.9) = 0.164747297068347$.
7. Error function
   $erf(1.89) = 0.9924793184331474$.
8. Fresnel integral
   $C(4.98) = 0.5605194154543555$.
9. Fresnel integral
   $S(1.96) = 0.3483829406343658$.
10. Sine integral
    $Si(10) = 1.658347594218874$.
11. Cosine integral
    $Ci(9) = 0.05534753133313368$.

The accuracy of evaluation of these functions has not been tested (P=0).

NOTE: Because the subroutines for the elementary functions in double precision are programmed, not built-in, the evaluation lasts 1 to 4 minutes if P=0.

P806 Bessel Functions $J_\nu(x)$, $Y_\nu(x)$, $K_\nu(x)$ (ν Is Any Real Number $>-\frac{1}{2}$), and Spherical Bessel Functions $j_n(x)$, $y_n(x)$ (n Is Zero or Any Positive Integer) in Single Precision

```
10 CLS:PRINT"P806 EVALUATES BESSEL FUNCTIONS Jn,Yn,Kn (n IS ANY REAL"
20 PRINT"NUMBER>-1/2), AND SPHERICAL BESSEL FUNCTIONS jn, yn (n IS"
30 PRINT"ZERO OR ANY POSITIVE INTEGER) AT A GIVEN X"
40 PRINT:PRINT"THE INPUT:"
50 PRINT"SINGLE PRECISION ONLY (DOUBLE PRECISION IN P807)."
60 PRINT"IF PRINTER NOT USED O2=0, IF USED O2>0."
70 DEFINT I-O:INPUT"O2";O2:CLS
80 PRINT"MAXIMUM RELATIVE ERROR IN THE EVALUATION: P"
90 PRINT"IF P=0 THE ACCURACY OF EVALUATION NOT TESTED."
100 INPUT"P";P:CLS
110 IF O2=0 THEN GOTO 130
120 LPRINT "P806: O2=";O2;"P=";P
130 DIM C(15),D(15),U(16),W(16),V(15),Z(15)
140 PRINT"READING DATA"
150 REM READ ABSCISSAS AND WEIGHTS FOR GAUSSIAN AND LAGUERRE...
160 REM ...QUADRATURE AND COEFFICIENTS FOR GAMMA FUNCTION
170 FOR I=1 TO 15
180 READ U(I),W(I),V(I),Z(I),C(I)
190 NEXT I
200 READ U(16),W(16)
210 LET D(15)=C(15)
220 CLS
230 PRINT"O3 IS AN INTEGER VARIABLE DETERMINING THE FUNCTION."
240 PRINT"INDEPENDENT VARIABLE: X; ORDER: n":PRINT
250 PRINT "BESSEL FUNCTION Jn: O3=1"
260 PRINT "BESSEL FUNCTION Yn: O3=4"
270 PRINT "MODIFIED BESSEL FUNCTION Kn: O3=3"
280 PRINT "SPHERICAL BESSEL FUNCTION jn: O3=2"
290 PRINT "SPHERICAL BESSEL FUNCTION yn: O3=5"
300 PRINT:PRINT"EVALUATE jn IN P808 FOR X>=5, n=0-9."
310 PRINT"EVALUATE yn IN P808 FOR n=0-9."
320 PRINT"IF O3>2 CHOOSE THE LIMIT OF GAUSSIAN QUADRATURE B=5-8."
330 PRINT"TO ACHIEVE MAXIMUM ACCURACY CHOOSE P>0 AND B SO HIGH"
340 PRINT"THAT Q(LAGUERRE) DIFFERS IN NO MORE THAN LAST TWO"
350 PRINT"SIGNIFICANT DIGITS FROM THE PRECEDING Q(GAUSS)."
360 IF O2>0 THEN LPRINT
370 INPUT"O3";O3
380 IF O3<1 THEN GOTO 420
390 IF O3>5 THEN GOTO 420
400 INPUT"X";X9:INPUT"n";S0
410 CLS:GOTO 430
420 PRINT"ERROR IN O3":GOTO 370
430 IF O3>2 THEN INPUT"B";B0
440 IF O2=0 THEN GOTO 500
450 IF O3=1 THEN LPRINT "BESSEL FUNCTION Jn, X=";X9;"n=";S0
460 IF O3=4 THEN LPRINT "BESSEL FUNCTION Yn, X=";X9;"n=";S0
470 IF O3=3 THEN LPRINT "MODIFIED BESSEL F. Kn, X=";X9;"n=";S0
480 IF O3=2 THEN LPRINT "SPHERICAL BESSEL F. jn, X=";X9;"n=";S0
490 IF O3=5 THEN LPRINT "SPHERICAL BESSEL F. yn, X=";X9;"n=";S0
500 LET X8=X9/2
510 LET S2=S0+S0
520 IF O3=3 THEN GOTO 890
530 LET B=1.570796
540 IF O3=2 THEN GOTO 780
550 IF O3=5 THEN GOTO 780
560 LET S9=1.128379
570 LET M0=INT(S0+.5)
580 LET M1=INT(S0)
```

```
590 IF S0<0 THEN LET M1=0
600 LET Z0=S0+.5-M0
610 IF Z0=0 THEN GOTO 700
620 FOR I=14 TO 1 STEP -1
630 LET J=I+1
640 LET D(I)=D(J)*Z0+C(I)
650 NEXT I
660 LET S9=S9*Z0*D(1)
670 IF M1>0 THEN GOTO 710
680 IF M0=1 THEN GOTO 750
690 IF M0=0 THEN GOTO 760
700 IF M1=0 THEN GOTO 760
710 FOR I=1 TO M1
720 LET S9=S9*X8/(S0+.5-I)
730 NEXT I
740 IF Z0=0 THEN GOTO 760
750 IF M0>M1 THEN LET S9=S9/Z0
760 LET S9=S9*(X8[(S0-M1))
770 GOTO 840
780 LET S9=1
790 IF S0=0 THEN GOTO 830
800 FOR I=1 TO S0
810 LET S9=S9*X8/I
820 NEXT I
830 LET S2=S2+1
840 IF O3>3 THEN GOTO 870
850 DEF FNF(X)=COS(X9*COS(X))*(SIN(X)[S2)
860 GOTO 930
870 DEF FNF(X)=SIN(X9*SIN(X))*(COS(X)[S2)
880 GOTO 930
890 LET S9=.5
900 DEF FNF(X)=(1+EXP(-S2*X))*EXP(S0*X-X8*(EXP(X)+EXP(-X)))
910 LET B=B0
920 REM GAUSSIAN QUADRATURE
930 LET L=1
940 LET K=0
950 LET M=1
960 LET H=B
970 LET H2=H/2
980 LET Q=0
990 LET A0=-H2
1000 FOR J=1 TO M
1010 LET X0=H*J
1020 FOR I=1 TO 16
1030 LET X=H2*U(I)
1040 LET X1=A0+X0+X
1050 LET X2=A0+X0-X
1060 LET Q=Q+W(I)*(FNF(X1)+FNF(X2))
1070 NEXT I
1080 NEXT J
1090 LET Q=Q*H2
1100 IF P=0 THEN GOTO 1230
1110 IF K=4 THEN GOTO 1230
1120 REM TESTING THE ACCURACY OF INTEGRATION
1130 PRINT "Q(GAUSS)=";Q*S9
1140 IF K=0 THEN GOTO 1170
1150 LET D0=ABS((Q-Q0)/Q)
1160 IF D0<P THEN GOTO 1230
1170 LET Q0=Q
1180 LET K=K+1
1190 LET M=M+M
1200 LET H=H2
1210 LET H2=H2/2
1220 GOTO 980
```

```
1230 IF O3<3 THEN GOTO 1400
1240 IF O3=3 THEN GOTO 1320
1250 IF L=2 THEN GOTO 1320
1260 LET Q1=Q
1270 DEF FNF(X)=EXP(S2*X-X8*(EXP(X)-EXP(-X)))*(((1+EXP(-X-X))/2)[S2)
1280 LET B=B0
1290 LET L=2
1300 GOTO 940
1310 REM LAGUERRE QUADRATURE
1320 FOR I=1 TO 15
1330 LET X=B+V(I)
1340 LET Q=Q+Z(I)*FNF(X)
1350 NEXT I
1360 IF P>0 THEN PRINT "Q(LAGUERRE)=";Q*S9
1370 IF O3=3 THEN GOTO 1400
1380 LET Q=Q1-Q
1390 REM PRINTING THE RESULT
1400 LET Q=Q*S9
1410 PRINT "VALUE OF THE FUNCTION=";Q
1420 IF P>0 THEN PRINT "NUMBER OF SUBINTERVALS M=";M
1430 IF O2=0 THEN GOTO 1470
1440 IF O3<3 THEN LPRINT "VALUE OF THE FUNCTION=";Q
1450 IF O3>2 THEN LPRINT "VALUE OF THE FUNCTION=";Q;"B=";B
1460 LPRINT
1470 PRINT:PRINT"TO CONTINUE PRESS ENTER, TO STOP PRESS BREAK"
1480 INPUT A$:CLS:GOTO 230
1490 DATA .0483077,.09654009,.0933078,.2395782,1
1500 DATA .1444720,.09563872,.4926917,.5601008,.5772157
1510 DATA .2392874,.0938444,1.215995,.8870083,-.6558781
1520 DATA .3318686,.09117388,2.269950,1.223664,-.04200264
1530 DATA .4213513,.08765209,3.667623,1.574449,.1665386
1540 DATA .5068999,.08331192,5.425337,1.944752,-.04219773
1550 DATA .5877158,.07819390,7.565916,2.341502,-.009621972
1560 DATA .6630443,.07234579,10.12023,2.774042,.007218943
1570 DATA .7321821,.06582222,13.13028,3.255643,-.001165168
1580 DATA .7944838,.05868409,16.65441,3.806312,-.0002152417
1590 DATA .8493676,.05099806,20.77648,4.458478,.0001280503
1600 DATA .8963212,.04283590,25.62389,5.270018,-2.0135E-05
1610 DATA .9349061,.03427386,31.40752,6.359563,-1.250E-06
1620 DATA .9647623,.02539207,38.53068,8.031788,1.133E-06
1630 DATA .9856115,.01627439,48.02609,11.52777,-2.056E-07
1640 DATA .9972639,.00701861
1650 END
```

Examples

1. Bessel function of first kind
    $J_0(9.4)=-0.176772$, P=0.
    $J_{1/4}(2.2)=0.296224$, P=0.
    $J_{-1/4}(6.2)=0.27443$,   P=0,   but $J_{-1/4}(6.2)=0.2803$ if P=$4*10^{-7}$; the correct value is 0.2823.
2. Bessel function of second kind
    $Y_0(0.1)=-1.53424$, B=8, P=0.
    $Y_2(0.3)=-14.4801$, B=8, P=0.
3. Modified Bessel function of second kind
    $K_1(5.4)=2.59663*10^{-3}$, B=8, P=0.
    $K_2(17.8)=6.12208*10^{-9}$, B=6, P=0.
4. Spherical Bessel function of first kind
    $j_2(5)=0.134730$, P=0.
    The evaluation is faster in P808 for x>=5, n=0-9.
5. Spherical Bessel function of second kind
    $y_{10}(1)=-6.75993*10^8$, B=8, P=0, but the function is evaluated correctly, $y_{10}(1)=-6.72214*10^8$ if B=8, P=$4*10^{-7}$.

NOTE: To achieve maximum accuracy choose P>0.

P807 Bessel Functions $J_\nu(x)$, $Y_\nu(x)$, $K_\nu(x)$ (ν Is Any Real Number $>-\frac{1}{2}$), and Spherical Bessel Functions $j_n(x)$, $y_n(x)$ (n Is Zero or Any Positive Integer) in Double Precision

```
10 CLS:PRINT"P807 EVALUATES BESSEL FUNCTIONS Jn,Yn (n IS ANY REAL"
20 PRINT"NUMBER>-1/2), MODIFIED BESSEL FUNCTION Kn, AND"
30 PRINT"SPHERICAL BESSEL FUNCTIONS jn,yn (n IS ZERO OR ANY"
40 PRINT"POSITIVE INTEGER) AT A GIVEN X."
50 PRINT:PRINT"THE INPUT:"
60 PRINT"DOUBLE PRECISION ONLY (SINGLE PRECISION IN P806)."
70 PRINT"IF PRINTER NOT USED O2=0, IF USED O2>0."
80 DEFINT I-O:DEFDBL A-H,P-Z:INPUT"O2";O2:CLS
90 PRINT"MAXIMUM RELATIVE ERROR IN THE EVALUATION: P"
100 PRINT"IF P=0 THE ACCURACY OF EVALUATION NOT TESTED."
110 INPUT"P";P:CLS
120 IF O2=0 THEN GOTO 140
130 LPRINT "P807: O2=";O2;"P=";P
140 DIM D(26),D0(26),U(16),W(16),V(15),Z(15)
150 PRINT"READING DATA"
160 READ P0,P1,P2,P3
170 REM READ ABSCISSAS AND WEIGHTS FOR GAUSSIAN AND LAGUERRE...
180 REM ...QUADRATURE AND COEFFICIENTS FOR GAMMA FUNCTION
190 FOR I=1 TO 15
200 IF I>13 THEN GOTO 250
210 LET I2=I+I
220 LET I1=I2-1
230 READ U(I),W(I),V(I),Z(I),D0(I1),D0(I2)
240 GOTO 260
250 READ U(I),W(I),V(I),Z(I)
260 NEXT I
270 READ U(16),W(16)
280 LET D(26)=D0(26)
290 CLS
300 PRINT"O3 IS AN INTEGER VARIABLE DETERMINING THE FUNCTION."
310 PRINT"INDEPENDENT VARIABLE: X; ORDER: n":PRINT
320 PRINT "BESSEL FUNCTION Jn: O3=1"
330 PRINT "BESSEL FUNCTION Yn: O3=4"
340 PRINT "MODIFIED BESSEL FUNCTION Kn: O3=3"
350 PRINT "SPHERICAL BESSEL FUNCTION jn: O3=2"
360 PRINT "SPHERICAL BESSEL FUNCTION yn: O3=5"
370 PRINT:PRINT"EVALUATE jn IN P808 FOR X>=5, n=0-9."
380 PRINT"EVALUATE yn IN P808 FOR n=0-9."
390 PRINT"IF O3>2 CHOOSE THE LIMIT OF GAUSSIAN QUADRATURE B=5-8."
400 PRINT"TO ACHIEVE MAXIMUM ACCURACY CHOOSE P>0 AND B SO HIGH"
410 PRINT"THAT Q(LAGUERRE) DIFFERS IN NO MORE THAN LAST TWO"
420 PRINT"SIGNIFICANT DIGITS FROM THE PRECEDING Q(GAUSS)."
430 IF O2>0 THEN LPRINT
440 INPUT"O3";O3
450 IF O3<1 THEN GOTO 490
460 IF O3>5 THEN GOTO 490
470 INPUT"X";X3:INPUT"n";S0
480 CLS:GOTO 500
490 PRINT"ERROR IN O3":GOTO 440
500 IF O3>2 THEN INPUT"B";B0
510 IF O2=0 THEN GOTO 570
520 IF O3=1 THEN LPRINT "BESSEL FUNCTION Jn, X=";X9;"n=";S0
530 IF O3=4 THEN LPRINT "BESSEL FUNCTION Yn, X=";X9;"n=";S0
540 IF O3=3 THEN LPRINT "MODIFIED BESSEL F. Kn, X=";X9;"n=";S0
550 IF O3=2 THEN LPRINT "SPHERICAL BESSEL F. jn, X=";X9;"n=";S0
560 IF O3=5 THEN LPRINT "SPHERICAL BESSEL F. yn, X=";X9;"n=";S0
570 PRINT"COMPUTING":PRINT
580 PRINT"IF P=0 EVALUATION OF A FUNCTION LASTS 4-12 MINUTES":PRINT
```

```
590 LET X4=X3/2
600 LET S2=S0+S0
610 IF O3=3 THEN GOTO 980
620 LET B=P2
630 IF O3=2 THEN GOTO 910
640 IF O3=5 THEN GOTO 910
650 LET S9=1.1283791670955126
660 LET M0=INT(S0+.5)
670 LET M1=INT(S0)
680 IF S0<0 THEN LET M1=0
690 LET Z0=S0+.5-M0
700 IF Z0=0 THEN GOTO 790
710 FOR I=25 TO 1 STEP -1
720 LET J=I+1
730 LET D(I)=D(J)*Z0+D0(I)
740 NEXT I
750 LET S9=S9*Z0*D(1)
760 IF M1>0 THEN GOTO 800
770 IF M0=1 THEN GOTO 840
780 IF M0=0 THEN GOTO 850
790 IF M1=0 THEN GOTO 850
800 FOR I=1 TO M1
810 LET S9=S9*X4/(S0+.5-I)
820 NEXT I
830 IF Z0=0 THEN GOTO 850
840 IF M0>M1 THEN LET S9=S9/Z0
850 LET X=X4
860 LET X8=S0-M1
870 GOSUB 3230
880 LET S9=S9*E
890 LET X8=S2
900 GOTO 1010
910 LET S9=1
920 IF S0=0 THEN GOTO 960
930 FOR I=1 TO S0
940 LET S9=S9*X4/I
950 NEXT I
960 LET X8=S2+1
970 GOTO 1010
980 LET S9=.5
990 LET B=B0
1000 REM GAUSSIAN QUADRATURE
1010 LET L=1
1020 LET K=0
1030 LET M=1
1040 LET H=B
1050 LET H2=H/2
1060 LET Q=0
1070 LET A0=-H2
1080 FOR J=1 TO M
1090 LET X1=A0+H*J
1100 FOR I=1 TO 16
1110 LET X2=H2*U(I)
1120 LET X=X1+X2
1130 LET O8=0
1140 LET F=0
1150 IF O3=3 THEN GOTO 1420
1160 IF L=2 THEN GOTO 1310
1170 GOSUB 2060
1180 IF O3>3 THEN GOTO 1250
1190 LET X=S
1200 GOSUB 3230
1210 LET X=X3*C
1220 GOSUB 2040
```

PROGRAM P807

```
1230 LET F=F+C*E
1240 GOTO 1520
1250 LET X=C
1260 GOSUB 3230
1270 LET X=X3*S
1280 GOSUB 2020
1290 LET F=F+S*E
1300 GOTO 1520
1310 LET X=-X
1320 GOSUB 2380
1330 LET F1=E
1340 LET X=X4*(E-1/E)-X8*X
1350 GOSUB 2380
1360 LET F2=E
1370 LET X=(1+F1*F1)/2
1380 GOSUB 3230
1390 LET F1=F2*E
1400 IF L=3 THEN GOTO 1860
1410 GOTO 1510
1420 LET X5=-S2*X
1430 GOSUB 2380
1440 LET X=S0*X-X4*(E+1/E)
1450 GOSUB 2380
1460 LET F1=E
1470 LET X=X5
1480 GOSUB 2380
1490 LET F1=F1*(1+E)
1500 IF L=2 THEN GOTO 1860
1510 LET F=F+F1
1520 IF O8=1 THEN GOTO 1560
1530 LET O8=1
1540 LET X=X1-X2
1550 GOTO 1150
1560 LET Q=Q+W(I)*F
1570 NEXT I
1580 NEXT J
1590 LET Q=Q*H2
1600 IF P=0 THEN GOTO 1730
1610 IF K=4 THEN GOTO 1730
1620 REM TESTING THE ACCURACY OF INTEGRATION
1630 PRINT "Q(GAUSS)=";Q*S9
1640 IF K=0 THEN GOTO 1670
1650 LET D0=ABS((Q-Q0)/Q)
1660 IF D0<P THEN GOTO 1730
1670 LET Q0=Q
1680 LET K=K+1
1690 LET M=M+M
1700 LET H=H2
1710 LET H2=H2/2
1720 GOTO 1060
1730 IF O3<3 THEN GOTO 1920
1740 IF O3=3 THEN GOTO 1780
1750 IF L=2 THEN GOTO 1780
1760 LET Q1=Q
1770 LET B=B0
1780 LET L=L+1
1790 IF L=3 THEN GOTO 1820
1800 IF O3>3 THEN GOTO 1020
1810 REM LAGUERRE QUADRATURE
1820 FOR I=1 TO 15
1830 LET X=B+V(I)
1840 IF O3=3 THEN GOTO 1420
1850 GOTO 1310
1860 LET Q=Q+Z(I)*F1
```

```
1870 NEXT I
1880 IF P>0 THEN PRINT "Q(LAGUERRE)=";Q*S9
1890 IF O3=3 THEN GOTO 1920
1900 LET Q=Q1-Q
1910 REM PRINTING THE RESULT
1920 LET Q=Q*S9
1930 PRINT "VALUE OF THE FUNCTION=";Q
1940 IF P>0 THEN PRINT "NUMBER OF SUBINTERVALS M=";M
1950 GOTO 1990
1960 IF O3<3 THEN LPRINT "VALUE OF THE FUNCTION=";Q
1970 IF O3>2 THEN LPRINT "VALUE OF THE FUNCTION=";Q;"B=";B
1980 LPRINT
1990 PRINT:PRINT"TO CONTINUE PRESS ENTER, TO STOP PRESS BREAK"
2000 INPUT A$:CLS:GOTO 300
2010 REM SUBROUTINE FOR S=SIN(X), C=COS(X) IN DOUBLE PRECISION
2020 LET O9=0
2030 GOTO 2070
2040 LET O9=1
2050 GOTO 2070
2060 LET O9=2
2070 LET X0=ABS(X)
2080 LET O4=0
2090 IF X0<=P0 THEN GOTO 2110
2100 LET X0=X0-P0*INT(X0/P0)
2110 IF X<=P2 THEN GOTO 2160
2120 LET O4=INT(X0/P2)
2130 IF O4=1 THEN LET X0=P1-X0
2140 IF O4=2 THEN LET X0=X0-P1
2150 IF O4=3 THEN LET X0=P0-X0
2160 LET X9=X0*X0
2170 IF O9=1 THEN GOTO 2270
2180 LET S=1-X9/600
2190 FOR I9=12 TO 1 STEP -1
2200 LET K9=(I9+I9)*(I9+I9+1)
2210 LET S=1-S*X9/K9
2220 NEXT I9
2230 LET S=S*X0
2240 IF O4>1 THEN LET S=-S
2250 IF X<0 THEN LET S=-S
2260 IF O9=0 THEN RETURN
2270 LET C=1-X9/650
2280 FOR I9=12 TO 1 STEP -1
2290 LET K9=(I9+I9)*(I9+I9-1)
2300 LET C=1-C*X9/K9
2310 NEXT I9
2320 IF O4=1 THEN LET C=-C
2330 IF O4=2 THEN LET C=-C
2340 RETURN
2350 DATA 6.283185307179586,3.141592653589793
2360 DATA 1.570796326794897
2370 REM SUBROUTINE FOR E=EXP(X) IN DOUBLE PRECISION
2380 IF X>-88 THEN GOTO 2410
2390 LET E=0
2400 RETURN
2410 LET X0=ABS(X)
2420 LET N9=0
2430 IF X0<P3 THEN GOTO 2470
2440 LET X0=X0/P3
2450 LET N9=INT(X0)
2460 LET X0=(X0-N9)*P3
2470 LET E=1+X0/23
2480 FOR I9=22 TO 1 STEP -1
2490 LET E=1+E*X0/I9
2500 NEXT I9
```

P. 264 PROGRAM P807

```
2510 LET E0=1
2520 IF N9=0 THEN GOTO 2560
2530 FOR I9=1 TO N9
2540 LET E0=10*E0
2550 NEXT I9
2560 LET E=E*E0
2570 IF X<0 THEN LET E=1/E
2580 RETURN
2590 DATA 2.302585092994046
2600 REM SUBROUTINE FOR G=LOG(X) IN DOUBLE PRECISION
2610 LET X0=X
2620 IF X0<1 THEN LET X0=1/X0
2630 LET N9=0
2640 IF X0<10 THEN GOTO 2680
2650 LET N9=N9+1
2660 LET X0=X0/10
2670 IF X0>10 THEN GOTO 2650
2680 LET H8=(X0-1)/2
2690 LET H9=(X0+1)/2
2700 LET G=0
2710 FOR I9=1 TO 16
2720 LET X9=H8*U(I9)
2730 LET G=G+W(I9)/(H9-X9*X9/H9)
2740 NEXT I9
2750 LET G=2*G*H8+N9*P3
2760 IF X<1 THEN LET G=-G
2770 RETURN
2780 DATA .04830766568773832,.09654008851472780
2790 DATA .093307812017,.239578170311
2800 DATA 1,.5772156649015329
2810 DATA .14447196158279649,.09563872007927486
2820 DATA .492691740302,.560100842793
2830 DATA -.6558780715202538,-.0420026350340952
2840 DATA .23928736225213707,.09384439908080457
2850 DATA 1.215595412071,.887008262919
2860 DATA .1665386113822915,-.0421977345555443
2870 DATA .33186860228212765,.09117387869576388
2880 DATA 2.269949526204,1.22366440215
2890 DATA -.0096219715278770,.0072189432466630
2900 DATA .42135127613063535,.08765209300440381
2910 DATA 3.667622721751,1.57444872163
2920 DATA -.0011651675918591,-.0002152416741149
2930 DATA .50689990893222939,.08331192422694676
2940 DATA 5.425336627414,1.94475197653
2950 DATA .0001280502823882,-.0000201348547807
2960 DATA .58771575724076233,.07819389578707031
2970 DATA 7.565916226613,2.34150205664
2980 DATA -.0000012504934821,.0000011330272320
2990 DATA .66304426693021520,.07234579410884851
3000 DATA 10.120228568019,2.77404192683
3010 DATA -.0000002056338417,.0000000061160950
3020 DATA .73218211874028968,.06582222277636185
3030 DATA 13.130282482176,3.25564334640
3040 DATA .0000000050020075,-.0000000011812746
3050 DATA .79448379596794241,.05868409347853555
3060 DATA 16.654407708330,3.80631171423
3070 DATA .0000000001043427,.0000000000077823
3080 DATA .84936761373256997,.05099805926237618
3090 DATA 20.776478709449,4.45847775384
3100 DATA -.0000000000036968,.0000000000005100
3110 DATA .89632115576605212,.04283589802222668
3120 DATA 25.623894226729,5.27001778443
3130 DATA -.000000000000206,-.000000000000054
3140 DATA .93490607593773969,.03427386291302143
```

```
3150 DATA 31.407519169754,6.35956346973
3160 DATA .000000000000014,.000000000000001
3170 DATA .96476225558750643,.02539206530926206
3180 DATA 38.530683306486,8.03178763212
3190 DATA .98561151154526834,.01627439473090567
3200 DATA 48.026085572686,11.5277721009
3210 DATA .99726386184948156,.0070186100094701D
3220 REM SUBROUTINE FOR E=X[X8 IN DOUBLE PRECISION
3230 GOSUB 2610
3240 LET X=X8*G
3250 GOSUB 2380
3260 RETURN
3270 REM SUBROUTINE FOR R=SQR(X) IN DOUBLE PRECISION
3280 LET R=SQR(X)
3290 LET X9=R/1.D16
3300 LET X8=R
3310 LET R=(R+X/R)/2
3320 IF ABS(R-X8)>X9 THEN GOTO 3300
3330 RETURN
3340 END
```

Examples

1. Bessel function of first kind

    $J_0(9.4) = -0.1767715727515078$.

    $J_{1/4}(2.2) = 0.296223934947109$.

    $J_{1/3}(7.8) = 0.279895323483816$.

    $J_{2/3}(3.2) = 0.07763199722135054$.

2. Bessel function of second kind

    $Y_0(0.1) = -1.534238651351449$, $B=8$.

    $Y_0(14.9) = 0.2065464347072768$, $B=8$.

    $Y_2(0.3) = -14.48009$, $B=8$ (only this accuracy achieved).

3. Modified Bessel function of second kind

    $K_1(5.4) = 2.596627039039027 \times 10^{-3}$, $B=8$.

    $K_2(17.8) = 6.122076018068536 \times 10^{-9}$, $B=5$.

4. Spherical Bessel function of first kind

    $j_2(5) = 0.1347312100851266$.

    $j_4(10) = -0.105589285117679$.

    Evaluate $j_n(x)$ in P808 for $x \geq 5$, $n=0-9$.

5. Spherical Bessel function of second kind

    $y_{10}(10) = -0.17245$, $B=7$ (only this accuracy achieved).

    Evaluate $y_n(x)$ in P808 for $n=0-9$.

PROGRAM P807

The accuracy of evaluation of these functions has not been tested (P=0). The input data were chosen to make comparison with the data from mathematical tables easy. Use this program to evaluate the functions for arguments not contained in the tables but first check the accuracy by comparing the values from the tables with the computed ones.

NOTE: Because the subroutines for the elementary functions in double precision are programmed, not built-in, the evaluation lasts 4 to 12 minutes if P=0.

P808 Spherical Bessel Functions $j_n(x)$, $y_n(x)$ of Order $n = 0, \ldots, 9$

```
10 CLS:PRINT"P808 EVALUATES SPHERICAL BESSEL FUNCTIONS jn,yn (n=0-9)"
20 PRINT"AT A GIVEN X."
30 PRINT:PRINT"INPUT:"
40 PRINT"IF SINGLE PRECISION O1=1, IF DOUBLE O1=2."
50 DEFINT I-O:INPUT"O1";O1:CLS
60 IF O1=2 THEN DEFDBL A-H,P-Z
70 PRINT"IF PRINTER NOT USED O2=0, IF USED O2>0."
80 INPUT"O2";O2:CLS
90 PRINT"BECAUSE OF ROUNDOFF ERRORS THE VALUES OF jn FOR X<5"
100 PRINT"ARE VERY INACCURATE. EVALUATE THEM IN P806 (SINGLE"
110 PRINT"PRECISION) OR IN P807 (DOUBLE PRECISION).":PRINT
120 IF O2=0 THEN GOTO 140
130 LPRINT "P808: O1=";O1;"O2=";O2
140 LET P=6.283185307179586
150 LET P1=3.141592653589793
160 LET P2=1.570796326794897
170 IF O2>0 THEN LPRINT "SPHERICAL BESSEL FUNCTIONS OF ORDER N"
180 PRINT"ORDER OF THE FUNCTION: N; INDEPENDENT VARIABLE: X":PRINT
190 INPUT"N";N:INPUT"X";X
200 LET Y=1/X
210 LET Y2=Y*Y
220 LET N1=N+1
230 ON N1 GOTO 240,270,300,330,360,390,420,450,480,510
240 LET Z1=1
250 LET Z2=0
260 GOTO 550
270 LET Z1=Y
280 LET Z2=1
290 GOTO 550
300 LET Z1 =-1+3*Y2
310 LET Z2=3*Y
320 GOTO 550
330 LET Z1=Y*(-6+15*Y2)
340 LET Z2=-1+15*Y2
350 GOTO 550
360 LET Z1=1+Y2*(-45+105*Y2)
370 LET Z2=Y*(-10+105*Y2)
380 GOTO 550
390 LET Z1=Y*(15+Y2*(-420+945*Y2))
400 LET Z2=1+Y2*(-105+945*Y2)
410 GOTO 550
420 LET Z1=-1+Y2*(210+Y2*(-4725+10395*Y2))
430 LET Z2=Y*(21+Y2*(-1260+10395*Y2))
440 GOTO 550
450 LET Z1=Y*(-28+Y2*(3150+Y2*(-62370+135135*Y2)))
460 LET Z2=-1+Y2*(378+Y2*(-17325+135135*Y2))
470 GOTO 550
480 LET Z1=1+Y2*(-630+Y2*(51975+Y2*(-945945+2027025*Y2)))
490 LET Z2=Y*(-36+Y2*(6930+Y2*(-270270+2027025*Y2)))
500 GOTO 550
510 LET Z1=945945+Y2*(-16216200+34459425*Y2)
520 LET Z1=Y*(45+Y2*(-13860+Y2*Z1))
530 LET Z2=135135+Y2*(-4729725+34459425*Y2)
540 LET Z2=1+Y2*(-990+Y2*Z2)
550 IF O1=2 THEN GOTO 650
560 LET S=SIN(X)
570 LET C=COS(X)
580 LET F1=Y*(Z1*S-Z2*C)
590 LET F2=-Y*(Z1*C+Z2*S)
600 PRINT "N=";N;"X=";X;"j=";F1;"y=";F2
610 IF O2>0 THEN LPRINT "N=";N;"X=";X;"j=";F1;"y=";F2
620 PRINT:PRINT"TO CONTINUE PRESS ENTER, TO STOP PRESS BREAK"
```

PROGRAM P808

```
630 INPUT A$:CLS:GOTO 180
640 REM COMPUTATION OF SIN(X) AND COS(X) IN DOUBLE PRECISION
650 LET X0=ABS(X)
660 LET O4=0
670 IF X0<=P THEN GOTO 690
680 LET X0=X0-P*INT(X0/P)
690 IF X<=P2 THEN GOTO 740
700 LET O4=INT(X0/P2)
710 IF O4=1 THEN LET X0=P1-X0
720 IF O4=2 THEN LET X0=X0-P1
730 IF O4=3 THEN LET X0=P-X0
740 LET X2=X0*X0
750 LET S=1-X2/600
760 FOR I=12 TO 1 STEP -1
770 LET K=(I+I)*(I+I+1)
780 LET S=1-S*X2/K
790 NEXT I
800 LET S=S*X0
810 IF O4>1 THEN LET S=-S
820 IF X<0 THEN LET S=-S
830 LET C=1-X2/650
840 FOR I=12 TO 1 STEP -1
850 LET K=(I+I)*(I+I-1)
860 LET C=1-C*X2/K
870 NEXT I
880 IF O4=1 THEN LET C=-C
890 IF O4=2 THEN LET C=-C
900 GOTO 580
910 END
```

Examples

1. Spherical Bessel functions in single precision (only the data not affected by roundoff errors are included):

$j_0(0.5)=0.958851,$ $y_0(0.5)=-1.75516.$
$j_1(0.5)=0.162537,$ $y_1(0.5)=-4.46918.$
$j_2(0.5)=0.01637,$ $y_2(0.5)=-25.0599.$
$j_3(0.5)=1.16,$ $y_3(0.5)=-246.13.$
$y_4(0.5)=-3420.76.$
$y_5(0.5)=-61327.6.$
$y_6(0.5)=-1.34579*10^6.$
$y_7(0.5)=-3.49291*10^7.$
$y_8(0.5)=-1.04653*10^9.$
$y_9(0.5)=-3.5547*10^{10}.$

2. Spherical Bessel functions in double precision:

$j_0(1)=0.8414709848978965$, $y_0(1)=-0.5403023058681397$.
$j_1(1)=0.3011686789397568$, $y_1(1)=-1.381773290676036$.
$j_2(1)=0.06203505201137383$, $y_2(1)=-3.605017566159969$.
$j_3(1)=9.006581117112389*10^{-3}$, $y_3(1)=-16.64331454012381$.
$j_4(1)=1.01101580841334*10^{-3}$, $y_4(1)=-112.8981842147067$.
$j_5(1)=9.256115860267756*10^{-5}$, $y_5(1)=-999.4403433922364$.
$j_6(1)=7.156936*10^{-6}$, $y_6(1)=-10880.94559309989$.
$j_7(1)=4.79013*10^{-7}$, $y_7(1)=-140452.8523669064$.
$j_8(1)=2.82*10^{-8}$, $y_8(1)=-2095911.89910496$.
$y_9(1)=-35490048.42611152$.

NOTE: Use P806 or P807 to evaluate $j_n(x)$ if x<5. On the other hand, the greater the argument x the more accurate the values this program yields.

P809 Gamma Function of a Real Argument

```
10 CLS:PRINT"P809 EVALUATES GAMMA FUNCTION AT A GIVEN X."
20 PRINT:PRINT"THE INPUT:"
30 PRINT"IF SINGLE PRECISION O1=1, IF DOUBLE O1=2."
40 DEFINT I-O:INPUT"O1";O1:CLS
50 IF O1=2 THEN DEFDBL A-H,P-Z
60 PRINT"IF PRINTER NOT USED O2=0, IF USED O2>0."
70 INPUT"O2";O2:CLS
80 IF O2=0 THEN GOTO 110
90 LPRINT "P809: O1=";O1;"O2=";O2
100 LPRINT "GAMMA FUNCTION:"
110 DIM B(26),C(25)
120 PRINT"READING DATA"
130 FOR I=1 TO 25
140 READ C(I)
150 NEXT I
160 READ B(26)
170 CLS:PRINT"INDEPENDENT VARIABLE: X"
180 INPUT"X";X
190 IF X=0 THEN GOTO 470
200 IF X<0 THEN GOTO 370
210 LET M=INT(X)
220 LET Z=X-M
230 IF Z=0 THEN GOTO 300
240 GOSUB 510
250 IF M=0 GOTO 440
260 FOR I=1 TO M
270 LET G=G*(X-I)
280 NEXT I
290 GOTO 440
300 LET G=1
310 IF M=1 THEN GOTO 440
320 LET M=M-1
330 FOR I=1 TO M
340 LET G=G*I
350 NEXT I
360 GOTO 440
370 IF (X+INT(-X))=0 THEN GOTO 470
380 LET M=1+INT(ABS(X))
390 LET Z=M+X
400 GOSUB 510
410 FOR I=1 TO M
420 LET G=G/(Z-I)
430 NEXT I
440 PRINT "X=";X;"G=";G
450 IF O2>0 THEN LPRINT "X=";X;"G=";G
460 GOTO 480
470 PRINT "SINGULAR POINT"
480 PRINT:PRINT"TO CONTINUE PRESS ENTER, TO STOP PRESS BREAK"
490 INPUT A$:GOTO 170
500 REM SUBROUTINE FOR 1/G(Z)
510 FOR I=25 TO 1 STEP -1
520 LET J=I+1
530 LET B(I)=B(J)*Z+C(I)
540 NEXT I
550 LET G=1/(Z*B(1))
560 RETURN
570 DATA 1,.5772156649015329,-.6558780715202538
580 DATA -.0420026350340952,.1665386113822915
590 DATA -.0421977345555443,-.0096219715278770
600 DATA .0072189432466630,-.0011651675918591
610 DATA -.0002152416741149,.0001280502823882
620 DATA -2.01348547807D-05,-1.2504934821D-06
```

P. 270

```
630 DATA 1.1330272320D-06,-2.056338417D-07
640 DATA 6.1160950D-09,5.0020075D-09,-1.1812746D-09
650 DATA 1.043427D-10,7.7823D-12,-3.6968D-12,5.100D-13
660 DATA -2.06D-14,-5.4D-15,1.4D-15,1.D-16
670 END
```

Examples

1. Gamma function G in single precision:
    G(1)=1,
    G(6)=120,
    G(0.5)=1.77345,
    G(1.265)=0.903436,
    G(34.5)=5.04462*10^{37},
    G(-0.5)=-3.54491,
    G(-2.0001)=-5004.68.

2. Gamma function G in double precision:
    G(1.575)=0.8909447685952175,
    G(1.985123)=0.9938011230938238,
    G(-2.0000000001)=-5000005136.957153,
    G(-10.000000000001)=-275671.1074815212,
    G(-10.5)=-2.640121820547716*10^{-7},
    G(-2.0001)=-4999.538701487421.

NOTE: There is a difference in the value of gamma function G(-2.0001) when computed in single and in double precision. To eliminate possible roundoff errors of this kind compute in double precision and round to the desired accuracy.

P810 Gamma Function of Argument $n+1/2$, n an Arbitrary Integer

```
10 CLS:PRINT"P810 EVALUATES GAMMA FUNCTION OF ARGUMENT (N+1/2),"
20 PRINT"N BEING AN ARBITRARY INTEGER."
30 PRINT:PRINT"THE INPUT:"
40 PRINT"IF SINGLE PRECISION O1=1, IF DOUBLE O1=2."
50 DEFINT I-O:INPUT"O1";O1:CLS
60 IF O1=2 THEN DEFDBL G
70 PRINT"IF PRINTER NOT USED O2=0, IF USED O2>0."
80 INPUT"O2";O2:CLS
90 IF O2=0 THEN GOTO 120
100 LPRINT "P810: O1=";O1;"O2=";O2
110 LPRINT "GAMMA FUNCTION G OF ARGUMENT (N+1/2):"
120 LET G0=1.772453850905516
130 PRINT"INDEPENDENT INTEGER VARIABLE: N"
140 PRINT"GAMMA FUNCTION OF ARGUMENT (N+1/2): G"
150 INPUT"N";N
160 LET G=1
170 IF N=0 THEN GOTO 280
180 IF N<0 THEN GOTO 250
190 LET N0=N-1
200 LET I=0
210 LET G=G*(I+.5)
220 IF I=N0 THEN GOTO 280
230 LET I=I+1
240 GOTO 210
250 FOR I=1 TO ABS(N)
260 LET G=G/(.5-I)
270 NEXT I
280 LET G=G0*G
290 PRINT "N=";N;"G=";G
300 IF O2>0 THEN LPRINT "N=";N;"G=";G
310 PRINT:PRINT"TO CONTINUE PRESS ENTER, TO STOP PRESS BREAK"
320 INPUT A$:CLS:GOTO 130
330 END
```

Examples

1. Gamma function G of argument (N+½) in single precision:
   N=0,   G=1.77245,
   N=3,   G=3.32335,
   N=34,   G=5.04462*10^{37},
   N=-11, G=-2.64012*10^{-7}.

2. Gamma function in double precision:
   N=0,   G=1.772453850905516,
   N=10,   G=1133278.388948786.
   N=-11, G=-2.640121820547716*10^{-7},
   N=-23, G=-5.859755192540227*10^{-22},
   N=-24, G=2.493512847889458*10^{-23}.

P811   Beta Function

```
10 CLS:PRINT"P811 EVALUATES BETA FUNCTION AT ANY GIVEN PAIR (X,Y)."
20 PRINT:PRINT"THE INPUT:"
30 PRINT"IF SINGLE PRECISION O1=1, IF DOUBLE O1=2."
40 DEFINT I-O:INPUT"O1";O1:CLS
50 IF O1=2 THEN DEFDBL A-H,P-Z
60 PRINT"IF PRINTER NOT USED O2=0, IF USED O2>0."
70 INPUT"O2";O2:CLS
80 IF O2=0 THEN GOTO 110
90 LPRINT "P811: O1=";O1;"O2=";O2
100 LPRINT "BETA FUNCTION:"
110 DIM A(26),C(25)
120 PRINT"READING DATA"
130 FOR I=1 TO 25
140 READ C(I)
150 NEXT I
160 READ A(26)
170 CLS:PRINT"INDEPENDENT VARIABLES: X,Y"
180 PRINT"BETA FUNCTION: B"
190 INPUT"X";X:INPUT"Y";Y
200 IF X<=Y THEN GOTO 240
210 LET Z=X
220 LET X=Y
230 LET Y=Z
240 LET M1=0
250 LET M2=0
260 LET M3=1
270 LET Z=X+1
280 IF (Z-INT(Z))=0 THEN LET M1=1
290 LET Z=Y+1
300 IF (Z-INT(Z))=0 THEN LET M2=2
310 LET M3=M1+M2+M3
320 ON M3 GOTO 340  ,620  ,650  ,620
330 REM COMPUTATION OF BETA IF NEITHER X NOR Y ARE INTEGER
340 LET M1=INT(X)
350 LET M2=INT(Y)
360 LET Z0=X+Y
370 LET M3=INT(Z0)
380 LET Z=X-M1
390 GOSUB 850
400 LET B=G
410 IF M1=0 THEN GOTO 450
420 FOR I=1 TO M1
430 LET B=B*(X-I)*(Y-I)/(Z0-I)
440 NEXT I
450 LET Z=Y-M2
460 GOSUB 850
470 LET B=B*G
480 IF M2=M1 THEN GOTO 520
490 FOR I=(M1+1) TO M2
500 LET B=B*(Y-I)/(Z0-I)
510 NEXT I
520 LET Z=Z0-M3
530 IF Z=0 THEN GOTO 740
540 GOSUB 850
550 LET B=B/G
560 IF M3=M2 THEN GOTO 800
570 FOR I=(M2+1) TO M3
580 LET B=B/(Z0-I)
590 NEXT I
600 GOTO 800
610 REM COMPUTATION OF BETA IF X OR Y OR X AND Y ARE INTEGER
620 LET M1=X-1
```

```
630 LET Z=Y
640 GOTO 670
650 LET M1=Y-1
660 LET Z=X
670 LET B=1/Z
680 IF M1=0 THEN GOTO 800
690 FOR I=1 TO M1
700 LET B=B*I/(Z+I)
710 NEXT I
720 GOTO 800
730 REM COMPUTATION OF BETA IF ONLY X+Y IS INTEGER
740 LET M1=M3-M2-1
750 IF M1<2 THEN GOTO 800
760 FOR I=2 TO M1
770 LET B=B/I
780 NEXT I
790 GOTO 800
800 PRINT "X=";X;"Y=";Y;"B=";B
810 IF O2>0 THEN LPRINT "X=";X;"Y=";Y;"B=";B
820 PRINT:PRINT"TO CONTINUE PRESS ENTER, TO STOP PRESS BREAK"
830 INPUT A$:GOTO 170
840 REM SUBROUTINE FOR 1/G(Z)
850 FOR I=25 TO 1 STEP -1
860 LET J=I+1
870 LET A(I)=A(J)*Z+C(I)
880 NEXT I
890 LET G=1/(Z*A(1))
900 RETURN
910 DATA 1,.5772156649015329,-.6558780715202538
920 DATA -.0420026350340952,.1665386113822915
930 DATA -.0421977345555443,-.0096219715278770
940 DATA .0072189432466630,-.0011651675918591
950 DATA -.0002152416741149,.0001280502823882
960 DATA -2.01348547807D-05,-1.2504934821D-06
970 DATA 1.1330272320D-06,-2.056338417D-07
980 DATA 6.1160950D-09,5.0020075D-09,-1.1812746D-09
990 DATA 1.043427D-10,7.7823D-12,-3.6968D-12,5.100D-13
1000 DATA -2.06D-14,-5.4D-15,1.4D-15,1.D-16
1010 END
```

Examples

1. Beta function B(x,y) in single precision:
    B(2.25,3.75)=0.041605,
    B(1,3.25)=0.307692,
    B(12,17)=2.73927*10^{-9}.

2. Beta function B(x,y) in double precision:
    B(2.5,7.25)=7.406949344527435*10^{-3},
    B(2,3)=0.08333333333333334,
    B(3.5,32.5)=1.490197821488766*10^{-5}.

P812 Gudermannian, Its Inverse, and Hyperbolic Functions in Single Precision

```
10 CLS:PRINT"P812 EVALUATES GUDERMANNIAN, ITS INVERSE AND"
20 PRINT"HYPERBOLIC FUNCTIONS AT A GIVEN X."
30 PRINT:PRINT"WARNING: BECAUSE OF ROUNDOFF ERRORS THE VALUES"
40 PRINT"OF SINH(X) AND COSH(X) ARE MORE AND MORE INACCURATE"
50 PRINT"IF X>3. IN THIS CASE USE PA1 (SINGLE PRECISION) OR"
60 PRINT"PA2 (DOUBLE PRECISION). PA1 AND PA2 COMPUTE ALSO"
70 PRINT"INVERSE HYPERBOLIC FUNCTIONS."
80 PRINT:PRINT"THE INPUT:"
90 PRINT"SINGLE PRECISION ONLY."
100 PRINT"IF PRINTER NOT USED O2=0, IF USED O2>0."
110 DEFINT I-O:INPUT"O2";O2:CLS
120 IF O2=0 THEN GOTO 140
130 LPRINT "P812: O2=";O2
140 PRINT"O3 IS AN INTEGER VARIABLE DETRMINING THE FUNCTION."
150 PRINT"INDEPENDENT VARIABLE: X":PRINT
160 PRINT "GUDERMANNIAN: O3=1"
170 PRINT "INVERSE GUDERMANNIAN: O3=2"
180 PRINT "SINH(X): O3=3"
190 PRINT "COSH(X): O3=4"
200 PRINT "TANH(X): O3=5"
210 PRINT "TANH(X/2): O3=6"
220 PRINT "COTH(X): O3=7"
230 PRINT "COTH(X/2): O3=8"
240 DEF FNG(X)=2*ATN(EXP(X))-1.570796
250 INPUT"O3";O3:INPUT"X";X
260 IF O3<1 THEN GOTO 290
270 IF O3>8 THEN GOTO 290
280 PRINT:ON O3 GOTO 300,350,390,430,470,510,550,590
290 CLS:PRINT"ERROR IN O3":GOTO 140
300 LET F=FNG(X)
310 IF F>1.570796 THEN LET F=1.570796
320 PRINT "X=";X;"GD(X)=";F
330 IF O2>0 THEN LPRINT "X=";X;"GD(X)=";F
340 GOTO 620
350 LET F=LOG(TAN(X/2+.7853982))
360 PRINT "X=";X;"INVERSE GD(X)=";F
370 IF O2>0 THEN LPRINT "X=";X;"INVERSE GD(X)=";F
380 GOTO 620
390 LET F=TAN(FNG(X))
400 PRINT "X=";X;"SINH(X)=";F
410 IF O2>0 THEN LPRINT "X=";X;"SINH(X)=";F
420 GOTO 620
430 LET F=1/COS(FNG(X))
440 PRINT "X=";X;"COSH(X)=";F
450 IF O2>0 THEN LPRINT "X=";X;"COSH(X)=";F
460 GOTO 620
470 LET F=SIN(FNG(X))
480 PRINT "X=";X;"TANH(X)=";F
490 IF O2>0 THEN LPRINT "X=";X;"TANH(X)=";F
500 GOTO 620
510 LET F=TAN(FNG(X)/2)
520 PRINT "X=";X;"TANH(X/2)=";F
530 IF O2>0 THEN LPRINT "X=";X;"TANH(X/2)=";F
540 GOTO 620
550 LET F=1/SIN(FNG(X))
560 PRINT "X=";X;"COTH(X)=";F
570 IF O2>0 THEN LPRINT "X=";X;"COTH(X)=";F
580 GOTO 620
590 LET F=1/TAN(FNG(X)/2)
600 PRINT "X=";X;"COTH(X/2)=";F
```

```
610 IF O2>0 THEN LPRINT "X=";X;"COTH(X/2)=";F
620 PRINT:PRINT"TO CONTINUE PRESS ENTER, TO STOP PRESS BREAK"
630 INPUT A$:CLS:GOTO 140
640 END
```

Examples

Gudermannian $gd(x)$, its inverse $gd^{-1}(x)$ and hyperbolic functions in single precision:

$gd(0.47)=0.453595$,
$gd^{-1}(1.57)=7.82881$,
$sinh(1.54)=2.22511$,
$cosh(0.43)=1.09388$,
$tanh(0.43)=0.405322$,
$tanh(5/2)=0.986615$,
$coth(0.21)=4.83169$,
$coth(0.42/2)=4.83169$.

CHAPTER 9: SELECTED PROBLEMS OF MATHEMATICAL STATISTICS

P901 Permutations, Variations, and Combinations

```
10 CLS:PRINT"P901 COMPUTES PERMUTATIONS, PERMUTATIONS WITH"
20 PRINT"REPETITION, VARIATIONS, VARIATIONS WITH REPETITION,"
30 PRINT"COMBINATIONS, COMBINATIONS WITH REPETITION."
40 PRINT:PRINT"THE INPUT:"
50 PRINT"DOUBLE PRECISION ONLY."
60 PRINT"IF PRINTER NOT USED O2=0, IF USED O2>0."
70 DEFINT I-O:DEFDBL P:INPUT"O2";O2:CLS
80 IF O2=0 THEN GOTO 110
90 LPRINT "P901: O2=";O2
100 LET L=0
110 PRINT"O3 IN AN INTEGER VARIABLE DETERMINING THE ARRANGEMENT."
120 PRINT
130 PRINT "PERMUTATIONS: O3=1"
140 PRINT "PERMUTATIONS WITH REPETITION: O3=2"
150 PRINT "VARIATIONS: O3=3"
160 PRINT "VARIATIONS WITH REPETITION: O3=4"
170 PRINT "COMBINATIONS: O3=5"
180 PRINT "COMBINATIONS WITH REPETITION: O3=6"
190 INPUT"O3";O3
200 IF O3<1 THEN GOTO 250
210 IF O3>6 THEN GOTO 250
220 IF O2>0 THEN LPRINT
230 CLS:PRINT"NUMBER OF ELEMENTS: N"
240 ON O3 GOTO 260,350,730,830,930,930
250 PRINT"ERROR IN O3":GOTO 190
260 INPUT"N";N
270 LET P=1
280 FOR I=2 TO N
290 LET P=P*I
300 NEXT I
310 PRINT "N=";N;"NUMBER OF PERMUTATIONS=";P
320 IF O2=0 THEN GOTO 1240
330 LPRINT "N=";N;"NUMBER OF PERMUTATIONS=";P
340 GOTO 1240
350 PRINT"THERE ARE M DIFFERENT ELEMENTS. THE I-TH ELEMENT"
360 PRINT"OCCURS p(I)-TIMES.":PRINT
370 PRINT"IF PERMUTATIONS WITH REPETITION ARE TO BE COMPUTED"
380 PRINT"AGAIN WITH M GREATER THAN M IN THE FIRST RUN PRESS"
390 PRINT"BREAK AND RUN AGAIN."
400 INPUT"M";M
410 IF L=0 THEN DIM K0(M)
420 LET L=1
430 FOR I=1 TO M
440 PRINT "I=";I
450 INPUT"p(I)";K0(I)
460 NEXT I
470 LET P=0
480 FOR I=1 TO M
490 LET P=P+K0(I)
500 NEXT I
510 LET N=P
520 LET P=1
530 FOR I=2 TO N
540 LET P=P*I
550 NEXT I
560 LET Q=1
570 FOR I=1 TO M
580 FOR J=1 TO K0(I)
```

P. 278   PROGRAM P901

```
590 LET Q=Q*J
600 NEXT J
610 NEXT I
620 LET P=INT(P/Q)
630 PRINT "N=";N;"M=";M;"NUMBER OF PERMUTATIONS WITH REPET.=";P
640 IF O2=0 THEN GOTO 660
650 LPRINT "N=";N;"M=";M;"NUMBER OF PERMUTATIONS WITH REPET.=";P
660 PRINT "THE I-TH ELEMENT OCCURS p(I)-TIMES:"
670 IF O2>0 THEN LPRINT "THE I-TH ELEMENT OCCURS p(I)-TIMES:"
680 FOR I=1 TO M
690 PRINT "I=";I;"p(I)=";K0(I)
700 IF O2>0 THEN LPRINT "I=";I;"p(I)=";K0(I)
710 NEXT I
720 GOTO 1240
730 PRINT"NUMBER OF ELEMENTS IN EACH SEQUENCE: K"
740 INPUT"N";N:INPUT"K";K
750 LET P=1
760 FOR I=(N-K+1) TO N
770 LET P=P*I
780 NEXT I
790 PRINT "N=";N;"K=";K;"NUMBER OF VARIATIONS=";P
800 IF O2=0 THEN GOTO 1240
810 LPRINT "N=";N;"K=";K;"NUMBER OF VARIATIONS=";P
820 GOTO 1240
830 PRINT"NUMBER OF ELEMENTS IN EACH SEQUENCE: K"
840 INPUT"N";N:INPUT"K";K
850 LET P=N
860 FOR I=2 TO K
870 LET P=P*N
880 NEXT I
890 PRINT "N=";N;"K=";K;"NUMBER OF VARIATIONS WITH REPET.=";P
900 IF O2=0 THEN GOTO 1240
910 LPRINT "N=";N;"K=";K;"NUMBER OF VARIATIONS WITH REPET.=";P
920 GOTO 1240
930 PRINT"NUMBER OF ELEMENTS IN EACH GROUP: K"
940 INPUT"N";N:INPUT"K";K
950 IF O3=6 THEN GOTO 1050
960 IF K<=N THEN GOTO 990
970 PRINT"ERROR IN K; K<=N.":GOTO 940
980 IF O3=6 THEN GOTO 1050
990 IF K<N THEN GOTO 1020
1000 LET P=1
1010 GOTO 1170
1020 LET N0=N
1030 LET M=N-K
1040 GOTO 1070
1050 LET N0=N+K-1
1060 LET M=N-1
1070 LET P=1
1080 IF K>N0/2 THEN GOTO 1130
1090 FOR I=1 TO K
1100 LET P=P*(I+M)/I
1110 NEXT I
1120 GOTO 1160
1130 FOR I=1 TO M
1140 LET P=P*(I+K)/I
1150 NEXT I
1160 IF O3=6 THEN GOTO 1210
1170 PRINT "N=";N;"K=";K;"NUMBER OF COMBINATIONS=";P
1180 IF O2=0 THEN GOTO 1240
1190 LPRINT "N=";N;"K=";K;"NUMBER OF COMBINATIONS=";P
1200 GOTO 1240
1210 PRINT "N=";N;"K=";K;"NUMBER OF COMBINATIONS WITH REPET.=";P
1220 IF O2=0 THEN GOTO 1240
```

```
1230 LPRINT "N=";N;"K=";K;"NUMBER OF COMBINATIONS WITH REPET.=";P
1240 PRINT:PRINT"TO CONTINUE PRESS ENTER, TO STOP PRESS BREAK"
1250 INPUT A$:CLS:GOTO 110
1260 END
```

Examples

1. Permutations of N different elements:

    N=5,   P=120,

    N=10, P=3828800.

2. Permutations of N elements with repetitions:

    N=6, $P_1=2$, $P_2=1$, $P_3=3$, $P_R=60$,

    N=9, $P_1=4$, $P_2=5$, $P_R=126$.

3. Variations of N elements, each sequence containing K elements:

    N=12, K=6,   V=665280,

    N=5, K=3,   V=60.

4. Variations of N elements with repetitions, each sequence containing K elements:

    N=12, K=6, $V_R=2985984$,

    N=5, K=3, $V_R=125$.

5. Combinations of N elements, each group containing K elements:

    N=12, K=7,   C=792,

    N=50, K=25,   C=126410606437752.

6. Combinations of N elements with repetitions, each group containing K elements:

    N=6, K=3, $C_R=56$,

    N=9, K=4, $C_R=495$.

P902 Arithmetic, Geometric, and Harmonic Mean Values, Variances S^2, s^2, Standard Deviations S, s, Standard Errors E, e for a Sample of n Data

```
10 CLS:PRINT"P902 COMPUTES THE BASIC STATISTICAL QUANTITIES:"
20 PRINT"ARITHMETIC, GEOMETRIC, HARMONIC MEAN VALUES; VARIANCES"
30 PRINT"S2,s2; STANDARD DEVIATIONS S,s; STANDARD ERRORS E,e."
40 PRINT:PRINT"THE INPUT:"
50 PRINT"SINGLE PRECISION ONLY."
60 PRINT"IF PRINTER NOT USED O2=0, IF USED O2>0."
70 DEFINT I-O:INPUT"O2";O2:CLS
80 PRINT"NUMBER OF OBJECTS IN A SAMPLE: N"
90 INPUT"N";N:PRINT
100 PRINT"SAMPLE VALUES: X(I)"
110 IF O2=0 THEN GOTO 130
120 LPRINT "P902: O2=";O2;"N=";N
130 DIM X(N)
140 REM INPUT OF DATA
150 FOR I=1 TO N
160 PRINT "I=";I
170 INPUT"X(I)";X(I)
180 NEXT I
190 CLS:PRINT "CHECK THE INPUT:"
200 FOR I=1 TO N
210 PRINT "I=";I;"X(I)=";X(I)
220 PRINT"IF NO ERROR INPUT 0"
230 PRINT"IF ERROR INPUT ANY NUMBER>0 AND THEN THE CORRECT X(I)"
240 INPUT"ERROR";K
250 IF K>0 THEN INPUT"X(I)";X(I)
260 CLS:NEXT I
270 IF O2=0 THEN GOTO 340
280 LPRINT "THE INPUT:"
290 FOR I=1 TO N
300 LPRINT "I=";I;"X(I)=";X(I)
310 NEXT I
320 LPRINT
330 REM COMPUTATION OF MEAN VALUES
340 LET S0=0
350 LET S1=1
360 LET S2=0
370 FOR I=1 TO N
380 LET S0=S0+X(I)
390 LET S1=S1*X(I)
400 LET S2=S2+1/X(I)
410 NEXT I
420 LET S0=S0/N
430 LET S1=S1[ (1/N)
440 LET S2=N/S2
450 REM COMPUTATION OF VARIANCES, STANDARD DEVIATIONS AND ERRORS
460 LET S3=0
470 FOR I=1 TO N
480 LET S3=S3+X(I)*X(I)
490 NEXT I
500 LET S3=S3-N*S0*S0
510 LET S4=S3/(N-1)
520 LET S3=S3/N
530 LET S5=SQR(S3)
540 LET S6=SQR(S4)
550 LET S9=SQR(N)
560 LET S7=S5/S9
570 LET S8=S6/S9
580 PRINT "ARITHMETIC MEAN=";S0
590 PRINT "GEOMETRIC MEAN=";S1
```

```
600 PRINT "HARMONIC MEAN=";S2
610 PRINT "VARIANCE S2=";S3
620 PRINT "VARIANCE s2=";S4
630 PRINT "STANDARD DEVIATION S=";S5
640 PRINT "STANDARD DEVIATION s=";S6
650 PRINT "STANDARD ERROR E=+-";S7
660 PRINT "STANDARD ERROR e=+-";S8
670 PRINT:PRINT"DIVISOR IS N WHEN S2,S,E COMPUTED."
680 PRINT"DIVISOR IS (N-1) WHEN s2,s,e COMPUTED."
690 IF O2=0 THEN STOP
700 LPRINT "ARITHMETIC MEAN=";S0
710 LPRINT "GEOMETRIC MEAN=";S1
720 LPRINT "HARMONIC MEAN=";S2
730 LPRINT "VARIANCE S2=";S3
740 LPRINT "VARIANCE s2=";S4
750 LPRINT "STANDARD DEVIATION S=";S5
760 LPRINT "STANDARD DEVIATION s=";S6
770 LPRINT "STANDARD ERROR E=+-";S7
780 LPRINT "STANDARD ERROR e=+-";S8
790 LPRINT:LPRINT"DIVISOR IS N WHEN S2,S,E COMPUTED."
800 LPRINT"DIVISOR IS (N-1) WHEN s2,s,e COMPUTED."
810 END
```

Example

A sample consists of 10 values x_i:

1.001, 1.024, 0.987, 0.976, 1.002,

1.010, 0.990, 0.973, 1.007, 1.003.

Arithemetic mean: 0.9973.

Geometric mean: 0.997188.

Harmonic mean: 0.997075.

Variance S^2: $2.24018*10^{-4}$.

Variance s^2: $2.48909*10^{-4}$.

Standard deviation S: 0.0149672.

Standard deviation s: 0.0157768.

Standard error E: $+-4.73306*10^{-3}$.

Standard error e: $+-4.98908*10^{-3}$.

P903 Arithmetic Mean Value, Variances S^2, s^2, Standard Deviations S, s, and Standard Errors E, e for a Sample of Grouped Data

```
10 CLS:PRINT"P903 COMPUTES THE BASIC STATISTICAL QUANTITIES"
20 PRINT"FOR GROUPPED DATA: ARITHMETIC MEAN VALUE; VARIANCES"
30 PRINT"S2,s2; STANDARD DEVIATIONS S,s; STANDARD ERRORS E,e."
40 PRINT:PRINT"THE INPUT:"
50 PRINT"SINGLE PRECISION ONLY."
60 PRINT"IF PRINTER NOT USED O2=0, IF USED O2>0."
70 DEFINT I-O:INPUT"O2";O2:CLS
80 PRINT"NUMBER OF GROUPS IN A SAMPLE: N"
90 INPUT"N";N:PRINT
100 PRINT"SAMPLE VALUES X(I) OCCUR WITH FREQUENCIES F(I)."
110 IF O2=0 THEN GOTO 130
120 LPRINT "P903: O2=";O2;"N=";N
130 DIM X(N),F(N)
140 REM INPUT OF DATA
150 FOR I=1 TO N
160 PRINT "I=";I
170 INPUT"X(I)";X(I):INPUT"F(I)";F(I)
180 NEXT I
190 CLS:PRINT "CHECK THE INPUT:"
200 FOR I=1 TO N
210 PRINT "I=";I;"X(I)=";X(I);"F(I)=";F(I)
220 PRINT"IF NO ERROR IN X(I) INPUT 0"
230 PRINT"IF ERROR INPUT ANY NUMBER>0 AND THEN THE CORRECT X(I)"
240 INPUT"ERROR";K
250 IF K>0 THEN INPUT"X(I)";X(I)
260 PRINT:PRINT"IF NO ERROR IN F(I) INPUT 0"
270 PRINT"IF ERROR INPUT ANY NUMBER>0 AND THEN THE CORRECT F(I)"
280 INPUT"ERROR";K
290 IF K>0 THEN INPUT"F(I)";F(I)
300 CLS:NEXT I
310 IF O2=0 THEN GOTO 380
320 LPRINT "THE INPUT:"
330 FOR I=1 TO N
340 LPRINT "I=";I;"X(I)=";X(I);"F(I)=";F(I)
350 NEXT I
360 LPRINT
370 REM COMPUTATION OF MEAN VALUES
380 LET S0=0
390 LET F0=0
400 FOR I=1 TO N
410 LET S0=S0+X(I)*F(I)
420 LET F0=F0+F(I)
430 NEXT I
440 LET S0=S0/F0
450 REM COMPUTATION OF VARIANCES, STANDARD DEVIATIONS AND ERRORS
460 LET S3=0
470 FOR I=1 TO N
480 LET S3=S3+X(I)*X(I)*F(I)
490 NEXT I
500 LET S3=S3-S0*S0*F0
510 LET S4=S3/(F0-1)
520 LET S3=S3/F0
530 LET S5=SQR(S3)
540 LET S6=SQR(S4)
550 LET S9=SQR(N)
560 LET S7=S5/S9
570 LET S8=S6/S9
580 PRINT "NUMBER OF ALL OBJECTS IN THE SAMPLE=";F0
590 PRINT "ARITHMETIC MEAN=";S0
```

```
600 PRINT "VARIANCE S2=";S3
610 PRINT "VARIANCE s2=";S4
620 PRINT "STANDARD DEVIATION S=";S5
630 PRINT "STANDARD DEVIATION s=";S6
640 PRINT "STANDARD ERROR E=+-";S7
650 PRINT "STANDARD ERROR e=+-";S8
660 PRINT:PRINT"DIVISOR IS N WHEN S2,S,E COMPUTED."
670 PRINT"DIVISOR IS (N-1) WHEN s2,s,e COMPUTED."
680 IF O2=0 THEN STOP
690 LPRINT "NUMBER OF ALL OBJECTS IN THE SAMPLE=";F0
700 LPRINT "ARITHMETIC MEAN=";S0
710 LPRINT "VARIANCE S2=";S4
720 LPRINT "VARIANCE s2=";S4
730 LPRINT "STANDARD DEVIATION S=";S5
740 LPRINT "STANDARD DEVIATION s=";S6
750 LPRINT "STANDARD ERROR E=+-";S7
760 LPRINT "STANDARD ERROR e=+-";S8
770 LPRINT:LPRINT"DIVISOR IS N WHEN S2,S,E COMPUTED."
780 LPRINT"DIVISOR IS (N-1) WHEN s2,s,e COMPUTED."
790 END
```

Example

A sample consists of 10 groups (x_j, f_j) in each of which a value x_j occurs f_j times:

(1.0021, 7), (0.9786, 6), (1.0001, 21), (0.934, 1), (0.910, 1),

(1.1, 2), (1.004, 27), (0.9991, 17), (0.1003, 4), (0.9931, 3).

Number of all objects in the sample: 89.

Arithmetic mean: 0.959614.

Variance S^2: 0.035551.

Variance s^2: 0.035551.

Standard deviation S: 0.187487.

Standard deviation s: 0.188550.

Standard error E: +-0.0592887.

Standard error e: +-0.0596247.

P904 Linear, Power, Exponential, and Logarithmic Curve Fit

```
10 CLS:PRINT"P904 COMPUTES THE LINEAR, POWER, EXPONENTIAL,"
20 PRINT"LOGARITHMIC CURVE FIT."
30 PRINT:PRINT"THE INPUT:"
40 PRINT"SINGLE PRECISION ONLY."
50 PRINT"IF PRINTER NOT USED O2=0, IF USED O2>0."
60 DEFINT I-O:INPUT"O2";O2:CLS
70 PRINT"NUMBER OF PAIRS X(I),Y(I) IN A SAMPLE: N"
80 INPUT"N";N:CLS
90 IF O2=0 THEN GOTO 110
100 LPRINT "P904: O2=";O2;"N=";N
110 DIM X(N),Y(N)
120 PRINT"INPUT OF DATA:"
130 FOR I=1 TO N
140 PRINT "I=";I
150 INPUT"X(I)";X(I):INPUT"Y(I)";Y(I)
160 NEXT I
170 CLS:PRINT "CHECK THE INPUT:"
180 FOR I=1 TO N
190 PRINT "I=";I;"X(I)=";X(I);"Y(I)=";Y(I)
200 PRINT"IF NO ERROR IN X(I) INPUT 0"
210 PRINT"IF ERROR INPUT ANY NUMBER>0 AND THEN THE CORRECT X(I)"
220 INPUT"ERROR";K
230 IF K>0 THEN INPUT"X(I)";X(I)
240 PRINT:PRINT"IF NO ERROR IN Y(I) INPUT 0"
250 PRINT"IF ERROR INPUT ANY NUMBER>0 AND THEN THE CORRECT Y(I)"
260 INPUT"ERROR";K
270 IF K>0 THEN INPUT"Y(I)";Y(I)
280 CLS:NEXT I
290 IF O2=0 THEN GOTO 350
300 LPRINT "THE INPUT:"
310 FOR I=1 TO N
320 LPRINT "I=";I;"X(I)=";X(I);"Y(I)=";Y(I)
330 NEXT I
340 LPRINT
350 CLS:PRINT"O3 IS AN INTEGER VARIABLE DETERMINING THE CURVE FIT"
360 PRINT"SEVERAL OR ALL CURVE FITS MAY BE RUN TO FIND THE BEST"
370 PRINT"CURVE FIT FOR THE GIVEN DATA.":PRINT
380 PRINT "LINEAR CURVE FIT: O3=1"
390 PRINT "POWER CURVE FIT: O3=3"
400 PRINT "EXPONENTIAL CURVE FIT: O3=4"
410 PRINT "LOGARITHMIC CURVE FIT: O3=2"
420 INPUT"O3";O3
430 IF O3<1 THEN GOTO 460
440 IF O3>4 THEN GOTO 460
450 CLS:GOTO 480
460 PRINT"ERROR IN O3":GOTO 420
470 REM COMPUTATION OF REGRESSION AND DETERMINATION COEFFICIENTS
480 LET X0=0
490 LET Y0=0
500 LET X2=0
510 LET X3=0
520 LET Y2=0
530 FOR I=1 TO N
540 ON O3 GOTO 550  ,580  ,610  ,640
550 LET C=X(I)
560 LET D=Y(I)
570 GOTO 660
580 LET C=LOG(X(I))
590 LET D=Y(I)
600 GOTO 660
610 LET C=LOG(X(I))
620 LET D=LOG(Y(I))
```

```
630 GOTO 660
640 LET C=X(I)
650 LET D=LOG(Y(I))
660 LET X0=X0+C
670 LET Y0=Y0+D
680 LET X2=X2+C*C
690 LET X3=X3+C*D
700 LET Y2=Y2+D*D
710 NEXT I
720 LET X0=X0/N
730 LET Y0=Y0/N
740 LET B=(X3-N*X0*Y0)/(X2-N*X0*X0)
750 LET A=Y0-B*X0
760 IF O3<3 THEN GOTO 780
770 LET A=EXP(A)
780 LET R2=B*B*(X2-N*X0*X0)/(Y2-N*Y0*Y0)
790 REM COMPUTATION OF THE SUM OF SQUARES OF RESIDUALS
800 LET S=0
810 FOR I=1 TO N
820 ON O3 GOTO 830  ,850  ,870  ,890
830 LET S0=A+B*X(I)-Y(I)
840 GOTO 900
850 LET S0=A+B*LOG(X(I))-Y(I)
860 GOTO 900
870 LET S0=A*(X(I)[B)-Y(I)
880 GOTO 900
890 LET S0=A*EXP(B*X(I))-Y(I)
900 LET S=S+S0*S0
910 NEXT I
920 PRINT"REGRESSION COEFFICIENTS: A,B"
930 PRINT"COEFFICIENT OF DETERMINATION: R2"
940 PRINT"IF R2=1 THE CURVE FIT IS PERFECT."
950 PRINT"SUM OF SQUARES OF RESIDUALS: S"
960 PRINT"IF S=0 THE CURVE GOES THROUGH ALL THE GIVEN POINTS."
970 PRINT
980 PRINT "O3=";O3;"A=";A;"B=";B;"R2=";R2;"S=";S
990 IF O2=0 THEN GOTO 1070
1000 IF O3=1 THEN LPRINT "LINEAR CURVE FIT:"
1010 IF O3=3 THEN LPRINT "POWER CURVE FIT:"
1020 IF O3=4 THEN LPRINT "EXPONENTIAL CURVE FIT:"
1030 IF O3=2 THEN LPRINT "LOGARITHMIC CURVE FIT:"
1040 LPRINT "REGRESSION COEFFICIENTS: A=";A;"B=";B
1050 LPRINT "DETERMINATION COEFFICIENT R2=";R2
1060 LPRINT "SUM OF SQUARES OF RESIDUALS=";S
1070 PRINT"THE VALUES Y ON THE REGRESSION CURVE ARE COMPUTED"
1080 PRINT"FOR ANY NUMBER OF VALUES OF THE VARIABLE X."
1090 INPUT"X";X1
1100 ON O3 GOTO 1110,1130,1150,1170
1110 LET Y1=A+B*X1
1120 GOTO 1180
1130 LET Y1=A+B*LOG(X1)
1140 GOTO 1180
1150 LET Y1=A*(X1[B)
1160 GOTO 1180
1170 LET Y1=A*EXP(B*X1)
1180 PRINT "O3=";O3;"X=";X1;"Y=";Y1
1190 IF O2>0 THEN LPRINT "O3=";O3;"X=";X1;"Y=";Y1
1200 PRINT:PRINT"IF THE SAME CURVE FIT O4=0, IF ANOTHER O4>0."
1210 PRINT"TO STOP PRESS BREAK"
1220 INPUT"O4";O4
1230 IF O4=0 THEN GOTO 1090
1240 IF O2>0 THEN LPRINT
1250 GOTO 350
1260 END
```

Examples

1. Kepler's data (Epitome of Copernican Astronomy, Bk. iv, Part 1) of relative distances x_i of 6 planets and their relative periods y_i (for Earth $x_3=1$, $y_3=1$):

(0.45, 0.241), (0.75, 0.616), (1,1),

(1.5, 1.882), (4.5, 12), (9,30).

Linear curve fit:
Regression coefficients: A=-2.48295, B=3.52539.
Determination coefficient: R^2=0.9931.
Sum of squares of residuals: 4.83207.
For x=1, y=1.04244.

Logarithmic curve fit:
Regression coefficients: A=2.83917, B=9.5028.
Determination coefficient: R^2=0.832258.
Sum of squares of residuals: 117.471.
For x=1, y=2.83917.

Power curve fit:
Regression coefficients: A=0.95659, B=1.61468.
Determination coefficient R^2=0.998126.
Sum of squares of residuals: 11.7527.
For x=1, y=0.95659.

Exponential curve fit:
Regression coefficients: A=0.500314, B=0.509657.
Determination coefficient R^2=0.862171.
Sum of squares of residuals: 416.244.
For x=1, y=0.832882.

The power curve fit is the best fit with the highest value of determination coefficient. The regression coefficient B=1.61 is the exponent. Kepler approximated this value as 3/2 when he formulated his third law.

2. Present data for 6 planets known to Kepler:

(0.387098, 0.241), (0.723333, 0.615), (1,1),

(1.5237, 1.881), (5.2037, 11.865), (9.5803, 29.65).

Program P904 yields the following values of the determination coefficient for these data:
Linear curve fit: R^2=0.947812.
Logarithmic curve fit: R^2=0.800982.
Power curve fit: R^2=1.
Exponential curve fit: R^2=0.882829.

In the power curve fit the regression coefficient B, the exponent, equals 1.49981, and for x=1, y=1.00008. Thus the present data confirm Kepler's third law with a far higher precision.

P905   Binomial-Power Curve Fit

```
10 CLS:PRINT"P905 COMPUTES THE BINOMIAL POWER CURVE FIT."
20 PRINT:PRINT"THE INPUT:"
30 PRINT"SINGLE PRECISION ONLY."
40 PRINT"IF PRINTER NOT USED O2=0, IF USED O2>0."
50 DEFINT I-O:INPUT"O2";O2:CLS
60 PRINT"NUMBER OF PAIRS X(I),Y(I) IN A SAMPLE: N"
70 INPUT"N";N:CLS
80 PRINT"GIVEN EXPONENTS IN THE BINOMIAL CURVE: C,D"
90 INPUT"C";C:INPUT"D";D:CLS
100 IF O2=0 THEN GOTO 120
110 LPRINT "P905: O2=";O2;"N=";N;"C=";C;"D=";D
120 DIM X(N),Y(N)
130 PRINT"INPUT OF DATA:"
140 FOR I=1 TO N
150 PRINT "I=";I
160 INPUT"X(I)";X(I):INPUT"Y(I)";Y(I)
170 NEXT I
180 CLS:PRINT "CHECK THE INPUT:"
190 FOR I=1 TO N
200 PRINT "I=";I;"X(I)=";X(I);"Y(I)=";Y(I)
210 PRINT"IF NO ERROR IN X(I) INPUT 0"
220 PRINT"IF ERROR INPUT ANY NUMBER>0 AND THEN THE CORRECT X(I)"
230 INPUT"ERROR";K
240 IF K>0 THEN INPUT"X(I)";X(I)
250 PRINT:PRINT"IF NO ERROR IN Y(I) INPUT 0"
260 PRINT"IF ERROR INPUT ANY NUMBER>0 AND THEN THE CORRECT Y(I)"
270 INPUT"ERROR";K
280 IF K>0 THEN INPUT"Y(I)";Y(I)
290 CLS:NEXT I
300 IF O2=0 THEN GOTO 360
310 LPRINT "THE INPUT:"
320 FOR I=1 TO N
330 LPRINT "I=";I;"X(I)=";X(I);"Y(I)=";Y(I)
340 NEXT I
350 LPRINT
360 CLS:PRINT "BINOMIAL POWER CURVE FIT:"
370 REM COMPUTATION OF REGRESSION COEFFICIENTS
380 LET X2=0
390 LET X3=0
400 LET X4=0
410 LET X5=0
420 LET X6=0
430 FOR I=1 TO N
440 LET C0=X(I)[C
450 LET D0=X(I)[D
460 LET X2=X2+C0*C0
470 LET X3=X3+D0*D0
480 LET X4=X4+C0*D0
490 LET X5=X5+C0*Y(I)
500 LET X6=X6+D0*Y(I)
510 NEXT I
520 LET B=(X6*X2-X5*X4)/(X2*X3-X4*X4)
530 LET A=(X5-B*X4)/X2
540 REM COMPUTATION OF THE SUM OF SQUARES OF RESIDUALS
550 LET S=0
560 FOR I=1 TO N
570 LET S0=A*(X(I)[C)+B*(X(I)[D)-Y(I)
580 LET S=S+S0*S0
590 NEXT I
600 PRINT"REGRESSION COEFFICIENTS: A,B"
610 PRINT"SUM OF SQUARES OF RESIDUALS: S"
620 PRINT"IF S=0 THE CURVE GOES THROUGH ALL THE GIVEN POINTS."
```

PROGRAM P905

```
630 PRINT "A=";A;"B=";B;"S=";S:PRINT
640 IF O2=0 THEN GOTO 680
650 LPRINT "BINOMIAL POWER CURVE FIT: C=";C;"D=";D
660 LPRINT "REGRESSION COEFFICIENTS: A=";A;"B=";B
670 LPRINT "SUM OF SQUARES OF RESIDUALS=";S
680 PRINT"THE VALUES Y ON THE REGRESSION CURVE MAY BE COMPUTED"
690 PRINT"FOR ANY NUMBER OF VALUES OF THE VARIABLE X."
700 PRINT"TO STOP PRESS BREAK."
710 INPUT"X";X1
720 LET Y1=A*(X1[C)+B*(X1[D)
730 PRINT "X=";X1;"Y=";Y1
740 IF O2>0 THEN LPRINT "X=";X1;"Y=";Y1
750 GOTO 710
760 END
```

Example

A curve is given by 4 points

(1, 9), (4, -44), (9, -699), (16, -4056).

It is approximated by the binomial-power curve fit

$$y = Ax^C + Bx^D$$

with C=.5 and D=3. The regression coefficients are

A=9.99994,   B=-1,

and the sum of squares of residuals equals 3.25512×10^{-7}.
For x=13.67,   y=-2517.53.

x=9,   y=-699.001.

x=6.5,   y=-249.13.

P906 Parabolic Curve Fit

```
10 CLS:PRINT"P906 COMPUTES THE PARABOLIC CURVE FIT."
20 PRINT:PRINT"THE INPUT:"
30 PRINT"SINGLE PRECISION ONLY."
40 PRINT"IF PRINTER NOT USED O2=0, IF USED O2>0."
50 DEFINT I-O:INPUT"O2";O2:CLS
60 PRINT"NUMBER OF PAIRS X(I),Y(I) IN A SAMPLE: N"
70 INPUT"N";N:CLS
80 IF O2=0 THEN GOTO 100
90 LPRINT "P906: O2=";O2;"N=";N
100 DIM X(N),Y(N)
110 PRINT"INPUT OF DATA:"
120 FOR I=1 TO N
130 PRINT "I=";I
140 INPUT"X(I)";X(I):INPUT"Y(I)";Y(I)
150 NEXT I
160 CLS:PRINT "CHECK THE INPUT:"
170 FOR I=1 TO N
180 PRINT "I=";I;"X(I)=";X(I);"Y(I)=";Y(I)
190 PRINT"IF NO ERROR IN X(I) INPUT 0"
200 PRINT"IF ERROR INPUT ANY NUMBER>0 AND THEN THE CORRECT X(I)"
210 INPUT"ERROR";K
220 IF K>0 THEN INPUT"X(I)";X(I)
230 PRINT:PRINT"IF NO ERROR IN Y(I) INPUT 0"
240 PRINT"IF ERROR INPUT ANY NUMBER>0 AND THEN THE CORRECT Y(I)"
250 INPUT"ERROR";K
260 IF K>0 THEN INPUT"Y(I)";Y(I)
270 CLS:NEXT I
280 IF O2=0 THEN GOTO 340
290 LPRINT "THE INPUT:"
300 FOR I=1 TO N
310 LPRINT "I=";I;"X(I)=";X(I);"Y(I)=";Y(I)
320 NEXT I
330 LPRINT
340 CLS:PRINT "PARABOLIC CURVE FIT:"
350 REM COMPUTATION OF REGRESSION COEFFICIENTS
360 LET X2=0
370 LET X3=0
380 LET X4=0
390 LET X5=0
400 LET Y2=0
410 LET Y3=0
420 LET Y4=0
430 FOR I=1 TO N
440 LET X0=X(I)
450 LET Y0=Y(I)
460 LET D=X0*X0
470 LET X2=X2+X0
480 LET X3=X3+D
490 LET X4=X4+X0*D
500 LET X5=X5+D*D
510 LET Y2=Y2+Y0
520 LET Y3=Y3+X0*Y0
530 LET Y4=Y4+D*Y0
540 NEXT I
550 LET F1=N*X3-X2*X2
560 LET F2=N*X4-X2*X3
570 LET F3=F1*(N*Y4-Y2*X3)
580 LET F4=F2*(N*Y3-Y2*X2)
590 LET C=(F3-F4)/(F1*(N*X5-X3*X3)-F2*F2)
600 LET B=(N*Y3-X2*Y2-C*F2)/F1
610 LET A=(Y2-B*X2-C*X3)/N
620 REM COMPUTATION OF THE SUM OF SQUARES OF RESIDUALS
```

```
630 LET S=0
640 FOR I=1 TO N
650 LET S0=A+X(I)*(B+C*X(I))-Y(I)
660 LET S=S+S0*S0
670 NEXT I
680 PRINT"REGRESSION COEFFICIENTS: A,B,C"
690 PRINT"SUM OF SQUARES OF RESIDUALS: S"
700 PRINT"IF S=0 THE CURVE GOES THROUGH ALL THE GIVEN POINTS."
710 PRINT "A=";A;"B=";B;"C=";C;"S=";S:PRINT
720 IF O2=0 THEN GOTO 760
730 LPRINT "PARABOLIC CURVE FIT:"
740 LPRINT "REGRESSION COEFFICIENTS: A=";A;"B=";B;"C=";C
750 LPRINT "SUM OF SQUARES OF RESIDUALS=";S
760 PRINT"THE VALUES Y ON THE REGRESSION CURVE ARE COMPUTED"
770 PRINT"FOR ANY NUMBER OF VALUES OF THE VARIABLE X."
780 PRINT"TO STOP PRESS BREAK."
790 INPUT"X";X1
800 LET Y1=A+X1*(B+C*X1)
810 PRINT "X=";X1;"Y=";Y1
820 IF O2>0 THEN LPRINT "X=";X1;"Y=";Y1
830 GOTO 790
840 END
```

Example

A curve is given by 5 points

(−2, 10.92), (−1, 1.15), (0, 2.3), (1, 9.17), (2, 23.13).

It is approximated by a parabolic curve fit

$y = A + bx + Cx^2$.

The regression coefficients are

A=1.73686,　B=3.24400,　C=3.79857,

and the sum of squares of residuals equals 2.08353.

For x=−2,　y=10.4431,

x=0,　y=1.73686,

x=2,　y=23.4191.

P907 Binomial and Cumulative Binomial Distribution

```
10 CLS:PRINT"P907 COMPUTES BINOMIAL AND CUMULATIVE BINOMIAL"
20 PRINT"DISTRIBUTION."
30 PRINT:PRINT"THE INPUT:"
40 PRINT"SINGLE PRECISION ONLY."
50 PRINT"IF PRINTER NOT USED O2=0, IF USED O2>0."
60 DEFINT I-O,X:INPUT"O2";O2:CLS
70 IF O2=0 THEN GOTO 100
80 LPRINT "P907: O2=";O2
90 LPRINT
100 PRINT"O3 IS AN INTEGER VARIABLE DETERMINING THE DISTRIBUTION."
110 PRINT:PRINT "BINOMIAL DISTRIBUTION: O3=1"
120 PRINT "CUMULATIVE BINOMIAL DISTRIBUTION: O3=2"
130 INPUT"O3";O3:PRINT
140 PRINT"NUMBER OF TRIALS: N"
150 INPUT"N";N
160 PRINT"NUMBER OF OCCURENCES OF THE EVENT: X"
170 INPUT"X";X
180 PRINT"PROBABILITY: P (0<P<1)"
190 INPUT"P";P:CLS
200 LET Q=1-P
210 LET S0=N*P
220 LET S=S0*Q
230 IF O3=2 THEN GOTO 480
240 LET F=1
250 LET M=N-X
260 LET P1=P*Q
270 IF X>N/2 THEN GOTO 360
280 FOR I=1 TO X
290 LET F=F*P1*(M+I)/I
300 NEXT I
310 IF X=N/2 THEN GOTO 420
320 FOR I=X+1 TO N-X
330 LET F=F*Q
340 NEXT I
350 GOTO 420
360 FOR I=1 TO M
370 LET F=F*P1*(X+I)/I
380 NEXT I
390 FOR I=M+1 TO X
400 LET F=F*P
410 NEXT I
420 PRINT "N=";N;"X=";X;"P=";P;"BINOMIAL DISTRIBUTION=";F
430 PRINT "MEAN VALUE=";S0;"   VARIANCE=";S
440 IF O2=0 THEN GOTO 580
450 LPRINT "N=";N;"X=";X;"P=";P;"BINOMIAL DISTRIBUTION=";F
460 LPRINT "MEAN VALUE=";S0;"   VARIANCE=";S
470 GOTO 580
480 LET F=Q[N
490 LET C=F
500 LET Q=P/Q
510 FOR I=1 TO X
520 LET F=F*Q*(N+1-I)/I
530 LET C=C+F
540 NEXT I
550 PRINT "N=";N;"X=";X;"P=";P;"CUMULATIVE BINOMIAL DISTR.=";C
560 IF O2=0 THEN GOTO 580
570 LPRINT "N=";N;"X=";X;"P=";P;"CUMULATIVE BINOMIAL DISTR.=";C
580 PRINT:PRINT"TO CONTINUE PRESS ENTER, TO STOP PRESS BREAK"
590 INPUT A$:CLS:GOTO 100
600 END
```

Examples

1. A binomial distribution with N=6 independent trials and probability P=0.2. The probability that the event occurs in N trials exactly X=4-times is 0.01536. The mean value is 1.2, and the variance 0.96.

2. A cumulative binomial distribution for the same data takes the value 0.998401.

P908 Negative Binomial and Cumulative Negative Binomial Distribution

```
10 CLS:PRINT"P908 COMPUTES NEGATIVE BINOMIAL AND CUMULATIVE"
20 PRINT"NEGATIVE BINOMIAL DISTRIBUTION."
30 PRINT:PRINT"THE INPUT:"
40 PRINT"SINGLE PRECISION ONLY."
50 PRINT"IF PRINTER NOT USED O2=0, IF USED O2>0."
60 DEFINT I-O,X:INPUT"O2";O2:CLS
70 IF O2=0 THEN GOTO 100
80 LPRINT "P908: O2=";O2
90 LPRINT
100 PRINT"O3 IS AN INTEGER VARIABLE DETERMINING THE DISTRIBUTION."
110 PRINT:PRINT "NEGATIVE BINOMIAL DISTRIBUTION: O3=1"
120 PRINT "CUMULATIVE NEGATIVE BINOMIAL DISTRIBUTION: O3=2"
130 INPUT"O3";O3:PRINT
140 PRINT"NUMBER OF TRIALS: N+X"
150 PRINT"NUMBER OF OCCURRENCES OF THE EVENT: N"
160 INPUT"N";N:INPUT"X";X
170 PRINT"PROBABILITY: P (0<P<1)"
180 INPUT"P";P:CLS
190 LET Q=1-P
200 LET S0=N*Q/P
210 LET S=S0/P
220 LET M=N-1
230 IF O3=2 THEN GOTO 490
240 LET F=P
250 LET P1=P*Q
260 IF M>X THEN GOTO 350
270 FOR I=1 TO M
280 LET F=F*P1*(X+I)/I
290 NEXT I
300 IF M=X THEN GOTO 410
310 FOR I=M+1 TO X
320 LET F=F*Q
330 NEXT I
340 GOTO 410
350 FOR I=1 TO X
360 LET F=F*P1*(M+I)/I
370 NEXT I
380 FOR I=X+1 TO M
390 LET F=F*P
400 NEXT I
410 PRINT "N=";N;"X=";X;"P=";P
420 PRINT "NEGATIVE BINOMIAL DISTRIBUTION=";F
430 PRINT "MEAN VALUE=";S0;"   VARIANCE=";S
440 IF O2=0 THEN GOTO 600
450 LPRINT "N=";N;"X=";X;"P=";P
460 LPRINT "NEGATIVE BINOMIAL DISTRIBUTION=";F
470 LPRINT "MEAN VALUE=";S0;"   VARIANCE=";S
480 GOTO 600
490 LET F=P[N
500 LET C=F
510 FOR I=1 TO X
520 LET F=F*Q*(M+I)/I
530 LET C=C+F
540 NEXT I
550 PRINT "N=";N;"X=";X;"P=";P
560 PRINT "CUMULATIVE NEGATIVE BINOMIAL DISTRIBUTION=";C
570 IF O2=0 THEN GOTO 600
580 LPRINT "N=";N;"X=";X;"P=";P
590 LPRINT "CUMULATIVE NEGATIVE BINOMIAL DISTRIBUTION=";C
600 PRINT:PRINT"TO CONTINUE PRESS ENTER, TO STOP PRESS BREAK"
610 INPUT A$:CLS:GOTO 100
620 END
```

Examples

1. A negative binomial distribution with N=4, probability P=0.9, and X=2. The probability that the event occurs N=4 times in (N+X) trials is given by the negative binomial distribution 0.06561. The mean value is 0.444445, and the variance 0.493827.

2. A cumulative negative binomial distribution for the same data takes the value 0.984151.

P909 Hypergeometric and Cumulative Hypergeometric Distribution

```
10 CLS:PRINT"P909 COMPUTES HYPERGEOMETRIC AND"
20 PRINT"CUMULATIVE HYPERGEOMETRIC DISTRIBUTION."
30 PRINT:PRINT"THE INPUT:"
40 PRINT"SINGLE PRECISION ONLY."
50 PRINT"IF PRINTER NOT USED O2=0, IF USED O2>0."
60 DEFINT I,K-O,X:DEFDBL J:INPUT"O2";O2:CLS
70 IF O2=0 THEN GOTO 100
80 LPRINT "P909: O2=";O2
90 LPRINT
100 PRINT"O3 IS AN INTEGER VARIABLE DETERMINING THE DISTRIBUTION."
110 PRINT:PRINT "HYPERGEOMETRIC DISTRIBUTION: O3=1"
120 PRINT "CUMULATIVE HYPERGEOMETRIC DISTRIBUTION: O3=2"
130 INPUT"O3";O3:PRINT
140 PRINT"TOTAL NUMBER OF OBJECTS: N"
150 PRINT"TOTAL NUMBER OF OBJECTS WITH PROPERTY A: M"
160 INPUT"N";N:INPUT"M";M
170 PRINT"PROBABILITY THAT THERE ARE X OBJECTS WITH PROPERTY A"
180 PRINT"IN A SAMPLE OF N0 OBJECTS: F"
190 PRINT"PROBABILITY F IS GIVEN BY HYPERGEOMETRIC DISTRIBUTION."
200 INPUT"X";X:INPUT"N0";N0:CLS
210 IF O3=2 THEN GOTO 450
220 LET S0=N0*M/N
230 LET S=S0*(N-M)*(N-N0)/(N*(N-1))
240 LET K=M
250 LET L=X
260 GOSUB 670
270 LET F=J
280 LET K=N
290 LET L=N0
300 GOSUB 670
310 LET F=F/J
320 LET K=N-M
330 LET L=N0-X
340 GOSUB 670
350 LET F=F*J
360 GOTO 370
370 PRINT "N=";N;"M=";M;"N0=";N0;"X=";X
380 PRINT "HYPERGEOMETRIC DISTRIBUTION=";F
390 PRINT "MEAN VALUE=";S0;"    VARIANCE=";S
400 IF O2=0 THEN GOTO 640
410 LPRINT "N=";N;"M=";M;"N0=";N0;"X=";X
420 LPRINT "HYPERGEOMETRIC DISTRIBUTION=";F
430 LPRINT "MEAN VALUE=";S0;"    VARIANCE=";S
440 GOTO 640
450 LET K=N-M
460 LET L=N0
470 GOSUB 670
480 LET F=J
490 LET K=N
500 LET L=N0
510 GOSUB 670
520 LET F=F/J
530 LET C=F
540 IF X=0 THEN GOTO 590
550 FOR I=1 TO X
560 LET F=F*(I-M-1)*(I-N0-1)/(I*(N-M-N0+I))
570 LET C=C+F
580 NEXT I
590 PRINT "N=";N;"M=";M;"N0=";N0;"X=";X
600 PRINT "CUMULATIVE HYPERGEOMETRIC DISTRIBUTION=";C
610 IF O2=0 THEN GOTO 640
```

P. 296   PROGRAM P909

```
620 LPRINT "N=";N;"M=";M;"N0=";N0;"X=";X
630 LPRINT "CUMULATIVE HYPERGEOMETRIC DISTRIBUTION=";C
640 PRINT:PRINT"TO CONTINUE PRESS ENTER, TO STOP PRESS BREAK"
650 INPUT A$:CLS:GOTO 100
660 REM SUBROUTINE COMPUTING BINOMIAL COEFFICIENTS
670 LET K0=K-L
680 LET J=1
690 IF L=0 THEN RETURN
700 IF L>K/2 THEN GOTO 750
710 FOR I=1 TO L
720 LET J=J*(K0+I)/I
730 NEXT I
740 RETURN
750 FOR I=1 TO K0
760 LET J=J*(L+I)/I
770 NEXT I
780 RETURN
790 END
```

Examples

1. In a sample of N objects there are M objects with a certain property A. The probability that there will be precisely X objects with the property A in drawing a sample of N0 objects is given by a hypergeometric distribution. For N=20, M=8, N0=6, X=5 it takes the value 0.0173375. The mean value is 2.4, and the variance 1.06105.

2. A cumulative hypergeometric distribution for N=20, M=8, N0=6, and X=0 is 0.023839. For the same data but X=3, it takes the value 0.862745.

P910   Poisson and Cumulative Poisson Distribution

```
10 CLS:PRINT"P910 COMPUTES POISSON AND"
20 PRINT"CUMULATIVE POISSON DISTRIBUTION."
30 PRINT:PRINT"THE INPUT:"
40 PRINT"SINGLE PRECISION ONLY."
50 PRINT"IF PRINTER NOT USED O2=0, IF USED O2>0."
60 DEFINT I,X:INPUT"O2";O2:CLS
70 IF O2=0 THEN GOTO 100
80 LPRINT "P910: O2=";O2
90 LPRINT
100 PRINT"O3 IS AN INTEGER VARIABLE DETERMINING THE DISTRIBUTION."
110 PRINT:PRINT "POISSON DISTRIBUTION: O3=1"
120 PRINT "CUMULATIVE POISSON DISTRIBUTION: O3=2"
130 INPUT"O3";O3:PRINT
140 PRINT"MEAN VALUE: M"
150 PRINT"NUMBER OF OCCURRENCES OF THE EVENT: X"
160 INPUT"M";M:INPUT"X";X:CLS
170 LET F=EXP(-M)
180 IF O3=2 THEN GOTO 270
190 IF X=0 THEN GOTO 230
200 FOR I=1 TO X
210 LET F=F*M/I
220 NEXT I
230 PRINT "M=";M;"X=";X;"POISSON DISTRIBUTION=";F
240 IF O2=0 THEN GOTO 350
250 LPRINT "M=";M;"X=";X;"POISSON DISTRIBUTION=";F
260 GOTO 350
270 LET C=F
280 FOR I=1 TO X
290 LET F=F*M/I
300 LET C=C+F
310 NEXT I
320 PRINT "M=";M;"X=";X;"CUMULATIVE POISSON DISTRIBUTION=";C
330 IF O2=0 THEN GOTO 350
340 LPRINT "M=";M;"X=";X;"CUMULATIVE POISSON DISTRIBUTION=";C
350 PRINT:PRINT"TO CONTINUE PRESS ENTER, TO STOP PRESS BREAK"
360 INPUT A$:CLS:GOTO 100
370 END
```

Examples

1. If the mean value for the occurrence of an event A is M, the probability that the event occurs exactly X times is given by the Poisson distribution.
For M=0.5, X=2, the probability f(X) is 0.0758163,
   M=3.2, X=0, f(X)=0.0407622,
   M=3.2, X=7, f(X)=0.0277893.

2. A cumulative Poisson distribution for M=0.5, X=2 is 0.985612, for M=3.2, X=7 it is 0.98317.

P911 Normal Distribution, Chi-Square Distribution, t Distribution, F Distribution

```
10 CLS:PRINT"P911 COMPUTES NORMAL DISTRIBUTION, CHI-SQUARE"
20 PRINT"DISTRIBUTION, t-DISTRIBUTION, F-DISTRIBUTION."
30 PRINT:PRINT"THE INPUT:"
40 PRINT"SINGLE PRECISION ONLY."
50 PRINT"IF PRINTER NOT USED O2=0, IF USED O2>0."
60 DEFINT I-O:INPUT"O2";O2:CLS
70 PRINT"MAXIMUM RELATIVE ERROR IN GAUSSIAN INTEGRATION: P0"
80 PRINT"IF P0=0 THE ACCURACY NOT TESTED."
90 PRINT"F-DISTRIBUTION MUST BE RUN WITH P>0 WHEN N1>1,N2=1."
100 INPUT"P0";P:CLS
110 IF O2=0 THEN GOTO 130
120 LPRINT "P911: O2=";O2;"P=";P
130 DIM U(16),W(16),X3(2),Y(2)
140 PRINT"READING DATA"
150 REM READ ABSCISSAS AND WEIGHTS FOR GAUSSIAN QUADRATURE
160 FOR I=1 TO 16
170 READ U(I),W(I)
180 NEXT I
190 CLS:PRINT"O3 IS AN INTEGER VARIABLE DEFINING THE DISTRIBUTION."
200 PRINT
210 PRINT "NORMAL DISTRIBUTION: O3=1"
220 PRINT "CHI-SQUARE DISTRIBUTION: O3=2"
230 PRINT "t-DISTRIBUTION: O3=3"
240 PRINT "F-DISTRIBUTION: O3=4"
250 IF O2>0 THEN LPRINT
260 INPUT "O3";O3
270 IF O3<1 THEN GOTO 310
280 IF O3>4 THEN GOTO 310
290 CLS
300 ON O3 GOTO 320,430,590,810
310 PRINT"ERROR IN O3":GOTO 260
320 PRINT"LIMITS OF THE INTEGRAL: (A,B)"
330 INPUT"A";A:INPUT"B";B
340 PRINT "THE MEAN :M":INPUT"M";S0
350 PRINT"STANDARD DEVIATION: S":INPUT"S";S1
360 IF O2=0 THEN GOTO 380
370 LPRINT "NORMAL DISTRIBUTION, A=";A;"B=";B;"M=";S0;"S=";S1
380 LET S9=.5641896
390 LET A=.7071068*(A-S0)/S1
400 LET B=.7071068*(B-S0)/S1
410 DEF FNF(X)=EXP(-X*X)
420 GOTO 1370
430 INPUT"CHI2";B:INPUT"N";N0
440 IF O2=0 THEN GOTO 460
450 LPRINT "CHI-SQUARE DISTRIBUTION, CHI2=";B;"N=";N0
460 LET A=0
470 IF ((N0/2)-INT(N0/2))=0 THEN GOTO 540
480 LET S9=.7978846
490 LET B=SQR(B)
500 IF N0=1 THEN DEF FNF(X)=EXP(-X*X/2)
510 LET N1=(N0-1)/2
520 LET S0=.5
530 GOTO 1370
540 LET S9=.5
550 IF N0=2 THEN DEF FNF(X)=EXP(-X/2)
560 LET N1=N0/2-1
570 LET S0=0
580 GOTO 1370
590 INPUT"t";B:INPUT"N";N0
600 IF O2=0 THEN GOTO 620
610 LPRINT "t-DISTRIBUTION, t=";B;"N=";N0
```

```
620 LET A=0
630 IF ((N0/2)-INT(N0/2))=0 THEN GOTO 720
640 LET S0=N0/2
650 LET N1=(N0-1)/2
660 LET S9=.6366198/SQR(N0)
670 IF N0=1 THEN GOTO 780
680 FOR I=1 TO N1
690 LET S9=S9*I/(I-.5)
700 NEXT I
710 GOTO 780
720 LET N1=N0/2-1
730 LET S9=1/SQR(N0)
740 IF N1=0 THEN GOTO 780
750 FOR I=1 TO N1
760 LET S9=S9*(I+.5)/I
770 NEXT I
780 LET S0=-(1+N0)/2
790 DEF FNF(X)=(1+X*X/N0)[S0
800 GOTO 1370
810 INPUT"F";B:INPUT"N1";N1:INPUT"N2";N2
820 IF O2=0 THEN GOTO 840
830 LPRINT "F-DISTRIBUTION, F=";B;"N1=";N1;"N2=";N2
840 LET A=0
850 IF N1>1 THEN GOTO 890
860 LET B=SQR(B)
870 LET N0=N2
880 GOTO 630
890 LET S1=N1/2
900 LET S2=N2/2
910 IF (S1-INT(S1))=0 THEN GOTO 1180
920 IF (S2-INT(S2))>0 THEN GOTO 990
930 LET S9=S1
940 IF N2=2 THEN GOTO 1350
950 FOR I=1 TO S2-1
960 LET S9=S9*(S1+I)/I
970 NEXT I
980 GOTO 1350
990 LET S9=.3183099
1000 IF N1>1 THEN GOTO 1050
1010 FOR I=1 TO (N2-1)/2
1020 LET S9=S9*I/(I-.5)
1030 NEXT I
1040 GOTO 1350
1050 IF N2>1 THEN GOTO 1100
1060 FOR I=1 TO (N1-1)/2
1070 LET S9=S9*I/(I-.5)
1080 NEXT I
1090 GOTO 1350
1100 FOR I=1 TO (N1-1)/2
1110 LET S9=S9*I/(I-.5)
1120 NEXT I
1130 LET N3=(N1-1)/2
1140 FOR I=1 TO (N2-1)/2
1150 LET S9=S9*(I+N3)/(I-.5)
1160 NEXT I
1170 GOTO 1350
1180 IF (S2-INT(S2))=0 THEN GOTO 1250
1190 LET S9=S2
1200 IF N1=2 THEN GOTO 1350
1210 FOR I=1 TO S1-1
1220 LET S9=S9*(S2+I)/I
1230 NEXT I
1240 GOTO 1350
1250 IF S1>1 THEN GOTO 1280
```

P. 300   PROGRAM P911

```
1260 LET S9=S2
1270 GOTO 1350
1280 IF S2>1 THEN GOTO 1310
1290 LET S9=S1
1300 GOTO 1350
1310 LET S9=S1+S2-1
1320 FOR I=1 TO S2-1
1330 LET S9=S9*(I+S1-1)/I
1340 NEXT I
1350 DEF FNF(X)=((N1*X/(N1*X+N2))[S1)*((N2/(N1*X+N2))[S2)/X
1360 REM GAUSSIAN QUADRATURE
1370 LET H=B-A
1380 LET H2=H/2
1390 LET K=0
1400 LET M=1
1410 LET Q=0
1420 LET A0=A-H2
1430 FOR J=1 TO M
1440 LET X0=H*J
1450 FOR I=1 TO 16
1460 LET X=H2*U(I)
1470 LET X1=A0+X0+X
1480 LET X2=A0+X0-X
1490 IF O3<>2 THEN GOTO 1640
1500 IF N0<3 THEN GOTO 1640
1510 LET X3(1)=X1/2
1520 LET X3(2)=X2/2
1530 IF S0=0 THEN GOTO 1560
1540 LET X3(1)=X3(1)*X1
1550 LET X3(2)=X3(2)*X2
1560 FOR L=1 TO 2
1570 LET Y(L)=EXP(-X3(L))
1580 FOR L0=1 TO N1
1590 LET Y(L)=Y(L)*X3(L)/(L0-S0)
1600 NEXT L0
1610 NEXT L
1620 LET Q=Q+W(I)*(Y(1)+Y(2))
1630 GOTO 1650
1640 LET Q=Q+W(I)*(FNF(X1)+FNF(X2))
1650 NEXT I
1660 NEXT J
1670 LET Q=Q*H2
1680 IF P=0 THEN GOTO 1830
1690 IF K=4 THEN GOTO 1820
1700 REM TESTING THE ACCURACY OF INTEGRATION
1710 PRINT "P=";Q*S9;"- COMPUTING MORE ACCURATE P"
1720 IF K=0 THEN GOTO 1750
1730 LET D=ABS((Q-Q0)/Q)
1740 IF D<P THEN GOTO 1820
1750 LET Q0=Q
1760 LET K=K+1
1770 LET M=M+M
1780 LET H=H2
1790 LET H2=H2/2
1800 GOTO 1410
1810 REM PRINTING THE RESULT
1820 PRINT
1830 LET Q=Q*S9
1840 IF O3=3 THEN GOTO 1890
1850 PRINT "VALUE OF THE DISTRIBUTION: P=";Q
1860 LET Q1=1-Q
1870 PRINT "VALUE OF THE DISTRIBUTION: Q=1-P=";Q1
1880 GOTO 1920
1890 PRINT "VALUE OF THE DISTRIBUTION: A=";Q
```

```
1900 LET Q1=(Q+1)/2
1910 PRINT "VALUE OF THE DISTRIBUTION: A'=(1+A)/2=";Q1
1920 IF P>0 THEN PRINT "NUMBER OF SUBINTERVALS M=";M
1930 IF O2=0 THEN GOTO 2000
1940 IF O3=3 THEN GOTO 1980
1950 LPRINT "VALUE OF THE DISTRIBUTION: P=";Q
1960 LPRINT "VALUE OF THE DISTRIBUTION: Q=1-P=";Q1
1970 GOTO 2000
1980 LPRINT "VALUE OF THE DISTRIBUTION: A=";Q
1990 LPRINT "VALUE OF THE DISTRIBUTION: A'=(1+A)/2=";Q1
2000 PRINT:PRINT"TO CONTINUE PRESS ENTER, TO STOP PRESS BREAK"
2010 INPUT A$:GOTO 190
2020 DATA .0483077,.09654009
2030 DATA .1444720,.09563872
2040 DATA .2392874,.0938444
2050 DATA .3318686,.09117388
2060 DATA .4213513,.08765209
2070 DATA .5068999,.08331192
2080 DATA .5877158,.07819390
2090 DATA .6630443,.07234579
2100 DATA .7321821,.06582222
2110 DATA .7944838,.05868409
2120 DATA .8493676,.05099806
2130 DATA .8963212,.04283590
2140 DATA .9349061,.03427386
2150 DATA .9647623,.02539207
2160 DATA .9856115,.01627439
2170 DATA .9972639,.00701861
2180 END
```

Examples

1. Normal distribution:

    A=0.6, B=1, M=0.8, S=0.2:   P=0.682689, Q=0.317311.

2. Chi-square distribution:

    CHI2=9.21, N=2:   P=0.989998, Q=0.0100015.

    CHI2=59.2, N=90:   P=5.0053*10^{-3}, Q=0.994995.

    CHI2=140.2, N=100:   P=0.995026, Q=4.9746*10^{-3}.

3. t-distribution:

    t=3.08, N=1:   A=0.800141, A'=0.900071.

    t=2.53, N=20:   A=0.980087, A'=0.990044.

    t=3.13, N=200:   A=0.997993, A'=0.998997.

4. F-distribution:
   F=4.21, N_1=7, N_2=6,   P=0.950087, Q=0.0499128.
   F=0.09, N_1=2, N_2=100,   P=0.0859956, Q=0.914005.
   F=1.39, N_1=100, N_2=100, P=0.949369, Q=0.0506309.

All the preceding data are computed with P0=0, no testing of accuracy of Gaussian integration. When F-distribution is computed with N_2=1 or 2, the integration should be carried out with P0>0, say P0=$2*10^{-4}$.

5. F-distribution, P0=$2*10^{-4}$:
   F=200, N_1=2, N_2=1,   P=0.950064, Q=0.049936.
   F=241, N_1=9, N_2=1,   P=0.950272, Q=0.049728.
   F=19.4, N_1=9, N_2=2,   P=0.950038, Q=0.0499618.
   F=19.5, N_1=100, N_2=2, P=0.950049, Q=0.049951.

P912 Inverse Normal Distribution

```
10 CLS:PRINT"P912 COMPUTES INVERSE NORMAL DISTRIBUTION."
20 PRINT:PRINT"THE INPUT:"
30 PRINT"SINGLE PRECISION ONLY."
40 PRINT"IF PRINTER NOT USED O2=0, IF USED O2>0."
50 DEFINT I-O:INPUT"O2";O2:CLS
60 IF O2=0 THEN GOTO 80
70 LPRINT "P912: O2=";O2
80 PRINT"MEAN VALUE: M"
90 INPUT"M";S0
100 PRINT"STANDARD DEVIATION: S"
110 INPUT"S";S1
120 PRINT"PROBABILITY: Q (0<Q<1/2)"
130 INPUT"Q";Q:PRINT
140 PRINT"LOWER LIMIT OF THE INTEGRAL: X"
150 LET T=SQR(-2*LOG(Q))
160 LET A=2.515517+T*(.802853+.010328*T)
170 LET A=T-A/(1+T*(1.432788+T*(.189269+.001308*T)))
180 LET X=S1*A+S0
190 PRINT "M=";S0;"S=";S1;"Q=";Q"X=";X
200 IF O2>0 THEN LPRINT "M=";S0;"S=";S1;"Q=";Q;"X=";X
210 PRINT:PRINT"TO CONTINUE PRESS ENTER, TO STOP PRESS BREAK"
220 INPUT A$:CLS:GOTO 80
230 END
```

Example

A normal distribution is given by its mean value M and standard deviation S. The inverse normal distribution determines the lower limit X of a random variable for which the probability that the variable is greater, or equal to, this limit is Q.
For M=0, S=1, Q=0.30854 the limit X=0.499583,
M=0, S=1, Q=1.35*10^{-3} the limit X=3.00029.

APPENDIX

PA1 Derived Elementary Functions in Single Precision

```
10 CLS:PRINT"PA1 EVALUATES DERIVED ELEMENTARY FUNCTIONS"
20 PRINT"IN SINGLE PRECISION."
30 PRINT:PRINT"THE INPUT:"
40 PRINT"SINGLE PRECISION ONLY (DOUBLE PRECISION IN PA2)."
50 PRINT"IF PRINTER NOT USED O2=0, IF USED O2>0."
60 DEFINT I-O:INPUT"O2";O2:CLS
70 IF O2=0 THEN GOTO 90
80 LPRINT "PA1: O2=";O2
90 PRINT"O3 IS AN INTEGER VARIABLE DETERMINING THE FUNCTION."
100 PRINT"INDEPENDENT VARIABLE: X"
110 LET C=1.570796
120 PRINT "COT(X): O3=1"
130 PRINT "ARCSIN(X), ABS(X)<=1: O3=2"
140 PRINT "ARCCOS(X), ABS(X)<=1: O3=3"
150 PRINT "ARCCOT(X): O3=4"
160 PRINT "SINH(X): O3=5"
170 PRINT "COSH(X): O3=6"
180 PRINT "TANH(X): O3=7"
190 PRINT "COTH(X): O3=8"
200 PRINT "ARSINH(X): O3=9"
210 PRINT "ARCOSH(X), X>=1: O3=10"
220 PRINT "ARTANH(X), ABS(X)<1: O3=11"
230 PRINT "ARCOTH(X), ABS(X)>1: O3=12"
240 PRINT "COMMON LOG(X), X>0: O3=13"
250 INPUT"O3";O3:INPUT"X";X:IF O3<1 THEN GOTO 290
260 PRINT:ON O3 GOTO 300,340,380,420,460,510,560,610,660
270 LET O3=O3-9
280 ON O3 GOTO 700,740,780,820
290 PRINT"ERROR IN O3; CHOOSE O3 AGAIN":GOTO 100
300 LET Y=1/TAN(X)
310 PRINT "X=";X;"COT(X)=";Y
320 IF O2>0 THEN LPRINT "X=";X;"COT(X)=";Y
330 GOTO 850
340 LET Y=ATN(X/SQR(1-X*X))
350 PRINT "X=";X;"ARCSIN(X)=";Y
360 IF O2>0 THEN LPRINT "X=";X;"ARCSIN(X)=";Y
370 GOTO 850
380 LET Y=C-ATN(X/SQR(1-X*X))
390 PRINT "X=";X;"ARCCOS(X)=";Y
400 IF O2>0 THEN LPRINT "X=";X;"ARCCOS(X)=";Y
410 GOTO 850
420 LET Y=C-ATN(X)
430 PRINT "X=";X;"ARCCOT(X)=";Y
440 IF O2>0 THEN LPRINT "X=";X;"ARCCOT(X)=";Y
450 GOTO 850
460 LET A=EXP(X)
470 LET Y=(A-1/A)/2
480 PRINT "X=";X;"SINH(X)=";Y
490 IF O2>0 THEN LPRINT "X=";X;"SINH(X)=";Y
500 GOTO 850
510 LET A=EXP(X)
520 LET Y=(A+1/A)/2
530 PRINT "X=";X;"COSH(X)=";Y
540 IF O2>0 THEN LPRINT "X=";X;"COSH(X)=";Y
550 GOTO 850
560 LET A=EXP(-X-X)
570 LET Y=(1-A)/(1+A)
580 PRINT "X=";X;"TANH(X)=";Y
590 IF O2>0 THEN LPRINT "X=";X;"TANH(X)=";Y
```

```
600 GOTO 850
610 LET A=EXP(-X-X)
620 LET Y=(1+A)/(1-A)
630 PRINT "X=";X;"COTH(X)=";Y
640 IF O2>0 THEN LPRINT "X=";X;"COTH(X)=";Y
650 GOTO 850
660 LET Y=LOG(X+SQR(X*X+1))
670 PRINT "X=";X;"ARSINH(X)=";Y
680 IF O2>0 THEN LPRINT "X=";X;"ARSINH(X)=";Y
690 GOTO 850
700 LET Y=LOG(X+SQR(X*X-1))
710 PRINT "X=";X;"ARCOSH(X)=+-";Y
720 IF O2>0 THEN LPRINT "X=";X;"ARCOSH(X)=+-";Y
730 GOTO 850
740 LET Y=.5*LOG((1+X)/(1-X))
750 PRINT "X=";X;"ARTANH(X)=";Y
760 IF O2>0 THEN LPRINT "X=";X;"ARTANH(X)=";Y
770 GOTO 850
780 LET Y=.5*LOG((X+1)/(X-1))
790 PRINT "X=";X;"ARCOTH(X)=";Y
800 IF O2>0 THEN LPRINT "X=";X;"ARCOTH(X)=";Y
810 GOTO 850
820 LET Y=.4342945*LOG(X)
830 PRINT "X=";X;"COMMON LOG(X)=";Y
840 IF O2>0 THEN LPRINT "X=";X;"COMMON LOG(X)=";Y
850 PRINT"TO CONTINUE PRESS ENTER; TO STOP PRESS BREAK"
860 INPUT A$:CLS:GOTO 90
870 END
```

Examples

```
cot(0.983)=0.666369,    cot(-0.25)=-3.916632
arcsin(0.453)=0.470128,    arcsin(-0.623)=-0.672572
arccos(0.721)=0.765552,    arccos(-0.347)=1.92517
arccot(0.931)=0.821116,    arccot(-0.213)=1.78066
sinh(4.1)=30.1619,    sinh(-0.57)=-0.601371
cosh(0.12)=1.00721,    cosh(-2.7)=7.47347
tanh(0.5)=0.462117,    tanh(-7)=-0.999998
coth(0.3)=3.43274,    coth(-0.9)=-1.39607
arsinh(0.21)=0.208486,    arsinh(-0.61)=-0.577381
arcosh(1.14)=+-0.523164,    arcosh(1.86)=+-1.23207
artanh(0.14)=0.140926,    artanh(-0.94)=+-1.73805
arcoth(2.5)=0.423649,    arcoth(-4.7)=-0.216067
```

$\log_{10}(2)=0.30103,\quad \log_{10}(1*10^{-7})=-7$

PA2 Elementary Functions in Double Precision

```
10 CLS:PRINT"PA2 EVALUATES ELEMENTARY FUNCTIONS IN DOUBLE PRECISION."
20 PRINT:PRINT"THE INPUT:"
30 PRINT"DOUBLE PRECISION ONLY (SINGLE PRECISION IN PA1)."
40 PRINT"IF PRINTER NOT USED O2=0, IF USED O2>0."
50 DEFINT I-O:INPUT"O2";O2:CLS
60 DEFDBL A-H,P-Z
70 PRINT "READING DATA"
80 IF O2=0 THEN GOTO 100
90 LPRINT "PA2: O2=";O2
100 DIM U(16),W(16)
110 LET P=6.283185307179586
120 LET P1=3.141592653589793
130 LET P2=1.570796326794897
140 LET P3=2.302585092994046
150 LET P4=.4342944819032518
160 FOR I=1 TO 16
170 READ U(I),W(I)
180 NEXT I
190 CLS
200 PRINT"O3 IS AN INTEGER VARIABLE DETERMINING THE FUNCTION."
210 PRINT"INDEPENDENT VARIABLE: X"
220 PRINT "SIN(X): O3=1          ARCCOT(X): O3=10"
230 PRINT "COS(X): O3=2          SINH(X): O3=11"
240 PRINT "EXP(X): O3=3          COSH(X): O3=12"
250 PRINT "LOG(X), O3=4          TANH(X): O3=13"
260 PRINT "COMMON LOG(X): O3=4"
270 PRINT "ARCTAN(X): O3=5       COTH(X): O3=14"
280 PRINT "TAN(X): O3=6          ARSINH(X): O3=15"
290 PRINT "COT(X): O3=7          ARCOSH(X): O3=16, X>1"
300 PRINT "ARCSIN(X): O3=8       ARTANH(X): O3=17"
310 PRINT "      ABS(X)<=1            ABS(X)<1"
320 PRINT "ARCCOS(X): O3=9       ARCOTH(X): O3=18"
330 PRINT "      ABS(X)<=1            ABS(X)>1"
340 INPUT"O3";O3:INPUT"X";X0:IF O3<1 THEN GOTO 380
350 PRINT:ON O3 GOTO 400,400,790,1040,1280,1460,1510,1560,1640
360 LET O5=O3-9
370 ON O5 GOTO 1730,1780,1830,1880,1940,2000,2080,2160,2230
380 CLS:PRINT "ERROR IN O3; CHOOSE O3 AGAIN":GOTO 210
390 REM COMPUTATION OF SIN(X) AND COS(X)
400 LET X=ABS(X0)
410 LET O4=0
420 IF X<=P THEN GOTO 440
430 LET X=X-P*INT(X/P)
440 IF X<=P2 THEN GOTO 490
450 LET O4=INT(X/P2)
460 IF O4=1 THEN LET X=P1-X
470 IF O4=2 THEN LET X=X-P1
480 IF O4=3 THEN LET X=P-X
490 LET X2=X*X
500 IF O3=2 THEN GOTO 660
510 LET Y=1-X2/600
520 FOR I=11 TO 1 STEP -1
530 LET K=(I+I)*(I+I+1)
540 LET Y=1-Y*X2/K
550 NEXT I
560 LET Y=Y*X
570 IF O4>1 THEN LET Y=-Y
580 IF X0<0 THEN LET Y=-Y
590 IF O3<6 THEN GOTO 630
600 IF O3>7 THEN GOTO 630
610 LET Y1=Y
620 GOTO 660
```

```
630 PRINT "X=";X0;"SIN(X)=";Y
640 IF O2>0 THEN LPRINT "X=";X0;"SIN(X)=";Y
650 GOTO 2290
660 LET Y=1-X2/650
670 FOR I=12 TO 1 STEP -1
680 LET K=(I+I)*(I+I-1)
690 LET Y=1-Y*X2/K
700 NEXT I
710 IF O4=1 THEN LET Y=-Y
720 IF O4=2 THEN LET Y=-Y
730 IF O3=6 THEN GOTO 1470
740 IF O3=7 THEN GOTO 1520
750 PRINT "X=";X0;"COS(X)=";Y
760 IF O2>0 THEN LPRINT "X=";X0;"COS(X)=";Y
770 GOTO 2290
780 REM COMPUTATION OF EXP(X)
790 LET X=ABS(X0)
800 LET N=0
810 IF X<P3 THEN GOTO 850
820 LET X=X/P3
830 LET N=INT(X)
840 LET X=(X-N)*P3
850 LET Y=1+X/23
860 FOR I=22 TO 1 STEP -1
870 LET Y=1+Y*X/I
880 NEXT I
890 LET Y0=1
900 IF N=0 THEN GOTO 940
910 FOR I=1 TO N
920 LET Y0=10*Y0
930 NEXT I
940 LET Y=Y*Y0
950 IF O3>12 THEN GOTO 980
960 IF X0<0 THEN LET Y=1/Y
970 GOTO 990
980 IF X0>0 THEN LET Y=1/Y
990 IF O3>5 THEN RETURN
1000 PRINT "X=";X0;"EXP(X)=";Y
1010 IF O2>0 THEN LPRINT "X=";X0;"EXP(X)=";Y
1020 GOTO 2290
1030 REM COMPUTATION OF LOG(X)
1040 LET X=X0
1050 IF X<1 THEN LET X=1/X
1060 LET N=0
1070 IF X<10 THEN GOTO 1110
1080 LET N=N+1
1090 LET X=X/10
1100 IF X>10 THEN GOTO 1080
1110 LET H=(X-1)/2
1120 LET H0=(X+1)/2
1130 LET Y=0
1140 FOR I=1 TO 16
1150 LET X1=H*U(I)
1160 LET Y=Y+W(I)/(H0-X1*X1/H0)
1170 NEXT I
1180 LET Y=2*Y*H+N*P3
1190 IF X0<1 THEN LET Y=-Y
1200 IF O3>5 THEN RETURN
1210 PRINT "X=";X0;"LOG(X)=";Y
1220 PRINT "X=";X0;"COMMON LOG(X)=";Y*P4
1230 IF O2=0 THEN GOTO 2290
1240 LPRINT "X=";X0;"LOG(X)=";Y
1250 LPRINT "X=";X0;"COMMON LOG(X)=";Y*P4
1260 GOTO 2290
```

P. 308 **PROGRAM PA2**

```
1270 REM COMPUTATION OF ARCTAN(X)
1280 LET X=ABS(X0)
1290 IF X>1 THEN LET X=1/X
1300 LET H=X/2
1310 LET Y=0
1320 FOR I=1 TO 16
1330 LET X3=H*U(I)
1340 LET X1=H+X3
1350 LET X2=H-X3
1360 LET Y=Y+W(I)/(X1*X1+1)+W(I)/(X2*X2+1)
1370 NEXT I
1380 LET Y=Y*H
1390 IF ABS(X0)>1 THEN LET Y=P2-Y
1400 IF X0<0 THEN LET Y=-Y
1410 IF O3>5 THEN RETURN
1420 PRINT "X=";X0;"ARCTAN(X)=";Y
1430 IF O2>0 THEN LPRINT "X=";X0;"ARCTAN(X)=";Y
1440 GOTO 2290
1450 REM COMPUTATION OF DERIVED FUNCTIONS
1460 GOTO 400
1470 LET Y=Y1/Y
1480 PRINT "X=";X0;"TAN(X)=";Y
1490 IF O2>0 THEN LPRINT "X=";X0;"TAN(X)=";Y
1500 GOTO 2290
1510 GOTO 400
1520 LET Y=Y/Y1
1530 PRINT "X=";X0;"COT(X)=";Y
1540 IF O2>0 THEN LPRINT "X=";X0;"COT(X)=";Y
1550 GOTO 2290
1560 LET X7=1-X0*X0
1570 LET X5=X0
1580 GOSUB 2320
1590 LET X0=X0/Z
1600 GOSUB 1280
1610 PRINT "X=";X5;"ARCSIN(X)=";Y
1620 IF O2>0 THEN LPRINT "X=";X5;"ARCSIN(X)=";Y
1630 GOTO 2290
1640 LET X7=1-X0*X0
1650 LET X5=X0
1660 GOSUB 2320
1670 LET X0=X0/Z
1680 GOSUB 1280
1690 LET Y=P2-Y
1700 PRINT "X=";X5;"ARCCOS(X)=";Y
1710 IF O2>0 THEN LPRINT "X=";X5;"ARCCOS(X)=";Y
1720 GOTO 2290
1730 GOSUB 1280
1740 LET Y=P2-Y
1750 PRINT "X=";X0;"ARCCOT(X)=";Y
1760 IF O2>0 THEN LPRINT "X=";X0;"ARCCOT(X)=";Y
1770 GOTO 2290
1780 GOSUB 790
1790 LET Y=(Y-1/Y)/2
1800 PRINT "X=";X0;"SINH(X)=";Y
1810 IF O2>0 THEN LPRINT "X=";X0;"SINH(X)=";Y
1820 GOTO 2290
1830 GOSUB 790
1840 LET Y=(Y+1/Y)/2
1850 PRINT "X=";X0;"COSH(X)=";Y
1860 IF O2>0 THEN LPRINT "X=";X0;"COSH(X)=";Y
1870 GOTO 2290
1880 LET X=ABS(X0+X0)
1890 GOSUB 800
1900 LET Y=(1-Y)/(1+Y)
```

```
1910 PRINT "X=";X0;"TANH(X)=";Y
1920 IF O2>0 THEN LPRINT "X=";X0;"TANH(X)=";Y
1930 GOTO 2290
1940 LET X=ABS(X0+X0)
1950 GOSUB 800
1960 LET Y=(1+Y)/(1-Y)
1970 PRINT "X=";X0;"COTH(X)=";Y
1980 IF O2>0 THEN LPRINT "X=";X0;"COTH(X)=";Y
1990 GOTO 2290
2000 LET X7=X0*X0+1
2010 GOSUB 2320
2020 LET X5=X0
2030 LET X0=X0+Z
2040 GOSUB 1040
2050 PRINT "X=";X5;"ARSINH(X)=";Y
2060 IF O2>0 THEN LPRINT "X=";X5;"ARSINH(X)=";Y
2070 GOTO 2290
2080 LET X7=X0*X0-1
2090 GOSUB 2320
2100 LET X5=X0
2110 LET X0=X0+Z
2120 GOSUB 1040
2130 PRINT "X=";X5;"ARCOSH(X)=+-";Y
2140 IF O2>0 THEN LPRINT "X=";X5;"ARCOSH(X)=+-";Y
2150 GOTO 2290
2160 LET X5=X0
2170 LET X0=(1+X0)/(1-X0)
2180 GOSUB 1040
2190 LET Y=Y/2
2200 PRINT "X=";X5;"ARTANH(X)=";Y
2210 IF O2>0 THEN LPRINT "X=";X5;"ARTANH(X)=";Y
2220 GOTO 2290
2230 LET X5=X0
2240 LET X0=(X0+1)/(X0-1)
2250 GOSUB 1040
2260 LET Y=Y/2
2270 PRINT "X=";X5;"ARCOTH(X)=";Y
2280 IF O2>0 THEN LPRINT "X=";X5;"ARCOTH(X)=";Y
2290 PRINT"TO CONTINUE PRESS ENTER; TO STOP PRESS BREAK"
2300 INPUT A$:CLS:GOTO 200
2310 REM SUBROUTINE FOR SQUARE ROOT IN DOUBLE PRECISION
2320 LET Z=SQR(X7)
2330 LET X9=Z/1.D16
2340 LET X8=Z
2350 LET Z=(Z+X7/Z)/2
2360 IF ABS(Z-X8)>X9 THEN GOTO 2340
2370 RETURN
2380 DATA .04830766568773832,.09654008851472780
2390 DATA .14447196158279649,.09563872007927486
2400 DATA .23928736225213707,.09384439908080457
2410 DATA .33186860228212765,.09117387869576388
2420 DATA .42135127613063535,.08765209300440381
2430 DATA .50689990893222939,.08331192422694676
2440 DATA .58771575724076233,.07819389578707031
2450 DATA .66304426693021520,.07234579410884851
2460 DATA .73218211874028968,.06582222277636185
2470 DATA .79448379596794241,.05868409347853555
2480 DATA .84936761373256997,.05099805926237618
2490 DATA .89632115576605212,.04283589802222668
2500 DATA .93490607593773969,.03427386291302143
2510 DATA .96476225558750643,.02539206530926206
2520 DATA .98561151154526834,.01627439473090567
2530 DATA .99726386184948156,.00701861000947010
2540 END
```

PROGRAM PA2

Examples

sin(-0.5)=-0.479425538604203,   sin(6)=-0.2794154981989252
cos(24)=0.4241790073369995,   cos(-0.5)=0.8775825618903728
exp(1.25)=3.490342957461842,   exp(-7)=9.118819655545171*10^{-4}
log(3)=1.09861228866811,   log$_{10}$(3)=0.4771212547196624
log(0.5)=-0.6931471805599453,   log$_{10}$(0.5)=-0.3010299956639812
log(1*10^{-8})=-18.42068074395237,   log$_{10}$(1*10^{-8})=-8
arctan(-0.1)=-0.09966865249116203,   arctan(20)=1.520837931072954
tan(1.65)=-12.59926478946572,   tan(-0.5)=-0.5463024898437905
cot(0.5)=1.830487721712452,   cot(-7)=-1.147515422405134
arcsin(0.1)=0.1001674211615598,   arcsin(-0.75)=0.8480620789814814
arccos(0.1)=1.470628905633337,   arccos(-0.99)=3.000053180265367
arccot(0.1)=1.471127674303735,   arccot(-40)=3.116597859970874
sinh(0.25)=0.2526123168081683,   sinh(-9)=-4051.541902082786
cosh(0.25)=1.031413099879573,   cosh(-9)=4051.54202549259
tanh(0.25)=0.2449186624037092,   tanh(-5)=-0.9999092042625952
coth(0.1)=10.03331113225399,   coth(-5)=-1.000090803982019
arsinh(0.25)=0.2474664615472653,   arsinh(-5)=-2.312438341272753
arcosh(1.25)=+-0.6931471805599453,   arcosh(5)=+-2.292431669561178
artanh(0.2)=0.2027325540540822,   artanh(-0.8)=-1.09861228866811
arcoth(1.25)=1.09861228866811,   arcoth(5)=0.2027325540540822

PA3 Subroutines for Programming in Double Precision

```
10 CLS:PRINT"PA3: SUBROUTINES FOR PROGRAMMING IN DOUBLE PRECISION"
20 PRINT"SOME OF THE FOLLOWING SUBROUTINES MUST BE MERGED WITH"
30 PRINT"PROGRAMS OF CHAPTERS 3,4,6 WHEN THE INTEGRAND OR THE"
40 PRINT"DIFFERENTIAL EQUATION OR THE TRANSCENDENTAL EQUATION,"
50 PRINT"RESPECTIVELY, SHOULD BE RUN IN DOUBLE PRECISION AND"
60 PRINT"CONTAIN SOME OF THE FOLLOWING FUNCTIONS: SIN, COS,"
70 PRINT"EXP, LOG, X[X8, SQR, ATN"
80 STOP
4992 REM CONSTANTS P0,P1,P2 (STATEMENTS 4995, 5340-5350)
4994 REM REQUIRED FOR THE SUBROUTINE FOR SIN(X) AND COS(X)
4995 READ P0,P1,P2
5000 REM SUBROUTINE FOR S=SIN(X), C=COS(X) IN DOUBLE PRECISION
5010 LET O9=0:REM ENTRY FOR SIN(X)
5020 GOTO 5060
5030 LET O9=1:REM ENTRY FOR COS(X)
5040 GOTO 5060
5050 LET O9=2:REM ENTRY FOR SIN(X) AND COS(X)
5060 LET X0=ABS(X)
5070 LET O4=0
5080 IF X0<=P0 THEN GOTO 5100
5090 LET X0=X0-P0*INT(X0/P0)
5100 IF X<=P2 THEN GOTO 5150
5110 LET O4=INT(X0/P2)
5120 IF O4=1 THEN LET X0=P1-X0
5130 IF O4=2 THEN LET X0=X0-P1
5140 IF O4=3 THEN LET X0=P0-X0
5150 LET X9=X0*X0
5160 IF O9=1 THEN GOTO 5260
5170 LET S=1-X9/600
5180 FOR I9=12 TO 1 STEP -1
5190 LET K9=(I9+I9)*(I9+I9+1)
5200 LET S=1-S*X9/K9
5210 NEXT I9
5220 LET S=S*X0
5230 IF O4>1 THEN LET S=-S
5240 IF X<0 THEN LET S=-S
5250 IF O9=0 THEN RETURN
5260 LET C=1-X9/650
5270 FOR I9=12 TO 1 STEP -1
5280 LET K9=(I9+I9)*(I9+I9-1)
5290 LET C=1-C*X9/K9
5300 NEXT I9
5310 IF O4=1 THEN LET C=-C
5320 IF O4=2 THEN LET C=-C
5330 RETURN
5340 DATA 6.283185307179586,3.141592653589793
5350 DATA 1.570796326794897
5492 REM CONSTANT P3 (STATEMENTS 5495, 5730) REQUIRED
5494 REM FOR THE SUBROUTINE FOR EXP(X)
5495 READ P3
5500 REM SUBROUTINE FOR E=EXP(X) IN DOUBLE PRECISION
5510 IF X>-87 THEN GOTO 5550
5520 LET E=0
5530 RETURN
5540 REM COMPUTATION OF EXP(X) FOR X>-87
5550 LET X0=ABS(X)
5560 LET N9=0
5570 IF X0<P3 THEN GOTO 5610
5580 LET X0=X0/P3
5590 LET N9=INT(X0)
5600 LET X0=(X0-N9)*P3
5610 LET E=1+X0/23
```

```
5620 FOR I9=22 TO 1 STEP -1
5630 LET E=1+E*X0/I9
5640 NEXT I9
5650 LET E0=1
5660 IF N9=0 THEN GOTO 5700
5670 FOR I9=1 TO N9
5680 LET E0=10*E0
5690 NEXT I9
5700 LET E=E*E0
5710 IF X<0 THEN LET E=1/E
5720 RETURN
5730 DATA 2.302585092994046
5990 REM CONSTANTS P3,U(I),W(I) (STATEMENTS 5495, 5730;
5991 REM 5994-5997, 6180-6330) REQUIRED FOR THE SUBROUTINE
5992 REM FOR LOG (X)
5994 DIM U(16),W(16)
5995 FOR I9=1 TO 16
5996 READ U(I9),W(I9)
5997 NEXT I9
6000 REM SUBROUTINE FOR G=LOG(X) IN DOUBLE PRECISION
6010 LET X0=X
6020 IF X0<1 THEN LET X0=1/X0
6030 LET N9=0
6040 IF X0<10 THEN GOTO 6080
6050 LET N9=N9+1
6060 LET X0=X0/10
6070 IF X0>10 THEN GOTO 6050
6080 LET H8=(X0-1)/2
6090 LET H9=(X0+1)/2
6100 LET G=0
6110 FOR I9=1 TO 16
6120 LET X9=H8*U(I9)
6130 LET G=G+W(I9)/(H9-X9*X9/H9)
6140 NEXT I9
6150 LET G=2*G*H8+N9*P3
6160 IF X<1 THEN LET G=-G
6170 RETURN
6180 DATA .04830766568773832,.09654008851472780
6190 DATA .14447196158279649,.09563872007927486
6200 DATA .23928736225213707,.09384439908080457
6210 DATA .33186860228212765,.09117387869576388
6220 DATA .42135127613063535,.08765209300440381
6230 DATA .50689990893222939,.08331192422694676
6240 DATA .58771575724076233,.07819389578707031
6250 DATA .66304426693021520,.07234579410884851
6260 DATA .73218211874028968,.06582222277636185
6270 DATA .79448379596794241,.05868409347853555
6280 DATA .84936761373256997,.05099805926237618
6290 DATA .89632115576605212,.04283589802222668
6300 DATA .93490607593773969,.03427386291302143
6310 DATA .96476225558750643,.02539206530926206
6320 DATA .98561151154526834,.01627439473090567
6330 DATA .99726386184948156,.00701861000947010
6490 REM SUBROUTINES FOR LOG(X) AND EXP(X) WITH CONSTANTS
6491 REM P3,U(I),W(I) (STATEMENTS 5495,5730; 5994-5997,
6492 REM 6180-6330) REQUIRED FOR THE SUBROUTINE FOR X[X8
6500 REM SUBROUTINE FOR E=X[X8 IN DOUBLE PRECISION
6510 GOSUB 6010
6520 LET X=X8*G
6530 GOSUB 5550
6540 RETURN
7000 REM SUBROUTINE FOR R=SQR(X) IN DOUBLE PRECISION
7010 LET R=SQR(X)
7020 LET X9=R/1.D16
```

```
7030 LET X8=R
7040 LET R=(R+X/R)/2
7050 IF ABS(R-X8)>X9 THEN GOTO 7030
7060 RETURN
7490 REM CONSTANTS U(I),W(I) (STATEMENTS 5994-5997,
7491 REM 6180-6330) REQUIRED FOR THE SUBROUTINE FOR ATN(X)
7500 REM SUBROUTINE FOR Y=ATN(X0) IN DOUBLE PRECISION
7510 LET X=ABS(X0)
7520 IF X>1 THEN LET X=1/X
7530 LET H=X/2
7540 LET Y=0
7550 FOR I=1 TO 16
7560 LET X3=H*U(I)
7570 LET X1=H+X3
7580 LET X2=H-X3
7590 LET Y=Y+W(I)/(X1*X1+1)+W(I)/(X2*X2+1)
7600 NEXT I
7610 LET Y=Y*H
7620 IF ABS(X0)>1 THEN LET Y=P2-Y
7630 IF X0<0 THEN LET Y=-Y
7640 RETURN
7650 END
```

Notes

When the integrand or the differential equations or the transcendental equation in programs P305 to P310, P401 to P407, or P605, respectively, are to be run in double precision and contain some of the elementary functions $\sin(x)$, $\cos(x)$, $\exp(x)$, $\log(x)$, x^y, $x^{1/2}$, $\arctan(x)$, or an elementary function derived from them, one or more sections of program PA3 must be merged with the one of the above-mentioned programs as follows:

1. Choose the section or sections to be merged.
2. Load program PA3.
3. Delete the lines that are not to be merged.
4. Name the remaining section of the program, say, AUX and save it in ASCII format on the diskette with the master program in BASIC.
5. Load the main program.
6. Merge the AUX program.
7. Decide whether this program should be stored on a special diskette to be run again. Do not save it on the original diskette with programs of Part Two.

This kind of programming (applied in Programs P805 and P807) requires a certain amount of skill and experience.

Index

References to programs indicate that the quantity referred to is computed, or problem solved, in the program.

Algebra:
  of complex numbers, 6.4 to 6.5
  of matrices, 1.5
Algebraic eigenvalue problems, 7.1 to 7.6
  Cayley-Hamilton theorem, 7.2
  characteristic values, 7.2
  characteristic vectors, 7.2
  eigenvalues of, 7.3 to 7.5, P.221
  eigenvectors of, 7.2, 7.5, P.229
    orthonormal, 7.5, P.229
  Gershgorin theorem, 7.2, P.213
  iterative methods, 7.5
  Krylov method, 7.4, P.216
  proper values, 7.2
  proper vectors, 7.2
  Schur theorem, 7.2
  secular equation, 7.2 to 7.4
  spectral norm, 7.3
  spectral radius, 7.3
Approximations:
  by Chebyshev polynomials, 2.17, P.68
  choosing method of, 2.30
  of even periodic function, 2.1, P.100
  least-squares techniques of, 2.15
  modified Padé approximation, 2.25, P.114
  of odd periodic function, 2.11, P.104
  by orthogonal polynomials, 2.19, P.73, P.79, P.84, P.90
  Padé approximation, 2.24, P.111

Approximations (*Cont.*):
  of periodic function, 2.11, P.95
  by rational function, 2.24 to 2.27, P.111, P.114, P.118

Boundary-value problems, 5.1 to 5.12
  analytical approach to, 5.1 to 5.6
  characteristic functions, 5.3
  characteristic values, 5.3
  defining orthogonal functions, 5.6 to 5.11
    associated Legendre, 5.9
    Bessel, 5.8
    Chebyshev, 5.10
    for harmonic oscillator, 5.6 to 5.7
    Hermite, 5.11
    Laguerre, 5.10
    Legendre, 5.9
    spherical Bessel, 5.8
  Dirichlet condition, 5.3
  eigenfunctions, 5.3 to 5.6
    normalized, 5.4
    orthonormal, 5.4
  eigenvalues, 5.3 to 5.6
  Neumann condition, 5.3
  norm of orthogonal functions, 5.4, 5.7 to 5.11
  numerical approach to, 5.11 to 5.12, P.183

Boundary-value problems (*Cont.*):
  periodicity condition, 5.3
  proper functions, 5.3
  proper values, 5.3
  square of norm, 5.4
  Sturm-Liouville equation, 5.2, 5.4 to 5.6
    Green function of, 5.6
    homogeneous equation, 5.2, 5.4
    nonhomogeneous equation, 5.5
    orthogonal eigenfunctions of, 5.4 to 5.6
  Sturm-Liouville operator, 5.4

Cardano's formulas, 6.3
Cauchy principal value, 3.10
Cayley-Hamilton theorem, 7.2
Characteristic functions, 5.3
Characteristic values, 5.3, 7.2
Characteristic vector, 7.2
Chebyshev abscissas, 2.3, 2.4, 2.6, 2.19
Chebyshev polynomials, 2.17 to 2.18, 2.23
Coefficient of correlation, 9.5
Coefficient of determination, 9.6, 9.7, P.284
Column vector, 1.3
Combinations, 9.2, P.277
Combinatorial analysis, 9.1, P.277
  combinations, 9.2, P.277
  permutations, 9.1, P.277
  variations, 9.2, P.277
Complete pivoting, 1.9
Condition indicator, 1.17, P.7
Condition number, 1.15
Correlation analysis, 9.4
Covariance, 9.5
Cramer's rule, 1.8 to 1.9, 4.9
Curve fitting, 2.15, 9.5 to 9.9
  binomial-power-curve fit, 9.8, P.287
  coefficient of determination, 9.6, 9.7, P.284
  exponential-curve fit, 9.6, P.284
  linear regression, 9.5, P.284
  logarithmic-curve fit, 9.7, P.284
  parabolic-curve fit, 9.8, P.289
  power-curve fit, 9.6, P.284
  regression coefficients, 9.4 to 9.7

Definite integrals, 3.1 to 3.12
  Chebyshev integration, 3.2
  Chebyshev-Gauss integration, 3.11, P.149
  composite corrected trapezoidal rule, 3.4, 3.5, P.132, P.134
  composite Simpson's rule, 3.4, P.128
  composite trapezoidal rule, 3.3
  corrected trapezoidal rule, 3.2

Definite integrals (*Cont.*):
  over finite intervals, 3.4, 3.5, P.128, P.130, P.132, P.134, P.136, P.147, P.149, P.150
  gaussian integration, 3.2, 3.3, 3.5, P.136
  Gauss-Laguerre integration, 3.7, 3.8, P.140, P.144
  Hermite integration, 3.2, 3.7, P.143
  over infinite intervals, 3.7, P.143, P.144
  Laguerre integration, 3.6, P.138
  Lobatto's formula, 3.3
  with logarithmic singularity, 3.12, P.150
  methods of numerical integration, 3.1 to 3.4
  modified composite Simpson's rule, 3.4, 3.5, P.130
  Newton-Cotes formulas, 3.1, 3.2, 4.25, 4.26
  with numerically defined integrand, 3.4, P.128, P.130, P.132, P.134
  Radau's formula, 3.3
  Richardson's extrapolation, 3.4, 3.9
  Romberg integration, 3.4, 3.8, P.147
  over semi-infinite intervals, 3.6, P.138, P.140
  Simpson's rule, 3.2
  with singular integrand, 3.10 to 3.12, P.149, P.150
  trapezoidal rule, 3.1
Determinant, 1.5 to 1.6, 1.8, 1.11, 1.13, 1.17, P.11, P.17, P.21
  Laplace expansion, 1.7, 1.8
Differential equations, 4.1 to 4.33
  analytical methods of integration, 4.1 to 4.12
  Bernoulli equation, 4.5
  canonical form of, 4.11
  Clairaut equation, 4.5
  confluent hypergeometric, 4.15
  defining orthogonal polynomials, 4.15 to 4.16
    Chebyshev, 4.16
    Gegenbauer ultraspherical, 4.16
    generalized Laguerre, 4.16
    Hermite, 4.16
    Jacobi, 4.16
    Laguerre, 4.16
    Legendre, 4.16
  defining special functions, 4.12 to 4.16
    associated Legendre, 4.14
    Bessel, 4.12
    confluent hypergeometric, 4.15
    hyperbolic Bessel, 4.13
    hypergeometric series, 4.14
    Kummer, 4.15
    Legendre, 4.13
    modified Bessel, 4.13
    spherical Bessel, 4.13
    Struve, 4.13

Differential equations (*Cont.*):
  exact equation, 4.4
  of first order, 4.4, 4.5
  Frobenius method, 4.12
  general solution, 4.1, 4.6, 4.9 to 4.11
  homogeneous equation, 4.1, 4.6 to 4.7, 4.12
  hypergeometric, 4.14
  integrating factor, 4.4
  irregular singular point, 4.10
  linear equation, 4.1
    with constant coefficients, 4.8
    of first order, 4.5
    of nth order, 4.6
    of second order, 4.10 to 4.12, 5.1
  nonhomogeneous equation, 4.1, 4.6 to 4.9, 4.11, 5.1
  nonlinear, 4.1, 4.5 to 4.6
  numerical methods of integration, 4.16 to 4.29
    Adams-Bashforth formulas, 4.25, 4.26
    Adams-Moulton formulas, 4.25 to 4.27
    choosing method, 4.23
    Euler method, 4.16, 4.27
    Gill's method, 4.22, P.164
    Hamming's corrector, 4.26
    improved Euler method, 4.21
    Milne's corrector, 4.25, 4.26
    Milne's predictor, 4.25, 4.26
    modified Euler method, 4.21
    predictor-corrector methods, 4.24 to 4.30, P.168, P.173, P.178
    Runge-Kutta methods, 4.20 to 4.24, P.160, P.164
    stability and accuracy of, 4.31
    standard Runge-Kutta method, 4.22, P.160
    Taylor-series methods, 4.17 to 4.20, P.151, P.156
  order of, 4.1
  ordinary point, 4.10
  particular integral, 4.1
  reduction of order of, 4.2 to 4.3
  regular singular point, 4.10
  Riemann, 4.14
  singular integral, 4.2, 4.5 to 4.6
  Sturm-Liouville equation (*see* Boundary-value problems, Sturm-Liouville equation)
  variation of constants, 4.7, 4.8, 4.10 to 4.11
  wronskian, 4.7, 4.11
    Abel's formula for, 4.11
Discretization, 2.28
Discretization error, 2.28, 4.2
Dispersion, 9.3

Distributions, 9.9 to 9.15
  continuous, 9.11 to 9.15
    chi-square, 9.13, P.298
    F distribution, 9.14, P.298
    inverse normal, 9.13, P.303
    normal, 9.11, P. 298
    t distribution, 9.14, P.298
  discrete, 9.9 to 9.11
    binomial, 9.9, P.291
    cumulative binomial, 9.9, P.291
    cumulative hypergeometric, 9.10, P.295
    cumulative negative binomial, 9.10, P.293
    cumulative Poisson, 9.11, P.297
    hypergeometric, 9.10, P.295
    negative binomial, 9.10, P.293
    Poisson, 9.11, P.297
Divided differences, 2.6, 2.7
  table of, 2.7, 2.8

Economization of power series, 2.23, P.108
Eigenvalue problems:
  algebraic (*see* Algebraic eigenvalue problems)
  of differential equation (*see* Boundary-value problems)
Eigenvectors, 7.2, 7.5, P.229
Error analysis:
  for linear algebraic equations, 1.13 to 1.20
  for nonlinear equations, 6.12
Error vector, 1.15

Fourier series, 2.21, P.95
  for even function, 2.21, P.100
  for odd function, 2.21, P.104
Frobenius method, 4.12
Functions:
  Bessel functions, 8.6 to 8.7, P.246, P.250, P.256, P.260
  beta function, 8.9, P.273
  complete elliptic integrals, 8.5, 8.6
  confluent hypergeometric function, 8.3 to 8.5, P.242
    for incomplete gamma function, 8.5
    for Laguerre polynomials, 8.5
  cosine integral, 8.11, P.246, P.250
  elementary functions, P.304, P.306, P.311
  error function, 8.10, P.246, P.250
  exponential integral, 8.12, P.246, P.250
  Fresnel integrals, 8.11, P.246, P.250
  gamma function: of argument ($\frac{1}{2} + n$), 8.9, P.272
    of real argument, 8.8, P.270

Functions (*Cont.*):
  gudermannian, 8.13, P.275
    inverse gudermannian, 8.13, P.275
    relation to hyperbolic functions, 8.13, P.275
  hypergeometric series, 8.3 to 8.5, P.242
    for Chebyshev polynomials, 8.5
    for complete elliptic integrals, 8.5
    for Gegenbauer ultraspherical polynomials, 8.5
    for Jacobi polynomials, 8.5
    for Legendre polynomials, 8.4
  incomplete elliptic integrals: of first kind, 8.5, P.246, P.250
    of second kind, 8.5, P.246, P.250
    of third kind, 8.5, P.246, P.250
  incomplete gamma function, 8.9, P.246, P.250
  logarithmic integral, 8.12, P.246, P.250
  modified Bessel functions, 8.6, 8.7, P.246, P.250, P.256, P.260
  orthogonal polynomials: Chebyshev, 8.2 to 8.3, P.238
    Hermite, 8.2, P.238
    Laguerre, 8.2, P.238
    Legendre, 8.2, P.238
    zeros of Chebyshev polynomials, 8.3, P.238
  polynomial, 8.1, P.235
  sine integral, 8.11, P.246, P.250
  spherical Bessel functions, 8.7, P.256, P.260, P.267

Gershgorin theorem, 7.2, P.213

Interpolation, 2.1 to 2.15
  Chebyshev abscissas, 2.3, 2.4, 2.6
  choosing method of, 2.30
  cubic-spline, 2.10, P.55, P.60
    boundary conditions, 2.12 to 2.13
  divided differences, 2.6, 2.7
    table of, 2.7, 2.8
  Hermite, 2.8, P.44, P.48, P.52
  inverse, 2.15
  Lagrange, 2.3, P.31, P.34
  Lagrange functions, 2.2, 2.3, 2.14
  Newton, 2.5, P.37, P.41
  techniques of, 2.1
  trigonometric, 2.14, P.65

Krylov method, 7.4, P.216

Lagrange functions, 2.2, 2.3, 2.14
Least-squares approximations, 2.15
Least-squares technique, 1.20, 2.1, 2.15
Linear algebraic equations, 1.1 to 1.23
  complete pivoting, 1.9
  condition indicator, 1.17, 1.19, P.7
  condition number, 1.15, 1.17
  Cramer's rule, 1.8 to 1.9
  direct methods, 1.8 to 1.13
    Choleski, 1.11
    Doolittle, 1.10, 7.4, P.11, P.216
    gaussian elimination, 1.9
    for pentadiagonal matrices, 1.12, P.21
    for tridiagonal matrices, 1.11, P.17
  divided differences, 2.6, 2.7
    table of, 2.7, 2.8
  error analysis, 1.13 to 1.20
  homogeneous, 1.8
  indicator, 1.17
  iterative methods, 1.21 to 1.23, P.27
    Gauss-Seidel, 1.21, P.27
    Jacobi, 1.21, P.27
    SOR (successive-overrelaxation) method, 1.21, P.27
  Laplace expansion, 1.8
  LU decomposition, 1.6, 1.8, 1.9, 1.11
  nonhomogeneous, 1.8
  nontrivial solution, 1.8
  norm: euclidean, 1.13, 1.14
    of matrix, 1.14, 7.3
    matrix-bound, 1.14
    maximum, 1.14, 1.15, 1.17, 7.3
    spectral, 1.15, 7.3
    of vector, 1.13 to 1.14
    weighted, 1.14
  normal equations, 1.20 to 1.21, P.25
  partial pivoting, 1.9 to 1.11, 1.20, 1.21
  pivotal element, 1.9
  relaxation parameter, 1.21
  residual vector, 1.13, 1.15, 1.20
  systems of, 1.1
    determined, 1.2, P.11
    ill-conditioned, 1.16 to 1.20, 1.22
    overdetermined, 1.2, 1.20, P.25
    pentadiagonal, 1.12, P.21
    tridiagonal, 1.11, P.17
    underdetermined, 1.2
    well-conditioned, 1.19

Linear algebraic equations (*Cont.*):
  trivial solution, 1.8
LU decomposition, 1.6, 1.8, 1.9, 1.11

Matrix, 1.1 to 1.8
  adjoint, 1.3
  algebra of, 1.5
    addition, 1.5
    equality, 1.5
    multiplication, 1.5
    subtraction, 1.5
  antisymmetric, 1.4
  basis of singular matrix, 1.7, 1.8
  cofactor of, 1.7
  commuting, 1.5
  complex-conjugate, 1.3
  diagonal, 1.2, 1.3
  full, 1.3
  hermitian, 1.4
  identity, 1.2, 1.3
  inverse, 1.2, 1.5 to 1.8, 1.16, 1.17
  invertible, 1.7
  lower-triangular, 1.3, 1.6, 1.7, 1.11 to 1.12
  LU decomposition, 1.6, 1.8, 1.9, 1.11
  minor of, 1.7
  nonsingular, 1.7, 1.8
  norm, 1.14
    maximum, 1.14
    spectral, 1.15, 7.3
  null, 1.3
  order of, 1.3
  orthogonal, 1.4
  pentadiagonal, 1.3, 1.12 to 1.13
  pure imaginary, 1.4
  rank of, 1.7, 1.8
  real, 1.4
  rectangular, 1.1, 1.7
  self-adjoint, 1.4
  single-column, 1.2
  single-row, 1.2
  singular, 1.7 to 1.9
  skew-hermitian, 1.4
  skew-symmetric, 1.4
  sparse, 1.3, 1.21
  square, 1.3, 1.5, 1.8
  strictly row dominant, 1.4
  Sylvester's criterion, 1.4
  symmetric, 1.4
  symmetric positive definite, 1.4, 1.11
  trace of, 1.5
  transposed, 1.3
  tridiagonal, 1.3, 1.11, 1.13

Matrix (*Cont.*):
  unit, 1.3
  unitary, 1.4
  upper-triangular, 1.3, 1.6, 1.7, 1.11 to 1.12
  zero, 1.3
Mean value (*see* Statistics, mean value)

Nonlinear equations (*see* Roots)
Norm:
  euclidean, 1.13, 1.14
  of matrix, 1.14, 7.3
  matrix-bound, 1.4
  maximum, 1.14, 1.15, 1.17, 7.3
  of orthogonal eigenfunctions, 5.4, 5.7 to 5.11
  spectral, 1.15, 7.3
  of vector, 1.13 to 1.14
  weighted, 1.14
Numerical differentiation, 2.1, 2.27 to 2.30,
        P.37, P.41, P.44, P.48, P.52, P.55, P.60,
        P.68, P.73, P.79, P.84, P.90, P.95,
        P.100, P.104, P.111, P.114, P.118, P.124
Numerical integration (*see* Definite integrals;
        Differential equations)

Orthogonal eigenfunctions (*see* Boundary-
        Value problems)

Partial pivoting, 1.9 to 1.11, 1.20, 1.21
Permutations, 9.1, P.277
Pivotal element, 1.9
Pivoting:
  complete, 1.9
  partial, 1.9 to 1.11, 1.20, 1.21
Polynomial equation (*see* Roots)

Quadrature (*see* Definite integrals)

Regression (*see* Curve fitting)
Regression analysis, 9.4
Regression coefficients, 9.4 to 9.7
Residual vector, 1.13, 1.15, 1.20
Richardson's extrapolation, 3.4, 3.9
Roots:
  of algebraic equation, 6.1 to 6.4, 6.9, 6.11,
        P.186, P.188, P.190, P.196, P.201
  analysis of errors in computation of, 6.12
  of biquadratic equation, 6.3, P.190
  Cardano's formulas, 6.3

Roots (*Cont.*):
  complex, 6.11, P.201
  of cubic equation, 6.2, P.188
  deflation technique, 6.10, 6.11
  direct methods, 6.1 to 6.5
  iterative methods, 6.5 to 6.12
    bisection, 6.6
    for complex roots, 6.11, P.201
    Laguerre, 6.8
    modified regula falsi, 6.6
    Muller, 6.8
    Newton, 6.7, P.196
    for real roots, 6.9, P.196
    regula falsi, 6.6
    secant, 6.6
  of polynomial equation, 6.1 to 6.4, 6.9, 6.11, P.186, P.188, P.190, P.196, P.201
  preliminary location of, 6.5, P.193, P.201
  of quadratic equation, 6.1, P.186
  real, 6.9, P.196
  synthetic division, 6.10
  of transcendental equation, 6.9, P.196
  of two equations, 6.13 to 6.15
    fixed-point iteration, 6.13, P.209
    modified Newton method, 6.14, P.211
    Newton method, 6.14, P.211
Row vector, 1.2

Schur theorem, 7.2
Secular equation, 7.2 to 7.4
Singular integral:
  of definite integral, 3.10, 3.12, P.149, P.150
  of differential equation, 4.2, 4.5 to 4.6

Special functions (*see* Functions)
Spectral norm, 1.15, 7.3
Spectral radius, 1.15, 1.22, 7.3
Standard deviations, 9.3, P.280, P.282
Standard error, 9.3, P.280, P.282
Statistics, 9.2 to 9.5
  coefficient of correlation, 9.5
  correlation analysis, 9.4
  covariance, 9.5
  dispersion, 9.3
  grouped data, 9.3 to 9.4, P.282
  mean value: arithmetic, 9.2, P.280, P.282
    geometric, 9.3, P.280
    harmonic, 9.3, P.280
  regression analysis, 9.4
  regression coefficients, 9.4 to 9.7
  standard deviation, 9.3, P.280, P.282
  standard error, 9.3, P.280, P.282
Sturm-Liouville equation, 5.2, 5.4 to 5.6
Sturm-Liouville operator, 5.4
Sylvester's criterion, 1.4
Synthetic division, 6.10, 8.2

Trace of matrix, 1.5
Transcendental equation, roots of, 6.9, P.196

Variances, 2.17, P.68, P.73, P.79, P.84, P.90, P.95, P.100, P.104
Variations, 9.2, P.277

Wronskian, 4.7, 4.11
  Abel's formula for, 4.11

About the Author

Jaroslav Pachner, a native of Czechoslovakia, received his doctorates from Technical University in Prague (Dr. Ing.) and from Charles University in Prague (Dr. Sc.). He immigrated to Canada in 1968. After a year as visiting professor of physics at the University of Alberta, Edmonton, Dr. Pachner was appointed professor of physics at the University of Regina, Saskatchewan, where he taught numerical analysis, advanced analysis, and theoretical physics, including general relativity, until his retirement in 1981.

Dr. Pachner has published more than 60 research papers in scientific journals and proceedings of conference in Czechoslovakia, the United States, Canada, Germany, England, Poland, Israel, and India. He is a member of the International Astronomical Union, International Society on General Relativity and Gravitation, American Physical Society, New York Academy of Sciences, and Einstein Foundation International.